Practical Aspects of Computational Chemistry IV

Jerzy Leszczynski · Manoj K. Shukla
Editors

Practical Aspects of Computational Chemistry IV

Springer

Editors
Jerzy Leszczynski
Department of Chemistry
Jackson State University
Jackson, MS
USA

Manoj K. Shukla
US Army Engineer Research
and Development
Vicksburg, MS
USA

ISBN 978-1-4899-7697-0 ISBN 978-1-4899-7699-4 (eBook)
DOI 10.1007/978-1-4899-7699-4

Library of Congress Control Number: 2016935391

Printed on acid-free paper

This Springer imprint is published by Springer Nature
The registered company is Springer Science+Business Media LLC New York

Preface

You are reading the next, most current volume of "Practical Aspects of Computational Chemistry" book series. Four volumes published since 2012 augmented by an initial book: the "Practical Aspects of Computational Chemistry" published in 2009 have established five volume series devoted to the various aspects of computational chemistry that we (Jerzy Leszczynski and Manoj Shukla) are pleased to bring to you over the last few years. This volume covers recent developments and current applications covering nanomaterials, hydrogen-bonded clusters, semiempirical local coupled-cluster theory, charge-transfer coupling, ro-vibrational energy levels, relativistic effects and quantum electrodynamics in chemistry, mechanochemistry, passivation on metal oxide surfaces, and nano-QSAR. The state-of-the-science research reviews covering the current volume are distributed in 12 chapters.

Chapter 1 contributed by Barysz provides an overview of applications of relativistic effects and quantum electrodynamics in solving chemical problems such as predicting reliable X-ray spectra. Chapter 2 discusses an efficient algorithm to locate global energy minima of hydrogen-bonded clusters containing up to 55 water molecules and is written by Kazachenko and Thakkar. In Chap. 3, Zakharov et al. have discussed the development and application of semiempirical coupled-cluster theory to calculate optical parameters such as polarizabilities and hyperpolarizabilities of fragments of conjugated polymers. Ramos et al. have reviewed methods to compute charge-transfer couplings efficiently and accurately in Chap. 4. Carrington has reviewed methods to compute ro-vibrational energy levels of small polyatomic molecules in Chap. 5.

The effectively unpaired electron theory for singlet states and its application extending from diatomic to graphene nanoclusters have been reviewed by Luzanov in Chap. 6. Bobadilla and Seminario have discussed the application of computational chemistry methods in designing of carbon-based nanodevices in Chap. 7. There is a growing interest in the area of computational mechanochemistry and it has been reviewed by Dopieralski and Latajka in Chap. 8. The mechanisms of different types of Lewis acid–Lewis base interactions have been discussed in Chap. 9 by Grabowski and results of computational modeling of iodine-containing drugs have been reviewed by Yuldasheva et al. in Chap. 10. Rybakov et al. have discussed atomistic modeling of Si(110) passivation by atomic layer deposition of Al_2O_3 in Chap. 11. The last chapter contributed by Toropov et al. deals with the development of nano-QSAR.

We would like to take this opportunity to thank all contributors for devoting their time and hard work to make this project a success. We acknowledge the excellent support from the Presidium of the European Academy of Science as well as Editors at Springer. Of course, many thanks go to our family and friends without their support the realization of the book would not have been possible.

Jackson, MS, USA

Jerzy Leszczynski
Manoj K. Shukla

Contents

Chapter 1
Relativistic Effects and Quantum Electrodynamics in Chemistry

Maria Barysz

Abstract In this chapter I discuss some aspects of relativistic theory, the accuracy of the infinite order two-component relativistic IOTC method and its advantage over the infinite order Douglas-Kroll-Hess (DKHn) theory, in the proper description of the molecular spectroscopic parameters and the potential energy curves. Spin-free and spin dependent atomic mean filed (AMFI) two-component theories are presented. The importance of the quantum electrodynamics (QED) corrections and their role in the correct description of the spectroscopic properties of many-electron atoms for the X-ray spectra is discussed as well. Some examples of the molecular QED calculations will be discussed here as well.

1.1 Introduction

In the last decades we have seen intense development of experimental techniques to study molecular systems and this means that we get more accurate results of these experiments. This means that we also need more and more accurate theoretical methods to calculate these properties. The ab initio calculations based on Schrödinger equation have been a routine way to study a molecular electronic structure and properties in chemical laboratory or in biological systems. Nonrelativistic Schrödinger quantum mechanics provides a consistent picture on numerous phenomena in the atomic and molecular domain. However the description of physical phenomena based on nonrelativistic quantum mechanics is incomplete. The description of phenomena at high energies requires the investigation of relativistic wave equations. This means equations which are invariant under the Lorentz transformations. The central theme in relativity is that the speed of light, c, is constant in all inertia frames (coordinate systems that move with respect to each other). Together with the requirement that physical lows should be identical in such frames, this has a consequence that time and space coordinates become equivalent. The fundamental structure of the Schrödinger equation is not invariant under the Lorentz

M. Barysz (✉)
Institute of Chemistry, N. Copernicus University, Gagarina 7, 87100 Toruń, Poland
e-mail: teomjb@chem.uni.torun.pl

J. Leszczynski and M.K. Shukla (eds.), *Practical Aspects of Computational Chemistry IV*, DOI 10.1007/978-1-4899-7699-4_1

transformation and is therefore not relativistic correct. Traditionally, the relativistic theory is first developed for one electron (particle) and leads to what is known as the Dirac equation which is relevant substitute for the Schrödinger equation. Although Dirac introduced special relativity into quantum mechanics in 1929, and the extension to many-electron systems was made by Breit soon after, the importance of relativistic corrections in molecular quantum chemistry was not fully appreciated until the 1980s. The Dirac equation describes both electrons (positive energy solutions) and positrons (negative energy solutions) and explicitly includes spin. For a given potential and chosen charge $q = -e$, both positive and negative energy solutions of the Dirac equation correspond to the electronic and positronic states. The excitation energy between positive and negative states is of the order $2mc^2 \approx 1$ MeV. Thus in low energy processes, one can proceed by ignoring the lower continuum. In the nonrelativistic limit, the lower positive energy solutions go to zero, whereas the upper part reduce to a spin orbital in which the spatial part of the wave function solves the nonrelativistic Schrödinger equation. The upper and lower two components of the Dirac wave function are generally referred to as the large and small component, respectively [1]. The relativistic effects can be analysed in terms of direct and indirect effects. The direct relativistic effects originates in the immediate vicinity of the nucleus, the indirect relativistic effects is influenced by the outer core orbitals.The direct relativistic effects are responsible for the radial contraction and energetic stabilization of the $s_{1/2}$ and $p_{1/2}$ shells as well as for the spin-orbit splitting of shells $l > 0$ into sub shells with $j = l - 1/2$ and $j = l + 1/2$. The SO splitting for the same l decreases with increasing number of sub shells i.e., it is much stronger for inner shells than for outer shells. The SO splitting decreases with increasing l for the same principal quantum number i.e. the $np_{1/2} - np_{3/2}$ splitting is larger than the $nd_{3/2} - nd_{5/2}$ and both are larger than the $nf_{3/2} - nf_{5/2}$. The indirect relativistic effects are consequences of a more efficient screening of the d and f shells accompanied by their energetic destabilization (cited from [1–3]). The nature is relativistic and it does not distinguish between relativistic and nonrelativistic effects and only theory allows to distinguish these effects. The values of the relativistic effects for a given quantity X are usually calculated as a difference between the relativistic and nonrelativistic values of this quantity, i.e. $\Delta X = X_{rel} - X_{nrel}$ calculated at the same level of theory. Calculations using relativistic quantum mechanics are becoming more routine nowadays and theoretical methods more and more accurate.

However, already in 1930s deviations were observed between the results of precision spectroscopy and the Dirac theory for simple atomic systems, primarily for the hydrogen atom. The existence of negative-energy states in the solutions of Dirac equation is the mathematical but not the physical grounds of the existence of particles and antiparticles (electrons and positrons). Besides, the velocity of light is finite. For an complete model we must turn to quantum field theory and quantum electrodynamics (QED) [4].

Dirac himself was not convinced of the importance of relativistic effects, which he declared would be of no importance in the consideration of atomic and molecular structure and ordinary chemical reactions. We know that he wasn't right. Are QED effects important in chemistry. Full answer to this question is not yet known.

The chapter consists of two parts. In the first part I discuss some aspects of relativistic theory, the accuracy of the infinite order two-component relativistic IOTC method and its advantage over the infinite order Douglas-Kroll-Hess (DKHn) theory, in the proper description of the molecular spectroscopic parameters. Spin-free and spin dependent atomic mean filed (AMFI) theory is presented. Additionally, the accuracy of the relevant potential energy curves is discussed as well. In the second part I show the role of the QED corrections and that they are necessary for the correct description of the spectroscopic properties of atoms for the X-ray spectra. Some examples of the molecular QED calculations will be discussed here as well.

1.2 The Infinite Order IOTC and DKHn Theories

The four-component relativistic Dirac wave function contains information about positive and negative-energy states of the system. In chemical applications, one is usually concerned with the electronic (or positive-energy) states only. Therefore, some reduction of the four-component wave function seems to be preferred. The history of this reduction goes back to the period before the Dirac equation. The first step in this direction appears to have been made by Pauli in the form of a quasi-relativistic Hamiltonian known as the Pauli Hamiltonian. This Hamiltonian can be approximately derived from the Dirac Hamiltonian by using the fact that in the nonrelativistic limit the large (the upper) 2-vector part of the Dirac 4-spinor becomes the electronic (positive energy) solution with spin. On this basis one could expect that there is some representation of the Dirac Hamiltonian which completely separates the positive and negative energy spectra and permits to focus all attention on the electronic part only. The exact separation of the two spectra would be equivalent to the transformation of the Dirac 4-spinors into either electronic or positronic 2-spinors. Alternatively, this means that the 4×4 Dirac Hamiltonian is to be block-diagonalized, i.e. brought into the form of the Direct sum of 2×2 matrix Hamiltonian's one of them corresponding to the electronic spectrum and the other one referring solely to the positronic eigenvalues. Once this is achieved most problems of relativistic quantum chemistry can be formulated solely in terms of electronic 2-spinors. The positronic solutions can be simple abandoned. In practice it means that we have to find the unitary transformation which diagonalize Dirac Hamiltonian.

$$\mathrm{H}^{\mathcal{U}} = \mathcal{U}^{\dagger}\mathrm{H}_{Dirac}\mathcal{U}, \tag{1.1}$$

with

$$H_{Dirac} = \begin{pmatrix} V & c\boldsymbol{\sigma p} \\ c\boldsymbol{\sigma p} & V - 2c^2 \end{pmatrix}, \tag{1.2}$$

and

$$\mathrm{H}^{\mathcal{U}} = \begin{pmatrix} \mathrm{h}_+ & 0 \\ 0 & \mathrm{h}_- \end{pmatrix}. \tag{1.3}$$

The potential energy operator V is assumed here to correspond to the electron–nucleus Coulomb attraction. Atomic units are used throughout this paper, c is the velocity of light, $c = 137.03599$ a.u.

The unitary transformation U can be determined as the product of two transformations $U_0 U_1$. The first transformation U_0 is the free-particle Fouldy–Wouthuysen transformation and leads to the approximate separation of the electronic and positronic spectra $H_1 = U_0^\dagger H_0 U_0$. The second unitary transformation $U_1^\dagger H_1 U_1$ is based on the idea of Heully et al. and is determined in terms of the auxiliary operator R [5]. The infinite–order (or exact) solution of the block-diagonalization problem is then reduced to the solution of the following operator equation:

$$R = [(H_1)_{22}]^{-1}[-(H_1)_{21} + R(H_1)_{11} + R(H_1)_{12}R] \tag{1.4}$$

Once the solution R of Eq. (1.4) is known, the *exact* two-component "electronic" Hamiltonian h_+ becomes:

$$h_+ = \Omega_+^\dagger (H_1)_{11}\Omega_+ + \Omega_+^\dagger R^\dagger (H_1)_{21}\Omega_+ + \Omega_+^\dagger (H_1)_{12} R\Omega_+ + \Omega_+^\dagger R^\dagger (H_1)_{22} R\Omega_+ \tag{1.5}$$

where the Ω operator is defined through the R operator [6, 7].

The operator Eq. (1.4) is not easy to solve since it involves terms which are linear in $\boldsymbol{\sigma p}$ [7]. One of the possible way to solve the Eq. (1.4) is by means of some iterative scheme. It can be made through some odd powers of α, say α^{2k-1}, k = 2, 3, ... (with α denoting the fine structure constant, $\alpha = 1/c$). Then, the unitary transformation U will be exact through the same order in α. Simultaneously, this will lead to the approximate form h_{2k}, k = 2, 3, ... of h_+. Thus the method leads to a series of two-component relativistic Hamiltonians whose accuracy is determined by the accuracy of the iterative solution for R. In each step of the iteration the analytical form of the R operator (Eq. (1.4)) and the Hamiltonian h_+ (Eq. (1.5)) have to be derived.

The above idea was the basis of the BSS method formulated by Barysz and Sadlej [8]. The BSS method has its roots in the historically earlier Douglas-Kroll-Hess (DKH2 and DKH3) [9, 10] approximation. In the BSS approximation the fine structure constant α is the perturbative parameters and it differs from the DKH method where the potential V is used as the perturbation. Formally the BSS and DKH methods are of the infinite order in α or V. However, the necessity to define the analytical form of the R operator and the Hamiltonian h_+ in each step of the iteration, makes the accuracy of both methods limited to the lowest order in α or V.

The break through of the above limitation came in 2002 when Barysz and Sadlej [7] found the way to exact solution of the R operator equation Eq. (1.4). In this new two-component infinite-order (IOTC) method the analytical form of the R operator equation is formulated only once and the iterative procedure is defined inside of the molecular code. The solution is exact in the given basis set. This is the main advantage of the IOTC method in comparison to the BSS or DKH methods.

After the IOTC method has been formulated, the so called infinite order DKHn method has been also defined [11–13]. The DKHn approximation is the generalization of the original DKH theory which enables to achieve the higher orders of the h_+ Hamiltonian. Unfortunately, the necessity to formulate the analytical form of the h_+ Hamiltonian in each order of perturbation V is still the basis of the DKHn method. The order of the DKHn approximation must be defined prior to any quantum-chemical calculations. The DKHn method is very well defined but it is only the approximation of the IOTC method which is exact.

There are a lot of mistakes in understanding and distinctions among the BSS, the IOTC and the DKHn methods and that is the reason we make a comment on it in this chapter.

For the lightest elements all the methods should give very similar results. The differences will appear for the heavier elements.

In the present work we will focus mainly on the infinite order two-component method, IOTC. However, some comparison between the IOTC and DKHn methods will be also presented. So far the discussion has been focus on the block–diagonalization of the one–electron Dirac Hamiltonian. For the N electron system a Hamiltonian may be written as the sum of the one-electron transformed Dirac Hamiltonian plus the Coulomb electron-electron interaction and it is commonly used form of the relativistic Hamiltonian.

1.3 Potential Energy Curves in the Spin-Free Relativistic DKHn and IOTC Theory

The electronic and spectroscopic properties of small dimers have been the topics of many experimental and theoretical studies in recent years [14–17]. The intense theoretical and experimental activity in this area resulted in a huge amount of results on the spectroscopic properties and potential curves of many dimers. Investigation of such species provide details on the low- and high-lying electronic states and the nature of the atom-atom bonds. Theoretical ab initio calculations on such dimers are on the increase due to advances in theoretical methods and yield information of significance to experimentalists and are of great value in enhancing our knowledge of bonding in these systems. Theoretical calculations have also become valuable in designing new experiments. Due to intense development of relativistic methods in recent years, special interest are in dimers of heavy atoms. The nonrelativistic Schrödinger theory is not able to properly described such dimers. The relativistic theory has to be applied instead [18].

The nature is relativistic. One can not receive the correct potential energy curve without taking into account the relativistic effects. To show the difference between the nonrelativistic and relativistic theory, the results of theoretical calculations for the ground $^1\Sigma^+$ and two $^3\Pi_r$ and $^1\Pi$ excited states of $SiAu^+$ molecule are presented in Table 1.1. All calculations were carried out with the complete active space CASSCF method followed by the second-order single state multireference perturbation, CASPT2 scheme [19–22].

The POL and POL.DK Gaussian basis sets have been used in the nonrelativistic and spin-free relativistic CASSCF/CASPT2 calculations. Gaussian basis sets

Table 1.1 Calculated CASSCF/CASPT2 bond lengths, R_e (in pm), D_e dissociation energies (in eV) and ω_e spectroscopic constants (in cm^{-1}) for the ground and excited states of $SiAu^+$ cation, in the IOTC and DKH_n methods

Method	R_e	D_e	ω_e
$^1\Sigma^+$			
DKH_2	2.1876	3.7870	459
DKH_3	2.1856	3.8099	461
DKH_4	2.1859	3.7708	460
DKH_5	2.1858	3.8079	461
DKH_6	2.1858	3.8073	460
DKH_7	2.1858	3.8074	460
DKH_8	2.1858	3.8074	460
IOTC	2.1858	3.8076	460
NR	2.5214	2.9057	367
$^3\Pi_r$			
DKH_2	2.2772	1.9822	339
DKH_3	2.2749	1.9848	340
DKH_4	2.2752	1.9845	340
DKH_5	2.2749	1.9849	340
DKH_6	2.2751	1.9846	340
DKH_7	2.2751	1.9846	340
DKH_8	2.2751	1.9846	340
IOTC	2.2751	1.9848	340
NR	2.6607	1.3179	180
$^1\Pi$			
DKH_2	2.3480	1.4920	299
DKH_3	2.3470	1.4963	299
DKH_4	2.3471	1.4956	299
DKH_5	2.3471	1.4956	299
DKH_6	2.3471	1.4958	299
DKH_7	2.3471	1.4958	299
DKH_8	2.3471	1.4958	299
IOTC	2.3471	1.4960	299
NR	2.6813	1.5001	167

[13s10p4d/7s5p2d] and [21s17p11d9f/13s11p7d4f] have been used in the calculations for *Si*, and *Au* respectively [23]. The basis set used differ a bit from those used earlier in our calculations [24], and they lead to slightly different values of the total energy. All calculations have been carried out in C_{2v} symmetry. The partition of the orbital space used in CASSCF calculation is (0.0.0.0/20.10.10.4/3.1.1.0;5el). In the CASPT2 method the partition is (17.8.8.3/3.2.2.1/3.1.1.0;5el).

The spin-free IOTC results have been compared with different DKHn approximations $n = 2, 3, 4, \ldots 8$ and with the nonrelativistic (NR) data. Analyzing the results of the calculations, one can observe a quite large role of the relativistic effect. For example, the calculated IOTC dissociation energies for the ground $^1\Sigma^+$ and $^3\Pi_r$ excited states are 3.8076 and 1.9848 eV, the corresponding nonrelativistic values are 2.9057

Table 1.2 The comparison of the CASSCF/CASPT2 *DKHn* $(n - 2, 3, 4, \ldots 8)$ and the IOTC energies (in a.u.) at the equilibrium distance R_e for the ground and excited states of $SiAu^+$ molecule

State	Method	Energy
$^1\Sigma^+$	DKH_2	−19282.10072357
	DKH_3	−19301.90094796
	DKH_4	−19299.16552475
	DKH_5	−19300.60487817
	DKH_6	−19300.08950129
	DKH_7	−19300.24097799
	DKH_8	−19299.99193999
	IOTC	−19299.97359888
	NR	−18154.11091612
$^3\Pi_r$	DKH_2	−19282.03278623
	DKH_3	−19301.83227194
	DKH_4	−19299.09695943
	DKH_5	−19300.33822623
	DKH_6	−19299.82285485
	DKH_7	−19299.97432997
	DKH_8	−19299.92334410
	IOTC	−19299.90500105
	NR	−18154.10095727
$^1\Pi$	DKH_2	−19282.01523132
	DKH_3	−19301.81477198
	DKH_4	−19299.07945149
	DKH_5	−19300.32072159
	DKH_6	−19299.80534885
	DKH_7	−19299.95682437
	DKH_8	−19299.90583836
	IOTC	−19299.88749478
	NR	−18154.07492452

and 1.3179 eV. The relativistic IOTC equilibrium distances R_e are about 0.3–0.4∀ shorter than the nonrelativistic values for all states. Similarly, the significant differences can be observed for the harmonic frequency ω_e. The calculated spectroscopic parameters by the IOTC method and the DKHn approximations are similar in both methods, and starting from the $n = 3$–4 DKHn order they are practically the same.

To plot the whole potential energy curve one needs a total energy values. The relativistic spin-free IOTC and DKHn values of the total energies for the $^1\Sigma^+$, $^3\Pi_r$ and $^1\Pi$ states at the equilibrium distance R_e are shown in the Table 1.2 According to the calculations the total energies obtained in the Douglas-Kroll-Hess method strongly depends on the used approximations. The perturbative DKHn results ($n = 2, 3, 4 \ldots 8$) converge nicely to the IOTC energy. They go a little bellow the IOTC value. It should be noted however, that the parametrization implemented in the DKHn approximation does not affect the DKHn Hamiltonian up to the fourth order only. Therefore, as long as one runs calculations with DKHn Hamiltonian below 5th order may use any parametrization as they would all yield the same results. Higher order DKHn Hamiltonian depend slightly on the chosen parametrization of the unitary transformation applied to decouple the Dirac Hamiltonian. Nonrelativistic energy values are about 1146 a.u. above the relativistic values, for all states.

1.4 Electronic States in the Spin Dependent Relativistic Theory

We do not receive a full description of excited states and potential energy curves without the spin-orbit terms. Spin-orbit effect arises due to the interaction of the magnetic dipole of the electronic spin and the movement of electrons in its orbit. For the nonrelativistic case, angular momentum **l** and spin **s** are normal constants of motion and they both commute with the nonrelativistic Hamiltonian. For the relativistic case and the Dirac equation neither **s** nor **l** are normal constants of motion for this case, but the total angular momentum operator $j = l + s$ is.

For the orbital angular momentum **l**, the effect of a rotation by an angle ϕ about the z axis is the effect of a rotation by an angle ϕ about z axis is [25]

$$R_z(\phi)|l, m_l >= e^{im_l\phi}|l, m_l > \tag{1.6}$$

where $|l, m_l>$ is an eigenfunction of $\mathbf{l}^2$ and $\mathbf{l_z}$.

For the angular momentum j, the effect of a rotation by an angle ϕ about z axis is [25]

$$R_z(\phi)|j, m_j >= e^{im_j\phi}|j, m_j > \tag{1.7}$$

In particular, if we rotate through an angle of 2π, we get

$$R_z(2\pi)|j, m_j >= e^{im_j 2\pi}|j, m_j >= (-1)^{2m_j}|j, m_j > \quad (1.8)$$

For orbital momentum l, this result was not a problem because m_l is always integer, and thus a rotation by 2π brings any physical system back to itself, and amounts to be an identity operator. However, m_j may assume half integer, and a rotation by 2π changes the sign of the function.

In general, for functions describing systems with half integer spin, the function must change sign under a rotation by 2π. This operation is given a special symbol, R, and is interpreted as rotation by an angle 2π around an arbitrary axis. This has consequences in the description of the symmetry of atomic and molecular system [2, 26]. For molecules described in the nonrelativistic theory, the symmetry is described by the point groups, which do not contain the operation R. We would expect these groups to require the addition of R in order to describe the relativistic symmetry of molecules. In order to define the double groups, we introduce that the molecule is to go into itself not upon a rotation by 2π around an axis, but only upon a rotation by 4π. The double groups are usually denoted in terms of the nonrelativistic point group symmetry with some specific superscript. Throughout the manuscript the $*$ symbol is used. For example, the relativistic counterpart of the $C_1(E)$ and $C_2(E, C_2)$ point groups would be $C_1^*(E, R)$ and $C_2^*(E, C_2, R, RC_2)$ respectively [2, 25, 26].

The potential energy term in the Hamiltonian for an atom is spherically symmetric. The formation of a diatomic molecules leads to the lowering of the spherical symmetry and the Hamiltonian has only cylindrical symmetry and hence any angular momentum, rotational and electronic, are meaningful, but only the electronic angular momentum L can have a component along the bond axis. Since the energy of the molecule is the same regardless of whether the projection is positive or negative, we usually use the quantum number $\Lambda = |M_L|$, i.e. the absolute value of M_L to characterize the system. We can also define two further quantum numbers, $\Sigma = |M_S|$ and $\Omega = \Sigma + \Lambda$, which describe the projection of the spin angular momentum and total electronic angular momentum, respectively, onto the internuclear axis. The possible Ω states could be obtained by combining Λ with spin or directly from the individual j states, $\Omega = M_{J_1} + M_{J_2}$.

The correspondence of the notation of the irreducible representations of the double group $\Sigma^+, \Sigma^-, \Pi, \Delta, \Phi, \Gamma, E_{1/2}, E_{3/2}, E_{5/2}, E_{n/2}$ and the nonrelativistic Ω states are $0^+, 0^-, 1, 2, 3, 4, 1/2, 3/2, 5/2$ and $n/2$, respectively. The direct product for the irreducible representations of the double groups need to be defined so that one could use the double-group theory to derive relativistic electronic states from the nonrelativistic states. For example the direct product $\Sigma^+ \otimes \Sigma^+$ has irreducible representation Σ^+ and the corresponding Ω state is 0^+, the direct product $\Pi \otimes \Delta$ has irreducible representation $\Phi \otimes \Pi$ what corresponds to 3, 1 Ω states. Similarly, the direct product $\Sigma^+ \otimes \Pi$ gives the Π irreducible representation and corresponds to 1 Ω state. More details can be found in [2, 26].

The next thing one needs to establish is how the spin multiplets D^s (s here denotes the total spin of the state) correlate with the irreducible representation of the double group. The transformation of D^s is isomorphic with the transformation of the rotation state j, denoted as D^j, in molecular spectroscopy. The irreducible representation s spanned by the D^s representation of the spin multiplet with quantum number s, are determined once the characters for various operations in the double group are obtained. The appropriate formulas can be found in [2] One can find that in $D^*_{\infty h}$ or $C^*_{\infty v}$ double groups, D^0 (s = 0) corresponds to Σ^+ irreducible representation, $D^{1/2}$ and D^2 corresponds to $E_{1/2}$ and $(\Sigma^+ + \Pi + \Delta)$ irreducible representations, respectively.

The determination of a relativistic state arising from a given non-relativistic state involves two steps. Firstly, the irreducible representations spanned by the spin multiplets using double group correlation (as discussed above) are found out. These irreducible representations are then multiplied with the spatial symmetry of the non-relativistic state in the next step. The resulting set of the irreducible representations is then transformed to the Ω state. As an example, for the nonrelativistic $^3\Pi_r^+$ state of the studied $SiAu^+$ cation, $s = 1$ and hence D^1 corresponds to Σ^- and Π irreducible representations. The direct products:

$$\Sigma^-(spin) \otimes \Pi(spatial) = \Pi \tag{1.9}$$

$$\Pi(spin) \otimes \Pi(spin) = \Sigma^- + \Sigma^+ + \Delta \tag{1.10}$$

Thus the $^3\Pi_r^+$ nonrelativistic state yields Π, Σ^-, Σ^+, and Δ states and their assignments according to Ω quantum numbers are 1, 0^+, 0^-, and 2. The above discussion gives only some background to further studies.

1.5 Spin-Dependent Two-Component Theories

In the last two decades, the two-component approaches for treatment relativistic effects in atoms and molecules has proven to lead to highly accurate results compared both to experiment and four-component calculations. It is practical, and the usual way, to separate operators that do not explicitly depend on the spin and to define a one-component (scalar) relativistic approximation, resulting in one-component equations of the type used in the usual non-relativistic approach. This separation is possible in a rigorous manner for the Dirac-Coulomb equation. Spin-orbit interaction can either be neglected (which is often a good approximation) or be treated by means of perturbation theory. While most of the relativistic effects at the spin-free level are well described using only one-electron relativistic terms and the Coulomb interactions, this is not true for spin-orbit coupling. As example all splittings in $^3\Sigma$ states can only be described by second-order perturbation contribution. However, the computation of the two-electron contributions by the spin-orbit operator requires a substantial effort due to the lower symmetry compared to the standard two-electron integrals and the necessity to calculate three different spin-orbit two-electron files. The demands on both CPU time and storage facilities become considerable. A method which overcomes this problem has been proposed already in 1996–1998 and is known under the name 'spin-orbit mean field integrals' (AMFI) [27–29]. It is based on the Breit-Pauli Hamiltonian but for the study of spin-orbit contributions, a suitable starting point is obtained by retaining only those terms that involve direct coupling between spin and orbital motion. The Breit-Pauli spin-orbit Hamiltonian has the following form:

$$\begin{aligned}\mathcal{H}^{SO} = & \frac{1}{2^2c^2}\sum_i\sum_\alpha Z_\alpha \vec{s}_i\left(\frac{\vec{r}_{i\alpha}}{r_{i\alpha}^3}\times\vec{p}_i\right) \\ & -\frac{1}{2^2c^2}\sum_{i\neq j}\left(\frac{\vec{r}_{ij}}{r_{ij}^3}\times\vec{p}_i\right)\left(\vec{s}_i+2\vec{s}_j\right)\end{aligned} \tag{1.11}$$

The two-electron spin-orbit $\mathcal{H}^{SO}$ integrals contribute to the spin-orbit matrix element between Slater determinants which are singly or double excited relative to one another. The matrix elements between singly excited determinants can, just like in the Hartree-Fock equations, be written as a pseudo one-electron integral. One of the key aspects of a mean-field theory is to neglect interactions between double excited states and to include all two-electron integrals in pseudo one-electron integrals.

The matrix element of the spin-orbit operator between a pair of Slater determinants differing by a single valence spin orbital excitation $i \rightarrow j$ is given by

$$\begin{aligned}\left\langle \Phi|\mathcal{H}^{SO}|\Phi_i^j\right\rangle = & \left\langle i|\mathcal{H}^{SO}(1)|j\right\rangle \\ & +\frac{1}{2}\sum_k n_k\left(\left\langle ik|\mathcal{H}^{SO}(1,2)|jk\right\rangle - \left\langle ik|\mathcal{H}^{SO}(1,2)|kj\right\rangle - \left\langle ki|\mathcal{H}^{SO}(1,2)|jk\right\rangle\right)\end{aligned} \tag{1.12}$$

where n_k denotes the occupancy of orbitals common to the determinant on the left- and right-hand sides, k runs over all occupied spin orbitals common to the determinants.

In the independent-particle model, Eq. (1.12) describes valence electrons (from orbitals i and j) moving in a field generated by the electrons in orbitals k (which includes the valence space). In other words, Eq. (1.12) defines a matrix element of a Fock operator for a one-determinant approximation with a certain occupancy. Based on this observation, we define an approximate spin-orbit operator by

$$\begin{aligned}H_{ij}^{mean-field} = & \left\langle \Phi|\mathcal{H}^{SO}|\Phi_i^j\right\rangle \\ & +\frac{1}{2}\sum_k n_k\left(\left\langle ik|\mathcal{H}^{SO}(1,2)|jk\right\rangle - \left\langle ik|\mathcal{H}^{SO}(1,2)|kj\right\rangle - \left\langle ki|\mathcal{H}^{SO}(1,2)|jk\right\rangle\right)\end{aligned} \tag{1.13}$$

Table 1.3 Kr, Xe and Rn—$np_{\frac{1}{2}} - np_{\frac{3}{2}}$ splittings of the energy levels (eV)

Method	$2p_{\frac{1}{2}} - 2p_{\frac{3}{2}}$	$3p_{\frac{1}{2}} - 3p_{\frac{3}{2}}$	$4p_{\frac{1}{2}} - 4p_{\frac{3}{2}}$	$5p_{\frac{1}{2}} - 5p_{\frac{3}{2}}$	$6p_{\frac{1}{2}} - 6p_{\frac{3}{2}}$
Kr AMFI	51.00	7.48	0.62	–	–
Kr exp.	52.50	7.8	0.65	–	–
Xe AMFI	306.72	58.88	11.25	1.20	–
Xe exp.	320.15	62.12	12.37	1.33	–
Rn AMFI	2492.85	571.03	142.91	28.31	3.39
Rn exp.	–	–	–	–	–

The IOTC CASSCF/CASPT2 method

In Eq. (1.13) the orbitals $|k>$ are taken from spin-free atomic or molecular calculations. It was additionally assumed, in order to make the mean-field method efficient, that all two-electron integrals between basis functions centred on different centres could be neglected.

Since the mean-field spin-orbit integrals have been defined and implemented to molecular codes (Dalton, Molcas etc.), hundreds or more successful molecular and atomic applications have been performed with the accuracy comparable to the experimental results.

As an illustration of the performance of the AMFI approximation I present the results of calculations of the $np_{\frac{1}{2}}$ and $np_{\frac{3}{2}}$ atomic splittings of the noble gases *Kr*, *Xe* and *Rn* elements Table 1.3. The scalar spin-free IOTC complete active space CASSCF/CASPT2 method was used followed by the restricted active space state-interaction (RASSI) method with the use of the atomic mean-filed spin-orbit (SO) integrals AMFI.

1.6 Quantum Electrodynamic

In the previous section we discussed how to calculate the spin-orbit splittings of atomic and molecular states. The effect of spin-orbit coupling was introduced via the restricted active space interaction (RASSI) method with the use of the atomic mean-field AMFI integrals. It appears however, that the discrepancies between the experimental and calculated values of energies can be still quite big.

To explain these differences the quantum electrodynamic corrections have to be implemented. The velocity of light is finite and this means retardation of the inter-particle interactions. This means that the Dirac-Coulomb Hamiltonian has to be corrected by further expressions.

One of the most important physical corrections to the Dirac-Coulomb Hamiltonian is the replacement of the nonrelativistic Coulomb repulsion, $\frac{1}{r_{ij}}$ in with a covariant expression derived from QED [30–32].

The precise form of this correction depends upon the gauge condition used to describe the electromagnetic field. In the Coulomb gauge, which has been employed more often in relativistic atomic structure, the electron-electron interactions come from one-photon exchange process and is sum of instantaneous Coulomb interaction and the transverse photon interaction.

$$H_{ij} = \sum_{i<j}^{N} \frac{1}{r_{ij}} + H_{trans} \tag{1.14}$$

where

$$H_{trans} = -\sum_{i<j}^{N}[+\frac{\alpha_i\alpha_j}{r_{ij}} + (\alpha_i \cdot \nabla_i)((\alpha_j \cdot \nabla_j)\frac{cos\omega_{ij}r_{ij}}{\omega_{ij}^2 r_{ij}}] \tag{1.15}$$

ω_{ij} is the wave number of the photon being transferred, α_i is the Dirac alpha matrix for particle "i". In the low-frequency limit ($\omega_{ij} \rightarrow 0$) the Eq. (1.15) reduces to the Breit interaction,

$$\mathbf{H_{HBreit}} = -\sum_{i<j}^{N}[+\frac{\alpha_i\alpha_j}{2r_{ij}} + (\alpha_\mathbf{i} \cdot \mathbf{r_{ij}})(\alpha_j \cdot \mathbf{r_{ij}})\frac{1}{2r_{ij}^3}] \tag{1.16}$$

where the first term is the correction due to two-electron magnetic interaction and represents so called Gaunt interaction, The second term is the correction due to retardation resulting from the finite velocity propagation of the interaction.

Both corrections form the Breit interaction. The fact that this is the limit of (1.15) as $\omega_{ij} \rightarrow 0$ means that it must not be used for describing the interaction of orbitals with large energy differences. A method of derivation of the Breit interaction assumes $\frac{1}{c}Z \ll 1$, where c is the velocity of light.

The Dirac-Coulomb-Breit Hamiltonian is derived perturbationally and thus it is frequently suggested that the Breit correction to the Coulomb interaction should be considered in the perturbation framework and evaluated as the first–order contribution to the energy which follows from the Dirac-Coulomb calculations, and this is the way how it is done, as example, in the atomic GRASP2K package.

The Breit corrections are sometimes classified as nonradiative effects in contrary to the radiative affects which are 'true' quantum-electrodynamical effects due to the electron self energy and vacuum polarization [30–32].

The effect of the Breit and Gaunt interactions has been investigated in many atomic systems and as will be demonstrated later in this chapter, it is known that they are very important in high-resolution atomic X-ray spectroscopy.

Some examples of the molecular calculations are known as well. For instance, the effect of the magnetic electron-electron Gaunt interaction on bond length as been demonstrated in reference is: CH_4 [0.0 pm]; SiH_4 [0.0 pm]; GeH_4 [0.1 pm, 326 ppm];

SnH_4 [0.1 pm, 514 ppm]; PbH_4 [0.2 pm, 962 ppm]—the difference in bond lengths compared to a calculation with only Coulomb interaction energy operators is given in pm in brackets and the percentage of the change compared to the absolute values of the bond length is given in parts per million (ppm); the wave function was approximated by a singly determinant. The effect of the Gaunt interaction on binding energies in CH_4 is about 0.3 kJ mol^{-1} and increases in the case of PbH_4 to about 2.1 kJ mol^{-1}.

It is a common way to expand the Coulomb-Breit Hamiltonian in the perturbation series in $1/c$ parameter. Then the one- and two-electron Breit-Pauli Hamiltonian for the N-electron system is obtained and can be written as

$$\hat{H_{BP}} = H_{NR} + H_R + H_{FS} \tag{1.17}$$

and is the sum of the nonrelativistic many-electron Hamiltonian H_{NR}, the relativistic operator H_R and the fine-structure operator H_{FS} [33]. The relativistic operator can be written as

$$\hat{H}_R = H_{MC} + H_{D1} + H_{D2} + H_{OO} + H_{SSC} \tag{1.18}$$

where H_{MC} is the mass velocity term attributed to relativistic correction arising from the variation of the mass of the electron with its speed.

$$H_{MC} = -\frac{1}{8c^2} \sum_{i=1}^{N} \nabla_i^4 \tag{1.19}$$

and H_{D1} and H_{D2} are the one- and two-body Darwin Terms

$$H_{D1} = -\frac{Z}{8c^2} \sum_{i}^{N} \nabla_i^2 (\frac{1}{r_i}) \tag{1.20}$$

$$H_{D2} = -\frac{1}{4c^2} \sum_{i<j}^{N} \nabla_i^2 (\frac{1}{r_{ij}}) \tag{1.21}$$

The Darwin correction, is a relativistic correction attributed to the electron's Zitterbewegung. It arises from the smearing of the charge of the electron due to its relativistic motion.

H_{SSC} is the spin-spin contact term

$$H_{SSC} = -\frac{8\pi}{3c^2} \sum_{i<j}^{N} (\mathbf{s_i} \cdot \mathbf{s_j} \delta(r_i \cdot r_j)) \tag{1.22}$$

and finally H_{OO} is the orbit-orbit term

$$H_{OO} = -\frac{1}{2c^2}\sum_{i<j}^{N}[\frac{\mathbf{p_i}\cdot\mathbf{p_j}}{r_{ij}} + \frac{r_{ij}(r_{ij}\cdot p_i)p_j}{r_{ij}^3} \quad (1.23)$$

The fine-structure operator H_{FS} describes interactions between the spin and orbital angular momentum of the electrons. The fine structure operator consists of three terms

$$\hat{H_{FS}} = H_{SO} + H_{SOO} + H_{SS} \quad (1.24)$$

Here H_{SO} is the spin-orbit term describing the interaction of the electron spin with the magnetic field generated by its own movement.

$$\hat{H_{SO}} = \frac{Z}{2c^2}\sum_{i=1}^{N}\frac{1}{r_i^3}\mathbf{l_i}\cdot s_i \quad (1.25)$$

H_{SOO} is the spin-other-orbit term

$$H_{SOO} = -\frac{1}{2c^2}\sum_{i<j}^{N}\frac{r_{ij}\times p_i}{r_{ij}^3}(\mathbf{s_i} + 2s_j) \quad (1.26)$$

The last two terms H_{SO} and H_{SOO} have been discussed already earlier in this chapter. H_{SS} is the spin-spin term

$$\hat{H_{SS}} = -\frac{1}{c^2}\sum_{i<j}^{N}\frac{1}{r_{ij}^3}[\mathbf{s_i}\cdot s_j - 3\frac{(s_i\cdot r_{ij})(s_j\cdot r_{ij})}{r_{ij}^2}] \quad (1.27)$$

Already in the 1930s deviations were observed between the results of precision spectroscopy and the Dirac or Dirac-Coulomb-Breit theory for simply atomic systems. Primarily this deviation was observed for the hydrogen atom. This is due to the fact, that the Dirac equation for the hydrogen atom does not contain radiative and other corrections that are of quantum-electrodynamic origin. Exact Dirac equation of the hydrogen atom do not differentiate between $2S_{1/2}$ and $2P_{1/2}$ states of the hydrogen atom for a given n quantum number, thus the degeneracy is not completely removed. High-resolution experimental studies show that the energy levels corresponding to $2S_{1/2}$ and $2P_{1/2}$ atomic states are not identical. In Dirac's theory, pairs of levels with $l = j \pm \frac{1}{2}$ for the same j values are degenerate.

Emission and absorption of a virtual photon on the same electron is an effect that is not included in the Dirac theory and it is known as the electron self-energy (SE). This forms the major part of the Lamb shift, discovered experimentally by Lamb and Rutherford in 1947. This was the starting point for the development of modern quantum electrodynamic. The second most important part of the Lamb shift is the vacuum polarization (VP) [4]. The field near the atomic nucleus can give rise to a polarization effect in the form of the creation of electron-positron pairs, an effect referred as the vacuum polarization. The self-energy and the vacuum polarization effects give the leading contribution to the Lamb shift.

Both, SE and VP, corrections are calculated, in the GRASP2K code, as the first order contributions to the Dirac-Fock-Coulomb energy.

The corresponding expressions are given by (1.28) and (1.29), respectively,

$$E_{SE} = (Z^4\alpha^3/\pi)\sum_a F_a(Z\alpha)q_a/n_a^3 \tag{1.28}$$

$$E_{VP} = \sum_a q_a \int_0^\infty [P_a^2(r) + Q_a^2(r)]V_{VP}(r)dr \tag{1.29}$$

where the sums run over all occupied orbitals, α is the fine-structure constant, q_a is the occupation number of the orbital, n_a is its principle quantum number, $P_a(r)$ and $Q_a(r)$ are large and small radial components of Dirac orbital, and V_{VP} is the vacuum polarization potential described by Fullerton and Rinker. The values of $F_a(Z\alpha)$ are calculated using an interpolation procedure of Grant [31].

1.7 X-Ray Spectroscopy and the Relativistic and QED Corrections

A very good illustration of the importance of QED effects (and relativistic) is the X-ray spectroscopy. Although the data showed in this section refers to atomic spectra, the similar relationships will apply to molecular spectroscopy. X-ray photoelectron spectroscopy (XPS) called ESCA (Electron Spectroscopy for Chemical Analysis) is a good tool to study atomic and molecular structure. It provides an information on atomic core electron binding energies (CEBE) and is closely related to the local electronic structure at the atom on which the localized ionization takes place. Already in 1960s Kai Siegbahn et al. showed that inner shell ionization energies depend on the chemical state (environment) of the atom from which the electron is ionized [34, 35]. The photoelectron X-ray technique can also be used in the study of the electron properties of new materials.

The most advanced relativistic approach in relativistic calculations of X-ray spectra, is most likely that based on the Dirac-Coulomb-Breit Hamiltonian and quantum electrodynamic contributions accounted for. In addition, one should also carry out the corresponding correlated-level calculation within these relativistic formalism. To illustrate the role and size of relativistic and QED corrections the core and valence ionisation potentials and excitation energies of noble gases are shown. The relativistic IOTC CASSCF/CASPT2 method together with the restricted active space state interaction, RASSI [36], method combined with the atomic mean field integrals method, AMFI [37], has been used to take spin-orbit (SO) effects into account. One electron part of AMFI code has been used only. It gives approximate spin-dependent corrections and the $np_{\frac{1}{2}}$ and $np_{\frac{3}{2}}$ splittings. The two-electron part of SO corrections were included after adding the Breit expressions. The most important, the Breit and quantum electrodynamic (QED) corrections have been taken into account after the IOTC/RASSI calculations.

To calculate Breit and low order QED corrections is a rather routine procedure in atomic codes. In the molecular calculations it is not an easy task and it is practically impossible to do it. But the goal of this discussion is to show the accuracy of molecular methods and the estimation of the size of the QED corrections that are not included in these codes. The two-component infinite order, IOTC, method is designed for the molecular calculations and, has been implemented in the Molcas, Gamess and Dirac codes, and it does not allow to calculate the QED corrections, at present.

However, some assumption can be made. It is known, from the atomic physics that, firstly, the correlation effects are not important for the calculations of the QED corrections, and secondly, that the core atomic QED corrections do not change significantly while calculated for an isolated atom and an atom in a molecule, particularly for the heavy elements. These features can be exploit and they have encouraged us to add the QED corrections calculated with the atomic GRASP2K code [38] a posteriori to the relativistic IOTC data [39] calculated with the Molcas code [40, 41]. The results have been presented in Tables 1.4, 1.5, 1.6, 1.7 and 1.8 and they have been published earlier by us in [39].

In Tables 1.4, 1.5, 1.6, 1.7 and 1.8 some selected results of the IOTC CASPT2/RASSI valence and core ionization potentials and excitation energies of noble gases are presented. The results of these calculations have been compared with the X-ray ionization and transition energies calculated within the relativistic many-body perturbation theory (RMBPT) designed for the atomic calculations [42], which include the following effects: (i) nuclear size; (ii) relativistic effects (corrections to Coulomb energy, magnetic and retardation energy); (iii) Coulomb and Breit correlation; (iv) radiative (QED) corrections etc., for isolated atoms in vacuum. The comparison with experiment is also presented [39].

In Tables 1.4, 1.5, 1.6 and 1.7 and the valence and core ionization potentials and quantum electrodynamic corrections for the noble gas elements Ne, Ar Table 1.4, and Kr and Xe (Table 1.5) are shown. For the light elements Ne and Ar the QED corrections are very small, and for $2s$ and $2p$ IPs of Ne are almost zero. However for the $1s^{-1}$ states of Ne and Ar are not negligible. The total QED corrections are -0.456 and -3.348 eV, for Ne and Ar respectively. It was also pleasing to note that the IP values of Ne and Ar obtained within the present IOTC method are close to the experimental data and the RMBPT results.

For heavier elements Kr and Xe (Table 1.5) and for ns states, the QED corrections to ionization potentials are more important. The QED corrections to Kr are -32.994, -2.901 and -0.364 eV for $1s^{-1}$, $2s^{-1}$ and $3s^{-1}$ states, accordingly. The appropriate

Table 1.4 Ionization potentials of Ne and Ar (in eV) taken from [39]

Ne and Ar							
Method	$1s^{-1}$	$2s^{-1}$	$3s^{-1}$	$2p_{\frac{1}{2}}^{-1}$	$2p_{\frac{3}{2}}^{-1}$	$3p_{\frac{1}{2}}^{-1}$	$3p_{\frac{3}{2}}^{-1}$
IOTC PT2 Ne	870.78	48.70	–	21.42	21.56	–	–
IOTC PT2 Ar	3208.81	327.86	30.62	251.00	248.47	15.58	15.52
QED Corrections							
BREIT Ne	−0.332	−0.006	–	−0.011	−0.003	–	–
BREIT Ar	−2.342	−0.107	−0.004	−0.178	−0.100	−0.008	−0.001
SE Ne	−0.131	−0.008	–	+0.001	+0.001	–	–
SE Ar	−1.083	−0.091	−0.009	+0.016	+0.012	+0.002	+0.001
VP Ne	+0.007	+0.000	–	−0.000	−0.000	–	–
VP Ar	+0.077	+0.006	+0.006	−0.001	−0.001	−0.000	−0.000
SUM Ne	−0.456	−0.013	–	−0.010	−0.002	–	–
SUM Ar	−3.348	−0.192	−0.013	−0.163	−0.090	−0.006	+0.000
IOTC PT2 + QED Ne	870.33	48.68	–	21.41	21.56	–	–
IOTC PT2 + QED Ar	3205.46	327.67	30.61	250.85	248.39	15.57	15.52
RMBPT Ne	870.73	53.04	–	21.63	21.55	–	–
RMBPT Ar	3207.44	327.31		251.55	249.54		
EXP. Ne[a]	870.23 (866.90)	48.45	–	21.66	21.56	–	–
EXP. Ar	3206.14 (3202.93)[a]	326.32[a]	29.3[b]	250.57[a]	248.46[a]	15.9[c]	15.7[c]

[a]Reference [42]
[b]Reference [34]
[c]Reference [43]

values for Xe are −124.439, −13.178 and −2.198 eV. For each calculated state the IOTC CASSCF/CASPT2 ionization potential improve significantly after the QED corrections are added. The comparison with the experimental and RMBPT data show that for the $1s$ core level the calculated IPs of Kr and Xe differ by about 10 eV and 32 eV respectively. However, if the accuracy of the relativistic IOTC + QED results is measured in terms of ratios of the error with respect to the experimental data and the total experimental values, it is less than 0.1 %, and the performance of the method can be considered as exceptionally good.

Another information, as follows from the data in Tables 1.4 and 1.5 is the role of the individual QED terms. We see that the most important is the negative Breit correction. The self energy has about half of the Breit value and is also negative. The vacuum polarization has a positive value and is much smaller than Breit and SE corrections.

Table 1.5 Ionization potentials of Kr and Xe (in eV) taken from [39]

Kr and Xe				
Method	$1s^{-1}$	$2s^{-1}$	$3s^{-1}$	$4s^{-1}$
NR PT2 Kr	14106.23	1869.70	280.80	27.96
NR PT2 Xe	33260.20	5117.41	–	201.39
IOTC PT2 Kr	14351.41	1924.06	295.12	28.81
IOTC PT2 Xe	34657.03	5462.93	–	212.82
QED Corrections in GRASP				
BREIT Kr	−22.142	−1.684	−0.167	−0.007
BREIT Xe	−82.490	−7.737	−1.101	−0.150
SE Kr	−12.135	−1.348	−0.218	+0.023
SE Xe	−48.990	−6.260	−1.260	−0.264
VP Kr	+1.283	+0.131	+0.021	+0.002
VP Xe	+7.041	+0.819	+0.163	+0.034
SUM Kr	−32.994	−2.901	−0.364	−0.027
SUM Xe	−124.439	−13.178	−2.198	−0.380
IOTC PT2 + QED Kr	14318.42	1921.16	294.76	28.78
IOTC PT2 + QED Xe	34532.59	5449.75	–	212.79
RMBPT Kr	14328.06	1925.49	–	–
RMBPT Xe	34566.5	5453.7	–	–
EXP. Kr	14327.19 (14324.61)[a]	1920.4 (1916.3)[a]	292.8[b]	27.4[b]
EXP. Xe	34565.13 (34593.)[a]	5452.57 (5452.89)[a]	1148.7[b]	213.2[b]
Relative error Kr	0.06	0.04	0.91	4.62
Relative error Xe	0.09	0.05	–	0.15

[a]Reference [42]
[b]Reference [34]

The X-ray spectra gives the information about the ionization potentials of the core and valence states but also about the excitations between states.

In Tables 1.6 we present some selected IOTC CASPT2 + QED excitation energies of Ne and Ar, such as $1s^{-1} \rightarrow ns^{-1}$ and $1s^{-1} \rightarrow np^{-1}$ transitions. These excitations are from the $1s^{-1}$ state with one hole to another state with one hole in the higher *ns* or *np* levels. As it can be expected, the QED corrections are important for all excitations. The calculated excitation energies for the Ne and Ar elements entirely agree with the experimental data and are sometimes even better than the RMBPT results (KL1, $K\alpha_1$, $K\alpha_2$ of neon and $K\beta_1$, $K\beta_3$ of argon).

Tables 1.7 and 1.8 show the calculated KL1, KM1, KN1 and $K\beta_1$, $K\beta_3$, $K\beta_2^{II}$, $K\beta_2^{I}$ excitation energies of Kr and Xe. In the case of krypton the difference between the IOTC + QED energies and the experimental values do not exceed 12 eV and is the smallest for the KL1 line (3 eV). Similar agreement can be observed with the RMBPT data.

Table 1.6 Excitation energies of Ne and Ar ions (in eV) taken from [39]

Ne and Ar						
Method	KL1[a]	KM1[b]	Kα_1[c]	Kα_2[d]	Kβ_1[e]	Kβ_3[f]
IOTC PT2 Ne	822.08	–	849.36	849.23	–	–
IOTC PT2 Ar	2880.95	3178.19	2960.34	2957.81	3193.29	3193.23
QED corrections						
BREIT Ne	−0.326	–	−0.329	−0.321	–	–
BREIT Ar	−2.235	−2.338	−2.242	−2.164	−2.341	−2.334
SE Ne	−0.123	–	−0.130	−0.130	–	–
SE Ar	−0.991	−1.073	−1, 094	−1.099	−1.084	−1.084
VP Ne	+0.007	–	+0.007	+0.007	–	–
VP Ar	+0.071	+0.076	+0.076	+0.076	+0.077	+0.077
SUM Ne	−0.442	–	−0.454	−0.456	–	–
SUM Ar	−3.155	−3.335	−3.258	−3.185	−3.348	−3.341
IOTC PT2 + QED Ne	821.64	–	848.91	848.77	–	–
IOTC PT2 + QED Ar	2877.79	3174.86	2957.08	2954.63	3190.03	3190.05
RMBPT Ne	817.69	–	849.17	849.09	–	–
RMBPT Ar	2880.13	3177.4	2957.90	2955.89	3191.47	3191.31
EXP. Ne	822.07[h]	–	848.61[g]	848.61[g]	–	–
EXP. Ar	2879.6[h]	3176.12[h]	2957.68[g]	2955.57[g]	3190.49[g]	3190.49[g]

[a] $1s^{-1} \rightarrow 2s^{-1}$
[b] $1s^{-1} \rightarrow 3s^{-1}$
[c] $1s^{-1} \rightarrow 2p\frac{3}{2}$
[d] $1s^{-1} \rightarrow 2p\frac{1}{2}$
[e] $1s^{-1} \rightarrow 3p\frac{3}{2}$
[f] $1s^{-1} \rightarrow 3p\frac{1}{2}$
[g] Reference [42]
[h] Calculated as the difference between the appropriate experimental energies (Table 1.2)

For xenon the absolute error is larger and varies from about 7.3 eV for the Kβ_3 excitation until 44 eV for the Kβ_1 line. However, the relative error does not exceed 0.13 % of the experimental value.

Additionally to the relativistic calculations one may see the nonrelativistic results for the *ns* ionization potentials (Table 1.5) and KL1, KM1 and KN1 excitation energies (Table 1.6) of Kr and Xe, which show the importance of the relativistic effects on the calculated values. The main relativistic effect is obviously associated with the deepest core level and varies from about 245 eV in Kr (total 1s IOTC IP is about 14351 eV) to 1396 eV in Xe (total 1s IOTC IP is about 34657 eV). However, for

Table 1.7 Excitation energies of Kr and Xe ions (in eV) taken from [39]

Kr and Xe			
Method	KL1[a]	KM1[b]	KN1[c]
NR PT2 Kr	12236.53	13825.43	14078.27
NR PT2 Xe	28142.79	–	33058.81
IOTC PT2 Kr	12427.35	14056.29	14322.6
IOTC PT2 Xe	29194.1	–	34444.21
Corrections in GRASP			
BREIT Kr	−20.458	−21.975	−22.135
BREIT Xe	−74.754	–	−82.341
SE Kr	−10.787	−11.917	−12.113
SE Xe	−42.730	–	−48.726
VP Kr	+1.152	+1.262	+1.281
VP Xe	+6.222	–	+7.001
SUM Kr	−30.093	−32.630	−32.967
SUM Xe	−111.261	–	−124.060
IOTC PT2 + QED Kr	12409.22	14022.52	14289.63
IOTC PT2 + QED Xe	29082.84	–	34320.15
RMBPT Kr	12402.57	14034.9	14301.2
RMBPT Xe	29112.8	–	34353.4
EXP. Kr[d]	12406.79	14034.39	14299.79
EXP. Xe[d]	29112.56	–	34351.83
Relative error % Kr	0.02	0.08	0.07
Relative error % Xe	0.10	–	0.09

[a] $1s^{-1} \rightarrow 2s^{-1}$
[b] $1s^{-1} \rightarrow 3s^{-1}$
[c] $1s^{-1} \rightarrow 4s^{-1}$
[d] Calculated as the difference between the appropriate experimental energies (Table 1.4)

intermediate levels, in particular for higher s-type holes, the relativistic effect is also quite significant and its inclusion is crucial in the core X-ray spectra calculations.

Finally, we may estimate that the total calculated QED corrections, for most of the states, are usually about 10 % of the corresponding relativistic correction.

1.8 Concluding Remarks

The role and importance of relativistic effects in the chemistry is already widely acknowledged. The problem which remains is the choice of the best method for the calculation of these effects. In advanced calculations they need to be spin-free or spin-dependent algorithms. One of the most exact two-component method is the infinite order two-component IOTC theory implemented in its spin-free version into

Table 1.8 Excitation energies of Kr and Xe ions (in eV) taken from [39]

Kr and Xe				
Method	Kβ_1[a]	Kβ_3[b]	Kβ_2^{II}[c]	Kβ_2^{I}[d]
IOTC PT2 Kr	14137.56	14129.19	14337.68	14337.00
IOTC PT2 Xe	33716.06	33653.58	34520.96	34509.00
QED Corrections GRASP2K				
BREIT Kr	−21.967	−21.813	−22.141	−22.128
BREIT Xe	−81.283	−80.445	−82.35	−82.18
SE Kr	−12.148	−12.162	−12.137	−12.139
SE Xe	−48.994	−49.083	−49.001	−49.018
VP Kr	+1.285	+1.285	1.284	1.297
VP Xe	+7.054	+7.046	+7.045	+7.043
SUM Kr	−32.83	−32.69	−32.994	−32.970
SUM Xe	−123.22	−122.48	−124.30	−124.15
IOTC PT2 + QED Kr	14104.73	14096.50	14304.68	14304.03
IOTC PT2 + QED Xe	33592.84	33531.10	34396.66	34384.85
RMBPT Kr	14113.3	14105.7	14314.10	14313.0
RMBPT Xe	33624.6	33563.03	34408.9	34408.
EXP. Kr[e]	14112.82	14104.96	14315.0	14315.0
EXP. Xe[e]	33624.23	33563.20	34414.7	34414.7
Relative error % Kr	0.06	0.07	0.07	0.08
Relative error % Xe	0.09	0.10	0.05	0.09

[a] $1s^{-1} \rightarrow 3p_{\frac{3}{2}}$
[b] $1s^{-1} \rightarrow 3p_{\frac{1}{2}}$
[c] $1s^{-1} \rightarrow 4p_{\frac{3}{2}}$
[d] $1s^{-1} \rightarrow 4p_{\frac{1}{2}}$
[e] Reference [42]

the MOLCAS and GAMES (USA) codes. Spin-dependent properties should also be included in the calculations of many properties, however it is not easy to do it. One of the possibilities gives the atomic mean-filed approximation discussed in this chapter. Comparison of the results obtained by the IOTC method with one of the most commonly used, the Douglas-Kroll-Hess method (DKHn, ($n = 2, 3, \ldots 8$), show the very good agreement of the results. In these chapter some illustration of these data have been presented. In recent years, due to the strong development of experimental methods, we observe an increased interest in spectroscopy of small chemical molecules, and their potential energy curves. To get the good quality potential energy curve one needs the good total energy. It was not necessary when we were only interested in spectroscopic parameters of the molecules. It turns out, that for the description of the potential energy curves the IOTC method gives the results which can be obtained only in the high order DKHn approximations.

The are some experimental methods, such as X-ray spectroscopy, for which a description of the experimental results by the methods based on the relativistic quantum mechanics is not sufficient. These spectroscopic methods play an increasing role, recent years, in the search for new materials. To properly described the X-ray spectroscopy, the Breit interactions and the corrections resulting from quantum electrodynamics, such as the self energy and vacuum polarization must be added to the nonrelativistic results. The calculation of these terms is not an easy task to do, especially in the molecular codes. This chapter shows how the results of the spin-free CASSCF/CASPT2 IOTC calculations of the core and valence ionization and excitation energies change when we add to them Breit and quantum electrodynamics corrections. It also shows that the molecular relativistic IOTC method together with the electrodynamic expressions is able to give the X-ray parameters comparable with the experimental results.

References

1. Balasubramanian K (1990) Chem Rev 90:93
2. Balasubramanian K (1997) Relativistic effects in chemistry, part A: theory and techniques. Wiley, New York
3. Pershina V (2010) Challenges and advances in computational chemistry and physics. In: Barysz M, Ishikawa J (eds) Relativistic methods for chemists, vol 10. Springer, Dordrecht, pp 451–513
4. Lindgren I (2011) Relativistic many-body theory. New field-theoretical approach. In: Springer series on atomic, optical and plasma physics, vol 63. Springer, New York
5. Heully J-L, Lindgren I, Lindroth E, Lundquist S, Mårtenson-Pendril A-M (1986) J Phys B 19:2799
6. Barysz M, Ishikawa Y (eds) (2010) Challenges and advances in computational chemistry and physics., Relativistic methods for chemistsSpringer, London
7. Barysz M, Sadlej AJ (2002) J Chem Phys 116:2696
8. Barysz M, Sadlej AJ (1997) Theoret Chem Acc 97:260
9. Hess BA (1985) Phys Rev A 32:756
10. Hess BA (1986) Phys Rev A 33:3742
11. Wolf A, Reiher M, Hess BA (2002) J Chem Phys 117:9215
12. Wolf A, Reiher M, Hess BA (2002) In: Schwerdfeger P (ed) Relativistic electronic structure theory, part 1 fundamentals. Elsevier, Amsterdam, p 622
13. Reiher M, Wolf A (2004) J Chem Phys 121:2037
14. Pyykkö P (1986) Relativistic theory of atoms and molecules I - a bibligraphy 1916–1985., Lecture notes in chemistrySpringer, Berlin
15. Pyykkö P (1993) Relativistic theory of atoms and molecules II - a bibligraphy 1986–1992, vol 60., Lecture notes in chemistrySpringer, Berlin
16. Pyykkö P (2000) Relativistic Theory of atoms and molecules III - a bibligraphy 1993–1999, vol 76., Lecture notes in chemistrySpringer, Berlin
17. Pyykkö P (2006) Database 'RTAM' (relativistic quantum chemistry database 1915–1998). http://www.csc.fi/lul/rtam/rtamquerry.html
18. Ilias M, Kellö V, Urban M (2010) Acta Physica Slovaca 60:259
19. Andersson K, Malmqvist P-A, Roos BO, Sadlej AJ, Woliński K (1990) J Chem Phys 94:5483
20. Andersson K, Malmqvist P-A, Roos BO (1992) J Chem Phys 96:1218
21. Andersson K (1995) Theoret Chim Acta 91:31

22. Roos BO, Andersson K, Fülscher MP, Malmqvist P.-A, Pierloot K, Merchán M (1996). In: Prigogine I, Rice SA (eds) Advances in chemical physics: new methods in computational quantum mechanics, vol XCIII. Wiley, New York, p 219
23. Cernusak I, Kellö V (2003) Collect Czech Chem Commun 68:211
24. Turski P, Barysz M (2000) J Chem Phys 111:4654
25. Dyall KG, Faegri JRK (2007) Introduction to relativistic quantum chemistry. Oxford University Press, New York
26. Majumdar D, Sz Roszak, Leszczyński J (2010) Challenges and advances in computational chemistry and physics. In: Barysz M, Ishikawa J (eds) Relativistic methods for chemists, vol 10. Springer, Dordrecht, pp 373–406
27. Hess BA, Marian CM, Wahlgren U, Gropen O (1996) Chem Phys Lett 251:365
28. Schimmelpfennig B, Maron L, Wahlgren U, Teitchteil Ch, Fagerli A (1998) Gropen O 286:267
29. Marian Ch (2001) Reviews in computational chemistry. In: Lipkowitz K, Boyd D (eds) vol 17. Wiley-VCH, pp 99–204
30. Craig DP, Thirunamachandran T (1984) Molecular quantum electrodynamics. An introduction to radiation-molecule interaction. Academic Press, London
31. Grant IP (2007) Relativistic quantum theory of atoms and molecules., Theory and computationSpringer, New York
32. Lindgren I (2011) Relativistic many-body theory., A new field -theoretical approachSpringer, NewYork
33. Jensen F (2007) Introduction to Computational Chemistry. John Wiley and Sons, England
34. Siegbahn K, Nordling C, Johansson G, Hedman J, Hedén JPF, Hamrin K, Gelius U, Bergmark T, Werme LO, Manne R, Baer Y (1969) ESCA applied to free molecules. North-Holland Publishing Company, Amsterdam
35. Siegbahn K, Fahlman C, Nordberg R, Hamrin K, Hedman J, Johansson G, Bergamrk T, Karlsson S-E, Lindgren L, Lindberg B (1967) ESCA, atomic, molecular and solid state structure studied by means of electron spectroscopy. Almqvist and Wiksell, Uppsala
36. P-øA Malmqvist, Roos BO, Schimmelpfenning B (2002) Chem Phys Lett 357:230
37. Schimmelpfennong B, Maron L, Wahlgren U, Teichteil C, Fagerli H, Gropen O (1998) Chem Phys Lett 286:267
38. Jönson P, He X, Froese Fischer C, Grant IP (2007) Comput Phys Commun. 177:597
39. Barysz M, Syrocki Ł (2013) Mol Phys http://dx.doi.org/10.1080/0026897.2013.843033
40. Aquilante F, de Vico L, Ferre N, Ghigo G, Malmqvist P-øA, Neogrady P, Pedersen TB, Pitonak M, Reiher M, Roos BO, Serrano-Andres L, Urban M, Veryazov V, Lindh R (2009) Molcas Version 7.3. Lund University, Sweden
41. The IOTC method was implemented in the version of the Molcas.7.3 system of programs by Slovakia Group of Quantum Chemistry (2010). Comenius University, Bratislava, Slovakia. The corresponding patches for Molcas 7.3 releases of Molcas can be obtained directly from Barysz M (E–mail: temjb@chem.umk.pl)
42. Deslattes RD, Kessler JEG, Indelicato P, de Billy L, Lindroth E, Anton J (2003) Rev Mod Phys 75:35
43. Sur Ch, Chaudhuri RK (2004) J Phys B: Atom Mol Opt Phys 37:4127

Chapter 2
How Can One Locate the Global Energy Minimum for Hydrogen-Bonded Clusters?

Sergey Kazachenko and Ajit J. Thakkar

Abstract An important problem in many areas of chemistry and physics is finding the global energy minimum on a potential energy surface. The difficulty stems from the exponential increase in the number of local minima with the size of the system. An efficient algorithm to find the global minima of water clusters is described and tested. It works well for clusters containing up to about 55 water molecules. A generalization to other hydrogen-bonded clusters is outlined. Applications of this algorithm to water clusters and methanol clusters have already been reported in the literature.

2.1 Introduction

Material particles consisting of a few to a few thousand atoms are called clusters. Cluster properties can have dramatic size and shape dependence. Clusters can serve as building blocks for new materials and electronic devices. Hence clusters of metals, semiconductors, ionic solids, rare gases, and small molecules have been studied using both theoretical and experimental methods. Atomic and molecular clusters [1] are held together by hydrogen bonds [2] or by relatively weak intermolecular forces [3, 4].

In particular, small clusters of hydrogen-bonded water molecules have received a lot of attention; see, for example, spectroscopic work [5–16], and density functional theory and *ab initio* investigations [17–31]. Water is not a simple substance and has anomalous physical and chemical properties. More than a century of work has been devoted to modeling and understanding these properties. Nevertheless, many aspects of water remain unsolved puzzles and the development of water models continues

S. Kazachenko · A.J. Thakkar (✉)
Department of Chemistry, University of New Brunswick,
Fredericton, NB E3B 5A3, Canada
e-mail: ajit@unb.ca

S. Kazachenko
Present Address: Department of Chemistry, Queen's University,
Kingston, ON K7L 3N6, Canada

J. Leszczynski and M.K. Shukla (eds.), *Practical Aspects of Computational Chemistry IV*, DOI 10.1007/978-1-4899-7699-4_2

to attract much interest. Water models can use water clusters, $(H_2O)_n$, as building blocks [32–36].

An important part of cluster research is a characterization of the global minimum of the potential energy surface—the structure of the most stable cluster. A scholarly and comprehensive survey of the difficult problem of locating the global minimum on a potential energy surface is available [37]. Computational approaches to the determination of global minima are very attractive because the experimental determination of the structure of the ground state is extremely difficult for complex systems. Unfortunately, the computational solution is rather expensive because the number of local minima on the surface grows exponentially with the dimensionality of the surface or, in other words, with the number of atoms in the system. An algorithm searching for a global minimum can easily get trapped in one of these local minima because the traversal of many intermediate minima and the crossing of high barriers separating them may be required to find a lower local minimum. In high-dimensional cases, which are the norm rather than the exception in physically important clusters, a global minimum can only be discovered and verified after a sufficiently large number of low energy minima have been located and compared.

We have been studying water clusters for some time now. Hence we have been gradually developing and refining our own algorithm for locating a global energy minimum for water clusters. We previously described successive improvements to our algorithm in an incremental manner in a series of papers [38–41] and a thesis [42]. The purpose of this chapter is to try and describe the current state of our algorithm in a relatively self-contained and cohesive manner without the reader having to consult our previous work.

All algorithm tests on water clusters discussed in this chapter are done with the empirical TIP4P model [43], a reparametrization of the venerable Bernal-Fowler model [44]. The interactions are considered to be pair-wise additive and the water monomers are held rigid so that they do not vibrate. The interaction between a pair of water molecules is given by a Lennard-Jones (12,6) interaction between the oxygen atoms and electrostatic interactions between three point charges on each water molecule. There are positive charges on the hydrogen atoms and a balancing negative charge between the hydrogen atoms along the C_2 symmetry axis. The TIP4P model gives a reasonable thermodynamic description of liquid water. Many studies have been devoted to finding the global minima of TIP4P water clusters [38–40, 45–51]. Generally good agreement has been found [48] between its predictions for small clusters with $n \leq 12$ and both *ab initio* and experimental results. The TIP4P global minima for cluster sizes up to 47 are now firmly established [41] except for $n = 39$ and $n = 45$. Putative TIP4P global minima have been reported [41] for larger clusters with $n \leq 55$. Low-energy TIP4P structures for much larger clusters with selected sizes have been located [7, 49]. Pure water clusters based on the TIP4P model are now a benchmark for methods for global optimization of hydrogen-bonded clusters.

Some of the basic terminology commonly used in the global optimization literature is summarized in Sect. 2.2 which also contains a brief description of basin and minima hopping. Next, the long Sect. 2.3 details the various ways by which we optimize the topology of water clusters with a fixed oxygen skeleton. The overall algorithm is then described in Sect. 2.4 and its generalization to other hydrogen-bonded clusters is outlined in Sect. 2.5. A few concluding remarks are made in Sect. 2.6.

2.2 Basin Hopping and Minima Hopping

A discussion of search algorithms is facilitated by the introduction of some terminology that is widely used in the literature. A *basin* is a region of geometrical configuration space around a minimum on the potential energy surface. The basin contains all structures, or configurations, from which a search can reach this minimum using only small steps and downhill moves. A *super-basin* is the union of several neighboring basins. A *funnel* $\mathcal{F}$ is a super-basin with the property that starting at any point in $\mathcal{F}$ one can reach the lowest minimum in $\mathcal{F}$ without crossing barriers that are very high relative to the average energy difference between local minima in $\mathcal{F}$. Figure 2.1 is a schematic illustration of two super-basins, one of which is a funnel and the other one is not.

Use of the Boltzmann factor, $\exp(-\Delta E/kT)$, to control all the steps used to leave a basin makes the crossing of high barriers a rare event. Hence, global minimization methods based exclusively on thermodynamic principles can be extremely slow because they may find it hard to exit a funnel and may repeatedly visit neighboring configurations that are close in energy. Many standard algorithms such as simulated annealing [53–55] and basin hopping [46, 56–59] are based on thermodynamic principles. However, genetic algorithms [60–62] and minima hopping [52] are not.

The fundamental idea in basin hopping is that the potential energy surface is effectively transformed to a stepped surface [46, 56, 63]. This is done by ending each search step with a local optimization so that one effectively searches over local minima. How one starts the next search step distinguishes various algorithms based on this seminal idea. We chose minima hopping [52] as the global optimization

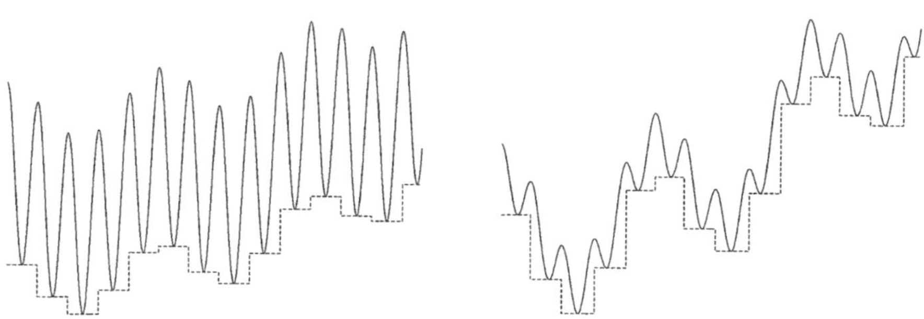

Fig. 2.1 Two super-basins: the one on the right is a funnel but the one on the left is not. Figure adapted from Goedecker [52]

algorithm in our preliminary work. Minima hopping can be thought of as a version of basin-hopping [46, 56, 63]. Its goal is to avoid revisiting previously located local minima without forbidding repeated passages through any transition basins that may separate many funnels. This is achieved in the minima-hopping method by maintaining a list of visited minima and using adaptive thresholds to leave a basin more and more vigorously each time the algorithm revisits it. Molecular dynamics (MD) steps are used as an efficient mechanism for crossing energy barriers.

The minima-hopping algorithm has an inner part for jumping to a new local minimum and an outer part for accepting or rejecting the new local minimum. In the inner part, one tries to escape the current minimum $\mathbf{M}_c$ by a short MD simulation in which the atoms have a Boltzmann velocity distribution such that their kinetic energy is fixed at E_{kin}. The simulation is stopped as soon as a minimum is encountered along the trajectory or the maximum number of MD steps is exceeded. Then one optimizes to the closest local minimum $\mathbf{M}$ using a suitable local optimization method. If this minimum has been visited previously, then multiply E_{kin} by $\beta > 1$ to make the next escape attempt more vigorous and repeat the inner part. If this is a newly found local minimum, divide E_{kin} by β to make the next escape attempt less vigorous and go to the outer part. The latter accepts or rejects the local minimum $E(\mathbf{M})$ as follows. If $E(\mathbf{M}) - E(\mathbf{M}_c) < E_{\mathrm{diff}}$, the minimum is accepted and E_{diff} is divided by $\alpha > 1$ to make the next acceptance more difficult. Otherwise it is rejected and E_{diff} is multiplied by α to make the next acceptance easier. We tried various values of α and β but were unable to improve upon the values of $\alpha = 1.02$ and $\beta = 1.05$ recommended by Goedecker [52]. The minima-hopping algorithm stops when E_{kin} reaches or exceeds a maximum value $E_{\mathrm{kin}}^{\mathrm{max}}$ or the number of inner steps exceeds a preset limit N_{iter}.

It is important to do local optimizations as efficiently as possible because they are a time-consuming part of minima hopping. In our implementation, we perform local minimization in two steps. The first optimization uses the limited memory L-BFGS method [64, 65] with a loose convergence threshold. The optimization is refined in the second step which uses Davidon's optimally conditioned variable metric method [66] and a more stringent convergence criterion.

It soon became apparent to us that minima hopping did not always succeed in finding the optimum hydrogen bond topology for water clusters [39]. Nevertheless, minima hopping served us as the basic algorithm upon which improved and specialized methods for water clusters were built. Special topology refining algorithms, described in the next section, are used both within the global search algorithm and independently for a refining step on the list of saved low energy minima. Although many features described here were designed to work only with water, the core algorithm is general and has been applied successfully to other hydrogen-bonded clusters including pure methanol clusters [67] and pure clusters of ethanol, n-propanol, and iso-propanol.

2.3 Finding the Optimal Hydrogen Bond Topology

2.3.1 Representation of H-Bonded Clusters

An algorithm for global optimization of hydrogen-bonded clusters requires a convenient description of a cluster and its features. At the most basic level, the cluster is defined by the Cartesian coordinates and types of the atoms it consists of. We use that information to derive some important properties and find a convenient way to represent them.

A basic property of any H-bonded cluster is the presence or absence of a hydrogen bond between a pair of given molecules. We use simple geometric criteria to decide whether there is an O–H···O hydrogen bond: the H···O distance should be less than 2.5 Å and the O–H···O angle should be greater than 90°. It is convenient to describe the H-bonds in a cluster of n molecules by a graph. The latter is represented as an $n \times n$ adjacency matrix **A** with elements obeying the usual rules: $A_{ij} = 1$ if there is a bond between molecules i and j and $A_{ij} = 0$ otherwise. By definition, $A_{ii} = 0$ because molecules are not connected to themselves and $A_{ij} = A_{ji}$ because the bond direction is not taken into consideration. The adjacency matrix allows one to calculate, for example, the total number of hydrogen bonds in a cluster as $N_{\text{bond}} = \frac{1}{2} \sum_{ij} A_{ij}$. One can also calculate the number of rings of a given size formed by connected molecules. We implemented the counting of rings using a backtracking algorithm based on the work of Franzblau [68]. There are several possible definitions of a ring in a graph. We count all rings in which each monomer is connected to exactly two other monomers belonging to the same ring.

The H-bond directionality is important in topology optimization. Keeping track of both the existence and direction of the H-bonds in a cluster can be accomplished by using a digraph. The latter can be represented by a directed adjacency matrix **D**. As with the adjacency matrix, the absence of a bond between monomers i and j is indicated by $D_{ij} = 0$ and so all diagonal elements vanish, $D_{ii} = 0$. If there is a bond in the $i \rightarrow j$ direction then $D_{ij} = 1$ and $D_{ji} = 0$. The **D** matrix contains, for example, information about the number of donor (don) and acceptor (acc) H-bonds for each molecule: $N^i_{\text{don}} = \sum_j D_{ji}$, $N^i_{\text{acc}} = \sum_j D_{ij}$. Note that the summation index depends on the definition of the H-bond direction. In a computer program, the sparsity of **A** and **D** can be exploited by storing them as linked lists of non-zero elements.

2.3.2 Why Is Topology Optimization Needed?

A simple characterization of a cluster of n water molecules is its skeleton or graph by which we mean the connectivity of the monomers described by an adjacency matrix **A**. The positions of the oxygen atoms and the skeleton define the shape of a cluster as in the left panel of Fig. 2.2. However, virtually all hydrogen bonds between water molecules can be assigned a direction, say from donor to acceptor. Hence,

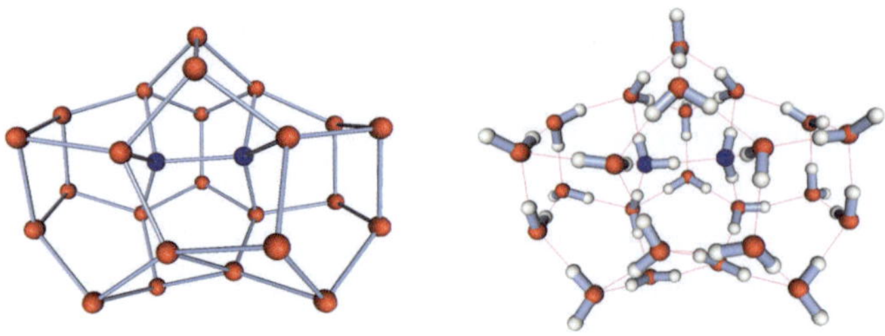

Fig. 2.2 An example of a framework and a complete cluster with one of the possible H-bond topologies for a $(H_2O)_{26}$ cluster. Two internal molecules are shown in *blue*

there can be many water cluster geometries with the same skeleton but different hydrogen bond topologies; that is to say, cluster structures with the same skeleton but different directions in one or more of the hydrogen bonds. In the terminology of Sect. 2.3.1, any given adjacency matrix **A** can correspond to many directed adjacency matrices $\mathbf{D}_k$. The oxygen framework and the H-bond topology together define a complete cluster as in the right panel of Fig. 2.2. The stability of a given framework can vary significantly depending on the hydrogen bond distribution [69–75].

Since each cluster framework maybe paired with a large number of H-bond topologies, the resulting cluster structures can have significant energy differences. Therefore, it is important that we are able to locate the one with the lowest energy. It might also be of interest to know how many topologies exist for a given framework and possibly separate them into categories. Locating the minimum energy topology, or topology optimization, can be done either as a separate procedure on selected water clusters or as a part of a global optimization.

The problem of finding the best hydrogen bond topology for water clusters has been studied extensively; see, for example, Refs. [49, 51, 69–74, 76–83]. Polyhedral and cubic water clusters were described in several publications using graph theory [69, 74, 78, 79]. The effects of H-bond topology on the stability and spectroscopic properties of water octamers were studied by Francisco et al. [73, 80]. The relation between the topology and interaction energy was studied for polyhedral clusters [72, 81] and for some other shapes and sizes [71, 81]. As a result, several formulas were developed to predict the relative energy of a water cluster based on its monomer connectivity. A different approach is to use a proton transfer to change the direction of hydrogen bonds and so sample a number of topologies using general optimization methods [49, 51]. There is also the brute-force approach of examining all possible topologies for an arbitrary cluster shape [77, 82, 83]. The last two approaches will be discussed later in more detail.

Unfortunately, most of the suggested methods were designed only for a particular oxygen framework or turned out to be inefficient. We created several algorithms suitable to our goal of reliably finding the lowest energy topology for a water cluster of an arbitrary shape. The algorithms were efficient at locating lowest-energy topologies when applied to clusters with no more than 55 molecules.

2.3.3 *Comparing and Storing Water Cluster Minima*

In our approach for H-bond topology optimization, a water cluster is defined as a structure with a particular oxygen framework. A framework is assumed to have a unique adjacency matrix **A**. No two frameworks will have the same matrix and, therefore, clusters can be compared by comparing their **A** matrices. All possible topologies for a framework are considered to be variations of the same cluster. Therefore, each cluster defined by its **A** matrix has a set of local minima corresponding to all possible H-bond topologies. Such separation of the framework and topology helps to reduce the number of local minima that are stored during a global minimum search. Only the version of a cluster with the best topology is kept.

How can we determine if two sets with an equal number of water molecules form clusters with an identical shape? **A** matrices can be compared directly if the position of the molecules has changed only slightly, i.e. after a small distortion. However, in general we want the comparison to be independent of the order of molecules in a coordinate list. In our case this is done in three steps which helps to reduce the number of time-consuming operations. First, the number of H-bonds must be the same, which is easy to calculate and to check. Next, we use the idea that the way molecules are connected has an effect on the number of rings formed by H-bonded monomers in a cluster. Moreover, the number of rings is independent of the order of molecules in a list. Therefore, as the second step, it is required that the number of rings of each size from 3 to 10 molecules must be the same. For those cases where the ring rule is also satisfied, an alignment of **A** matrices is used to perform the final check which must allow for the possibility that the order of the elements (monomers) is different. A backtracking procedure is used to examine the connectivity of each monomer and to try and match pairs of molecules from the clusters being compared. This procedure allows us to determine whether two matrices correspond to the same framework even when the order of the elements is not the same.

When the list of visited minima is large enough, it is no longer feasible to keep either the Cartesian coordinates or the adjacency matrix for each local minimum. In that case, the third step that includes alignment of **A** matrices can be omitted. Nine integers, the number of H-bonds and the numbers of three- through ten-membered rings, are saved for each local minimum and provide a robust and reliable way to compare cluster frameworks.

2.3.4 *Filters to Screen Topologies*

Finding the lowest-energy topology of a large cluster framework would be impossible if it required the local optimization of all or even most of its topologies. Fortunately, the number of expensive local optimizations required can be reduced by using simple filters or criteria based solely on geometrical considerations to weed out the topologies that are likely to have a high energy. Note that although the shape of a

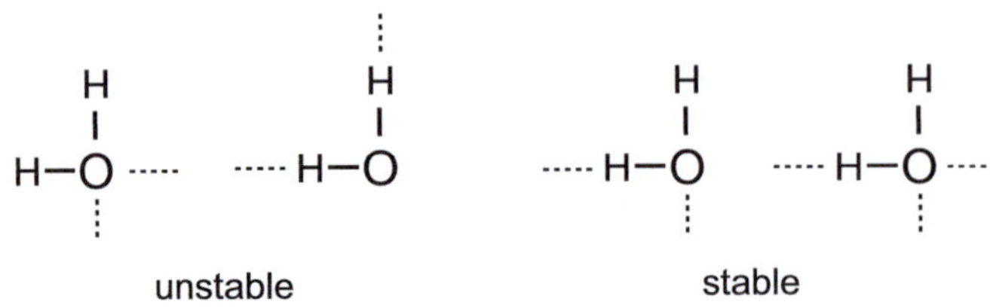

Fig. 2.3 Examples of less stable and more stable hydrogen bonding of a monomer

water cluster depends somewhat on its topology, a stable (i.e. low-energy) framework does not break or change when its topology is changed.

The first thing to do is to count the number of H-bonds and their types for each molecule. We define a donor bond as a bond created by a hydrogen atom of a given molecule and an acceptor bond as the one created by an oxygen atom of the molecule. We use a modification of the venerable Bernal-Fowler "ice rules" [44] and require that

$$1 \leq \text{number of donor bonds} \leq 2,$$

and

$$1 \leq \text{number of acceptor bonds} \leq 2 \text{ or } 3.$$

In words, each water molecule in a cluster must have at least one donor and one acceptor bond. Moreover, there can be at most two donor bonds (one for each hydrogen) and at most two acceptor bonds. These rules allow one to avoid unstable connectivity of molecules; see Fig. 2.3. In rare cases a water molecule in a cluster accepts three hydrogen bonds and it is necessary to account for such a possibility. However, allowing all molecules to accept three H-bonds would significantly increase the number of topology combinations. Thus, the algorithm detects penta-coordinated molecules in the input geometry and only those can have up to three acceptor bonds, while the rest are restricted to have no more than two acceptor bonds.

We found that in some cases changing the cluster topology leads to unrealistic monomer angles; see Fig. 2.4. Note that the molecules in a cluster shift somewhat during a local optimization of a new topology. Thus we must allow for a range of angles to be accepted. Test calculations suggest that a topology can reliably be

Fig. 2.4 An example of an unrealistic monomer angle in one of the H-bond topologies for an $(H_2O)_{12}$ cluster

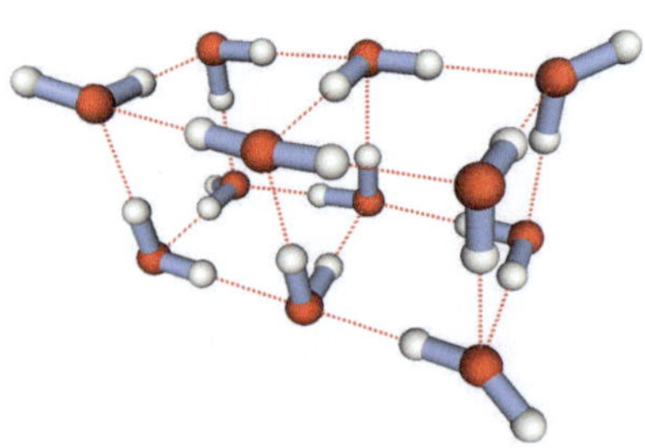

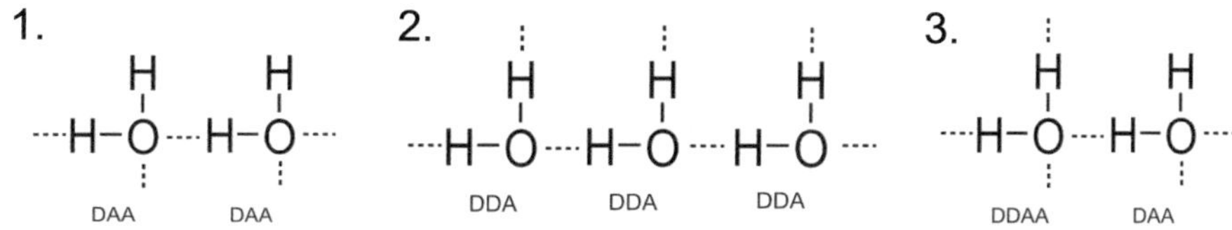

Fig. 2.5 Adjacent monomers with a connectivity that leads to higher energy (D—donor, A—acceptor)

considered unfavorable if it leads to a monomer angle larger than 150° or smaller than 65°. These values span a large enough range to ensure that no important topologies are missed because of the angle filter.

It is well-established that adjacent dangling hydrogen atoms lead to a higher cluster energy [69, 71–73, 79]. In addition, Anick found two other patterns leading to an increased energy [81]. These three motifs of connectivity are shown in Fig. 2.5. The types of H-bonds are labeled by D for donor and A for acceptor. The number of motifs 1 and 2 that occur in a cluster must be kept as low as possible. This means that one can safely discard all topologies with more such motifs than the lowest number found at any given stage. The third rule is not as strong. We found that a topology can be discarded safely only when the number of such motifs exceeds the lowest number found *plus two*.

2.3.5 Topology Optimization by Enumeration

A straightforward way of finding the best topology is to generate all possible H-bond distributions, do local optimizations on each one, and compare their energies. Miyake and Aida used adjacency and directed adjacency matrices (graphs and digraphs) to describe the framework and the topology of a cluster, respectively [82]. Knowing the digraph and the coordinates of the oxygen atoms, one can recreate the topology and then perform a local optimization to relax the cluster geometry. The use of adjacency matrices was a promising idea; however, in their algorithm all possible **D** matrices were generated first and only later were they checked to see if they were useful. This led to a large wasted computational effort and the largest cluster size they were able to study was limited to just eight water molecules. Vukičević et al. improved the performance of the method by eliminating the need to generate unrealistic matrices [77, 83] and analyzed the topology for clusters with up to 12 water molecules.

The ideas of using graphs and digraphs and eliminating undesired combinations on the fly were used in the creation of our method, called topology enumeration (NT). Consider the cage hexamer shown in Fig. 2.6a as an example. We start from the given framework and its adjacency matrix **A**. Each row and column of **A** with the same index corresponds to a molecule. Defining all connections of a molecule to the

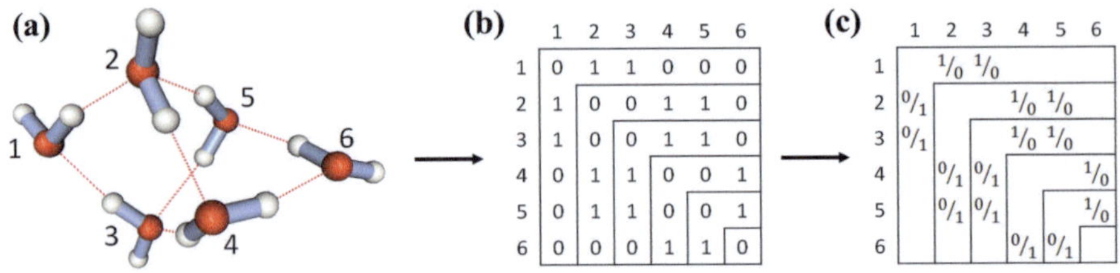

Fig. 2.6 Adjacency matrices for the enumeration method. **a** sample geometry; **b** adjacency matrix; **c** possible directed adjacency matrices

other molecules automatically defines the connections of all molecules to the current one. Therefore, it is sufficient to consider each row and column starting from the corresponding diagonal element; see Fig. 2.6b. The connectivity of the last molecule (number 6), as expected, is completely defined by the connectivity of the previous molecules.

The positions with ones in each row (or column) of **A** tell us which molecules are connected to a given molecule. Let us choose rows as a reference. Then, by going through all combinations of 1 and 0 for positions marked with "1/0" in Fig. 2.6c we can generate all possible H-bond directions for the given skeleton. This can be achieved by taking a bit representation of an integer value that ranges from 0 to $2^n - 1$. From the definition of **D**, the values in a column are the opposite of the values in a row (positions with "0/1" in Fig. 2.6c). H-bond directions are generated for each molecule in this manner using a backtracking loop. The backtracking allows one to apply geometry filters for each molecule right away without generating a complete matrix. The steps constituting our backtracking method are shown in Listing 2.1.

Listing 2.1 Backtracking algorithm for topology enumeration.

```
------------------------------------------------------------
Mark all bit combinations of all molecules as not used
Select molecule 1 as the current molecule 'M'
WHILE ('M' > 0)
  IF (all bit combinations for 'M' have been used) THEN
    Mark all bit combinations for 'M' as not used
    Set 'M' = 'M' - 1
  ELSE
    Choose next bit combination for 'M'
    Mark this bit combination as used
    IF (topology filters are satisfied) THEN
      Set 'M' = 'M' + 1
    END IF
    IF ('M' > number of molecules) THEN
      Create and save a cluster geometry for local optimization
      Set 'M' = 'M' - 1
    END IF
  END IF
END WHILE
------------------------------------------------------------
```

In the beginning it is important to sort molecules by their distance from the first one to allow the algorithm to backtrack from bad combinations sooner rather than later. The bicoordinated molecules are a special case because they can generally have two stable directions for the dangling hydrogen. This is accounted for by generating all combinations of the directions for bicoordinated molecules for each topology. A symmetrical framework is not considered as a special case because it is a rare case for water clusters. During creation of the geometry, hydrogen atoms are placed on the line between two corresponding oxygen atoms. In the case of rigid monomers (as in the TIP4P potential), and to save optimization time in the case of non-rigid ones, the monomer bond lengths and angles are adjusted to the equilibrium values of the given potential using a simple geometrical transformation.

A large number of trial topologies is usually generated. It is not possible to perform a tight local optimization for all topologies even with an inexpensive potential. Therefore, a local optimization is done in several steps of increasing precision. A single-point energy is calculated for the initial geometry of each topology. The geometries of the structures with the 5000 lowest single-point energies are then optimized until a 1.0 kcal/mol/Å gradient threshold is reached. Then the 500 best resulting geometries are further optimized with a 0.1 kcal/mol/Å threshold. Next, the 50 lowest-energy geometries that result are optimized using a 0.01 kcal/mol/Å threshold. Finally, a very tight optimization is performed at the best geometry.

The number of topologies generated for three cluster sizes and the effects of topology filters are shown in Table 2.1. Note that in these test cases the best topology found is the same no matter what filters are used; the difference is in the number of geometries that need to be locally optimized. The use of the modified ice rules still leads to a huge number of topologies left for energy calculations. Checking for monomer angle and adjacent dangling (non-H-bonded) hydrogen atoms further reduces the number of accepted topologies by three orders of magnitude. The next filter is not as effective but reduces the number of topologies in all cases. The final filter makes a difference only for the larger cluster with $n = 31$. Even with all the filters included the number of resulting topologies shows an exponential growth. Thus the method becomes increasingly inefficient for clusters with more than about 35 water molecules.

The backtracking loop for generating **D** matrices was also used in the routine that produces a random topology. Such a topology should be random but still satisfy

Table 2.1 The number of topologies to be locally optimized for a sample oxygen framework in $(H_2O)_n$

Filters used	$n = 23$	$n = 28$	$n = 31$
Modified ice rules	$\sim10^7$	$\sim10^8$	$\sim10^9$
+ HOH angle	$\sim10^6$	$\sim10^7$	$\sim10^8$
+ 2 adjacent DAA	11556	46922	403465
+ 3 adjacent ADD	6892	34971	337986
+ AADD – AAD	6892	34971	318728

The plus sign means that a filter is added to all previous ones topology filters. To achieve this, random bond directions are selected for each line of the **D** matrix. Then the backtracking search is performed starting from this random matrix until all topology filters are satisfied.

2.3.6 *Short Topology-Altering Optimization*

Kazimirski and Buch [49] foresaw and Takeuchi [51] refined a different approximate but fast H-bond topology optimization which can be used as part of a global minimum search. We now describe our version of this method which we call the "short topology-altering optimization".

Tri-coordinate water molecules in a cluster are either double-acceptor, single-donor (AAD) molecules or single-acceptor, double-donor (ADD) molecules. The basic H-bond topology-altering operation is the reversal of the direction of the H-bonds in a contiguous sequence of water molecules. It can also be viewed as a proton transfer. It is possible to perform such a reversal without altering the connectivity in rings and in those chain substructures which have an AAD water at one end and an ADD water at the other end. Examples of such a H-bond topology-altering operation are shown in Fig. 2.7.

Sometimes this operation has the effect of creating an extra pair of non-H-bonded or "dangling" hydrogen atoms on adjacent water molecules. However, as discussed earlier in Sect. 2.3.4, structures with adjacent AAD water molecules are energetically unfavorable [69, 71–73, 79]. To reduce the number of such structures generated, we introduce a split-chain H-bond topology-alteration. This new operation consists of splitting a chain substructure into two shorter chains between a pair of adjacent ADD and AAD waters, and performing the H-bond reversal in both subchains without altering the H-bond connecting the subchains. An example is shown in Fig. 2.8. Adding this type of operation helps to distribute "dangling" hydrogen atoms on the surface of a cluster.

A short topology-altering search consists of applying the operations described above to all rings containing no more than 10 monomers and all chain and split-chain substructures with no more than 5 monomers. Note that Takeuchi used only rings and size 2 chain substructures while Buch and Kazimirski used only chains and rings of size 4. New topologies are created by changing bond directions of one substructure at a time. If a new topology has a lower energy, it is accepted as a new

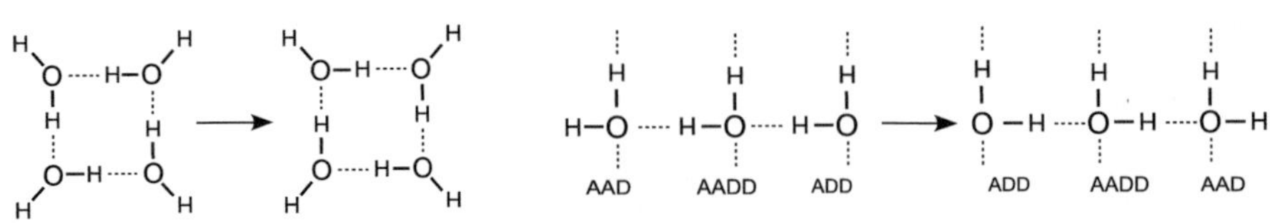

Fig. 2.7 Examples of a H-bond topology-altering operation applied to (*left*) a ring and (*right*) to a chain of water molecules

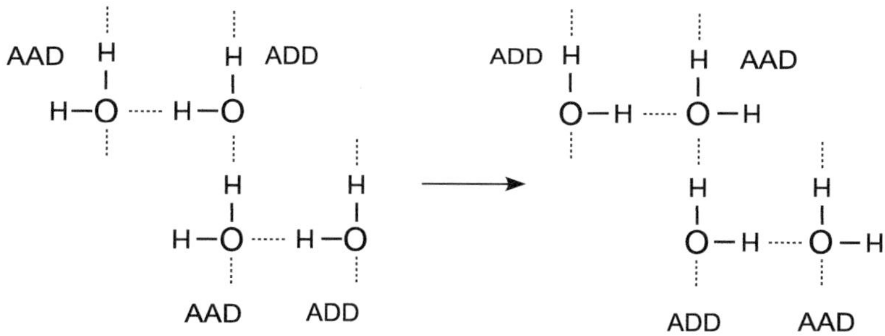

Fig. 2.8 A split-chain, H-bond topology-altering operation. In this example, the split is between the second and third waters

reference point and optimization returns to the smallest substructure. The search is allowed to move only in a downhill direction of the potential energy surface to limit the number of topologies considered. The framework of the initial cluster is allowed to change during optimization. In fact, a framework change does happen often when the input structure has a relatively high energy. The search is complete when no substructure alteration leads to energy lowering.

Since the algorithm goes through all possible substructure changes, it is likely that the same topology would be generated and locally optimized more than once. Saving optimized geometries in a list removes that problem at the cost of adding complexity to the algorithm. A list entry consists of an energy, Cartesian coordinates, and a **D** matrix. New structures are saved in the list. If a structure with a given **D** matrix is already in the list, the geometry and energy from the list are used instead of performing a local optimization. The list is created for a given framework. It can be limited to a single optimization call or can be global in case the parent process uses short topology-altering optimization as part of a larger algorithm as in Sect. 2.3.8. If the framework changes during the optimization, the list is either reset to be used with the new framework, or, in the case of a global list, the search is terminated. The algorithm can only go downhill in energy. Listing 2.2 describes the general flow of the ST algorithm.

Listing 2.2 Short topology-altering optimization (ST).

```
-----------------------------------------------------------------
Set substructure size M = 2
WHILE (M <= 10)
  FOR each substructure of type 'T'
    Find substructures 'T(M)' that allow altering H-bond direction
    FOR each substructure 'T(M)'
      Alter H-bond direction of the substructure
      IF (topology filters are satisfied) THEN
        Perform local optimization
        Keep the best topology generated for the current M
      END IF
    END FOR
  END FOR
```

```
    IF (lower energy geometry has been found) then
      M = 2
    else
      M = M + 1
    end if
  END WHILE
  ------------------------------------------------------------------
```

The effects of using topology filters and the minima list have been examined for three large cluster sizes: $n = 34$, 43, and 52. For each size, 1000 low energy clusters with randomized topologies were generated as a test set. Short topology-altering (ST) optimization was applied to each cluster, first with no filters or lists as a reference point, and then adding one filter at a time. The average speed up of run times was calculated for each case; see Table 2.2.

The direction of a search often changes when filters are added. In some cases, optimization finishes at a higher energy minimum than the one reached without the use of the additional filter. However, with all of the improvements, the energy lowering is the same on average while the run time is reduced by about 40 %. Note that the AADD–AAD pair motif and the three adjacent ADD motif were not used as filters because the improvement in speed was not significant.

This algorithm has also been implemented to work with mixed clusters containing many water molecules and a few other molecules. The latter are treated as a special case and only the H-bond network between water molecules is allowed to change. Of course, for such an optimization to make sense, there should be enough water molecules to form a topological network. The algorithm can be used to optimize an initial distribution of a large number of water molecules around a solvated molecule, for example a protein, before starting a molecular dynamics simulation.

2.3.7 Extended Topology-Altering Optimization

The short topology-altering optimization allows only downhill moves. However, it is desirable to extend it to cross energy barriers and explore larger areas of the potential energy surface. Using a Monte Carlo search in the manner of Buch and Kazimirski [7] is one option. We wanted, however, to keep the algorithm more

Table 2.2 Effect of topology filters shown as an average percentage improvement in calculation time relative to the trial with no filters

Features used	$(H_2O)_n$		
	$n = 34$	$n = 43$	$n = 52$
+ HOH angle	11.68	7.64	6.39
+ 2 adjacent DAA	41.82	37.18	31.31
+ list of minima	45.30	40.16	33.99

deterministic. Our algorithm, called extended topology-altering optimization ET(m), uses a population of m topologies that act as different solutions and are refined independently. This way some topologies with an energy higher than the best are saved as part of a population thus allowing the algorithm to search in several directions.

The algorithm is shown in Listing 2.3. A population member is described by Cartesian coordinates, energy, **A** and **D** matrices, and its substructure level—the size of the largest substructure for which the altering of H-bond directions has been performed. At the beginning, the population contains only one member, an initial structure with its substructure level set to 1. A single substructure level is considered during a search iteration. An H-bond alternation search is performed with the ST algorithm of Sect. 2.3.6 for all members whose substructure level is $M - 1$ in which M is the current level. In the case a new topology is accepted, it replaces the highest energy one in the population if its energy is lower and the population does not already have such a topology. The new member's substructure level is set to 1, and M is reset to 2. Thus, only the part of the population with smaller substructure levels is considered until all members have the same level again. The search stops when substructures of size 10 have been examined for all members. The lowest-energy member of the population is the result of the optimization.

Listing 2.3 Extended topology (ET) optimization.

```
----------------------------------------------------------------
Set current population size to 1
Use initial structure as the first member
Set the substructure level of the first member to 1
Set current substructure size M = 2
WHILE (M <= 10)
  Copy current population into parent population
  FOR all current population members with substructure level < M
    Increase substructure level by 1
  END FOR
  FOR all parent population members
    IF (the substructure level of the member < M) THEN
      FOR each substructure of type 'T'
        Find all substructures 'T(M)' that allow
           altering H-bond direction
        FOR each substructure 'T(M)'
          Alter H-bond direction of the substructure
          IF (topology filters are satisfied) then
            Local optimization
            IF (new topology is better than the worst in
                the current population) THEN
              Add structure if population is not full
                      or replace the worst
              Set the substructure level of a new member to 1
            END IF
          END IF
        END FOR
      END FOR
    END IF
```

```
    END FOR
    IF (current population has a new member) then
      Set M = 2
    else
      Set M = M + 1
    end if
  END WHILE
  -----------------------------------------------------------------
```

We can control the extent of the search by changing the size of the population. Using a larger population allows us to explore a larger part of the potential energy surface (PES) but with the computational effort being significantly larger. An ET optimization with a population of one is equivalent to an ST optimization. Since each population member has its own adjacency matrix, the framework is allowed to change during the optimization. Therefore, if the input structure is not sufficiently stable, it is possible that the population will contain several different frameworks. We have not implemented a list of visited geometries for this algorithm.

2.3.8 *Genetic Topology Optimization*

The enumeration algorithm is limited to clusters with up to 35 water molecules while the ST and ET algorithms that alter the H-bond directions of the substructures are not designed to perform an exhaustive search. We need a different approach to reliably find the best topology for larger clusters.

If we consider the positions of ones in an **A** matrix and the matrix dimensions, we can create a transformation rule that converts a corresponding **D** matrix into a string containing only information on the direction of existing bonds. In other words, all zeroes that indicate the absence of a H-bond and do not change when the topology is altered are removed. It suffices to use only half of the **D** matrix because D_{ji} can be determined from D_{ij}. Since the bond direction is encoded in a binary manner by 0 and 1, we can create a bit string that unambiguously represents a topology for a given oxygen framework; see Fig. 2.9.

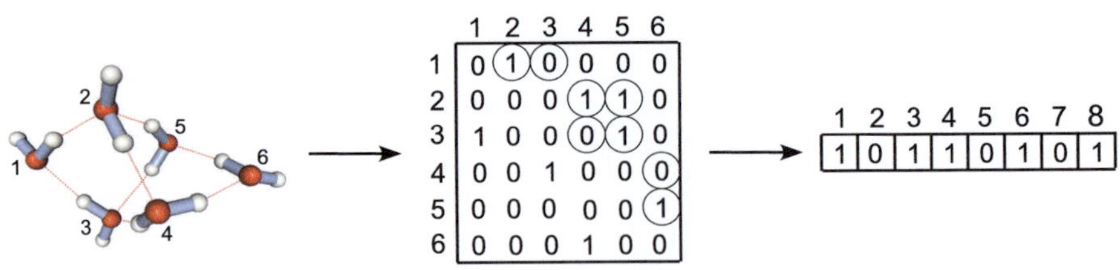

Fig. 2.9 Representation of a hydrogen bond topology using a bit string

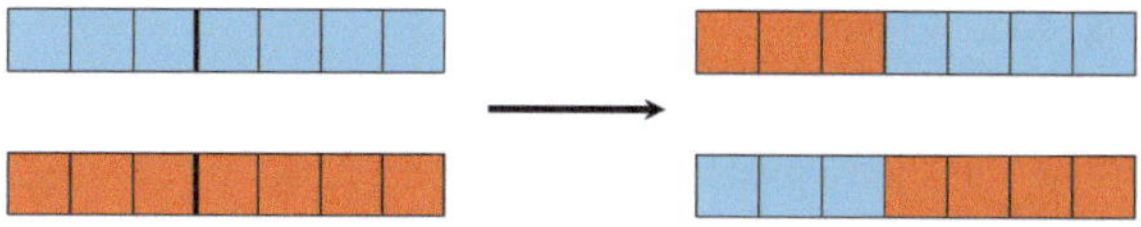

Fig. 2.10 Crossover operation

In the jargon of genetic algorithms, a new bit string can be created by a crossover operation on a pair of parent strings as in Fig. 2.10. The definition of a suitable crossover operation makes it possible to search over a large number of topologies and find the best one with genetic optimization methods.

Genetic topology optimization (GT) proceeds in the following manner. A list of visited topologies is maintained in the form of bit strings. A new generation is created using steps standard in genetic algorithms: elitism, crossover, and mutation. Elitism copies 20 % of the previous population into the next generation to keep several of the lowest energy topologies in the pool. During the crossover phase, all possible pairs of the population members exchange sequences of bits. All possible crossover positions and sequence lengths are examined. Each child that has not been visited previously is checked for satisfaction of the topology filters. Each accepted child is added to the list of visited topologies and processed using a multiple step procedure described in the next paragraph. Sometimes there are not enough new topologies generated by the crossover phase to fill the population. In that case, random topologies are added to the population during a mutation phase. The search stops either when the best energy remains unchanged after 20 generations or when neither crossover nor mutation can generate an unvisited topology.

When a new topology is accepted, it is processed in several steps. First, a local optimization is performed in two stages of increasing precision. After the first stage, the topology is discarded if the energy of a child is higher than the highest energy of the parent population by 3 kcal/mol or more. Otherwise, a tighter optimization is performed. The adjacency matrices of the initial and current geometries are compared at the second step to ensure that the oxygen framework has not changed. Allowing the framework to change during the genetic topology search would add unnecessary complications. Next, a short topology-altering (ST) optimization is applied. A crossover operation generates diverse starting points, sampling different areas of the potential energy surface, while an ST optimization allows the algorithm to reach a low energy minimum at the bottom of a funnel. The framework distortion is checked again. If the energy was lowered during the ST step, the new topology is checked against the visited list.

The initial population is generated as follows. Random topologies corresponding to the framework of the input structure are created using the enumeration (NT) method described in Sect. 2.3.5 and processed as described in the previous paragraph until the desired population size is obtained.

The genetic topology algorithm is successful in optimizing topologies of large clusters. However, a significant number of random topologies is often added to the population during optimization. Perhaps the random topologies are more important starting points for the ST optimization than the results of a crossover operation.

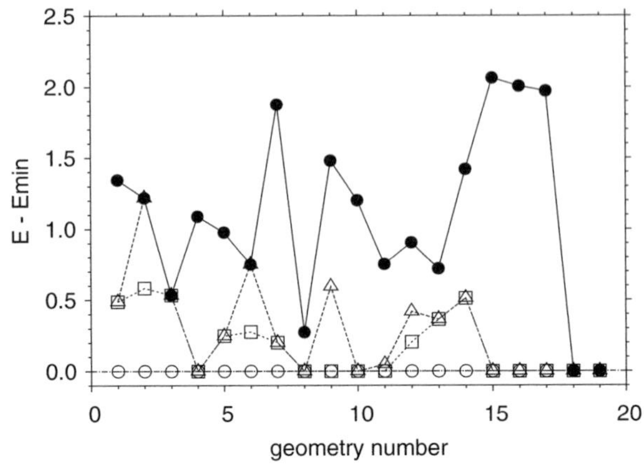

Fig. 2.11 Comparison of four topology optimization methods. Energy differences $E - E_{\min}$ are in kcal/mol. *Filled circles* ST, *triangles* ST + ET(10), *squares* ST + ET(20), *circles* ST + ET(10) + GT

A careful assessment of this issue would help to improve algorithm performance. However, one should remember that test runs over a large variety of structures are required to obtain information on the details of the algorithm performance, since topology optimization is much more difficult for some oxygen frameworks than others.

2.3.9 Comparison of Topology Optimization Methods

The ability of the ST, ET, and GT methods described in Sects. 2.3.6–2.3.8 to locate the best topology was tested. To perform a comparison, 19 low-energy $(H_2O)_{52}$ clusters with randomized topologies were subject to topology optimization. During topology optimization, we start with a faster algorithm and then refine results employing more precise methods. Therefore, different combinations of methods were examined to see whether the inclusion of more complex algorithms leads to a better topology.

The results are shown in Fig. 2.11. $E_{\min}$ is the lowest energy that has been found for a cluster whereas E is the cluster energy after a sequence of topology optimizations. The initial random topologies have an energy difference, $E - E_{\min}$, ranging from 5 kcal/mol to 25 kcal/mol. The short topology-altering optimization lowers the energy significantly, leading to $E - E_{\min} < 2.5$ kcal/mol. The addition of a more complex method helps to lower the energy further in all but two cases. Comparison of ET(10) and ET(20) shows that increasing the population size of ET(m) is advantageous for 5 geometries out of 19. In nine cases GT is necessary to find the best topology.

2.4 Global Optimization of Water Clusters

An improved parallel evolutionary algorithm [84] and basin hopping combined with vibrational modes [50] are examples of successful approaches to the problem of global optimization of water clusters. However, these methods are not able to find the lowest energy geometry for systems with more than 30 water molecules. A detailed comparison of results for each cluster size can be found in our publication that describes an application of minima hopping to water clusters [41].

2.4.1 Improved Minima Hopping

Goedecker's minima-hopping algorithm was shown to be efficient for atomic clusters [52, 85]. However, our test calculations revealed that the original minima-hopping algorithm does not have a satisfactory performance on water clusters with more than 25 molecules. To improve the algorithm, we decided [38] to add geometry altering operations that are different from molecular dynamics or Monte Carlo steps. These were allowed to be specific to a water cluster system in some cases. Some examples of operations that take into account the current energy or geometry of a cluster are direct mutation [84], internal, surface, and rotational operators [51], and cluster surface smoothing [86].

Our improved algorithm uses several types of geometry transformation steps which we call operators. An operator acts as a black box. It takes a cluster geometry, performs some transformations independent of the other operators, and returns a new geometry. For example, the MD simulation step in standard minima hopping [52] is considered to be an operator. The main purpose of the MD operator is to "hop" over higher energy minima, thereby sampling different areas of the PES. The other operators are designed to lower the energy of a cluster starting from a minimum obtained by MD steps. Four types of geometry transformations were added. The topology operations, ST and ET, described in Sect. 2.3 are used to lower a cluster's energy through topology variation. A translational operator (TRN) moves a molecule to a more favorable position in a cluster [38]. Lastly, a distortion operator (DST) changes the position of several molecules in a selected area of a cluster.

A global optimization using the improved minima-hopping method consists of many cycles. A cycle takes search parameters and cluster coordinates as an input. Each cycle combines operator calls in an optimal sequence. During a cycle, the output list and the visited minima list are updated and the final search parameters and coordinates are returned. To account for the different computational expense of the different operators, a cycle is divided into an initial exploration phase and a subsequent energy lowering phase. The exploration phase attempts to move the search to a slightly different area of the PES by crossing energy barriers using the MD and ST operators. We found that the ST operator is cheap enough to be applied to each accepted minimum during the search, significantly improving the algorithm performance. The other three operators are more expensive and hence they are used sparingly. In the second phase of the cycle, the energy of a structure is lowered using sequences of TRN, DST, and ET(4) operators. The aim of the energy lowering phase is to reach the bottom of a funnel. Each operator is followed by a tight local optimization and an acceptance or rejection of the resulting local minimum.

2.4.2 *Accept or Reject?*

The decision process closely follows the original minima-hopping [52] algorithm and its modification [85]. An essential item for deciding what to do with a new minimum is the list of visited minima which contains a description of each accepted minimum and a count of the number of times it has been visited. Our improved algorithm searches over unique cluster frameworks rather than unique topologies thus reducing the number of possible minima to be considered. To achieve this, the list contains not only the energy for each minimum but also its nine framework parameters, the number of hydrogen bonds and eight ring counts, as described in Sect. 2.3.3. The improved minima-hopping search is controlled by two adjustable energy criteria, E_{kinetic} and E_{diff}. Five parameters control the rate of change of these two thresholds: $\beta_1 = \beta_2 = \beta$, $\beta_3 = 1/\beta$, $\alpha_1 = \alpha$, $\alpha_2 = 1/\alpha$. Values of the parameters are $\alpha = 1.02$ and $\beta = 1.05$ as in Goedecker's work [52].

Detailed pseudo code for the processing of a newly found minimum is shown in Listing 2.4. The first section checks whether the minimum has not been changed by the operator or has been visited before. The rate of kinetic energy increase is modified by an *enhanced feedback* mechanism [85]; it depends on the number of visits to a minimum according to the formula $\beta_2' = \beta_2(1 + c \log N)$, where N is the number of visits and $c = 0.1$ is a feedback coefficient. With this modification, repeated visits to a minimum lead to a greater repulsion from that minimum. A special case arises when the topology of a minimum is improved but the framework remains the same. In such a case, the energy in the visited list is updated and the minimum is accepted. The second section of the processing checks the energy of the new minimum. The MD operator uses E_{diff} to determine whether to accept or reject the minimum. For all other operators a new minimum is accepted only if the energy is lowered. Finally, the accepted minimum is written to the output and visited minima lists.

Listing 2.4 Analysis of a new minimum. 'Mcurrent' is the current minimum and 'M' is a new minimum. 'M' is rejected unless said otherwise.

```
-----------------------------------------------------------------
---Checking for repeated visits---
if (energy('M') equals energy('Mcurrent')) then
  if (operator is MD) then
    Ekinetic=Ekinetic*beta1
  end if
  GOTO CHECKEND
```

```
end if
Look for 'M' in a list of visited minima
if ('M' matches minimum 'Mlist' in the list) then
  if (energy('M') larger than energy('Mlist')) then
    if (operator is MD) then
      Increase number of visits 'N' to 'Mlist' by one
      Ekinetic=Ekinetic*beta2*(1+c*log('N'))
    end if
    GOTO CHECKEND
  else
    energy('Mlist') is substituted by energy('M')
   end if
end if
---Checking energy---
if (operator is MD) then
  Ekinetic=Ekinetic*beta3
  if (energy('M') - energy('Mcurrent') < Ediff) then
    Ediff=Ediff*alpha1
    Accept 'M'
  else
    Ediff=Ediff*alpha2
  end if
else
  if (energy('M') < energy('Mcurrent')) then
    Accept 'M'
  end if
end if
CHECKEND
----------------------------------------------------------------
```

2.4.3 Operator Call Sequences

Next, we consider in greater detail the search during a cycle of our improved minima-hopping method. During the exploratory phase the MD operator is called repeatedly. The ST operator is used to relax the topology whenever MD leads to a new accepted minimum. To use the expensive energy lowering operators efficiently, the exploration phase must stop at an optimal minimum which is not too high in energy and preferably belongs to a different funnel of the PES. To achieve this, a change in geometry is first calculated for each new minimum. A shift of the molecules' centers of mass (CM) between the initial cluster geometry of a cycle and the current geometry is calculated as

$$S = \left[\frac{\sum_{i=1}^{N_{\text{mol}}} [\text{CM}_{\text{init}}(i) - \text{CM}_{\text{new}}(i)]^2}{N_{\text{mol}}} \right]^{1/2}.$$

The geometry is considered to be different when S is higher than an empirically selected threshold value of 0.95 Å. The exploration phase ends when the last three accepted minima satisfy the geometry condition, and the energy of the second minimum in this sequence is the lowest of the three. The latter condition prevents the search from stopping when the energy of the minima is changing monotonically. The lowest-energy minimum of these minima is the result of the exploratory phase.

The energy lowering operators take a significant amount of time. Often, the result of an operator is the same minimum or a minimum much higher in energy than the best one. The following procedure is used to avoid wasting time on expensive calculations far from the lowest energy. The energy of a minimum obtained in the exploration phase is compared with the best energy found so far. The translation and distortion operators are designed to have two different sets of parameters: one for a more thorough search and the other for a less thorough search. The settings for a more thorough search are used if the difference in energy between the initial minimum for the energy-lowering phase and the best energy $\Delta E = E_{\mathrm{new}} - E_{\mathrm{best}} < E_{\mathrm{high}}$, where $E_{\mathrm{high}} = 5$ kcal/mol. If $E_{\mathrm{high}} \leq \Delta E < E_{\mathrm{low}}$, where E_{low} is an automatically adjusted threshold, then the less thorough settings are used and E_{low} is decreased. If $\Delta E \geq E_{\mathrm{low}}$, then the energy lowering phase is not used at all, E_{low} is increased, and the cycle goes back to the beginning of the exploration phase. In the energy-lowering phase, operators are called in the sequence

$$\mathrm{TRN} \Rightarrow \mathrm{DST} \Rightarrow \mathrm{ETOP} \Rightarrow \mathrm{TRN} \Rightarrow \cdots$$

An operator is called again if it is successful in lowering the energy; otherwise, the next one is called. The energy-lowering phase is completed when application of all three operators in a row fails to lower the energy.

2.4.4 *How the Operators Work*

An effective operator should take into account the properties of a system and employ transformations that make use of the way molecules are connected. An efficient operator should use as few local optimizations as possible. This can be achieved by using geometry filters to predict high energy minima without geometry optimization, avoiding acceptance of the same geometry repeatedly by using a list of visited minima, and by doing local optimization in stages of increasing precision so that it can be stopped when it becomes clear that it is not needed.

2.4.4.1 The MD Operator

The basic operator of our algorithm is a short MD simulation as in standard minima hopping. How short should the simulation be? How many MD steps should be taken before continuing with a geometry relaxation? Three options were considered: (1) stop after a fixed number of steps, (2) stop at the first energy local minimum which is not too shallow, and (3) stop after a certain number of energy minima. Goedecker's minima hopping uses the third criterion [52]; however, there were no clear rules presented on how to choose the number of minima. In our first implementation we used a fixed number of steps for simplicity but this was not optimal. If we examine the usual energy trend during the MD steps, as shown in Fig. 2.12, we can see that there is a minimum within about 100 steps while other minima could occur much later. Since the primary object of the MD simulation is to distort the geometry and reach a new local minimum, it suffices to stop the MD trajectory at the first energy minimum. Therefore, we settled on the second option. The trajectory continues until a minimum is reached provided that the energy decreases during at least the last five steps before the minimum is reached. The latter restriction prevents termination of the MD simulation at a very shallow local minimum. Since this prescription could lead to a long MD simulation, the trajectory is terminated if such energy lowering is not found within the first 100 steps. Our algorithm also checks for the situation when monomers move too far away from the cluster since such distortions are not useful.

In principle, following the directions of soft vibrational normal modes should result in easier barrier crossing. Such a technique was used by Kabrede in a vibrational mode basin hopping study of water clusters [50]. A method of choosing a low curvature direction for minima hopping was suggested by Goedecker et al. [85]. However, they tested it only on atomic clusters. They also noted that excessive use of low curvature directions could reduce the randomness of the PES exploration. We tried several approaches to make use of low curvature directions for the initial MD velocities; however, nothing gave any improvement over randomly directed velocities.

2.4.4.2 The Translation (TRN) Operator

Translation of one or more molecules to a different part of a cluster provides a good addition to the MD moves in which all molecules are shifted by a relatively

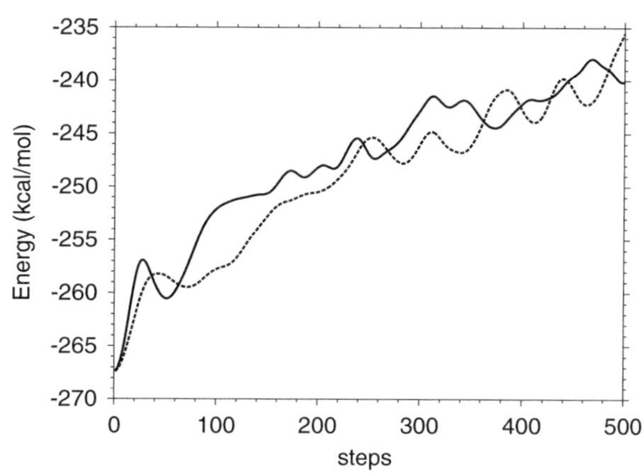

Fig. 2.12 Energies of two typical MD trajectories for a water cluster

small amount. The translation operator is general and can be applied with little modification to a cluster of any type. It was used in various forms in previous studies [38, 51, 84, 86, 87]. The purpose of a translational operator is to find a few molecules that contribute least to the cluster binding energy and put them in a different location that will lower the total energy. We use an algorithm that translates up to two molecules at once. While it can be easily extended to the simultaneous translation of three or more molecules, we came to the conclusion that going beyond two molecules increases the cost-efficiency ratio significantly. Further, only bi- and tri-coordinated water molecules are moved as they are the ones most likely to have higher energies.

Suppose that the cluster is composed of n monomers. First consider that only one monomer is to be moved. We begin by selecting a scaffold composed of $n-1$ monomers around which the remaining monomer is moved. This scaffold is chosen by requiring that it minimizes disruption of the stability of the cluster. Candidates for the scaffold are created as follows. Each bi- and tri-coordinated molecule is removed in turn from the cluster and the residual structure with one less molecule is locally optimized. The candidate with the lowest energy is selected to be the scaffold geometry around which the molecule is to be moved. Next, a grid of points is set up around the cluster. The grid size depends on the minimal distance that needs to be maintained between the centers of mass of the monomers. The grid has a rectangular prism shape to allow for different lengths of the cluster in the three principal directions. The nth monomer is moved to a grid point only if it is not too close to its original position and if it is within a threshold distance from any three other molecules. The nth molecule's center of mass is placed at an accepted grid point and the position and orientation of the molecule are then adjusted. This simple adjustment might involve several random rotations followed by a single-point energy calculation or a loose local optimization. A more sophisticated adjustment mechanism was developed for water clusters and it will be described below. For each grid point, the energy of the obtained structure is compared with the initial structure's energy. If the energy is not lowered by moving a single molecule, two molecules are moved simultaneously in an analogous fashion. In this case, the $n-2$ scaffold is created as follows. Each possible pair of bi- and tri-coordinated molecules is removed in turn, the resulting scaffold candidate is locally optimized, and the lowest-energy candidate is selected to be the scaffold geometry. As soon as the energy is improved, the operator stops and returns the structure without checking any unexplored grid points. This does not leave other grid points unexplored because an energy-lowering translation is always followed by another translation as the next operator in the minima-hopping cycle.

2.4.4.3 The Distortion (DST) Operator

The distortion operator attempts to improve a selected area of the cluster by moving several adjacent molecules a relatively short distance while the rest of the cluster remains intact. This operator is expected to be useful for sufficiently large clusters for which distortion of one section of the surface will not significantly perturb the rest of the cluster. The DST operator works as follows. The molecules are sorted by their contribution to the cluster binding energy. The half of the molecules with the smaller contributions serve as the centers of distortion. The molecules closest to a distortion center are shifted and the resulting geometry is analyzed. The number of molecules being shifted varies in turn from three to five. If energy lowering is found, the operator stops immediately. If the final energy is higher than the starting one, but within 0.5 kcal/mol, the operator is repeated from that higher energy point.

In our first attempt, the shifting was performed by short MD simulations with only the selected molecules allowed to move. This did not work well enough and a more efficient and complex distortion mechanism was created as described below. A stable hydrogen bond network in a water cluster creates high energy barriers. Unlike the case of an atomic cluster, it is not enough to just move a molecule; one also has to change the existing H-bond network in the area where the molecule is placed and in the area from which the molecule was removed. Initially, we left the creation of new hydrogen bonds to chance using random rotations. To improve upon this, we created a H-bond adjustment mechanism. The general idea is to remove H-bonds that create energy barriers, change the positions of the pertinent water molecules, and then recreate hydrogen bonds in a meaningful way for a new distribution. When we remove hydrogen bonds, we remove the hydrogen atoms from the cluster and switch to a representation of each water monomer as a bead or point particle. To reproduce reasonable water molecule positions without the presence of the hydrogen bonds we employ Molinero and Moore's mW coarse-grained model of water [88] based on the Stillinger-Weber silicon potential [89]. The energy of a system in the mW model depends on distances between pairs of beads and angles between triplets of beads. The short-ranged potential sets to zero the forces between beads farther apart than 4.32 Å. The mW model successfully reproduces bulk properties of water. For us, the important feature of the model is that it maintains a tetrahedral distribution of beads and a correct distance between them. The angle term is a three-body term which leads to more elaborate code than is needed for pairwise potentials.

The adjustment process consists of redistributing selected molecules and then recreating hydrogen bonds between them. The process starts by selecting a central point of distortion. For the DST operator, it is the geometrical center of the several molecules selected to be shifted. The cluster is converted into the bead representation and divided into two sections. The N_{free} molecules adjacent to the center of distortion are allowed to move during the local optimization and adjust their positions depending on a new distribution. There are $N_{\text{frozen}} = N_{\text{total}} - N_{\text{free}}$ molecules which have their coordinates frozen and their connectivity does not change. Next, the $N_{\text{move}} < N_{\text{free}}$ molecules closest to the center of distortion are selected to be relocated. A new distribution of beads is created by placing them at random positions very close to the central point of distortion and performing a local optimization using the mW model potential. During the optimization beads will move away from the central point and assume a tetrahedral distribution. N_{try} such attempts are made for each point of distortion. The mW energy is not reliable. Therefore, the N_{pos} lowest energy distributions with different connectivity are saved for analysis. Adjacency matrices of the mW geometries are compared to avoid saving identical distributions. In addition, bi- and penta-coordinated beads are avoided to reduce the number of possible distributions considered. A process of hydrogen bond restoration is performed for the saved geometries.

The process of creating new H-bonds for free molecules makes use of the backtracking algorithm described in Sect. 2.3.5. At first, possible bonds are determined and a directed adjacency matrix is created for a cluster using three bond categories: existing bonds, no bonds, and possible bonds. Existing and absent bonds belong to the frozen molecules and possible bonds to the free ones. Hydrogen atoms are placed back into the cluster in such a way that water molecules would form hydrogen bonds according to the modified ice-rules discussed in Sect. 2.3.4. A backtracking algorithm is used to analyze all possible combinations of bonds in an efficient way. The following filters are used to avoid high energy combinations: bicoordinated monomers, non-bonded molecules that are too close, three-membered rings, adjacent dangling hydrogen atoms, and unrealistic monomer angles. A single point energy for each accepted structure is calculated and local optimization is performed for the low energy ones. An ST optimization is used to relax the topology for several of the energetically-best geometries.

An example of distortion using the adjustment mechanism is shown in Fig. 2.13 which highlights in blue the molecules selected to be moved (1). The molecules are stripped of hydrogen atoms (2) and then placed close to each other (3). The next image depicts the result of mW optimization, where highlighted molecules and

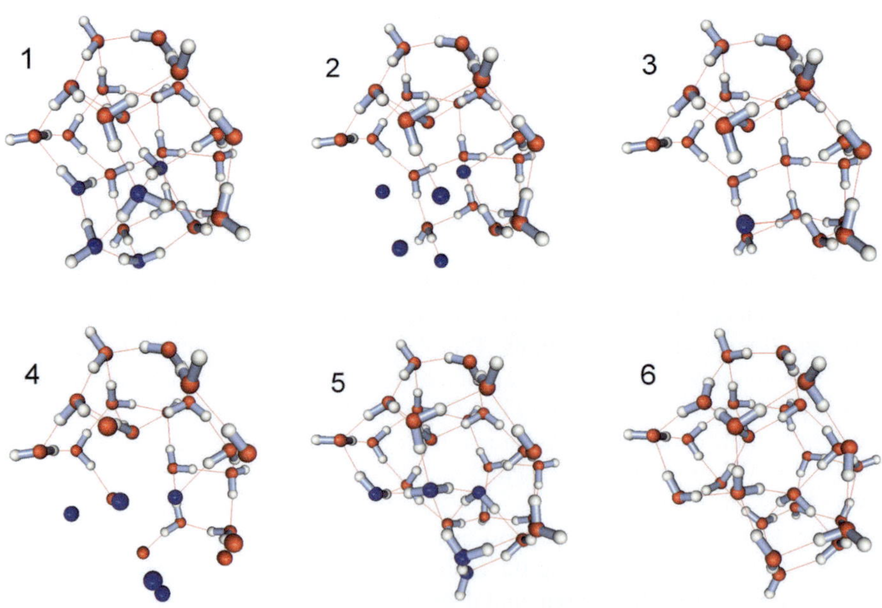

Fig. 2.13 An example of distortion using the adjustment mechanism

Table 2.3 Values of adjustment mechanism parameters for higher and lower accuracy operators

	TRN(high)	TRN(low)	DST(high)	DST(low)
N_{move}	2	2	3–5	4
N_{free}	12	8	12	8
N_{try}	4	2	10	4
N_{pos}	4	2	5	2

several adjacent ones were moved to new positions (4). Then hydrogen bonds are restored (5) and ST optimization is performed (6).

The translation and distortion operators become similar when the adjustment mechanism is used. In the translation, movement of a molecule is followed by an adjustment procedure. In the distortion, the adjustment is applied right away on a selected area of a cluster; however, more adjustment attempts are made in that case.

Different sets of parameters for the adjustment mechanism are used to create versions of operators with higher and lower accuracy. The values used are listed in Table 2.3.

2.4.5 *Performance Tests*

Test runs were made to check the effects of the various operators on the performance of our improved minima hopping. It is hard to judge the change in performance of the algorithm from a single run due to its random search nature. We need an average result from several global optimization runs on the same cluster size. The latter must be small enough to allow the algorithm to reach the global minimum in a reasonable time. Following these considerations, we performed 20 global optimization runs each on TIP4P $(H_2O)_n$ clusters with $n = 19$, 23, and 27. A run was stopped if a global minimum was not found within two days. The average time used to locate the global minimum framework in each cluster size was calculated. The operators with a simple adjustment mechanism were used.

Table 2.4 shows the results of the test runs. In four cases, not all of the 20 runs were able to locate the global minimum within two days and the number of successful runs is indicated instead of an average time. We can conclude that all operators improve the performance and are necessary to obtain the global minimum in a reasonable time for large clusters. The inclusion of the mW potential and an advanced adjustment mechanism led to improvement in performance. The time taken to locate the global minimum was almost halved from 54.9×10^3 to 29.1×10^3 s for $n = 27$. The algorithm performance using a simple list of visited minima was also assessed. Treating every topology as a unique minimum leads, as expected, to an increase in the average search time to 35.5×10^3 s.

Table 2.4 Average computer time (seconds) used to locate the global minima for $(H_2O)_n$ clusters using different sets of the search operators

n	MD	+ST	+ TRN	+ DST	+ ET(4)
19	1741	371	246	150	110
23	(6 of 20)	4199	2200	1336	1191
27	(0 of 20)	(6 of 20)	(11 of 20)	78775	54927

The operators with a simple adjustment mechanism were used. If the global minimum was not found in all 20 runs within two days, the number of successful runs is shown in parentheses instead of the time

Table 2.5 Operator success rate (%)

n	MD	ST	TRN	DST	ET(4)
44	25.63	72.73	36.67	17.94	31.33
45	25.63	72.51	36.93	17.90	31.59
46	25.64	72.31	36.40	17.46	31.87
47	25.64	72.02	37.22	17.52	31.86
48	25.64	71.89	37.43	18.26	31.70
49	25.65	71.85	38.00	17.38	32.15
50	25.65	71.56	38.01	17.48	32.38
51	25.64	71.52	36.47	16.76	32.42
52	25.66	71.18	36.30	16.94	32.63
53	25.66	71.11	35.76	16.07	32.67
54	25.64	70.88	36.07	16.31	32.70
55	25.64	70.88	35.17	15.94	32.89

Table 2.5 lists the percentage of calls that led to energy lowering during the search. Values for the TRN and DST operators combine both low and high precision calls. A 25 % rate for the MD operator follows from the minima-hopping logic with only 1/4 of generated minima being accepted. Although the other operators did not have any constraints, their success rate is consistent over a number of cluster sizes. The ST operator shows very high success rates showing its importance. The ET(4) success rate grows slightly as the cluster size increases while the ST success rate decreases. This is a result of the hydrogen bond topology becoming more complicated.

Despite obvious improvements, the average time to locate the global minimum grows rapidly. The algorithm can reliably locate the global minimum for $(H_2O)_n$ clusters up to $n = 35$. Reliability means that it always finds the same lowest energy minimum within several days of searching on a single CPU. With a significant amount of calculation, it was possible to find the same lowest energy minima at least twice for sizes up to $n = 47$, except for $n = 39$ and 45 [41]. The best minima for $n > 47$ reported earlier [41] were found only once.

2.5 General Version of Improved Minima Hopping

Our improved minima-hopping algorithm can be used with any type of cluster. In that case, all operators specific to water are disabled. By design, the MD operator can be used with any monomers and does not require changes. However, the improved minima-hopping cycle parameters should be adjusted for each type of cluster to achieve the best performance. The list of visited minima is simplified by using only the energy and the number of visits for each minimum as in standard minima hopping.

A general version of the translational operator was created as well. The same ideas of scaffold and grid points are used. The changes from the water version include moving only one molecule at a time because advanced adjustment cannot be used. Instead the highest and the second highest energy molecules are used. The adjustment mechanism consists of several random rotations. A collision check is performed after each rotation and a single point energy is calculated. Low energy trial structures are accumulated in a list and the best one is selected after a local optimization. As before, higher and lower quality settings are available.

The general version of the distortion operator uses several molecular dynamics steps to create a local distortion, followed by geometry relaxation. The operator uses several molecular dynamics steps in which only certain molecules are allowed to move. Such distortions are applied to several areas of a cluster, each followed by a geometry relaxation.

We have reported [67] application of the general version of improved minima hopping to methanol clusters $(CH_3OH)_n$ with $n \leq 15$ to generate a set of low-energy geometries for subsequent electronic structure calculations. In unpublished work, we have applied it to pure clusters of ethanol, n-propanol, and iso-propanol. In that work, a conformational rotation operator was added to speed up the optimization.

2.6 Concluding Remarks

Our improved minima-hopping algorithm has proved its worth in applications to water clusters using five different potential functions including two which had vibrating monomers [41]. Clusters sizes up to $n = 55$ were studied successfully although the minima for sizes with $n > 47$ must be considered putative since they were found only once. Our ambitious goal of global optimization of water clusters containing up to $n = 100$ monomers has not yet been reached and requires further development of the algorithm.

Acknowledgments The Natural Sciences and Engineering Research Council of Canada (NSERC) supported this work. SK is grateful to the University of New Brunswick for the award of a Frank J. and Norah Toole Graduate Scholarship.

References

1. Johnston RL (2002) Atomic and molecular clusters. CRC Press, Boca Raton
2. Scheiner S (1997) Hydrogen bonding: a theoretical perspective. Oxford, New York
3. Stone AJ (1996) The theory of intermolecular forces. Oxford, New York
4. Thakkar AJ (2001) In: Moore J, Spencer N (eds) Encyclopedia of chemical physics and physical chemistry, vol I. Fundamentals, chap. A1.5, pp 161–186. Institute of Physics Publishing, Bristol
5. Keutsch FN, Saykally RJ (2001) Proc Nat Acad Sci USA 98:10533
6. Keutsch FN, Cruzan JD, Saykally RJ (2003) Chem Rev 103:2533
7. Buch V, Bauerecker S, Devlin JP, Buck U, Kazimirski JK (2004) Int Rev Phys Chem 23:375
8. Prell JS, Williams ER (2009) J Am Chem Soc 131:4110
9. Hamashima T, Mizuse K, Fujii A (2011) J Phys Chem A 115:620
10. Sliter R, Gish M, Vilesov AF (2011) J Phys Chem A 115:9682
11. Nilsson A, Pettersson LGM (2011) Chem Phys 389:1
12. Ceponkus J, Uvdal P, Nelander B (2012) J Phys Chem A 116:4842
13. Pérez C, Muckle MT, Zaleski DP, Seifert NA, Temelso B, Shields GC, Kisiel Z, Pate BH (2012) Science 336:897
14. Pradzynski CC, Forck RM, Zeuch T, Slavíček P, Buck U (2012) Science 337:1529
15. Saykally RJ, Wales DJ (2012) Science 336:814
16. Richardson JO, Wales DJ, Althorpe SC, McLaughlin RP, Viant MR, Shih O, Saykally RJ (2013) J Phys Chem A 117:6960
17. Bulusu S, Yoo S, Aprà E, Xantheas S, Zeng XC (2006) J Phys Chem A 110:11781
18. Lenz A, Ojamäe L (2010) J Mol Struct (Theochem) 944:163
19. Loboda O, Goncharuk V (2010) Chem Phys Lett 484:144
20. Yoo S, Aprà E, Zeng XC, Xantheas SS (2010) J Phys Chem Lett 1:3122
21. Shanker S, Bandyopadhyay P (2011) J Phys Chem A 115:11866
22. Góra U, Podeszwa R, Cencek W, Szalewicz K (2011) J Chem Phys 135:224102
23. Temelso B, Archer KA, Shields GC (2011) J Phys Chem A 115:12034
24. Ramírez F, Hadad CZ, Guerra D, David J, Restrepo A (2011) Chem Phys Lett 507:229
25. Liu X, Lu WC, Wang CZ, Ho KM (2011) Chem Phys Lett 508:270
26. Bates DM, Tschumper GS (2011) J Chem Theory Comput 7:2753
27. Pruitt SR, Addicoat MA, Collins MA, Gordon MS (2012) Phys Chem Chem Phys 14:7752
28. Wang Y, Babin V, Bowman JM, Paesani F (2012) J Am Chem Soc 134:11116
29. Gillan MJ, Manby FR, Towler MD, Alfè D (2012) J Chem Phys 136:244105
30. Miliordos E, Apra E, Xantheas SS (2013) J Chem Phys 139:114302
31. Iwata S, Bandyopadhyay P, Xantheas SS (2013) J Phys Chem A 117:6641
32. Weinhold F (1998) J Chem Phys 109:373
33. Chaplin MF (2000) Biophys Chem 83:211
34. Ludwig RW (2007) ChemPhysChem 8:938
35. Bukowski R, Szalewicz K, Groenenboom GC, van der Avoird A (2007) Science 315:1249
36. Lenz A, Ojamäe L (2009) J Chem Phys 131:134302
37. Wales DJ (2003) Energy landscapes. Cambridge University Press, New York
38. Kazachenko S, Thakkar AJ (2009) AIP Conf Proc 1108:90
39. Kazachenko S, Thakkar AJ (2009) Chem Phys Lett 476:120
40. Kazachenko S, Thakkar AJ (2010) Mol Phys 108:2187
41. Kazachenko S, Thakkar AJ (2013) J Chem Phys 138:194302
42. Kazachenko S (2013) Global energy optimization of hydrogen-bonded clusters. PhD thesis, University of New Brunswick, Fredericton, Canada
43. Jorgensen WL, Chandrasekhar J, Madura JD, Impey RW, Klein ML (1983) J Chem Phys 79:926
44. Bernal JD, Fowler RH (1933) J Chem Phys 1:515
45. Tsai CJ, Jordan KD (1993) J Phys Chem 97:5208
46. Wales DJ, Hodges MP (1998) Chem Phys Lett 286:65

47. Hartke B (2000) Z Phys Chem 214:1251
48. Kabrede H, Hentschke R (2003) J Phys Chem B 107:3914
49. Kazimirski JK, Buch V (2003) J Phys Chem A 107:9762
50. Kabrede H (2006) Chem Phys Lett 430:336
51. Takeuchi H (2008) J Chem Inf Model 48:2226
52. Goedecker S (2004) J Chem Phys 120:9911
53. Kirkpatrick S, Gelatt CD Jr, Vecchi MP (1983) Science 220:671
54. Černý V (1985) J Optim Theory Appl 45:41
55. Salomon P, Sibani P, Frost R (2002) Facts, conjectures and improvements for simulated annealing. SIAM, Philadelphia
56. Li Z, Scheraga HA (1987) Proc Natl Acad Sci USA 84:6611
57. Wales DJ, Doye JPK (1997) J Phys Chem A 101:5111
58. Doye JPK, Wales DJ (1998) Phys Rev Lett 80:1357
59. Doye JPK, Wales DJ, Miller MA (1998) J Chem Phys 109:8143
60. Fraser AS (1959) Biometrics 15:158
61. Booker LB, Goldberg DE, Holland JH (1989) Artif Intell 40:235
62. Goldberg D (1989) Genetic algorithms in search, optimization, and machine learning. Addison-Wesley, New York
63. Wales DJ, Scheraga HA (1999) Science 285:1368
64. Nocedal J (1980) Math Comp 35:773
65. Liu DC, Nocedal J (1989) Math Prog 45:503
66. Davidon WC (1975) Math Prog 9:1
67. Kazachenko S, Bulusu S, Thakkar AJ (2013) J Chem Phys 138:224303
68. Franzblau DS (1991) Phys Rev B 44:4925
69. McDonald S, Ojamäe L, Singer SJ (1998) J Phys Chem A 102:2824
70. Kuo JL, Ciobanu CV, Ojamäe L, Shavitt I, Singer SJ (2003) J Chem Phys 118:3583
71. Lenz A, Ojamäe L (2005) Phys Chem Chem Phys 7:1905
72. Anick DJ (2003) J Chem Phys 119:12442
73. Belair SD, Francisco JS (2003) Phys Rev A 67:063206
74. Shi Q, Kais S, Francisco JS (2005) J Phys Chem A 109:12036
75. Khan A (2008) J Mol Struct (Theochem) 850:144
76. Tissandier MD, Singer SJ, Coe JV (2000) J Phys Chem A 104:752
77. Vukičević D, Grubeša T, Graovac A (2005) Chem Phys Lett 416:212
78. Anick DJ (2002) J Mol Struct (Theochem) 587:87
79. Anick DJ (2002) J Mol Struct (Theochem) 587:97
80. Suitte BP, Belair SD, Francisco JS (2005) Phys Rev A 71:043204
81. Anick DJ (2010) J Chem Phys 132:164311
82. Miyake T, Aida M (2002) Chem Phys Lett 363:106
83. Vukičević D, Graovac A (2008) Croat Chem Acta 81:347
84. Bandow B, Hartke B (2006) J Phys Chem A 110:5809
85. Schönborn SE, Goedecker S, Roy S, Oganov AR (2009) J Chem Phys 130:144108
86. Cheng L, Feng Y, Yang J, Yang J (2009) J Chem Phys 130:214112
87. Cheng L, Cai W, Shao X (2005) Chem Phys Lett 404:182
88. Molinero V, Moore EB (2009) J Phys Chem B 113:4008
89. Stillinger FH, Weber TA (1985) Phys Rev B 31:5262

Chapter 3
Optical Parameters of π-Conjugated Oligomer Chains from the Semiempirical Local Coupled-Cluster Theory

Anton B. Zakharov, Vladimir V. Ivanov and Ludwik Adamowicz

Abstract The π-electron semiempirical local coupled-cluster theory has been developed and used to calculate molecular optical parameters (polarizabilities and hyperpolarizabilities) of fragments of conjugated polymers. The method is based on using molecular π-orbitals of ethylene as an orbital basis for the conjugated systems. The method is termed the Covalent Unbonded molecules of Ethylene, *cue*. Based on the comparison of the calculations performed with higher levels of theory (especially with the full configuration interaction method), it has been demonstrated that for selected conjugated molecules the approach is accurate and capable of reproducing the available experimental data with good accuracy. The *cue*-CC results show qualitatively correct dependency on the sizes of π-systems.

3.1 Introduction

Static electrical parameters of molecules (polarizabilities and hyperpolarizabilities) are quantities which play an important role in the characterization of a wide spectrum of physical-chemistry properties of molecular systems and materials. Among the properties which are particularly relevant to this characterization one should mention the electric polarizability, the optical absorption characteristics, and the intermolecular dispersion interaction (molecule-molecule, molecule-surface, etc) [2, 9, 74].

The linear and nonlinear optic properties of organic materials, including π-conjugated molecular polymers, make it possible to develop numerous technological and industrial applications of these systems. The applications in the field

A.B. Zakharov (✉) · V.V. Ivanov
V. N. Karazin Kharkiv National University, Kharkiv, Ukraine
e-mail: abzakharov@karazin.ua

V.V. Ivanov
e-mail: vivanov@karazin.ua

L. Adamowicz
University of Arizona, Tucson, AZ, USA
e-mail: ludwik@u.arizona.edu

J. Leszczynski and M.K. Shukla (eds.), *Practical Aspects of Computational Chemistry IV*, DOI 10.1007/978-1-4899-7699-4_3

of electronics are particularly important. The conjugated molecular structure and the mobile nature of the π-electrons lead to a significant response of these systems to non-resonance optical excitations. Such a property is important in applications involving optical switches, in devices for second and third harmonic generation, in devices for generating strong electromagnetic fields, etc.

Another important area, where information about polarizability is used, involves statistical prognostic models and the description of the ligands-protein target interactions in the Quantity–Structure–Activity Relation (QSAR) method [27].

The quantum chemical calculations of optical and nonlinear optic molecular parameters are an important step in designing new materials. However, adequate description of molecular optical parameters presents a challenge for contemporary quantum chemistry. The main problem in such calculations is the necessity of accounting for a significant part of the electron correlation effects. In the last decade the density functional theory (DFT) has been used for (hyper)polarizability calculations (see for instance [45]). It allows the consideration of systems with extended sizes. However, the DFT calculations are known to produce significant errors in the evaluation of the optical properties of π-conjugated systems [8, 26, 83].

Among the methods which guarantee adequate inclusion of the electron correlation effects, the coupled cluster (CC) theory is one of the most effective [4, 5, 33, 39, 59, 65, 79]. The first CC calculations of molecular properties related to the interaction of the molecule with the electric field date back to the first years of using the CC theory in molecular calculations [6, 14]. However, the CC calculations, even those performed with the standard approach which includes single and double excitations from the reference wave function (CCSD), involve significant computational cost.

Certain improvement in the efficiency of the CC calculations can be achieved by using the idea of "local correlation effects" [18, 33, 64, 67, 69]. In that approach the CC reference function is built using localized molecular orbitals (LMO) [62]. The use of the LMOs enables to significantly reduce the computational cost of the CCSD calculations [1, 32, 44, 53, 66, 70, 84]. Also, the local CC approach has been implemented in the calculations of electronic excited states [29, 30, 37, 52]. The use of the local CCSD approach in the calculations of the (hyper)polarizabilities has been described in Refs. [28, 35, 36, 38, 66]. For a description of the calculations of the frequency-dependent polarizability and dispersion-coefficients see Ref. [85].

An adequate account of the electron correlation effects requires the use of basis sets of considerable sizes which include polarization and diffuse functions. The ab initio CC calculations with such basis sets are expensive. Thus more exact CC calculations of molecular optical parameters can be carried out only for relatively small molecules. In the calculations of polarizabilities and hyperpolarizabilities for larger π-systems more approximate methods need to be used. Such calculations still remain a difficult problem in quantum chemistry.

An alternative approach for predicting optical properties of larger molecular systems is to employ semiempirical methods which can be used to calculate large polymeric systems. As mentioned above, the description of the nonlinear molecular optical parameters strongly depends on the level of accounting for the electron correlation effects. Thus, the semiempirical wave function has to include these effects in order to provide adequate accuracy in the calculations of these parameters. In our works we have used a semiempirical version of the CCSD theory which is based on the Pople-Pariser-Parr (PPP) Hamiltonian [60, 63]. It should be noted that the PPP model has become increasingly popular in predicting the optical properties of large π-conjugated systems (see for instance Refs. [34, 73, 89] and references therein). Detailed investigations of the correlation effects in the (hyper)polarizability calculations of π-electron systems have been presented in a number of works [22, 23, 31, 42, 43, 47, 50, 61].

Notwithstanding the fact that the PPP-model is a relatively simple quantum chemistry approach, our PPP CCSD calculations have demonstrated that an adequate description of the molecular optical parameters can be obtained using this model. The π CCSD calculations have been done in two ways [31]. The first way involves a separate Hartree–Fock (HF) calculation for each value of the electric field strength followed by a CCSD calculation that uses the HF MOs generated for that field strength. The approach is termed a "relaxed" CCSD (r-CCSD). The second way is to perform the HF calculation only once at zero electric field strength and then use the HF orbitals in the CCSD calculations for all fields including zero field. The approach is termed "unrelaxed" CCSD (u-CCSD). As it was shown in Ref. [31], the u-CCSD and r-CCSD results for some test examples agree very well with the π-electron Full Configuration Interaction (FCI) results. For details of our π-electron FCI and CCSD calculations see Refs. [46, 49, 61]. Further development of the π-electron CCSD theory has involved using the "local correlation" idea. In our local π CCSD theory we employ a classical representation of a molecule with a conjugated system of π bonds involving a set of ethylene fragments. The molecular orbitals of the separate covalent unbonded ethylene fragments (*cue*) form the orbital basis for the CCSD calculation. Our local π-electron (PPP) CC approach is termed the cue-CCSD method [25, 86, 88]. The initial testing of the cue-CCSD approach was performed using some small π-conjugated systems [86, 87]. It was shown that the cue-CCSD calculations lead to results which are close to the FCI results and almost independent on the choice of the double-bond arrangement in the structure of the molecule. The cue-CCSD calculations of π-conjugated polymers enable the evaluation of the limit values (per unit cell) of the optical parameters corresponding to systems with an infinite size [25, 88]. The aim of the present article is to review the cue-CCSD results obtained in the (hyper)polarizabilities calculations for different π-conjugated polymers. The systems under consideration include classical polyenes, polyynes, different polycyclic hydrocarbons (nonalternants), as well as a set of nonclassical "Push-Pull" polymers.

3.2 Semiempirical Local cue-CCSD Theory

Our semiempirical π-electron CCSD theory is based on the standard CCSD scheme. The CCSD wave function has the form:

$$|\Psi_{CCSD}\rangle = e^{\hat{T}_1+\hat{T}_2}|0\rangle, \tag{3.1}$$

where the cluster operators, $\hat{T}_1$, and $\hat{T}_2$, generate all single and double electron excitations from the reference state, $|0\rangle$. In the unitary group approach the cluster operators have the following form:

$$\hat{T}_1 = \sum_{i,a} t_i^a \hat{E}_{ai}, \tag{3.2}$$

$$\hat{T}_2 = \sum_{(i,a)\geq(j,b)} t_{ij}^{ab} \hat{E}_{ai} \hat{E}_{bj}, \tag{3.3}$$

where the operators $\hat{E}_{ai}$ and $\hat{E}_{bj}$ are unitary group generators [54, 55].

Indices *i* and *j* correspond to the occupied orbitals in the reference state, $|0\rangle$, while *a* and *b* correspond to the vacant orbitals. Amplitude matrices t_i^a and t_{ij}^{ab} contain all information about the structure of the many-particle wave function in terms of electronic excitations. The working equations for energy E_{CCSD} and amplitudes t_i^a and t_{ij}^{ab} can be obtained by using projection of the Schrödinger equation onto the reference configuration, $|0\rangle$, and onto singly and doubly excited configurations, $|_i^a\rangle$ and $|_{ij}^{ab}\rangle$. The latter two sets of configurations can be generated by using the unitary group operators (3.2, 3.3).

$$t_i^a \Leftarrow \langle 0\hat{E}_{ia}|\hat{H}_0 - E_{CCSD}|\Psi_{CCSD}\rangle = 0, \tag{3.4}$$

$$t_{ij}^{ab} \Leftarrow \langle 0|(2\hat{E}_{ia}\hat{E}_{jb} + \hat{E}_{ja}\hat{E}_{ib})|\hat{H}_0 - E_{CCSD}|\Psi_{CCSD}\rangle = 0, \tag{3.5}$$

$$E_{CCSD} \Leftarrow \langle 0|\hat{H}_0 - E_{CCSD}|\Psi_{CCSD}\rangle = 0, \tag{3.6}$$

where $\hat{H}_0$ is PPP Hamiltonian of the system.

In the cue-CCSD method we developed [25, 86–88] the π-conjugated system is represented as a set of ethylene units. The acronym *cue* stands for "covalently unbonded molecules of ethylene". In the framework of this approach, an arbitrary π-conjugated molecule (for example, naphthalene) is first represented as a set of single and double bonds as shown in Fig. 3.1.

Then, the electronic wave function of the π-electron system is constructed from ethylene-like MOs. For example, for structure A of naphthalene the wave functions is:

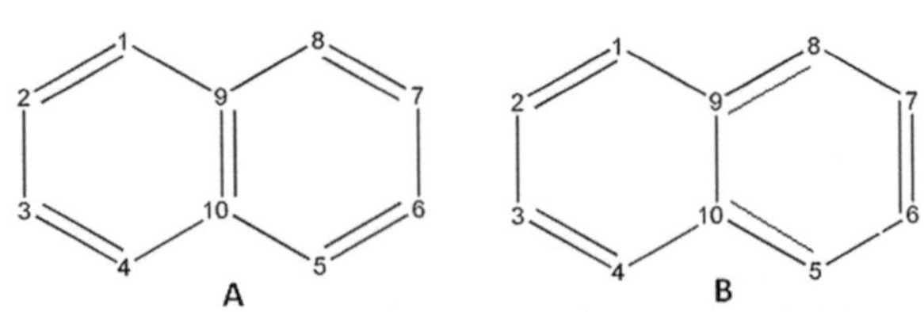

Fig. 3.1 Two ways of double-bonds distribution in the naphthalene molecule

$$|\varphi_1\rangle = \frac{1}{\sqrt{2}}(|\chi_1\rangle + |\chi_2\rangle), \quad |\varphi_1'\rangle = \frac{1}{\sqrt{2}}(|\chi_1\rangle - |\chi_2\rangle), \tag{3.7}$$

$$|\varphi_2\rangle = \frac{1}{\sqrt{2}}(|\chi_3\rangle + |\chi_4\rangle), \quad |\varphi_2'\rangle = \frac{1}{\sqrt{2}}(|\chi_3\rangle - |\chi_4\rangle), \tag{3.8}$$

$$|\varphi_3\rangle = \frac{1}{\sqrt{2}}(|\chi_5\rangle + |\chi_6\rangle), \quad |\varphi_3'\rangle = \frac{1}{\sqrt{2}}(|\chi_5\rangle - |\chi_6\rangle), \tag{3.9}$$

$$|\varphi_4\rangle = \frac{1}{\sqrt{2}}(|\chi_7\rangle + |\chi_8\rangle), \quad |\varphi_4'\rangle = \frac{1}{\sqrt{2}}(|\chi_7\rangle - |\chi_8\rangle), \tag{3.10}$$

$$|\varphi_5\rangle = \frac{1}{\sqrt{2}}(|\chi_9\rangle + |\chi_{10}\rangle), \quad |\varphi_5'\rangle = \frac{1}{\sqrt{2}}(|\chi_9\rangle - |\chi_{10}\rangle), \tag{3.11}$$

where the orbitals $|\varphi_1\rangle$, $|\varphi_2\rangle$, $|\varphi_3\rangle$, $|\varphi_4\rangle$, and $|\varphi_5\rangle$ are occupied MOs while the orbitals $|\varphi_1'\rangle$, $|\varphi_2'\rangle$, $|\varphi_3'\rangle$, $|\varphi_4'\rangle$, and $|\varphi_5'\rangle$ are vacant MOs. This representation corresponds to the case where each ethylene fragment provides an occupied orbital for the construction of the reference determinant, $|0\rangle$. Wave function $|0\rangle$ provides zero-order representation of the conjugated bond structure of naphthalene A. Hereinafter, for simplicity, we denote the molecular orbitals of the *cue* basis with the number of the ethylene fragments to which the basis corresponds. To indicate a vacant MO, we use prime symbol ($|\varphi_k'\rangle$).

It should be stressed that for condensed aromatic hydrocarbons, there are usually several ways for arranging the single and double bonds. This leads to some uncertainty in constructing the reference wave function. However, as our test calculations have shown [86], the differences in the values of the optical parameters obtained for different resonance structures of the system do not differ much even for small systems. For instance, for the naphthalene molecule (Fig. 3.1) there are two possible arrangement of the double bonds corresponding to two different symmetry groups, the D_{2h} and C_{2v} groups. The first case corresponds to structure A and the second one to structure B (see Fig. 3.1). The calculations show that the difference of the average 2nd hyperpolarizabilities of the two structures is insignificant (less than 1 %).

It is worth to note that a good agreement of the results for different resonance structures occurs only if the structures correspond to covalent arrangements of the π-electrons. In one of the previous works [86] we presented results for systems with different covalent coupling (including Dewar-benzene-like long-range covalent coupling) using the fulvene molecule as an example. In the case of non-classical covalent structures, the obtained values may differ by an order of magnitude and even in sign. However for structures with classical covalent arrangement of the double bonds (like the A and B structures shown in Fig. 3.1) the values of the optical parameters obtained in the calculations are very similar. It should be noted that the ambiguity in the double-bond distribution does not arise in the polyene molecules where the *cue* basis used in the calculation corresponds to the localized HF MO's.

The use of the *cue* basis in the CCSD method for calculating the energy of the unperturbed (i.e. zero-field case) system is justified by the Thouless theorem [78] which states that the transformation between two non-orthogonal determinants can be always performed with the use of an $e^{\hat{T}_1}$ operator with an appropriately selected single-excitation $\hat{T}_1$ operator. The operator of this type is included in the wave function in the standard method CCSD. The Euclidian norm of the amplitude vector, t_1, associated with the $\hat{T}_1$ operator can be used as a measure of the orbital relaxation [41].

Among the advantages of using the *cue* representation in the CCSD calculations are: a considerable simplification in the procedure for solving the CC amplitude equations, a clear physical interpretation of the results of the calculation, the topological notation of correlation effects ("correlation radius"),[1] the possibility of systematically improving the accuracy of the calculation by including in the CC cluster operator excitations corresponding to a longer correlation radius, and the simplicity of generating the reference state, $|0\rangle$. The latter means that it is not necessary to carry out an HF calculation and the MO localization prior to the CCSD calculation. This eliminates possible convergence problems in the HF procedure associated with a possible quasi-degeneracy of the HOMO-LUMO orbitals.

Thus the use of the *cue* basis leads to significant simplification in solving the CCSD equations. This is primarily due to the simplification of transformation of the two-electron integrals. The two-electron integrals (we use the Mulliken integral notation where indices denote spatial orbitals) in the zero-differential overlap (ZDO) approximation are calculated according to the following formulas:

$$[AB||CD] = 2[AB|CD] - [AD|CB], \tag{3.12}$$

$$[AB|CD] = \sum_{\mu}^{N} \sum_{\nu}^{N} C_{\mu A} C_{\mu B} C_{\nu C} C_{\nu D} \Gamma_{\mu\nu}, \tag{3.13}$$

where N is the number of the basis functions (in π-electron approximation N is also equal to the number of carbon atoms), $C_{\mu A}$ are the MO-LCAO coefficients, and $\Gamma_{\mu\nu} = [\mu\mu|\nu\nu]$ is a two-centered Coulomb integral.

The calculation of the two-electron integrals in the standard CCSD scheme, even within the ZDO approximation, involves a time-consuming calculation, which precludes the applicability of the CC theory to larger systems. The use of the *cue* approach in the evaluation of expression (3.13) in the CCSD scheme can be illustrated with the following example concerning polyene molecules (see Fig. 3.2). The calculation of the $[AA'|BB']$ integral using the *cue* basis is reduced to a simple expression:

$$[AA'|BB'] = \frac{1}{4}(\Gamma_{rt} - \Gamma_{ru} - \Gamma_{st} + \Gamma_{su}). \tag{3.14}$$

It is clear that the number of components in expression (3.14) does not depend on the system size. It is also worth mentioning that the only non-zero integrals are those with orbitals belonging to the same fragment. This leads to a significant simplifica-

[1] The "correlation radius" is the distance between two structural *cue* elements of the π-system whose electrons are correlated by including in the CC cluster operator electron excitations from both elements.

Fig. 3.2 A scheme describing the calculation of two-electron integrals of polyene systems

tion in the computation of the CCSD diagrams representing the components of the CCSD amplitude equation obtained by projecting the Schrödinger equation with the CC wave function onto singly and doubly excited determinants. This simplification can be illustrated using the diagram involved in the projection of the $\hat{T}_2\hat{T}_1^2$ term onto doubly excited determinant $\langle^{ab}_{ij}|$: $\langle^{ab}_{ij}|H|\hat{T}_2\hat{T}_1^2|0\rangle$.

$$\sum_{k,l,c,d}[kc||ld](t^{ab}_{kl}t^c_i t^d_j + t^{cd}_{ij}t^a_k t^b_l) = [kk'|ll']\left\{t^{ab}_{kl}\left(t^{k'}_i t^{l'}_j - t^{l'}_i t^{k'}_j\right) + 2t^{k'l'}_{ij}t^a_k t^b_l\right\}, \quad (3.15)$$

where k and k' (as well as l and l') correspond to occupied and vacant orbitals belonging to same ethylene fragment, respectively. The $[kk'|ll']$ and $[kl'|lk']$ integrals are the usual Coulomb and exchange parts of two-electron interaction $[kk'||ll']$.

An important element of the cue-CCSD theory is the topological interpretation of the correlation effects. The *cue* formalism allows building a hierarchy of approximations based on the classification of the double-excitation structure. The possible excitations for a polyene chain are shown in Fig. 3.3. Hereinafter use the following notation: cue$_{(l)}$-CCSD is the method that takes into account all excitations up to level l; thus cue-CCSD means that all possible excitations between the ethylene fragments are accounted for. In the figure, the amplitudes are represented as electronic transitions between different ethylene fragments. Therefore, $l = 1$ means including only local excitations, $l = 2$ means including also excitation between neighboring fragments (separated by one bond), and $l = 3$ means including also excitations between the second neighbors (separated by three bonds), etc. In the lth level of the theory all excitations between ethylene fragments separated by up to $(2l - 3)$ bonds are included.

The number of bonds which separate the fragments can be identified by the Floyd-Warshall procedure [3]. As a result of applying this procedure to a molecule a matrix containing information about the number of edges between the atomic orbitals (say

Fig. 3.3 *cue* excitations in polyenes

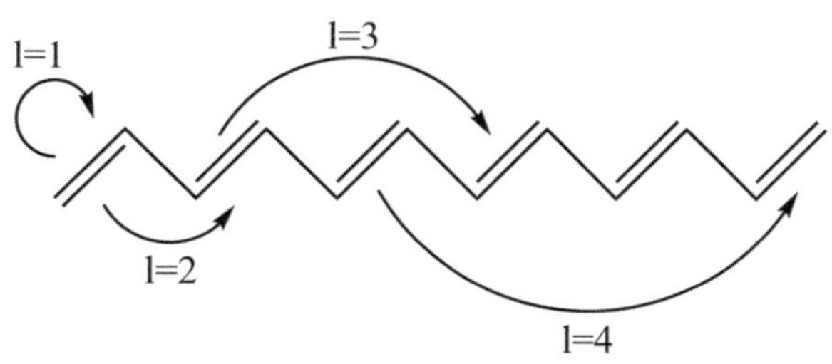

μ, ν), $W_{\mu\nu}^{(AO)}$, is obtained. Then the topological distance between two ethylene-like MOs is defined by the following expression:

$$\forall \mu \in i \cup \nu \in j : W_{ij}^{(MO)} = min\{W_{\mu\nu}^{(AO)}\}. \tag{3.16}$$

The classification of the cluster amplitudes is unambiguous for linear polyene-like systems (Fig. 3.3). For condensed (polyaromatic) hydrocarbons (i.e. polyacenes) there is some uncertainty in the topological evaluation of the distances between some double bonds. To explain this let us consider an example of a graphene fragment (see Fig. 3.4). The local ($l = 1$) and the first-neighbor ($l = 2$) excitations in this case are the same as for the polyene chain. However some excitations taken into account in the cue$_{(3)}$-CCSD approximation ($l = 3$) with the distances determined as the Euclidian distances between centroids of fragments separated by three bonds (see Fig. 3.4; excitations *a*) are greater than excitations *c* separated by four bonds. Thus excitations *c* need to be included in the cue$_{(3)}$-CCSD method. Analogically, the same rule is applied to all excitations of higher orders in the cue$_{(l)}$-CCSD method corresponding to approximation level l.

As mentioned, when all possible double excitations between the ethylene fragments are included in the wave function, the method is termed cue-CCSD. In this case l is equal to the number of ethylene units comprising the system, i.e. the system length in terms of the ethylene units. Also it should be noted that the restriction of including selected classes of excitations applies in the cue$_{(l)}$-CCSD method (and in the full cue-CCSD method) only to the $\hat{T}_2$ cluster operator and not to $\hat{T}_1$. All excitations generated by the $\hat{T}_1$ operator are included. This is done to properly account for the orbital relaxation effects. Such an approach differentiate our approach from the conventional approximate local CC theories where the restrictions are made in both $\hat{T}_1$ and $\hat{T}_2$ operators.

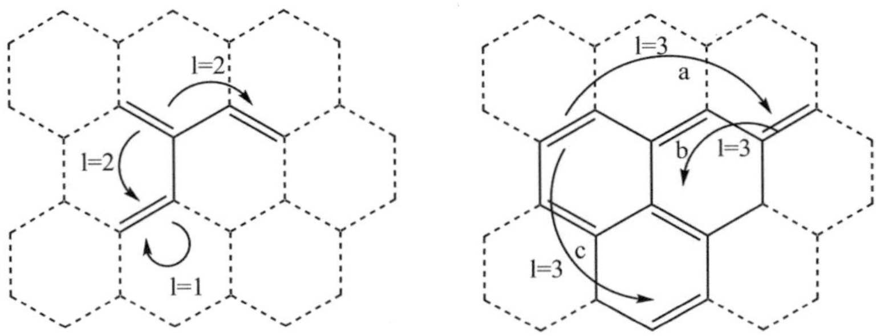

Fig. 3.4 *cue* excitations in polyacenes

3.3 Semiempirical cue-CCSD Calculations of (Hyper)polarizabilities

The most direct introduction of molecular nonlinear optic parameters is based on the Buckingham expansion [7]. According to this expansion the energy of a molecular system in the external (static) electric field ($\vec{\mathbf{F}} = F_x\vec{\mathbf{i}} + F_y\vec{\mathbf{j}} + F_z\vec{\mathbf{k}}$) can be described in the following form:

$$E(\vec{\mathbf{F}}) = E(0) - \mu_r E_r - \frac{1}{2}\alpha_{rs}F_rF_s - \frac{1}{3!}\beta_{rst}F_rF_sF_t - \frac{1}{4!}\gamma_{rstu}F_rF_sF_tF_u - \cdots \tag{3.17}$$

Einstein summation convention is assumed throughout this work. In Eq. 3.17 the indices r, s, t, u correspond to the cartesian coordinates ($r, s, t, u = x, y, z$). The coefficients in expansion (3.17) are: μ_s—the static dipole moment cartesian component, α_{rs}—the dipole polarizability components, and β_{rst} and γ_{rstu}—the first and second static dipole hyperpolarizabilities, respectively.

Also the corresponding coefficients can be introduced by using the induced dipole moment as:

$$\mu_r(\vec{\mathbf{F}}) = \mu_r(0) + \alpha_{rs}F_s + \frac{1}{2}\beta_{rst}F_sF_t + \frac{1}{3!}\gamma_{rstu}F_sF_tF_u + \cdots \tag{3.18}$$

The molecular optic parameters used in the above equations are determined as follows:

$$\alpha_{rs} = -\left.\frac{\partial^2 E(\vec{\mathbf{F}})}{\partial F_r \partial F_s}\right|_{\mathbf{F=0}} = \left.\frac{\partial \mu_r(\vec{\mathbf{F}})}{\partial F_s}\right|_{\mathbf{F=0}}, \tag{3.19}$$

$$\beta_{rst} = -\left.\frac{\partial^3 E(\vec{\mathbf{F}})}{\partial F_r \partial F_s \partial F_t}\right|_{\mathbf{F=0}} = \left.\frac{\partial^2 \mu_r(\vec{\mathbf{F}})}{\partial F_s \partial F_t}\right|_{\mathbf{F=0}}, \tag{3.20}$$

$$\gamma_{rstu} = -\left.\frac{\partial^4 E(\vec{\mathbf{F}})}{\partial F_r \partial F_s \partial F_t \partial F_u}\right|_{\mathbf{F=0}} = \left.\frac{\partial^3 \mu_r(\vec{\mathbf{F}})}{\partial F_s \partial F_t \partial F_u}\right|_{\mathbf{F=0}}. \tag{3.21}$$

The derivatives of energy in the above equations can be obtained by numerical differentiation (Finite Field approach, FF). For instance, to calculate the second- and fourth-order derivatives which correspond to polarizability and 2nd hyperpolarizability the following "seven point" formulaes can be used:

$$\alpha \approx \frac{2E_3 - 27E_2 + 270E_1 - 490E_0 + 270E_{-1} - 27E_{-2} + 2E_{-3}}{180\xi^2}, \tag{3.22}$$

$$\gamma \approx \frac{-E_3 + 12E_2 - 39E_1 + 56E_0 - 39E_{-1} + 12E_{-2} - E_{-3}}{6\xi^4}, \tag{3.23}$$

where ξ is differentiation step and $E_{\pm m}$ are energies of the system in static electric field with strength $\pm m\xi$. The energy of the system in field can be obtained by solution of corresponding Schrödinger equation with Hamiltonian:

$$\hat{H}(\pm m\xi) = \hat{H}_0 \pm m\xi\hat{R}, \quad (3.24)$$

where $\hat{H}_0$ is unperturbed PPP-Hamiltonian (without field) of the system, while $\hat{R}$ is the dipole moment operator.

The average value of the (hyper)polarizabilities can be represented by the corresponding invariants [7, 51]. These invariants are represented by the following expressions:

$$\langle\alpha\rangle = \frac{1}{3}(\alpha_{xx} + \alpha_{yy} + \alpha_{zz}). \quad (3.25)$$

$$\langle\gamma\rangle = \frac{1}{5}(\gamma_{xxxx} + \gamma_{yyyy} + \gamma_{zzzz} + 2\gamma_{xxyy} + 2\gamma_{zzyy} + 2\gamma_{xxzz}). \quad (3.26)$$

In all calculations described in this section "idealized" geometries for all considered π-systems are used. In these geometries all –C–C– bond lengths are equal to 1.4 Å. Geometries of all zig-zag *trans* carbon chains of polyene systems are planar. All –C–C–C– bond angles for *trans* polyenes are equal to 120°. All cyclic systems and polycyclic hydrocarbons are equilateral (regular) polygons.

The PPP-model of π-conjugated molecule assumes that each carbon atom contributes a $2p_z$ orbital and a single π-electron to the π-system. The PPP parametrization of the Hamiltonian matrix elements includes resonance (electron hopping) integrals for the connected carbon atoms μ and ν:

$$t_0 = \langle\mu|H|\nu\rangle = 2.274 \text{ eV}. \quad (3.27)$$

In the calculations of polyene molecules the bond alternation model (BAM) is used (see for instance [10, 74]). In our calculations in the BAM the values of two alternated-bond resonance integrals are equal to:

$$t^{\pm}_{\mu\nu} = (1 \pm 0.1)t_0, \quad (3.28)$$

where the '+' and '−' signs correspond to the hopping integrals for double and single bonds, respectively.

The one-center Coulomb integral for the carbon atom is equal to $\Gamma_{\mu\mu} = [\mu\mu|\mu\mu] = 11.13$ eV. The two-center Coulomb integrals are calculated by using the Onho empirical formula [57] (atom unit assumed):

$$\Gamma_{\mu\nu} = [\mu\mu|\nu\nu] = \frac{\zeta\, A_{\mu\nu}}{\sqrt{1 + (A_{\mu\nu}R_{\mu\nu})^2}}, \quad (3.29)$$

where $R_{\mu\nu}$ is the distance between the μ and ν atoms. Constant ζ guarantees the necessary physical dimension of $\Gamma_{\mu\nu}$. $A_{\mu\nu}$ is the average one-center integral for atoms μ and ν:

$$A_{\mu\nu} = \frac{1}{2}(\Gamma_{\mu\mu} + \Gamma_{\nu\nu}). \tag{3.30}$$

3.3.1 Justification of π-Electron CCSD Theory

An important question which can be raised is how the π-electron CCSD approximation in (hyper)polarizability calculations compare with the results of other methods, particularly with the ab initio and semiempirical all-valence methods. It is also important to compare obtained theoretical results with available experimental data.

To demonstrate adequacy of π-electron approximation in comparison with all-valence semiempirical AM1 method we carried out calculations of polarizability and 2nd hyperpolarizability of condensed and non-condensed polyfulvenes, (c-polyfulvene and nc-polyfulvene correspondently, see Fig. 3.18, Sect. 3.3.3.3). GAMESS package [68] is used in the AM1 (and further ab initio) calculations. First of all we are comparing AM1 and PPP variant of the HF method. Obtained results for 2nd hyperpolarizability presented on the Fig. 3.5. The values for two methods correlate for corresponding polymers with the coefficient greater than 0.999. It indicates that magnitudes under consideration do not demand more expensive calculation in

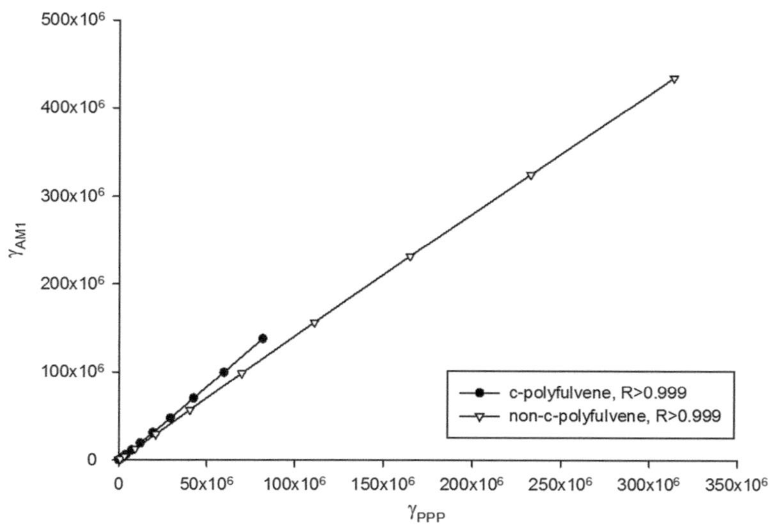

Fig. 3.5 Hartree-Fock AM1 versus PPP average 2nd hyperpolarizability of polyfulvenes. Reprinted with permission from Ref. [88]. Copyright 2014 American Chemical Society

the all-valence method due to additivity of σ-contribution for both polarizability and 2nd hyperpolarizability [25]. Such a result is typical and can be obtained for different conjugated polymers.

For the monomer units one can directly compare values of 2nd hyperpolarizability obtained in π-electron approximation and ab inito methods with augmented basis set, 6-31++G(d,p). We carried out calculations for monomer units of considered (in the Sect. 3.3.3.3) polyfulvenes at MP2 level of theory. Results are presented in the Table 3.1.

Difference in the results for non-condensed case are quite noticeable but it is not critical. In general the hyperpolarizabilities obtained at the MP2/ab initio and MP2/PPP levels are reasonably close to each other. With the growth of the length of the conjugation chain σ-contribution increases additively, unlike π-contribution.

First comparison of 2nd hyperpolarizability obtained in CCSD/PPP approximations were performed in Ref. [73]. The results of π-electron CCSD calculations and especially cue-CCSD calculations in comparison with experimental data [12] are presented in the Table 3.2. The Table 3.2 results reveals quite similar calculated values (r-CCSD and cue-CCSD) with experimental data. A detailed comparison of the results for different variants of CCSD theory with FCI values were performed in Ref. [31].

The values for first members of polyacene series obtained in cue-CCSD and ab initio variant of CCSD(T) methods (with extrapolation onto infinite basis set) as well as the experimental data are presented in the Table 3.3. For the sake of comparability with experimental data, average values for polarizability corrected by σ-electron contribution according to additive scheme [80, 82] $\alpha^{total} = \alpha^{\pi} + \alpha^{\sigma}$. For polyacene molecules cue-CCSD method reveals a perfect match. However, the cue-

Table 3.1 Average 2nd hyperpolarizability ($\langle\gamma\rangle/10^5$, a.u.) of monomer units polyfulvenes obtained in MP2/PPP approximation and in ab initio/6-31++G(d,p) level of theory

	MP2/PPP	MP2/6-31++(d,p)
nc-polyfulvene	1.71	0.64
c-polyfulvene	0.50	0.37

Table 3.2 Calculated and experimental average 2nd hyperpolarizabilities (a.u. 10^5) for linear *trans*-polyenes (N_C is the number of the carbon atoms in the molecule)

N_C	HF	MP2	r-CCSD	cue-CCSD	Experimental
8	0.791	0.966	0.963	1.016	0.993
10	2.07	2.44	2.06	2.26	1.97
12	4.29	4.97	3.65	4.11	3.57
14	7.59	8.72	5.67	6.49	5.56
16	12.0	13.8	8.00	9.31	7.94
18	17.5	20.0	10.5	12.43	11.3

Table 3.3 (Hyper)polarizabilities of polyacenes

	Benzene	Naphthalene	Anthracene	Tetracene
$\langle\alpha^{total}\rangle$ cue-CCSD	66.9	117.9	175.2	235.7
$\langle\alpha\rangle$ CCSD(T)/cc-pV∞Z [20]	N/A	116.7	176.2	239.7
$\langle\alpha\rangle$ CCSD(T)/aug-cc-pV∞Z [20]	N/A	115.5	164.9	N/A
$\langle\alpha\rangle$ experim. [11]	66.8	112.0	171	N/A
$\langle\gamma\rangle/10^4$ cue-CCSD	0.536	2.17	–	–
$\langle\gamma\rangle/10^4$ experim.	0.195 [40], 1.28 $\pm$ 0.06 [81]	6.19 $\pm$ 1.24 [81]	N/A	N/A

CCSD calculations of fullerene C_{60} demonstrated some discrepancies (experimental values are taken from [56]). Namely $\langle\alpha^{total}\rangle = 660$ a.u. while experimental value equal to 579 a.u. For the 2nd hyperpolarizability $\langle\gamma\rangle/10^4 = 11.1$ a.u. while corresponding experimental value evaluated as 9.3 ± 1.3 a.u.

3.3.2 *The Wave Function Structure in cue-CCSD Approach*

The use of *cue* basis in the CCSD calculation and the analysis of the resulting many-particle CCSD wave function give a unique possibility to interpret the correlation effects in the system in structural terms. Namely the transitions between the ethylene units of a conjugated molecule are the parameters which characterize the correlation effects in the system. The cue-CCSD approach differs from the standard ab initio CCSD approach due to the delocalization of the MOs forming the reference wave function in the latter approach. These MOs are obtained in the HF SCF calculation. However the first question is whether the *cue* reference is suitable for describing the electron-correlation effects. To determine that it is necessary to compare the values of the correlation energies (λ_{corr}) obtained in the cue-CCSD method with those obtained with other methods. As usual, λ_{corr} is evaluated as a difference of the total energy of the system obtained with the particular method (E_{method}) and the HF energy (E_{HF}). For polymeric systems it is convenient to calculate the specific λ_{corr}, i.e. the correlation energy per electron or per atom:

$$\lambda_{corr} = (E_{HF} - E_{method})/n, \tag{3.31}$$

where n is the number of π-electrons (or carbon atoms) in the system. In Table 3.4 specific correlation energies obtained using the MP2 and CCSD methods (in both the HF MOs are used) and the FCI method for polyenes with different lengths are shown. The largest polyene for which the FCI results have been obtained is $C_{16}H_{18}$. The percentage value in the parentheses shows the ratio of the correlation energy obtained with the particular method and the FCI correlation energy.

Table 3.4 Specific correlation energies (eV, per carbon atom) obtained with different methods for even-numbered polyene molecules, C_NH_{N+2}

Method	N					
	8	10	12	14	16	30
MP2	0.0713 (38.9)	0.0727 (39.3)	0.0738 (39.5)	0.0747 (39.8)	0.0754 (39.9)	0.0778
cue-CCSD	0.1807 (98.6)	0.1817 (98.1)	0.1824 (97.7)	0.1829 (97.3)	0.1833 (97.1)	0.1845
CCSD	0.1811 (98.9)	0.1821 (98.3)	0.1828 (97.9)	0.1833 (97.6)	0.1837 (97.3)	0.1849
FCI	0.1832	0.1852	0.1867	0.1879	0.1888	N/A

In the parentheses percentage of the particular correlation energy with respect to the correlation energy obtained with the FCI method is shown. All the methods use the π-electron PPP parametrization

The correlation energy values shown in Table 3.4 convincingly demonstrate that the use of the *cue* basis leads to very good results. The correlation energies obtained with cue-CCSD and with the complete CCSD method differ insignificantly. In addition, the differences between cue-CCSD and CCSD do not increase with the growth of the polyene chain. Also it should be noted that the correlation energies obtained with the MP2 method account only for about 40 % of the FCI results.

An important question which can be asked concerns the magnitude of the contributions of certain types of amplitudes which correspond to different classes of excitations (which are represented by l). To construct an effective local theory it is necessary to include excitations that give the main contributions to the total wavefunction for the system under study and to neglect those that have smaller influence on the results of the calculation. As the measure of the contribution of a particular l layer of particular amplitudes (excitations which correspond to the cue$_{(l)}$-CCSD level of theory; see Sect. 3.2) it is convenient to use the sum of squares of the amplitude matrix elements:

$$\varepsilon^{(l)}_{\|\cdot\|} = \sum_{\binom{ab}{ij}\in l} (t^{ab}_{ij})^2 / \sum_{\binom{ab}{ij}} (t^{ab}_{ij})^2. \tag{3.32}$$

$\varepsilon^{l}_{\|\cdot\|}$ can approximately describe the magnitude of the contribution resulting from amplitudes of a certain order as it provides complete information on the level of theory to achieve a certain accuracy threshold. The corresponding number of configurations is simply equal to:

$$\varepsilon^{(l)}_{N} = \frac{N^{(l)}}{N}, \tag{3.33}$$

where $N^{(l)}$ is the number of amplitudes corresponding to the l layer and N is the total number of the amplitudes corresponding to the doubly excited configurations. Hence the parameter:

$$\varepsilon^{(l)} = \varepsilon^{(l)}_{\|\cdot\|} / \varepsilon^{(l)}_{N}, \tag{3.34}$$

or the corresponding normalized parameter:

$$\overline{\varepsilon^{(l)}} = \varepsilon^{(l)} / \sum_{l} \varepsilon^{(l)}, \tag{3.35}$$

can be interpreted as a parameter that characterizes the effectiveness of the cue$_{(l)}$-CCSD models. This parameter can help to choose the right approach to use within the limits of the computational capabilities provided for the calculations. It allows determination fo the necessary level of theory without a direct calculation of the correlation energy.

In view of the obvious differences in the structure of the wave-functions for the linear and polycyclic conjugated polymers (linear polyacenes), let us analyze two representative members of both classes. In Table 3.5 values of the above described parameters are shown for polyene $C_{50}H_{52}$ and polyacene $C_{50}H_{28}$ that consists of 12 benzene rings. The values of λ_{corr} that describe the contributions to the correlation energy obtained from different l-layers in the cue-CCSD method are also shown in the table.

It can be seen from the table that the dominant part of the excitations involved in the calculation of the correlation energy comes from the local electron transitions ($l = 1, \varepsilon^{(l)}_{\|\cdot\|}$ value). The next largest contribution to the total cue-CCSD wave function comes from the excitations between adjacent double bonds ($l = 2$). Excitations from the $l = 3$ layer constitutes only 5.83 % of the total for polyenes, but they are still significant for polyacenes (24.87 %).

As can be seen from the data, the dominant contribution to the wave-function of cue-CCSD method are made by an extremely small number of amplitudes of the $\hat{T}_2$ operator matrix (the size of corresponding amplitude matrix, N, mainly determines the computational complexity of the method). Including l-layers with larger l (starting from the 4th for the polyenes, and 5th for polyacenes) involves larger amount of the matrix elements, but the contribution made by them is insignificant (less than 3 %). This clearly demonstrates the previously discussed fact of the correlation-effects locality.

It should be mentioned that the structure of the amplitude matrix for a polycyclic system is much more complex than for a polyene system due to the lack of excitations of certain kinds in the former (compare Figs. 3.3 and 3.4 concerning inter-bond electron jumping for $l = 3$). This leads to the contributions of some selected l-layers to the total cue-CCSD wave function of polyenes to be noticeably different from the contributions of the corresponding layers in polyacenes. For example, this is evident by examining the $l = 3$ contribution in polyenes and the $l = 4$ contribution in polyacenes. One can conclude that in the calculation of electro-optical properties of polycyclic systems the required minimal level of the theory is cue$_{(2)}$-CCSD. Such a choice guarantees a high level of accounting for the electron correlation effects (more than 99 %) for both polyenes and polyacenes.

Parameter $\overline{\varepsilon^{(l)}}$, which characterizes the "effectiveness" of the local theory, shows fast convergence with the increase of the polymeric chain. For example, for polyenes

Table 3.5 Effectivity parameters for different l layers of the cue-CCSD wave function for polyene and linear polyacene

l	Polyene, $C_{50}H_{52}$					Polyacene, $C_{50}H_{28}$				
	$\varepsilon^{(l)}_{\lVert\cdot\rVert}$, %	$\varepsilon^{(l)}_{N}$, %	$\varepsilon^{(l)}$	$\overline{\varepsilon^{(l)}}$, %	λ_{corr}, %	$\varepsilon^{(l)}_{\lVert\cdot\rVert}$, %	$\varepsilon^{(l)}_{N}$, %	$\varepsilon^{(l)}$	$\overline{\varepsilon^{(l)}}$, %	λ_{corr}, %
1	68.12	0.01	6810	95.16	95.19	36.99	0.01	3700	93.42	83.95
2	24.55	0.10	246	4.46	4.21	28.26	0.16	177	5.83	15.7
3	5.83	0.31	18.8	0.34	0.50	24.87	0.56	44.4	0.66	0.21
4	1.17	0.63	1.86	0.03	0.07	7.59	2.14	3.55	0.08	0.10
5	0.25	1.05	0.24	$\sim10^{-3}$	0.02	1.79	4.26	0.42	0.01	0.03
6	0.06	1.55	0.04	$\sim10^{-3}$	$\sim10^{-3}$	0.39	6.71	0.06	$\sim10^{-3}$	0.01
Rest	0.02	96.35	–	$\sim10^{-4}$	$\sim10^{-3}$	0.11	86.16	–	$\sim10^{-4}$	$\sim10^{-3}$

The percentage of the correlation, λ_{corr}, obtained with the particular cue$_{(l)}$-CCSD method relative to the correlation energy obtained with the full cue-CCSD method

$C_{30}H_{32}$, $C_{40}H_{42}$, and $C_{50}H_{52}$ the values of $\overline{\varepsilon^{(l)}}$ (for layers 1–4) differ by less than $\sim 10^{-2}$ %. Also the correlation energy converges faster with the increasing number of l layers included in the calculation, as evident from the data shown in Table 3.6. In the table the percentages of the account for the correlation effects for different methods with respect to the full cue-CCSD results are shown.

As expected, for polyenes even lowest level of theory allows for an adequate description of the correlation effects in contrast to the polyacene case. However, despite a low level of accounting for the electron correlation by the cue$_{(2)}$-CCSD method for polyacenes, this approach gives a qualitatively correct dependence of the values of the optical and non-linear optical properties on the system size (see the next section). We attribute this behavior to the size-extensivity of the CCSD method.

Another structural aspect of the CCSD wave function is related to its exponential form (3.1). The cumulative value of the configuration coefficient of a particular determinant is determined by expanding the exponent in a Taylor serious in terms of the singly and doubly excited CC operators and collecting terms generating this particular determinant. The configuration coefficient of a particular determinant (say, determinant $\langle (k)_\mu|$, where k indicates the level of excitation of the determinant with respect to the reference determinant and μ is the determinant number in the manifold of the k excited determinants) can be determined as:

$$C_{k\mu} = \langle (k)_\mu | \Psi_{CCSD} \rangle. \tag{3.36}$$

The contribution from the reference determinant, $|0\rangle$, according to the intermediate normalization condition, is assumed to be equal to unity:

$$C_0 = W_0 = \langle 0 | \Psi_{CCSD} \rangle = 1. \tag{3.37}$$

The total weight of all k-fold excited determinants in the wave function is:

$$W_k = \sum_\mu C_{k\mu}^2. \tag{3.38}$$

Table 3.6 Percentage accounting of the correlation energy (relative to full cue-CCSD) by the MP2 method and methods corresponding to different levels of the cue$_{(l)}$-CCSD method

Method	Polyene			Polyacene		
	$C_{30}H_{32}$	$C_{40}H_{42}$	$C_{50}H_{52}$	$C_{30}H_{18}$	$C_{42}H_{24}$	$C_{50}H_{28}$
MP2	42.2	42.5	42.7	50.3	50.4	50.5
cue$_{(2)}$-CCSD	81.6	80.7	80.2	30.2	28.2	27.4
cue$_{(3)}$-CCSD	94.1	93.7	93.5	79.4	77.0	76.2
cue$_{(4)}$-CCSD	98.2	98.0	97.9	95.6	94.6	94.2
cue$_{(5)}$-CCSD	99.5	99.4	99.3	99.2	98.8	98.7
cue$_{(6)}$-CCSD	99.8	99.8	99.8	99.9	99.7	99.7
cue-CCSD	0.1845	0.1849	0.1851	0.1677	0.1725	0.1744

In the last row the correlation energy in eV per carbon atom is shown

Table 3.7 Configurational weights, W_k, corresponding to different k values of the CCSD wave functions with the HF reference and the *cue* reference for polyene molecules

Weight	$C_{10}H_{12}$		$C_{12}H_{14}$		$C_{14}H_{16}$	
	HF	*cue*	HF	*cue*	HF	*cue*
W_1	0.493×10^{-3}	0.375	0.665×10^{-3}	0.475	0.844×10^{-3}	0.575
W_2	0.177	0.292	0.213	0.384	0.250	0.486
W_3	0.454×10^{-4}	0.565×10^{-1}	0.838×10^{-4}	0.101	0.136×10^{-3}	0.160
W_4	0.103×10^{-1}	0.229×10^{-1}	0.161×10^{-1}	0.418×10^{-1}	0.232×10^{-1}	0.696×10^{-1}
W_5	0.117×10^{-5}	0.221×10^{-2}	0.337×10^{-5}	0.629×10^{-2}	0.755×10^{-5}	0.141×10^{-1}
W_6	0.246×10^{-3}	0.673×10^{-3}	0.548×10^{-3}	0.181×10^{-2}	0.104×10^{-2}	0.414×10^{-2}
$\sum_k W_k$	0.19	0.74	0.23	1.01	0.27	1.31

The information about the total weights of excited determinants corresponding to different values of k for polyene molecules are shown in Table 3.7.

In the table, the results corresponding to the standard CCSD method are obtained with the HF MOs while the cue-CCSD results are obtained with the *cue* MO basis. As one can see, the contributions of higher excitations obtained with the MO basis are significant. In particular, the contributions from four- and six-fold excitations noticeably increase with the elongation of the carbon chain (notice that $W_1 \approx W_6$). The cue-CCSD wave function has different structure in terms of Ws. In the *cue* basis there is a significant relaxation of the reference state as evident from a larger value of W_1. The main contributions correspond to singly- and doubly-excited configurations. For instance, for polyene $C_{14}H_{16}$ the weights are $W_1 = 0.575$ and $W_2 = 0.486$. Also higher levels of excitations (e.g. four-fold and six-fold excitations) become important. The total weight, $\sum_k W_k$, characterize the complexity of the *cue* wave function. All presented data, including the total weights, indicate that the cue-CCSD wave function is significantly different from the reference determinant. The opposite can be said about the wave function obtained with the standard CCSD method based on the HF MOs.

3.3.3 (Hyper)polarizabilities of π-Conjugated Systems

3.3.3.1 Polyenes and Polyynes

In the field of nonlinear optics, probably the best known and studied systems are quasi-one-dimensional polymers such as polyenes (C_MH_{M+2}). In these polymers the correlation effects are relatively large. The importance of the correlation effect in these systems has been investigated using the semiempirical π-electron approach in Refs. [24, 31, 42, 43, 61, 73]. As shown earlier (Table 3.4), the limit values of the specific correlation energies for polyenes (per electron) are near 0.19 eV. As the quality of the description of nonlinear optical parameters significantly depends on the level of accounting for the electron correlation, one should expect that the calculated values of γ for long polyene chains are highly dependent on the level of the cue-CCSD theory. In Figs. 3.6 and 3.7 the dependency of the average specific (per electron) polarizability and 2nd hyperpolarizability for *trans*-polyenes on the level of theory used in the calulations are presented.

It can be seen in the figures that the results obtained with different variants of the CC theory are rather similar. Upon examination of the results, one can conclude that the specific values tend to converge with increase of the chain length. This observation has a clear physical explanation. With the increase of the number of electrons in the system the inclusion of an addition monomeric unit has increasingly smaller influence on the specific optical properties. The results of the calculations within the HF and MP2 frameworks show pronounced growth of the $\langle\alpha\rangle/N$ and $\langle\gamma\rangle/N$ values with increasing length of the polymer chain. Apparently these methods significantly overestimate the stability limits.

To investigate the influence of the conformation of the conjugated polymer chain on the optical parameter, we performed calculations for a number of regular linear polymers which different conformations. To describe the linear size of the system we use the following parameter:

$$\rho_\alpha = L_\alpha/N, \tag{3.39}$$

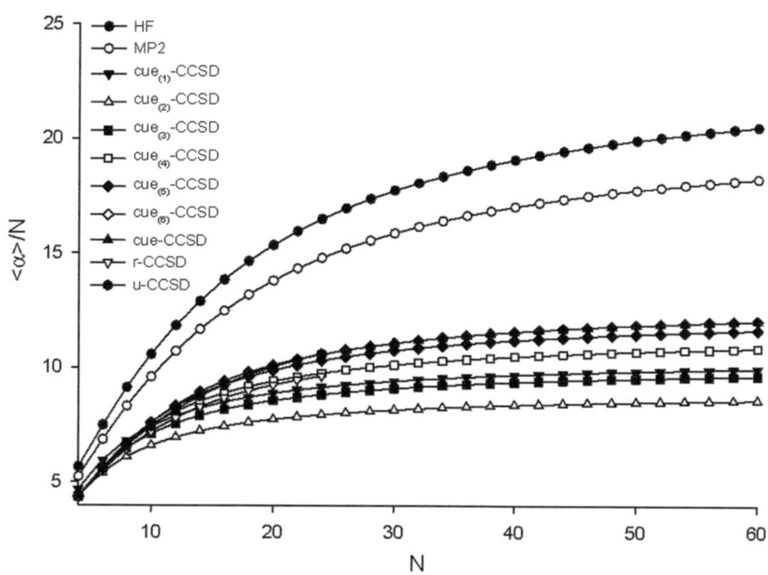

Fig. 3.6 Average specific (per electron) polarizabilities of polyenes obtained using different approaches

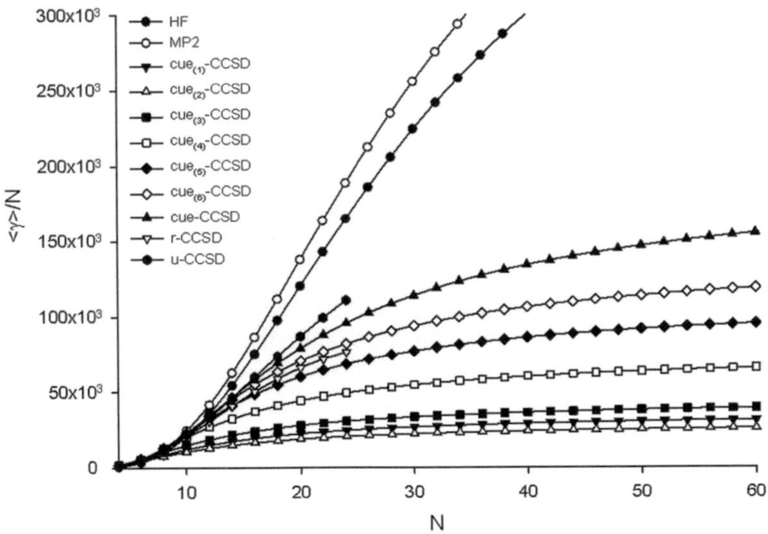

Fig. 3.7 Average specific (per electron) 2nd hyperpolarizabilities of polyenes

where $\alpha \in \{x, y, z\}$ are the Cartesian coordinates, L_α is the length of the polymere chain along a certain axis, and N is the number of π-electrons. The structures of the considered polyene are shown in Fig. 3.8 (A—*all-trans*, B—*all-cis*, C—combination of *trans/cis*, D—linear polyenes).

The linear size (say along the dominant "x"-direction) can be described as:

$$\rho_x = \rho_x^\infty + \rho_x^0/N, \tag{3.40}$$

where ρ_x^∞ is the limit (specific, per electron) length of the polymer and ρ_x^0 is the geometrical parameter responsible for the "end-effect" contribution. The corresponding

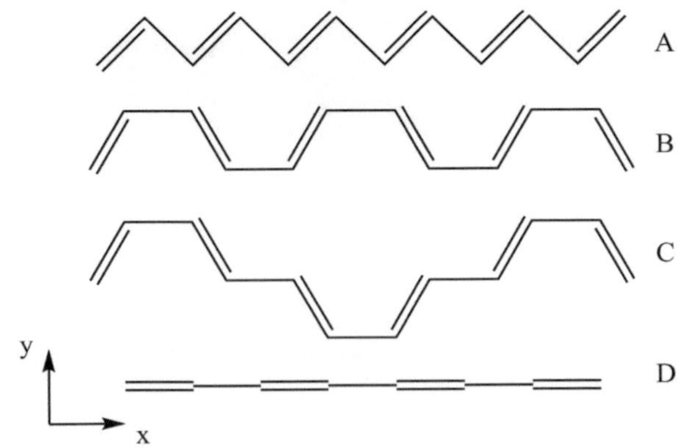

Fig. 3.8 Conformational models of linear polyenes

molecular optical parameters of π conjugated oligomers can be represented in an analogical form as:

$$\langle\alpha\rangle/N = \alpha^{\infty} + \alpha_0/N, \tag{3.41}$$

$$\langle\gamma\rangle/N = \gamma^{\infty} + \gamma_0/N, \tag{3.42}$$

where $\alpha^{\infty} = \langle\alpha\rangle/N$ and $\gamma^{\infty} = \langle\gamma\rangle/N$ are infinite-size-limit (when $N \to \infty$) quantities for the respective optical properties of the polymer. Obviously, the interval of N that corresponds to the region where the values α^{∞} and α_0 (as well as γ^{∞} and γ_0) are converging fits a linear relation.

By approximating this interval one can obtain the limit values of the specific α and γ. In Table 3.8 the parameters obtained using the Eq. (3.40) and the specific components of α and γ at the limit are shown. The extrapolation for obtaining the limit values for polyenes has been performed for systems with 50–60 atoms. For this 50–60 atom range the corresponding correlation coefficient is greater than 0.999. An important conclusion can be derived from the results shown in Table 3.8: in the limit of $N \to \infty$, the calculated values of the (hyper)polarizabilities for various polyenes closely correlate to the values of these properties for the linear-size polymer.

Optical parameters obtained with Eqs. (3.41) and (3.42) with different variants of cue-CCSD, as well as with the MP2 and HF methods, are presented in Table 3.9.

As expected, the results obtained with the HF and MP2 methods are significantly overestimated (the HF results are more than three times higher relative to the cue-CCSD results for the 2nd hyperpolarizability). Also these methods give qualitatively incorrect dependence of the studied properties on the polymer-chain length (see

Table 3.8 Linear sizes of different conformations of polyenes and limit values of the polarizability, α_{xx}, and the hyperpolarizability, γ_{xxxx}

Conformation	ρ_x^{∞}	ρ_x^0	α_{xx}^{∞}	$\gamma_{xxxx}^{\infty}/10^5$
A	1.21	−1.21	38.2	9.8
B	1.05	−1.40	24.2	4.9
C	1.05	−1.40	23.6	5.0
D	1.40	−1.40	57.9	19.4

Table 3.9 Limit values of (hyper)polarizabilities for *trans*-polyenes obtained using different many-body approaches

	HF	MP2	cue$_l$-CCSD						cue-CCSD
			$l=1$	$l=2$	$l=3$	$l=4$	$l=5$	$l=6$	
$\langle\alpha\rangle^{\infty}$	23.43	20.73	10.52	9.04	10.21	11.61	12.55	13.09	13.82
$\langle\alpha\rangle_0$	−174.4	−148.4	−33.21	−26.10	−33.41	−44.58	−54.10	−60.52	−75.10
$\langle\gamma\rangle^{\infty}/10^5$	5.92	6.53	0.36	0.29	0.45	0.77	1.14	1.45	1.99
$\langle\gamma\rangle_0/10^6$	−11.7	−12.7	−0.28	−0.22	−0.35	−0.67	−1.11	−1.54	−2.64

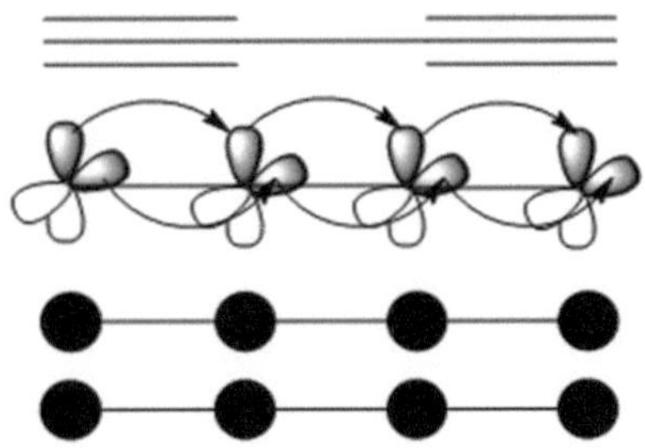

Fig. 3.9 Two orthogonal π-systems in polyynes. 1,3-butadyyne is used as an example

Figs. 3.6 and 3.7). Therefore the HF and MP2 methods are not suitable for describing the optical nonlinear properties for systems with strong electron correlation.

Based on the above considerations, one can expect that polyynes, $(C_M H_2)$, should exhibit significant optical nonlinearities. These polymers are linear and contain two orthogonal π-subsystems. The interaction of these two orthogonal subsystems and its impact on the optical properties is still a subject of studies. In the π-electron approximation, a triple bond is modeled as two double bonds with each carbon atom contributing two 2p-orbitals, one to each π-system (Fig. 3.9). For each pair of bonded atoms (contributing μ and v 2p-orbitals to the first π-system and μ' and v' 2p-orbitals the second π-system) two resonance integrals have to be defined ($t_{\mu v}$ and $t_{\mu' v'}$). The cross-resonance integrals (between different π-systems) are set to zero, $t_{\mu\mu'} = t_{vv'} = 0$. Originally the model was used in the frameworks of the Hückel theory [13]. We used this parametrization of triple bonds before to describe the spin distribution in dublet states [21] and in hyperpolarizability calculations of small acetilenic systems [31].

The limit values of optical properties obtained for polyynes by the extrapolation procedure (Eqs. 3.41 and 3.42) are presented in the Table 3.10.

By comparing the limit values for the longitudinal component of the 2nd hyperpolarizability for linear polyene and polyyne (19.4×10^5 a.u. and 10.8×10^5 a.u., respectively) one can evaluate the significance of the interaction between the orthogonal subsystems. The non-additive factor of 1.5 for the polarizability and 1.7 for the 2nd hyperpolarizability indicate that the cross-system interaction differently affects the different optical properties.

Table 3.10 Limit values of (hyper)polarizabilities for polyynes obtained with different many-body approaches

	HF	MP2	cue$_{(l)}$-CCSD						cue-CCSD
			$l = 1$	$l = 2$	$l = 3$	$l = 4$	$l = 5$	$l = 6$	
$\langle\alpha\rangle^{\infty}$	18.42	14.82	7.91	6.88	8.46	10.08	11.17	11.77	12.51
$\langle\alpha\rangle_0$	−97.09	−70.23	−57.77	−46.57	−67.08	−94.39	−116.8	−131.7	−156.8
$\langle\gamma\rangle^{\infty}/10^5$	4.38	5.62	0.29	0.20	0.35	0.71	1.16	1.54	2.15
$\langle\gamma\rangle_0/10^6$	−6.39	−7.87	−0.47	−0.31	−0.59	−1.35	−2.43	−3.49	−5.65

Fig. 3.10 A fragment of polydiacetylene polymer

It is also worth to note that methods with inaccurate accounting for the electron correlation effects cannot adequately describe the entire set of the electro-optical effects. Such methods show significant errors for the limit values of the optical parameters. For instance, the ratios $\langle\alpha\rangle^{\infty}_{HF}/\langle\alpha\rangle^{\infty}_{cue\text{-}CCSD} = 1.47$, $\langle\alpha\rangle^{\infty}_{MP2}/\langle\alpha\rangle^{\infty}_{cue\text{-}CCSD} = 1.18$, $\langle\gamma\rangle^{\infty}_{HF}/\langle\gamma\rangle^{\infty}_{cue\text{-}CCSD} = 2.04$, $\langle\gamma\rangle^{\infty}_{MP2}/\langle\gamma\rangle^{\infty}_{cue\text{-}CCSD} = 2.61$ demonstrate the fact that inclusion of MP2 correction improves values for polarizability but gives worse results for 2nd hyperpolarizability in comparison with cue-CCSD method.

Systematic improvement of the cue$_{(l)}$-CCSD theory with l up to $l = 6$ is still not enough to give results coinciding with the cue-CCSD results. However, unlike the MP2 approach, improvement of the quality of the wave-function with l leads to more accurate values for both polarizability and 2nd hyperpolarizability. Also, as a rule, the lower level of theory gives a lower bound for the calculated optical property.

Of considerable interest is the polydiacetylene polymer that can be viewed as a combination of a polyene and a polyyne. A fragment of the polydiacetylene molecule is shown in Fig. 3.10.

For comparison, in Table 3.11 some data for three iso-π-electron fragments: polyene, $C_{28}H_{30}$, polydiacetylene, $C_{18}H_{10}$, and polyyne $C_{14}H_2$ are presented.

The values shown in the table are smaller for polydiacetylene than for polyene and polyyne. This is likely due to the fact that addition of a non-conjugated π-bond makes an almost additive perturbation to the π-system and leads to a reduction of the specific (per atom) linear size. Also, as could be seen from Table 3.11, the largest values of the specific (per atom) nonlinear optical properties among all considered polymers are obtained for polyynes.

Analogically to the representation of the wave-function in structural terms, there is a way to separate (hyper)polarizabilities into the individual contributions from individual atoms. A method for such separation was developed by Brédas [15, 16] and is called the real-space finite-field method. The approach can be easily implemented for a post-Hartree-Fock method in the π-electron approximation due to the simplicity of the calculation of the one-electron reduced density matrix (RDM1) elements. In our calculations we are using a simple numerical-derivative "two-points" formula for RDM1 matrix elements (R_{rs}) [88] (see also [48]):

Table 3.11 Average (hyper)polarizabilities (per electron, N_{el}, or per atom, N_{at}) of different iso-π-electron systems

	$\langle\alpha\rangle/N_{el}$	$\langle\alpha\rangle/N_{at}$	$\langle\gamma\rangle/N_{el}$	$\langle\gamma\rangle/N_{at}$
$C_{28}H_{30}$	11.2	11.2	1.09	1.09
$C_{18}H_{10}$	6.2	9.7	0.35	0.55
$C_{14}H_2$	7.3	14.5	0.42	0.65

$$R_{rs} \approx \frac{1}{2-\delta_{rs}} \frac{E(h_{rs}+\xi)-E(h_{rs}-\xi)}{2\xi}, \tag{3.43}$$

where ξ is differentiation step, while h_{rs} is core Hamiltonian matrix element. For a detailed description of RDM1 CC calculations see Ref. [71].

A characteristic element of the Brédas method is the use of an expansion of the dipole moment in terms of powers of the applied electrostatic field.

$$\mu_x = \mu_x^0 + \alpha_{xx} F_x + \beta_{xxx} F_x^2 + \gamma_{xxxx} F_x^3 + \cdots. \tag{3.44}$$

The expression for the charge of an atom in the electrostatic field applied along a certain axis can be written as a sum of products of so-called α-, β-, and γ-charges (generally hypercharges) and the field strength, F_r, raised to the corresponding powers:

$$q_i = q_i^{(0)} + q_i^{(\alpha)} F_r + q_i^{(\beta)} F_r^2 + q_i^{(\gamma)} F_r^3 + ; \tag{3.45}$$

where $q_i^{(0)}$ is the charge on the ith atom in the absence of the field, $r \in \{x, y, z\}$.

Hence, for the 2nd hyperpolarizability one can write:

$$\gamma_{xxxx} = \frac{\partial^3 \mu_x}{\partial F_x^3} = \frac{\partial^3(\sum_{i=1}^N q_i x_i)}{\partial F_x^3} = \sum_{i=1}^N x_i \frac{\partial^3 q_i}{\partial F_x^3}, \tag{3.46}$$

where x_i is the corresponding Cartesian coordinate of the atom. Thus the $xxxx$ component of 2nd hyperpolarizability can be obtained as:

$$\gamma_{xxxx} = \sum_{i=1}^N x_i q_i^\gamma, \tag{3.47}$$

where q_i^γ is the atomic change of charge in the external field:

$$q_i^\gamma = \frac{\partial^3 q_i}{\partial F_x^3}. \tag{3.48}$$

In a similar way one can determine all other optical properties of interest:

$$\alpha_{xx} = \sum_{i=1}^N x_i \frac{\partial q_i}{\partial F_x} = \sum_{i=1}^N x_i q_i^\alpha, \tag{3.49}$$

$$\beta_{xxx} = \sum_{i=1}^N x_i \frac{\partial^2 q_i}{\partial F_x^2} = \sum_{i=1}^N x_i q_i^\beta, \tag{3.50}$$

The contributions to the particular property from the particular atom, which generally depend on the choice of the origin of the coordinate system, will be called α-, β-, γ-moments and denoted as:

$$\widetilde{\alpha} = x_i q_i^{(\alpha)}, \tag{3.51}$$

$$\widetilde{\beta} = x_i q_i^{(\beta)}, \tag{3.52}$$

$$\widetilde{\gamma} = x_i q_i^{(\gamma)}. \tag{3.53}$$

The histogram Fig. 3.11 shows the distribution of α- (black) and γ-charges (white) along the polyene $C_{34}H_{36}$ chain obtained with the cue-CCSD method ($x = 0$ corresponds to the center of the molecule).

Due to the presence of a center of inversion in the systems under study, β-charges and β-moments are equal to zero. Figure 3.12 shows the contributions from the ethylene units of the polyene to α_{xx} (black) and to γ_{xxxx} (white). The x-axis in the histograms correspond to the coordinates of the centroids representing the ethylene units. The coordinate origin is located at the center of inversion of the polyene. The polymer is placed along the x-axis. The distribution of the α-charge shows alternating signs and almost constant absolute values along the entire chain with only small deviations at the edge atoms. For the γ-charges (Fig. 3.11) at adjacent atoms, particularly close to the chain edges, the differences in the values (between atoms belonging to a single ethylene) are of the order of 0.6. The difference, as one moves closer to the center of the molecule, slowly decreases. The distribution of α- and γ-charges vary

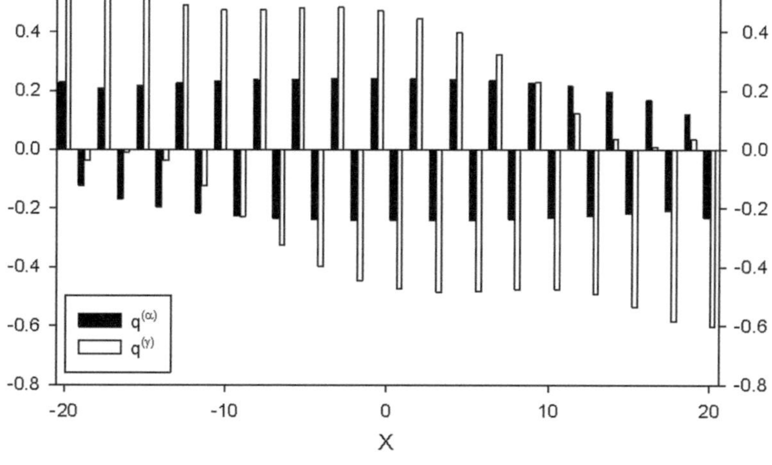

Fig. 3.11 Distribution of α and γ-charges in $C_{34}H_{36}$ *trans*-polyene chain

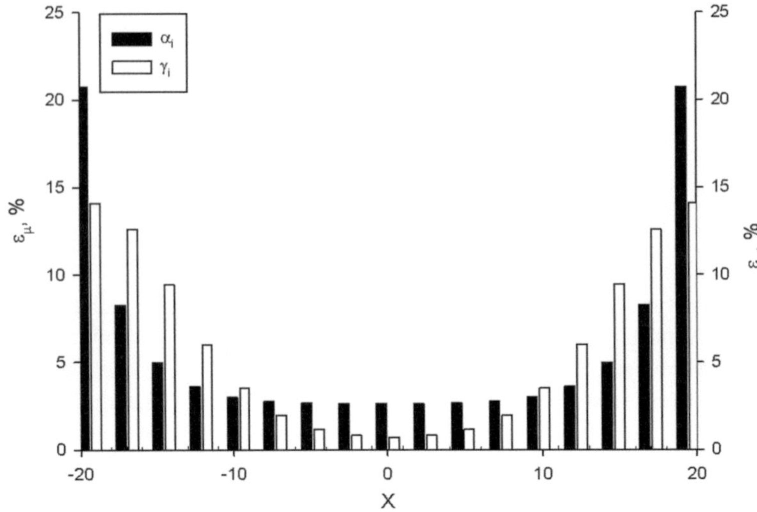

Fig. 3.12 Contribution to α_{xx} and γ_{xxxx} from ethylene units in *trans*-polyene, $C_{34}H_{36}$

slightly with the increase of the polymer chain. For polyenes $C_{22}H_{24}$ and $C_{32}H_{34}$, the total α- and γ-charges at the end ethylenes change in intervals of 0.107–0.108 and 0.536–0.566, respectively. Alternating signs of the hypercharges means that the field produces an oscillating wave of atomic charges. The charges on ethylenes, in turn, increase with the increase of the strength of the field applied along the chain. The largest contribution to the longitudinal components of the polarizability and the 2nd hyperpolarizability are made by the end ethylene fragments (for the above-defined arrangements of the molecules relatively to the origin). This is due to the fact that the atoms of these fragments are most distant from the center of the polymer (their charges are multiplied by the largest factors) and the charges on the atoms belonging to these fragments reach the maximum values for the chain. Thus for the $C_{34}H_{36}$ system, 42 % of the longitudinal polarizability component is due to the pair of the end ethylene fragments. For γ_{xxxx} the corresponding contribution is smaller—28 %, however it decreases less in moving to the center of the chain comparatively to the polarizability contribution. In the limit, the contribution of the end ethylene fragments are 40 and 25 % for α_{xx} and γ_{xxxx}, respectively. For all polyenes, starting with $C_{22}H_{24}$, the total moments for all ethylenes in the polymer chain (but not for separate atoms) are positive.

3.3.3.2 Polycyclic Hydrocarbons

In this section, we discuss various members of the family of polycyclic hydrocarbons (see Fig. 3.13). Among them there are linear (A) and angular (B) polyacenes,

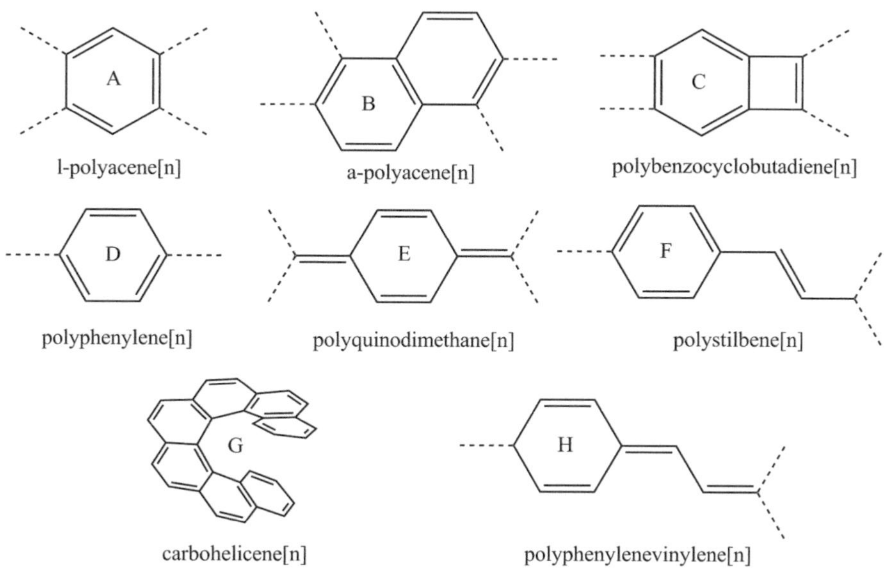

Fig. 3.13 Polycyclic Aromatic Hydrocarbons under study

helicenes (G) [72], polyphenylenes (D), poly-para-quinodimethanes (E), polybenzocyclobutadienes (C), poly-para-stilbenes (F), and poly-para-phenylene-vinylene (H).

The dihedral angles of the internal edge in the polyacene chain in helicene are equal to 25 °. For polyphenylene the dihedral angle for each two neighboring cycles is set to 45°. All presented systems contain conjugated six-membered cycles, so, in addition to other properties, one can consider the optical parameters in terms of benzene ring units. For the HF, MP2, and cue-CCSD methods, the corresponding values of the polarizability for the benzene ring are 25.94, 25.74, and 25.06 a.u., respectively. For 2nd hyperpolarizability the values are 844.9, 905.8, and 2603.3 a.u., respectively.

As noted in the Sect. 3.3.2, the polyacenes are characterized by a more complex structure of the wave function, and therefore, to adequately describe this structure a higher level of theory is needed. It is expected that for the polycyclic aromatic hydrocarbons discussed in this section the selection of an appropriate correlation radius is a very important aspect of the calculation. To study the effect of the level of accounting for the electron correlation effects for polyacenes, we have calculated the polarizability and 2nd hyperpolarizability values for different levels of the cue-CCSD theory. In Figs. 3.14 and 3.15 the dependencies of the specific values of these properties on the number of the π-electrons are shown.

The values obtained in the calculations, in comparison to those obtained for polyenes, show that a principal difference exists between the two sets of results. In the polyene case, an increase of the level of theory leads to a systematical improvement of the values of all studied properties relative to the results obtained with the

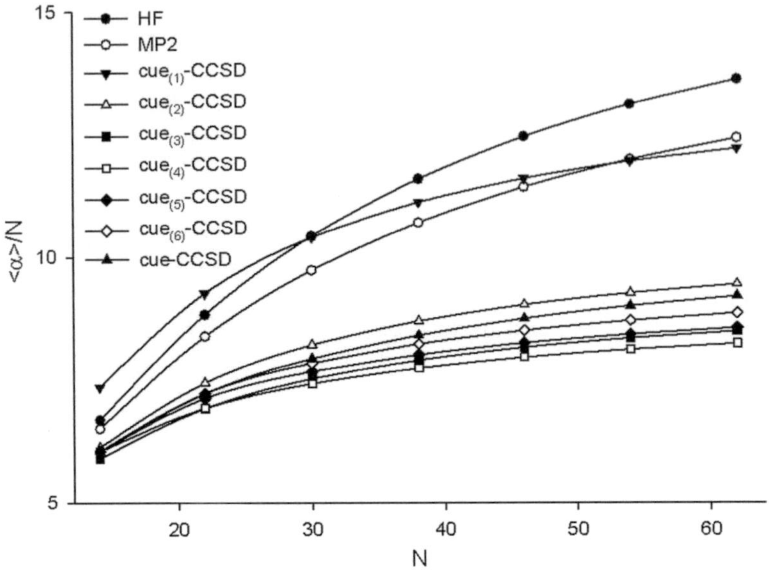

Fig. 3.14 Average specific (per electron) polarizabilities of linear polyacenes as a function of the number of the π-electrons

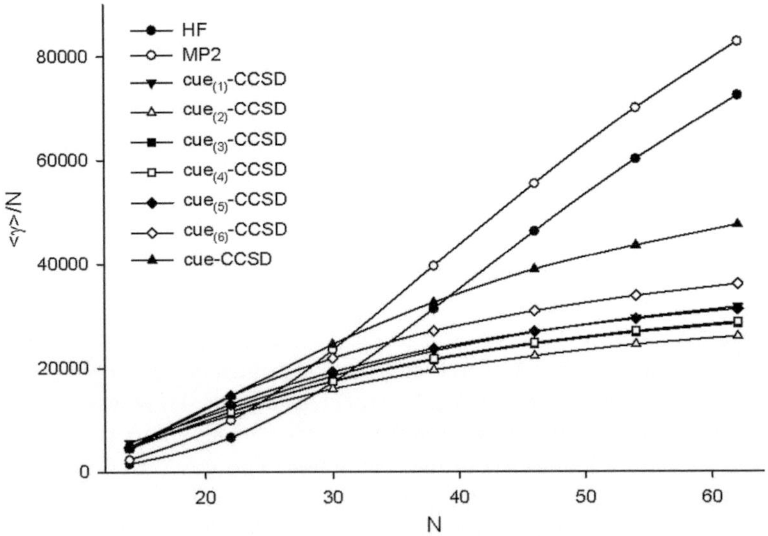

Fig. 3.15 Average specific (per electron) hyperpolarizabilities of linear polyacenes as a function of the number of the π-electrons

cue-CCSD method. In the case of polyacenes, for the polarizability, the approximation that takes into account only the local excitations, $l = 1$, shows a significant overestimation in comparison with the results obtained with other levels. However the results still reveal correct dependence on the number of π-electrons and they are more accurate than the HF and MP2 limit values. For the hyperpolarizability the picture is partially the same. The results calculated with the cue$_{(1)}$-CCSD method fall out from the general dependency but they are not overvalued relative to cue-CCSD. Starting with the $l = 2$ level corresponding to an enlargement of the excitation set, a consistent improvement of the calculated values is observed. The HF and MP2 hyperpolarizabilities show some analogy with the corresponding results for polyenes, namely the stabilization limit is overestimated, which can also be explained by an insufficient flexibility of the wave functions in these methods.

Let us now consider several possible isomers of the linear polyacene: angular polyacenes and helicenes (see Fig. 3.13). In Table 3.12 the results obtained for the $C_{22}H_{14}$–$C_{42}H_{24}$ systems (containing 5–10 benzene rings) are presented for structures corresponding to systems A, B, and G for different numbers, n, of the benzene rings in the chain.

Based on the results shown in the table, one can conclude that the largest optical nonlinearity occurs in the longest isomer, system A. This fact confirms the conclusion made for the linear model polyene (Fig. 3.8) and based on the comparison involving its different conformers. The conclusion was generalized for a larger class of condensed aromatic hydrocarbons.

Next, let us consider the optical properties of polymers presented in Fig. 3.13, which involve conjugated aromatic rings. In Table 3.13 the calculated polarizabilities and 2nd hyperpolarizabilities are shown for systems presented in Fig. 3.13 (A—linear polyacene $C_{30}H_{18}$, B—angular polyacene $C_{30}H_{18}$, C—polybenzocyclobutadiene $C_{42}H_{18}$, D—polyphenylene $C_{42}H_{30}$, polyquinodimethane $C_{44}H_{32}$, and G—helicene $C_{30}H_{18}$). The values corresponding to the benzene unit calculated with the different appropriate methods are also shown in the table. All systems contain seven benzene rings. The values demonstrate the non-additive character of the calculated

Table 3.12 Average values of (hyper)polarizabilities for polyacenes A, B, and G shown in Fig. 3.13

n	5	6	7	8	9	10
	$\langle\alpha\rangle$, a.u.					
A	158.96	198.09	238.03	278.57	319.58	360.93
B	148.25	184.34	221.18	258.72	296.64	334.91
G	137.85	162.00	186.32	209.95	233.36	254.23
	$\langle\gamma\rangle$, a.u.					
A	3.22	5.16	7.36	9.85	12.4	15.1
B	1.92	2.95	4.15	5.47	6.88	8.35
G	1.25	1.45	1.69	1.86	2.03	2.11

Table 3.13 Average values of (hyper)polarizabilities for polyacenes shown in Fig. 3.13

	A	B	C	D	E	G
	$\langle\alpha\rangle$, a.u.					
HF	312.91	236.78	373.49	262.79	N/A	189.90
MP2	292.06	234.57	385.60	258.73	N/A	191.17
cue-CCSD	238.03	221.18	371.85	249.44	1300.2	186.32
	$\langle\alpha\rangle/\langle\alpha\rangle_{benzene}$					
HF	12.1	9.1	14.4	10.1	N/A	7.3
MP2	11.3	9.1	15.0	10.1	N/A	7.4
cue-CCSD	9.5	8.8	14.8	10.0	51.9	7.4
	$\langle\gamma\rangle \times 10^5$, a.u.					
HF	5.15	2.79	11.6	4.02	N/A	0.863
MP2	7.04	3.52	15.8	4.28	N/A	1.07
cue-CCSD	7.36	4.15	16.4	3.99	858	1.69
	$\langle\gamma\rangle\ /\langle\gamma\rangle_{benzene}$					
HF	610	330	1370	476	N/A	102
MP2	777	389	1740	473	N/A	118
cue-CCSD	283	159	630	153	33000	65

properties. They show the failure of approaches based on the assumption of additivity in evaluating the studied nonlinear optical properties.

The data presented in Table 3.13 shows several important features. The first and most obvious concerns the degree of non-additivity of the polarizability and the 2nd hyperpolarizability. The ratio of $\langle\alpha\rangle$ and $\langle\alpha\rangle_{benzene}$ only marginally differ from the number of benzene rings in the oligomeric chain (the ratio is almost equal to that number for helicenes), while the $\langle\gamma\rangle/\langle\gamma\rangle_{benzene}$ ratio exceeds the number of the benzene rings by a factor of 2–3. This observation once again underlines the exceptional difficulty to accurately calculate the full set of the electro-optical properties.

The second noteworthy feature is the difference in the accuracy of the description of $\langle\alpha\rangle$ and $\langle\gamma\rangle$ in the HF and MP2 methods. In general, these methods are satisfactory in describing the values of the polarizabilities (except the case of linear polyacenes). At the same time, for the 2nd hyperpolarizabilities, there are significant deviations from the results obtained from the cue-CCSD calculations (especially for $\langle\gamma\rangle_{benzene}$). Therefore, a rough accounting for the electron correlation effects cannot guarantee a similarly accurate description of every optical property. As it will be shown below, this conclusion remains true for optical properties proportional to odd powers of the strength of the applied field.

The highest (hyper)polarizability values appear for poly-para-quinodimethane. The polarizability for this polymer is 5–7 times larger than those for the other systems and the 2nd hyperpolarizability is more than 100 times larger. The HF (and thus MP2) method for the chain poly-para-quinodimethane gives unstable results. Apparently, this system has an allowed long-wavelength transition which makes it behave like a metal and this, to a large extent, determines the value of its $\langle\gamma\rangle$. Despite this metallic-like behavior, the various approximate versions of the cue-CCSD theory provide rather stable results for the optical properties.

Table 3.14 shows limit values of the optical parameters calculated according to Eqs. 3.41 and 3.42 for systems A, C, E, F, and H shown in Fig. 3.13. Polystilbenes (F) are combinations of model planar polyphenylenes and *trans*-polyenes. System H can also be represented as a combination of poly-para-quinodimethane and trans-polyene.

Based on the presented data, it can be concluded that the highest hyperpolarizability occurs in polymers based on para-quinodimethane. The results for system E, shown in the table, give an estimation of the lower bound due to the fact that, for the specific $\langle\gamma\rangle$, a fast increase of the value with the number of the monomeric units is observed. Thus, one can expect even larger values for the linear poly-p-quinodimethane.

In many cases, the application of the HF and MP2 methods is hampered by convergence problems and by instabilities in the calculation to obtain the HF solution. The use of an orthogonalized *cue* basis set solves these problems ensuring smooth convergence even if the HOMO-LUMO gap is small. Also, according to the values of α_0 and γ_0, a more physically correct behavior of the specific values of optical properties depends on the size of the system as demonstrated by calculations performed with the cue$_{(l)}$-CCSD methods. Despite the limitations imposed on the excitation set, all levels (even the lowest, $l = 1$) demonstrate correct behavior for the limit α^∞ and γ^∞ values. The systematic improvement of the structure of the wave function equally affects the values of the polarizability and the 2nd hyperpolarizability. This happens due to the size consistency of the CCSD method.

Table 3.14 Limit average values of (hyper)polarizabilities for polyacenes shown in Fig. 3.13

		A	C	E	F	H
$\langle\alpha\rangle^\infty$	HF	17.28	11.17	N/A	14.85	N/A
	MP2	15.60	11.78	N/A	13.62	N/A
	cue-CCSD	10.53	10.58	>60	10.64	29.54
$\langle\alpha\rangle_0$	HF	−227.3	−98.34	N/A	−153.8	N/A
	MP2	−197.4	−112.1	N/A	−132.6	N/A
	cue-CCSD	−82.03	−73.21	N/A	−73.72	−430.8
$\langle\gamma\rangle^\infty/10^5$	HF	1.72	0.690	N/A	2.96	N/A
	MP2	1.86	1.02	N/A	3.33	N/A
	cue-CCSD	0.736	0.676	>110	1.27	19.8
$\langle\gamma\rangle_0/10^6$	HF	−6.22	−1.85	N/A	−8.27	N/A
	MP2	−6.47	−2.87	N/A	−9.19	N/A
	cue-CCSD	−1.62	−1.14	N/A	−2.32	−58.9

3.3.3.3 Nonalternant Hydrocarbons

In our previous work [88] we studied systems containing odd π-electron cycles. In the study, in addition to classical fulvalenes, we also considered systems where the cycles are connected by *trans*-butadiene and *para*-quinodimethane bonding fragments. In this section, only classical members of the fulvalene family, where each two cycles are connected with a double bond, are described. For convenience we use the notation shown in Fig. 3.16.

Systems considered in this section are characterized by extremely large π-electron dipole moments, while remaining entirely carbon-based. A large dipole moment of a system indicates that a significant charge transfer occurs between the cycles. This behavior also affects the polarizability and hyperpolarizability and makes these properties difficult to describe in quantum-chemical calculations. We have observed these difficulties in the calculations where different variations of the CCSD method were employed. More specifically, the CCSD hyperpolarizability calculations of some fulvalenes revealed a significant error (more than 30 %) in comparison with the results obtained in the exact π-FCI method. Also, considerable differences between the hyperpolarizability results obtained with different variants of the CCSD method were observed. Table 3.15 shows the optical properties for calicene [17, 19, 58] ("[5-3]" in foregoing notation) obtained with different methods including various CC approaches.

The results are compared with the results obtained with relaxed CC methods where an exact accounting for the triple (r-CCSDT) and quadruple (r-CCSDTQ) excitations is included. In this case, the almost exact match of the results of the MP2 calculations (for the second hyperpolarizability) with the FCI (exact) results is likely a coincidence. This can be concluded based on the results for other systems considered in Ref. [88], where the MP2 method, as well as the HF method, gives results for individual components and for average values with incorrect signs or even with incorrect order of magnitude. As can be seen from the data presented in Table 3.15, CCSD gives a noticeably large error (~10 %) for the 1st hyperpolarizability, and it incorrectly describes the 2nd hyperpolarizability (error ~30 % for the best of the results). The inclusion of the triply-excited configurations (r-CCSDT) results in a decrease of the error in up to 5 %, and using the method with quadruples, the r-CCSDTQ approach, leads to an almost exact value. It should be noted that such large errors

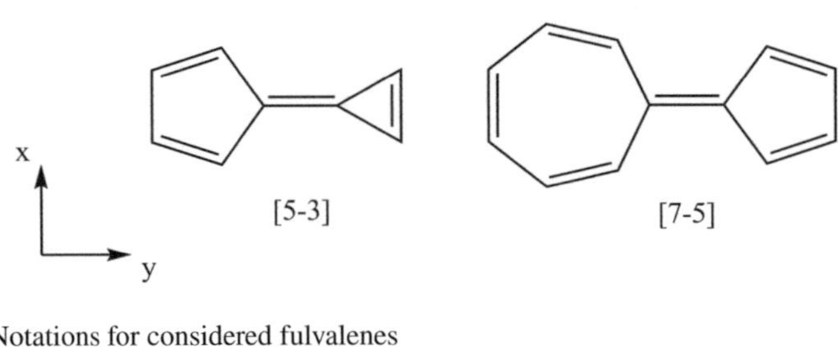

Fig. 3.16 Notations for considered fulvalenes

Table 3.15 Components of the optical parameters for [5-3] fulvalene obtained in different approaches (in a.u.)

[5-3]	α_{xx}	α_{yy}	β_{yyy}	$\gamma_{xxxx}/10^3$	$\gamma_{yyyy}/10^4$
HF	43.8 (9.3)	119.4 (−0.6)	−362 (−118.9)	8.05 (−15.8)	−4.74 (−20.9)
MP2	42.4 (5.7)	119.4 (−0.7)	159 (−91.7)	8.85 (−7.4)	−5.91 (−1.5)
cue-CCSD	40.2 (0.3)	123.8 (3.0)	1868 (−2.5)	9.37 (−2.0)	−8.23 **−(37.1)**
u-CCSD	40.5 (1.0)	122.9 (2.3)	1752 (−8.6)	9.11 (−4.7)	−9.07 **−(51.2)**
r-CCSD	40.3 (0.5)	121.8 (1.4)	1726 (−9.9)	9.07 (−5.1)	−7.85 **(30.8)**
r-CCSDT	40.2 (0.2)	120.5 (0.2)	1902 (−0.7)	9.37 (−2.0)	−6.31 (5.2)
r-CCSDTQ	40.1 (0.0)	120.2 (0.0)	1916 (0.0)	9.48 (−0.8)	−6.03 (0.5)
FCI	40.1	120.2	1916	9.56	−6.00

Errors relatively to FCI are presented in parenthesis. Reprinted with permission from Ref. [88]. Copyright 2014 American Chemical Society

Table 3.16 Classification of fulvalenes

Class	1	2	3
	$m = n$	$m > n$	
Condition	$n + m = 4k + 2$	$m + n = 4k$	$m + n = 4k + 2$

(~30 %) are obtained only for selected members of the family of the nonalternant systems. For almost all studied systems, the CCSD approach shows good agreement with more accurate methods. At the same time, for systems [5-5] and [7-3] r-CCSD, u-CCSD, and cue-CCSD methods give values that differ from the FCI results on the order of one percent. In the studies of the optical properties of fulvalenes we introduced a classification of these systems based on their sizes and on the number of cycles they contain. The classification uses the [m-n] symbols which are explained in Table 3.16.

Structurally, this classification is related to the tendency of the cycles contained in the system to form an aromatic bond arrangement. To explain this, let us consider one representative of each class. In Fig. 3.17 schemes of inter-cycle charge transfer for the [5-5], [5-3], and [7-3] systems are presented.

The systems where the charge transfer does not occur for the symmetry reason (m = n) belong to the first class.

For the systems that belong to the second class, a possibility appears to form two aromatic rings. In the case of the [5-3] system, this, however, requires that a tranfer of the electron density from the trinomial cycle to the 5-membered cycle uccurs. It leads to the formation of 6-electron-5-center and 2-electron-3-center systems. In this case, the trinomial cycle is an electron donor and the 5-member cycle is an electron acceptor.

For the third class, there is the possibility of the formation of one aromatic system and one antiaromatic system. Unlike the previous systems, for the systems belonging

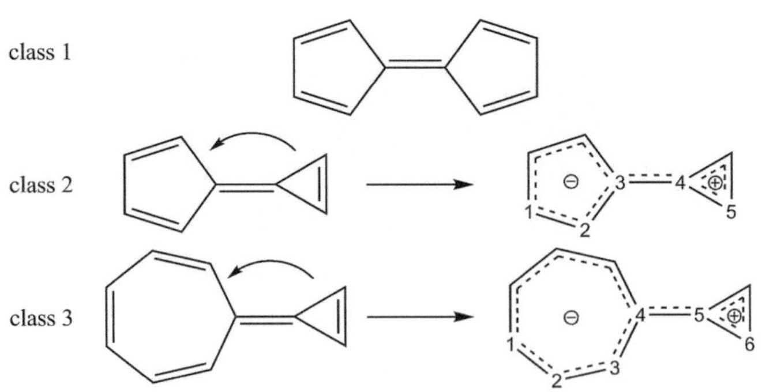

Fig. 3.17 Charge transfer in different classes of nonalternants

to the third class there is a competition between two cycles to acquire additional electron density to achieve full aromaticity.

According to our calculations, the direction of the charge transfer is determined by the electron-donor ability of the smaller cycle. If the two cycles are connected with a bridging molecular fragment, there are two possible variants of the charge distribution. If the hole/electron density of the bridging fragment is sufficient to provide enough electrons to make both cycles aromatic, a displacement of the electronic density in the direction from the bridge to the cycles is observed. Otherwise, when the electron-donor capacity of the bridging fragment is not sufficient, a transfer will still occur, but it will result in the formation of an aromatic ring only for the smaller cycle and there will be no electron transfer to the larger cycle. In [88], all possible variants of the intramolecular charge transfer were discussed for fulvalenes and fulvalene-like systems.

The first and the third class differ from each other because in the latter case, due to the difference in electron affinities, one cycle surpasses the other in the competition for the extra electron density obtained from the bridging molecular fragment.

In this section we will consider the charge distribution only for classical fulvalenes (without bridging fragments). In Table 3.17, the charges on atoms for the [5-3] and [7-3] systems obtained in the calculations are presented (the atom numbering shown in Fig. 3.17 is used). The charge transfer magnitude, $\delta^{(0)}$, is marked with a 0 index, as it characterizes the charge distribution at zero field.

According to the data presented in Table 3.17 one can see that a simple consideration based on the tendencies of the different cycles to form aromatic rings explains

Table 3.17 Charges and transferred charges between cycles in fulvalenes obtained using different methods

Atom	Hückel	HF	MP2	cue-CCSD	r-CCSD	FCI
	[5-3]					
1	0.281	0.145	0.141	0.129	0.125	0.124
2	0.262	0.156	0.131	0.088	0.087	0.081
3	−0.181	−0.150	−0.126	−0.085	−0.094	−0.089
4	−0.157	−0.088	−0.089	−0.090	−0.081	−0.085
5	−0.165	−0.060	−0.054	−0.040	−0.040	−0.039
$\delta^{(0)}$	0.824	0.445	0.412	0.346	0.336	0.330
	[7-3]					
1	−0.033	−0.002	0.000	0.001	0.001	0.002
2	−0.154	−0.060	−0.055	−0.043	−0.044	−0.045
3	0.087	0.057	0.053	0.042	0.041	0.043
4	−0.278	−0.182	−0.165	−0.128	−0.132	−0.129
5	0.080	−0.017	−0.035	−0.061	−0.052	−0.058
6	0.200	0.105	0.103	0.094	0.094	0.094
$\delta^{(0)}$	0.480	0.193	0.171	0.128	0.136	0.130

the direction of the charge transfer. It is worth noting that the simplest of the methods used in the calculations, namely the Hückel method, gives for system [5-3] the amount of the electron density transferred between the cycles of about 1 in units of the electron charge. The inclusion of the electron-electron interaction does not qualitatively change the picture of the charge distribution, but significantly reduces the amount of the electron density transferred (the ratio of the results obtained with the FCI and MO Hučkel methods is about 2.5 times). Also note that the cue-CCSD method correctly describes the total charge on each cycle, as well as the charges on the individual atoms.

A more complex issue is the behavior of the charges when the field is applied along the longitudinal axis of fulvalene. Using the partitioning scheme introduced in Sect. 3.3.3.1 and employed for describing the contributions to the (hyper) polarizability from different atoms, we can analyze the atomic changes in fulvalenes in the presence of the field. The intramolecular charge transfer at the zero field is described by the different variations of the CCSD method at a sufficient level of accuracy, so it cannot serve as an indicator of any incorrectness in the description of the hyperpolarizabilities of systems belonging to the second class. Therefore one needs to study the hypercharges and the moments of some corresponding quantities. In Table 3.18 the results concerning the (hyper)charge transfer and the components of the dipole moment and the (hyper)polarizability calculations are presented for three representative members of each considered class.

The values presented in Table 3.18 allow us to trace the relationship between the error in the description of the intramolecular transfer of the hypercharges and the accuracy of the values for the corresponding optical properties, i.e. polarizability, hyperpolarizability, etc. The most striking example is system [7-5]. For this system a wrong sign and the order of magnitude of $\delta^{(\gamma)}$ is obtained with the cue-CCSD method as compared with the FCI results. This results are in error for the longitudinal component of the 2nd hyperpolarizability by more than 65 % ! This exceeds the error for the previously described fulvalene [5-3] more than twice. At the same time, for the [9-3] system which is isoelectronic to the [7-5] system and belongs to the same class of systems, the error in the 2nd hyperpolarizability component is less than 1 %.

Performing a more detailed analysis of the hypercharges and the moments of the (hyper)polarizabilities for system [7-5] one can conclude that the distribution of the hypercharges obtained with the CCSD method is generally wrong (especially the charges corresponding to the first and second hyperpolarizabilities). In Table 3.19 the total charges, $q^{(\alpha)}$, $q^{(\beta)}$, and $q^{(\gamma)}$, for the cycles forming the systems and the corresponding total moments of the (hyper)polarizabilities are shown.

Very large errors are obtained for the two "end" atoms of the five-membered ring (error of more than 500 % for each), but the value of the charge is small, hence its contribution to the corresponding components is also small. The next largest error for $q^{(\gamma)}$ is for the pair of atoms forming the bond connecting the two rings (70 and 85 % for the atoms belonging to seven-membered and five-membered rings, respectively). For this pair of atoms the highest values of γ-hypercharge are observed, but, as they are located close to the origin, the contribution from the corresponding moment is also small. Nevertheless, the total moment for the seven-membered cycle is under-

Table 3.18 Charge transferred ($\delta^{(0)}$) and the contributions of the transferred charge corresponding to the polarizability, hyperpolarizability, etc, for fulvalenes

Class	m-n	$\delta^{(0)}$		$\delta^{(\alpha)}$		$\delta^{(\beta)}$		$\delta^{(\gamma)}$	
		cue	FCI	cue	FCI	cue	FCI	cue	FCI
1	3-3	0	0	0.2041	0.2042	0	0	0.0568	0.0651
	5-5	0	0	0.3226	0.2986	0	0	0.3264	0.3464
	7-7	0	0	0.4373	0.3726	0	0	1.0309	1.1180
2	5-3	0.3463	0.3295	0.3493	0.3412	0.0923	0.1052	−0.1778	−0.1485
	7-5	−0.2524	−0.2148	0.4890	0.4673	−0.3266	−0.4139	−0.0198	0.2142
	9-3	0.3041	0.2840	0.4289	0.4226	0.2522	0.2886	−0.0341	0.0027
3	7-3	0.1277	0.1295	0.3039	0.2885	0.0886	0.0842	0.2114	0.2207
	9-5	−0.0767	−0.0685	0.4267	0.3719	−0.1263	−0.1369	0.7702	0.8617
	11-3	0.1874	0.1781	0.3791	0.3497	0.1702	0.1721	0.3139	0.3655
Class	m-n	μ_y		α_{yy}		β_{yyy}		γ_{yyyy}	
		cue	FCI	cue	FCI	cue	FCI	cue	FCI
1	3-3	0	0	45.8	45.9	0	0	0.356	0.414
	5-5	0	0	154.8	148.9	0	0	2.712	2.758
	7-7	0	0	252.8	230.5	0	0	11.348	11.265
2	5-3	5.38	5.10	123.8	120.2	1867	1916	−0.823	−0.600
	7-5	−4.44	−3.73	244.1	234.8	−6591	−7537	1.027	2.921
	9-3	5.54	5.08	215.5	208.5	6562	6422	2.450	2.440
3	7-3	2.49	2.52	128.5	125.2	1540	1375	2.186	2.073
	9-5	−1.39	−1.25	254.9	235.9	−1662	−1863	9.223	9.222
	11-3	3.68	3.50	226.6	214.8	3703	3359	6.073	5.830

Table 3.19 (Hyper)polarizability moments for fulvalene [7-5]

		q_i^{α}	α_i	q_i^{β}	β_i	q_i^{γ}	$\gamma_i/10^4$
$\sum_7$	CUE	0.4890	110.3	−0.3266	−2992	−0.0198	9.08
	FCI	0.4673	106.2	−0.4139	−3173	0.2142	15.40
$\sum_5$	CUE	−0.4890	133.8	0.3266	−3600	0.0198	1.19
	FCI	−0.4673	128.7	0.4139	−4364	−0.2142	13.80
$\sum$	CUE	–	244.1	–	−6591	–	10.27
	FCI	–	234.8	–	−7537	–	29.21

estimated by 40 % and for the five-membered cycle by 90 %! As a result, the value of the longitudinal component of the 2nd hyperpolarizability obtained with the cue-CCSD method has an error of 65 %.

As mentioned above, for compounds belonging to the second class, the addition of a molecular fragment bridging the two cycles largely affects the intramolecular charge transfer. The presence of the bridging group also affects the charge distribution and the hypercharges. In the case of para-quinodimethane as the bridge, the change of the total (by cycle) charge is shown in Table 3.20. The insertion of the bridge between the cycles also leads to an increase of the longitudinal components of the hyperpolarizability by more than one order of magnitude, as determined with the FCI method. It also results in an incorrect order of magnitude for the first hyperpolarizability and an incorrect sign for the second hyperpolarizability calculated with the HF and MP2 methods.

Recalling the conclusions made in previous two sections and considering the results for the non-alternant hydrocarbons, we can conclude that the hyperpolarizability is a complex property not only for long π-conjugated systems but also for smaller systems containing several dozen atoms. Fulvalenes add a striking confirmation of this conclusion. To correctly describe the nonlinear optical properties even of systems as small as fulvalenes it is necessary to use high-level theories that accurately account for the electron correlation and include the third and higher order excitations from the reference wave function. Some of the methods widely used for molecular calculations (such as MP2 or DFT) usually do not guarantee sufficient accuracy of the calculations of the optical properties.

Table 3.20 FCI hypercharges of [5-3] fulvalene without and with para-quinodimethane bridge

	$q_i^{(\mu)}$	$q_i^{(\alpha)}$	$q_i^{(\beta)}$	$q_i^{(\gamma)}$
	[5-3]			
$\sum_5$	−0.329	0.341	0.105	−0.148
	[5-para-quinodimethane-3]			
$\sum_5$	−0.173	0.572	1.383	4.095
$\sum_{bf}$	0.104	−0.032	−0.186	−0.895

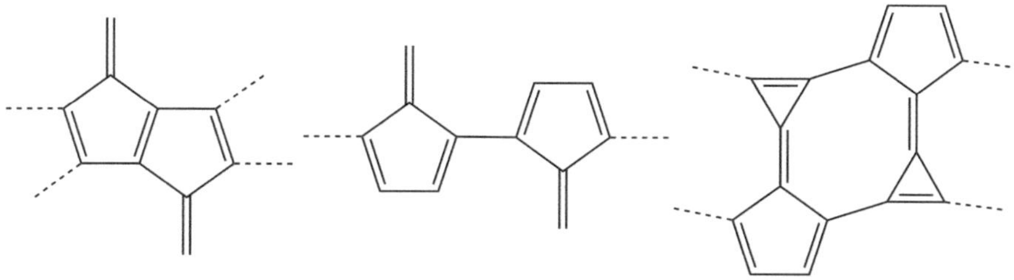

Fig. 3.18 Nonalternant polymers under consideration in this work

Among the nonalternant polymers, one should separate systems that are formed from nonalternant monomeric units from the structures with "alternant" chains and with "nonalternant" end groups. Both types were considered in our previous work [88]. Here we give only a general description of the systems and we present conclusions that have been derived from the analysis of the results obtained in the calculations.

The first group includes condensed (c-polyfulvenes) and non-condensed (nc-polyfulvenes) polyfulvenes and polycalicenes. The structures of these systems are shown in Fig. 3.18. The limit values of optical properties obtained by extrapolation for these systems are shown in Table 3.21. The corresponding specific values will be presented at end of section.

As one can see, the limit values for the condensed and uncondensed polyfulvenes are similar. In comparison with other polymers (represented in the two previous sections) these values are small, indicating unlikely use of these polymers as chromophores for generating the second and third harmonics.

Polycalicenes are characterized by significant limit hyperpolarizabilities that surpass the hyperpolarizabilities of *trans*-polyenes, but do not exceed those of some members of the polycyclic family. Some works [75–77] are devoted to the synthesis of polymeric calicene, which offer some potential for use as materials in optical devices.

Polymers belonging to the first group are characterized by very modest values of the first hyperpolarizability, therefore, are not suitable as chromophores for the second harmonic generation. For this purpose it is usually preferred to use polymers with the donor and acceptor groups separated with a long π-conjugated fragment. It is most common to use amino and nitro-groups as a pair of donor and acceptor.

Table 3.21 Limit values calculated for the polymers under consideration

	c-polyfulvene	nc-polyfulvene	polycalicene
$\langle\alpha\rangle^{\infty}$	7.0	6.6	20
$\langle\gamma\rangle^{\infty}/10^4$	5.8	5.2	39

In our previous work [88], the use of odd-electron π-conjugated cycles as bonding fragments that guarantee large values of the 1st hyperpolarizability was suggested. According to the results discussed in the section on fulvalenes, having different cycles forming compounds of the second class has some advantage. Therefore, a trinomial was selected as the donor and a five-membered ring as an acceptor. *trans*-polyenes, poly-para-quinodimethanes, and poly-para-phenylenevinylene were selected as the π-conjugate bonding fragments (corresponding polymers designated as pp-PE, pp-PQ, pp-PPV). The structures of the systems under investigation are shown in Fig. 3.19. This choice was made based on the data presented in the Table 3.22.

Limit values of optical properties for considered non-classical push-pull polymers are presented in Table 3.22. For these systems the dipole moment and first hyperpolarizability tend to converge with the chain length to limit values. This is why, when obtaining values for an infinite chain, the $|\mu|$ and $|\beta|$ values can be used directly. The fast and monotonic convergence of these values to the limit values can be explained by the shielding of the cycles by the π-conjugated chain. For a sufficiently long length of the polymer, the donor and acceptor groups do not interact with each other and the atoms located in the middle of the chain do not acquire any charge. The pp-PQ $|\beta|^{\infty}$ values are the largest among the three groups of polymers. Hence, these polymers have potential for being good materials for high-performance non-linear optical devices.

The specific limit $\langle\gamma\rangle$ for the push-pull variant of the poly-para-quinodimethane is found to be larger than that for the pure polymer. Clearly, the presence of the end

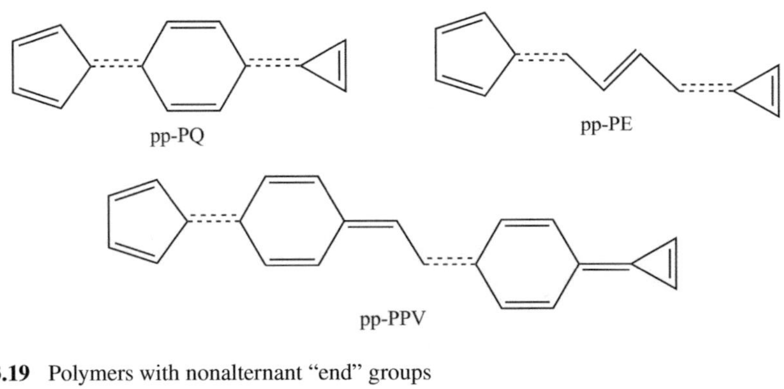

Fig. 3.19 Polymers with nonalternant "end" groups

Table 3.22 Limit values of (hyper)polarizabilities for polymers shown in Fig. 3.19

Bridge	$\lvert\mu\rvert$ (D)	$\langle\alpha\rangle^{\infty}/N_{el}$ (a.u.)	$\lvert\beta\rvert^{\infty}$ (a.u.)	$\langle\gamma\rangle^{\infty}/N_{el}$ (a.u.)
PE	5	14	2×10^4	2×10^5
PQ	14	>55	1×10^6	$>2\times10^7$
PPV	9	29	2×10^5	2×10^6

cycles affects not only the value of the first hyperpolarizability but also the value of the second hyperpolarizability. Just as in the case of the polymers without the donor and acceptor groups, the polarizability and the 2nd hyperpolarizability of pp-PQ are presented to estimate the lower-boundary values for the corresponding quantities, which exhibit intensive growth with the length of the system.

3.3.3.4 Nanosystems (Hyper)polarizability Calculations

Potential applications of π-conjugated nanosystems in molecular electronics and optics are currently of great interest. Among these systems, one should particularly mention fullerenes (see Sect. 3.3.1), graphene fragments, nanotubes, and their derivatives. The theoretical approaches most frequently used to investigate the properties of nanosystems are usually based on the concept of an infinite chain of periodic units with a particular topology. However, such approaches are not suitable for investigating systems with structural defects (partial hydrogenation atom substitution and other topological defects). The method we introduced allows computation of optical properties, i.e. the (hyper)polarizability, for aperiodic systems that contain thousands of atoms.

For the systems discussed earlier, the calculated specific optical properties were already converged to the limit values for a relatively small number of atoms (from several dozen up to a few hundreds). For nano-sized systems, the convergence requires from several hundred up to several thousand atoms. To demonstrate the ability of our method to describe nano-systems, we carried out calculations of a model macromolecular system with the structure presented in Fig. 3.20. The system represents a looped nanotube (nanotore). The looping of the nanotube allows for avoiding the "end" effects to appear in the calculations. The size of the nanotore is defined by the number of unit cells in the system. It can be expected that the C–C bonds

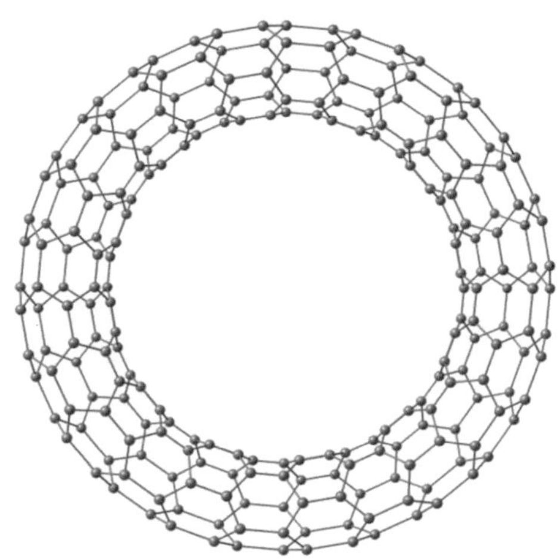

Fig. 3.20 The nanotore structure

Table 3.23 (Hyper)polarizabilities for nanotores with different numbers of unit cells, *n*

n	16	20	24	28	32	36
N_{el}	320	400	480	560	640	720
$\langle\alpha\rangle/10^3$	9.76	13.57	17.58	21.72	25.94	30.21
$\langle\gamma\rangle/10^7$	17.01	24.05	30.55	36.29	41.28	45.46
κ	1.39	1.31	1.26	1.22	1.19	1.17
size $T_2^{l=2}/10^4$	7.42	9.28	11.14	13.00	14.85	16.70
size $T_2^{l=3}/10^4$	1014	1267	1520	1774	2027	2280

The size of T_2 amplitude matrix, presented in the last two rows

located near the inner radius of the nanotore are somewhat shorter than the C–C bonds located near the outer radius. However with the growth of the tore radius this difference should be disappearing. The nanotore presented in Fig. 3.20 corresponds to the zig-zag nanotube with the (5,0) topology. The number *n* of the unit cells in our calculations varied in the interval from 16 to 36.

The difference in the C–C bond lengths for the nanotore can be expressed with the curvature parameter κ, which is defined as the ratio of the longest C–C bond length and the smallest one. It is clear that for $\lim_{n\to\infty} \kappa = 1$ and in the limit case nanotore becomes equivalent to an infinite nanotube without the "end" effect. The calculated values of the average polarizability and 2nd hyperpolarizability and a parameter of the nanotore curvature are presented in Table 3.23. The calculation of the parameters has been carried out with the cue$_{(2)}$-CCSD method. The use of higher levels of theory would require much more significant computational expenses. Also in Table 3.23, the size of the amplitude vector is presented for level $l = 2$ and $l = 3$. Based on the data presented in the table, one can see that extending the excitation set from the set used in cue$_{(2)}$-CCSD to the set used in cue$_{(3)}$-CCSD leads to an increase of the size the amplitude vector by more than two orders of magnitude. For the nanotore system, HF and MP2 results are not reported because they were not obtained due to the divergence of the SCF iteration process. The divergence is caused by quasi-degeneracy of the HOMO and LUMO that happens due to the high level of symmetry of the system.

3.4 Conclusion

The calculation of linear and nonlinear optical properties of conjugated systems is still a difficult problem for contemporary quantum chemistry. Wide use of conjugated systems, especially such nanosystems as carbon nanotubes, fullerenes, and graphenes in materials applications, require development of reliable theoretical methods for predicting the electronic and optical properties through quantum-mechanical calculations. The coupled cluster theory, which offers high accuracy in ab initio calculations, cannot be applied to investigate large molecular systems containing hundreds of atoms. Our semiempirical local coupled-cluster method (cue-CCSD) provides an alternative to the ab initio approach. It enables calculation of optical parameters of π-conjugated molecules and to produce results which qualitatively agree with the experiment. The results presented in this review clearly show the need to account for a significant part of the electronic correlation effects in the calculations in order to predict the molecular optical parameters with an adequate accuracy. This conclusion applies to both long polymers and small systems containing only a dozen atoms. A question which arises when the accuracy of a particular computational approach is evaluates, is what reference data one should use in such evaluation. In general, the optical parameters for short polymer fragments obtained in the cue-CCSD calculations are in good agreement with those obtained in the FCI calculations, as well as with the available experimental data. Thus, the FCI results can be used as the reference. The results obtained with the cue-CCSD method for quasi-one-dimensional conjugated systems with different lengths allow determination of limit values of the optical parameters. Some of these values are shown in this review. Certainly, one cannot expect the results obtained with the cue-CCSD method, which is a semiempirical approach, to be quantitatively accurate for a partic-

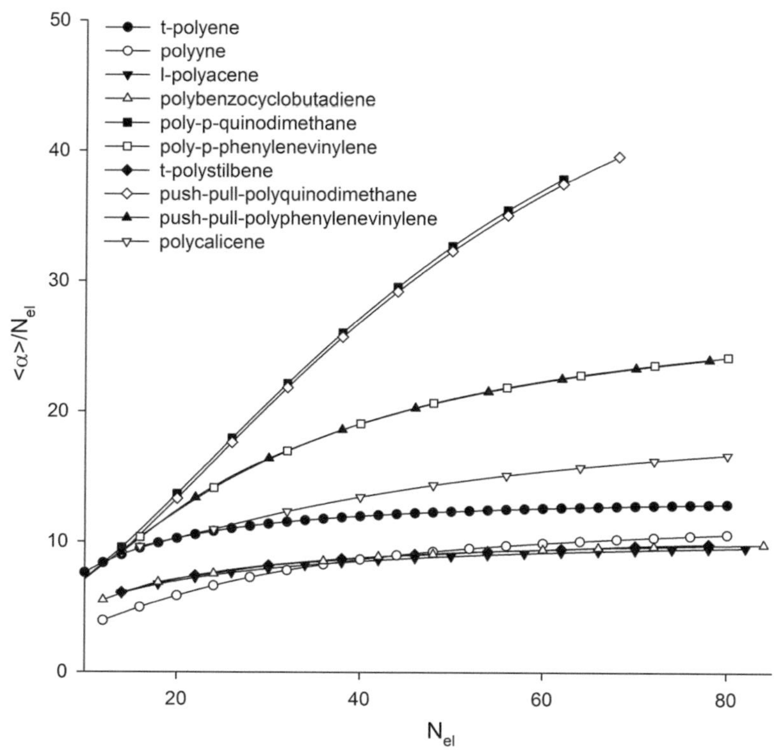

Fig. 3.21 Average specific (per electron) polarizabilities of different conjugated polymers in cue-CCSD approach

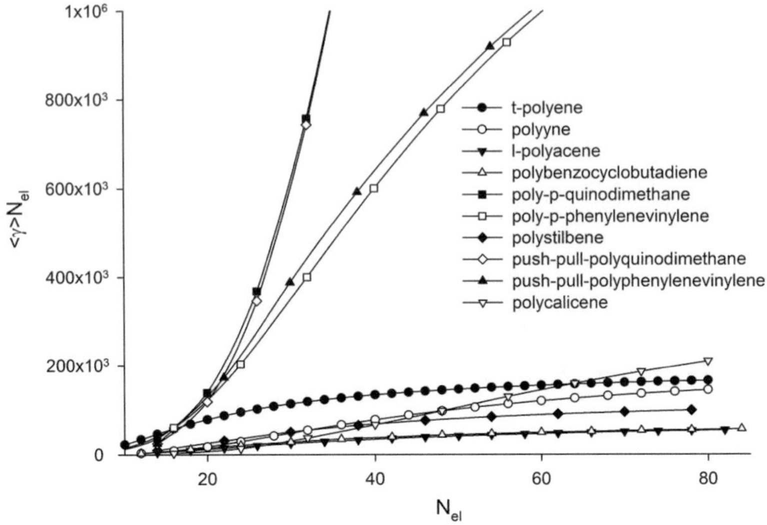

Fig. 3.22 Average specific (per electron) 2nd hyperpolarizabilities of different conjugated polymers in cue-CCSD approach

ular molecular system. However, the trends in the changes of the optical parameters with the lengths of the polymer are usually correctly described with the method. The same applies to polymers which are structurally similar. For example, the comparative analysis of the average (per electron) polarizability described in Fig. 3.21 and the 2nd average hyperpolarizability described in Fig. 3.22 for different systems (for a detailed description of the systems and their optical properties see Sects. 3.3.3.1–3.3.3.3) provides information about the optical response of the systems, which can be useful in predicting their optoelectronics applications.

References

1. Adler TB, Werner HJ (2011) J Chem Phys 135(144):117
2. Andre JM, Delhalle J (1991) Chem Rev 91:843
3. Bang-Jensen J, Gutin G (2007) Digraphs. Theory, algorithms and applications. Springer, London
4. Bartlett RJ (1981) Ann Rev Phys Chem 32:359
5. Bartlett RJ, Musiał M (2007) Rev Mod Phys 79:291
6. Bartlett RJ, Purvis-III GD (1979) Phys Rev A 20(4):1313
7. Buckingham AD (1967) In: Hirschfelder JO (ed) Advances in chemical physics, vol 12. Wiley, pp 107–142
8. Bulat FA, Toro-Labbè A et al (2005) J Chem Phys 123(014):319
9. Burland DM, Miller RD, Walsh CA (1994) Chem Rev 94:31
10. Champagne B, Perpète EA (1999) Int J Quant Chem 75:441
11. Cheng LT, Tam W et al (19991) SHS. J Phys Chem 95:10,631

12. Craig GSW, Cohen RE, Schrock RR, Silbey RJ, Puccetti G (1993) J Am Chem Soc 115:860
13. Cyvin SJ (1982) J Mol Struct (THEOCHEM) 86:315
14. Dalgaard E, Monkhorst HJ (1983) Phys Rev A 28(3):1217
15. Geskin VM, Brédas JL (1998) J Chem Phys 109(14):6163
16. Geskin VM, Lambert C, Brédas JL (2003) J Am Chem Soc 125(15):651
17. Ghigo G, Shahi ARM, Gagliardi L, Solstad LM, Cramer CJ (2007) J Org Chem 72:2823
18. Hampel C, Werner HJ (1996) J Chem Phys 104:6286
19. Hess Jr BA, Schaad LJ, Ewig CS, Carsky P (1983) J Comp Comp Chem 4:53
20. Huzak M, Deleuze MS (2013) J Chem Phys 138(024):319
21. Ivanov VV, Luzanov AV (1994) Ukr Chim J (in Russian) 60:11
22. Ivanov VV, Luzanov AV (1997) J Struct Chem (EnglTransl) 38(1):14
23. Ivanov VV, Boichenko IV, Luzanov AV (1998) J Struct Chem (EnglTransl) 41(4):457
24. Ivanov VV, Sleta LA, Klimenko TA (2007) Kharkov University Bulletin Chemical Series (in Russian) 770(15(38)):195
25. Ivanov VV, Zakharov AB, Adamowicz L (2013) Mol Phys 111(24):3779
26. Kamiya M, Sekino H, Tsuneda T, Hirao K (2005) J Chem Phys 122(234):111
27. Karelson M, Lobanov VS (1996) Chem Rev 96:1027
28. Kats D, Schütz M (2009) J Chem Phys 131(124):117
29. Kats D, Schütz M (2010) Z Phys Chem 224:601
30. Kats D, Korona T, Schütz M (2007) J Chem Phys 127(064):107
31. Klimenko TA, Ivanov VV, Adamowicz L (2009) Mol Phys 107(17):1729
32. Knab R, Förner W, Čížek J, Ladik J (1996) J Mol Struct 366:11
33. Knowles PJ, Schütz M, Werner HJ (2000) In: Grotendorst J (ed) Modern Methods and Algorithms of Quantum Chemistry, Proceedings, vol 3, John von Neumann Institute for Computing, Jülich, chap *Ab initio* Methods for Electron Correlation in Molecules, p 97
34. Kondayya G, Shukla A (2012) Comput Phys Comm 183(4):677
35. Korona T (2011) Theor Chem Acc 129:15
36. Korona T, Jeziorski B (2006) J Chem Phys 125(184):109
37. Korona T, Werner HJ (2003) J Chem Phys 118(7):3006
38. Korona T, Pflüger K, Werner HJ (2004) Phys Chem Chem Phys 6:2059
39. Kummel H (1960) Nucl Phys 17:477
40. Kurtz HA, Stewart JJP, Dieter KM (1990) J Comp Chem 11:82
41. Lee TJ, Rice JE, Scuseria GE, Schaefer HF (1989) Theor Chim Acta 75:81
42. Li Q, Chen L, Li Q, Shuai Z (2008) Chem Phys Lett 457:276
43. Li Q, Yi Y, Shuai Z (2008) J Comp Chem 29:1650
44. Li S, Ma J, Jiang Y (2002) J Comp Chem 23:237
45. Limacher PA, Mikkelsen KV, Lüthi HP (2009) J Chem Phys 130(194):114
46. Luzanov AV (1989) Theor Exp Chem (EnglTransl) 25(1):1
47. Luzanov AV, Ivanov VV (1997) J Struct Chem (Engl Transl) 38:863
48. Luzanov AV, Prezhdo OV (2006) J Chem Phys 124(224):109
49. Luzanov AV, Pedash YF, Ivanov VV (1989) J Struct Chem (EnglTransl) 30(5):701
50. Luzanov AV, Ivanov VV, Klimko GT, Mestechkin MM (1998) J Struct Chem (EnglTransl) 39(2):277
51. Bass M (eds) (2009) Handbook of Optics. Optical Properties of Materials, Nonlinear Optics, Quantum Optics. McGraw Hill
52. Mata RA, Stoll H (2011) J Chem Phys 134(034):122
53. Mata RA, Werner HJ, Schütz M (2008) J Chem Phys 128(144):106
54. Matsen FA (1978) The unitary group and the many-body problem. In: Löwdin PO (ed) Advances in Quantum Chemistry, Advances in Quantum Chemistry, vol II. Academic Press, p 223
55. Matsen FA (1987) Int J Quant Chem 32:87
56. Norman P, Luo Y, Jonsson D et al (1997) J Chem Phys 106:8788
57. Ohno K (1964) Theor Chim Acta 2:219

58. Oziminski WP, Palusiak M, Dominikowska J, Krygowski TM (2013) Phys Chem Chem Phys 15:3286
59. Paldus J, Čížek J, Shavitt I (1972) Phys Rev A 5:50
60. Pariser R, Parr RG (1953) J Chem Phys 21:466
61. Pedash YF, Ivanov VV, Luzanov AV (1989) Theor Exp Chem (EnglTransl) 25(6):607
62. Pipek J, Mezey PG (1989) J Chem Phys 90(9):4916
63. Pople JA (1953) Trans Faraday Soc 49:1375
64. Pulay P (1983) Chem Phys Lett 100(2):151
65. Schlegel PVR et al (eds) (2004) Encyclopedia of computational chemistry. Wiley, Chichester
66. Russ NJ, Crawford TD (2004) Chem Phys Lett 400:104
67. Saebø S, Pulay P (1993) Annu Rev Phys Chem 44:213
68. Schmidt MW, Baldridge KK, Boatz JA et al (1993) J Comput Chem 14:1347
69. Schütz M, Werner HJ (2001) J Chem Phys 114:661
70. Scuseria GE, Ayala PY (1999) J Chem Phys 111(18):8330
71. Shavitt I, Bartlett RJ (2009) Many-body methods in chemistry and physics. MBPT and coupled cluster theory. Cambridge University Press, Cambridge
72. Shen Y, Chen CF (2012) Chem Rev 112:1463
73. Shuai Z, Bredas JL (2000) Phys Rev B62(23):15,452
74. Spingborg M, Dong Y (2004) Conjugated polymers in external dc fields. In: Sabin JR, Brändas E (eds) Advances in quantum chemistry, vol 47. Elsevier Academic Press Inc., p 269
75. Sugimoto T, Yoshida Z (1990) Pure Appl Chem 62:551
76. Sugimoto T, Shibata M, Yoneda S, Yoshida Z et al (1986) J Am Chem Soc 108:7032
77. Sugimoto T, Shibata M et al (1991) Angew Chem 4:454
78. Thouless DJ (1960) Nucl Phys 21:225
79. Čížek J (1966) J Chem Phys 45(11):4256
80. Visotski YB, Briantsev BC (2004) Quantum chemistry of radicals and ion-radicals with conjugated bonds. DonGuet, Donetsk
81. Waite J, Papadopoulos MG (1983) J Comp Chem 4:578
82. Waite J, Papadopoulos MG (1989) J Chem Phys 91:791
83. Wergifosse M, Champagne B (2011) J Chem Phys 134(074):113
84. Werner HJ, Schütz M (2011) J Chem Phys 135(144):116
85. Wheatley RJ (2008) J Comput Chem 29(3):445
86. Zakharov AB, Ivanov VV (2011) J Struct Chem (Engl Transl) 52(4):645
87. Zakharov AB, Ivanov VV (2011) Kharkov University Bulletin Chemical Series 976(20(43)):9 (in Russian)
88. Zakharov AB, Ivanov VV, Adamowicz L (2014) J Phys Chem C 118:8111
89. Zhang D, Liu C (2011) J Chem Phys 135(134):117

Chapter 4
A Critical Look at Methods for Calculating Charge Transfer Couplings Fast and Accurately

Pablo Ramos, Marc Mankarious and Michele Pavanello

Abstract We present here a short and subjective review of methods for calculating charge transfer couplings. Although we mostly focus on Density Functional Theory, we discuss a small subset of semiempirical methods as well as the adiabatic-to-diabatic transformation methods typically coupled with wavefunction-based electronic structure calculations. In this work, we will present the reader with a critical assessment of the regimes that can be modelled by the various methods—their strengths and weaknesses. In order to give a feeling about the practical aspects of the calculations, we also provide the reader with a practical protocol for running coupling calculations with the recently developed FDE-ET method.

4.1 Introduction

Charge transfer (CT) between molecular species play vital roles in processes that occur in biology such as protein communication [1–5], respiratory systems in the mitochondria [6], oxidative damage on DNA [7–9], photosynthetic cycles [10, 11], as well as in materials science conduction in organic semiconductors [12, 13]. In order to achieve an accurate modeling of these processes in the simulations, one needs to include several levels of complexity, which in most instances lead to considering model systems featuring hundreds of atoms and an even larger number of electrons. The large system sizes preclude the use of high-level wavefunction-based quantum-chemical methods. For this reason, researchers worldwide have invested a great deal of effort in developing approximate, fast, yet still accurate methods for describing CT reactions. Methods based on Density-Functional Theory have in recent years become competitive in regards to the accuracy while still maintaining a generally low computational cost.

Marcus theory [14, 15] is perhaps the most applicable theory for modeling a CT process. This theory was originally derived under three main approximations. First, a CT event is thought of in terms of a two-dimensional basis set (donor and acceptor).

P. Ramos · M. Mankarious · M. Pavanello (✉)
Department of Chemistry, Rutgers University, Newark 07102, USA
e-mail: m.pavanello@rutgers.edu

J. Leszczynski and M.K. Shukla (eds.), *Practical Aspects of Computational Chemistry IV*, DOI 10.1007/978-1-4899-7699-4_4

The interaction matrix element of the Hamiltonian is the central quantity in determining the probability of a transition in populations from the basis function representing the donor state to one representing the acceptor state. This interaction is the electronic coupling V_{DA} of the two electronic states involved in the CT reaction [16, 17]. Second, it relies on the Condon approximation [18, 19], in which the electronic coupling is considered to be independent of the nuclear motion when the transfer occurs. Third, reactants and products are modeled as being enclosed by spheres on which the polarization of the solvent is represented as a dielectric continuum [15, 20–22]. Marcus theory can be summarized as [23]:

$$k_{CT} = \frac{2\pi}{\hbar}|V_{DA}|^2 \frac{e^{-\frac{(\Delta G+\lambda)^2}{4\lambda K_B T}}}{\sqrt{4\pi\lambda K_B T}}. \tag{4.1}$$

where λ is the reorganization energy, and V_{DA} is the electronic coupling.

States that most resemble the initial and final states of electron transfer have been often referred to as "diabatic states" [24, 25] and their corresponding wavefunctions "diabats". Although it is known that diabatic states have a formal definition [26, 27], it was shown [28] that charge-localized states satisfy the requirements for diabatic states for condensed phase electron transfer reactions.

Several approaches are available in the literature to generate and evaluate Hamiltonian matrix elements with wavefunctions of charge-localized, diabatic states. They differ in the level of theory used in the calculation and in the way localized electronic structures are created [15, 25, 26, 29–31]. When wavefunction-based quantum-chemical methods are employed, the framework of the generalized Mulliken-Hush method (GMH) [29, 32–34], is particularly successful. So far, it has been used in conjunction with accurate electronic structure methods for small and medium sized systems [35–37]. As an alternative to GMH and other derived methods [38, 39], additional methods have been explored for their applicability in larger systems such as constrained density functional method (CDFT) [25, 37, 40, 41], and fragmentation approaches [42–47], which also include the frozen density embedding (FDE) method [48, 49].

So far, we have mentioned methods that produce all-electron diabatic wavefunctions and corresponding Hamiltonian matrix elements. There are two other classes of methods which simplify the quantum problem by focusing on the wavefunction of the transferred charge: such as methods making use of the frozen core approximation Fragment Orbital methods (FO), and methods that assume the charge to be localized on single atomic orbitals [50]. In this work, we will also treat these computationally low-cost methods.

As our group is involved in the development of the Frozen Density Embedding (FDE) formulation of subsystem DFT, this chapter will pay particular attention to the FDE methodology. We believe FDE to be a very promising method capable of achieving a good description of the electronic coupling in CT reactions, while maintaining a low computational complexity.

This chapter is divided in two parts: the first part is devoted to the FDE method as well as other DFT-based alternatives. The second part covers more accurate methods (wavefunction-based). In each of the two parts, we discuss the numerical stability and accuracy of the methods in the generation of diabatic states with the overarching goal of obtaining reliable electronic couplings with a contained computational effort.

We will start with a description of FDE and its ability to generate diabats and to compute Hamiltonian matrix elements—the FDE-ET method (ET stands for Electron Transfer). In the subsequent section, we will present specific examples of FDE-ET computations to provide the reader with a comprehensive view of the performance and applicability of FDE-ET. After FDE has been treated, four additional methods to generate diabatic states are presented in order of accuracy: CDFT, FODFT, AOM, and Pathways. In order to output a comprehensive presentation, we also describe those methods in which wavefunctions methods can be used, in particular GMH and other adiabatic-to-diabatic diabatization methods. Finally, we provide the reader with a "protocol" for running FDE-ET calculations with the only available implementation of the method in the Amsterdam Density Functional software [51]. In closing, we outline our concluding remarks and our vision of what the future holds for the field of computational chemistry applyed to electron transfer.

4.2 DFT Based Methods

4.2.1 The Frozen Density Embedding Formalism

The frozen density-embedding (FDE) formalism [52] developed by Wesolowski and Warshel [52–54] has been applied to a plethora of chemical problems, for instance, solvent effects on different types of spectroscopy [55–57], magnetic properties [58–62], excited states [55, 63–66], charge transfer states [49, 67, 68]. Computationally, FDE is available for molecular systems in ADF [51, 69], Dalton [70, 71], Q-Chem [72, 73], and Turbomole [74–76] packages, as well as for molecular periodic systems in CP2K [77, 78] and fully periodic systems (although in different flavors) in CASTEP [79, 80], Quantum Espresso [81–83], and Abinit [84, 85].

FDE prescribes that the total electron density should be expressed as the sum of subsystem electron densities [53, 86–89], this is based on the idea that a molecular system can be more easily approached if it is subdivided into many smaller subsystems. Namely,

$$\rho_{tot}(\mathbf{r}) = \sum_{I=1}^{\#\,of\ subsystems} \rho_I(\mathbf{r}). \tag{4.2}$$

As in regular DFT calculations, the electron density of each subsystem is computed by solving selfconsistently a Kohn–Sham (KS) like equation per subsystem. These KS like equations read as:

$$\left[\frac{-\nabla^2}{2} + v^I_{KS}(\mathbf{r}) + v^I_{emb}(\mathbf{r})\right] \phi_{(i)I}(\mathbf{r}) = \epsilon_{(i)I}(\mathbf{r})\phi_{(i)I}(\mathbf{r}). \tag{4.3}$$

where $\phi_{(i)I}(\mathbf{r})$, $\epsilon_{(i)I}$ are the molecular orbitals and orbital energies of subsystem I. In (4.3) we have augmented the Kohn–Sham single particle Hamiltonian by an embedding potential, v^I_{emb}, in which are encoded the interactions with the other subsystems. In the following, $v^I_{emb}(\mathbf{r})$ is the embedding potential acting on subsystem I:

$$\begin{aligned} v^I_{emb}(\mathbf{r}) = \sum_{J \neq I}^{N_S} \left[\int \frac{\rho_J(\mathbf{r}')}{|\mathbf{r} - \mathbf{r}'|} d\mathbf{r}' - \sum_{\alpha \in J} \frac{Z_\alpha}{|\mathbf{r} - \mathbf{R}_\alpha|} \right] + \\ + \frac{\delta T_s[\rho]}{\delta \rho(\mathbf{r})} - \frac{\delta T_s[\rho_I]}{\delta \rho_I(\mathbf{r})} + \frac{\delta E_{xc}[\rho]}{\delta \rho(\mathbf{r})} - \frac{\delta E_{xc}[\rho_I]}{\delta \rho_I(\mathbf{r})}. \end{aligned} \tag{4.4}$$

In the above, T_s, E_{xc} and Z_α are kinetic and exchange-correlation energy functionals, and the nuclear charge, respectively, and N_S is the total number of subsystems considered. In practical FDE calculations, the kinetic energy is calculated in terms of orbital free semilocal functionals. This approximation is ultimately the biggest difference between an FDE and a full KS-DFT calculation of the supersystem [90–93]. As a consequence, the embedding potential becomes inaccurate when the subsystems feature a large overlap between their electron densities [83, 94, 95] (this is because the larger the density overlap is, the larger the magnitude of the nonadditive potentials become). In FDE, the subsystem KS equations are left to converge to selfconsistency with respect to each other. This is often achieved by employing the so-called freeze-and-thaw procedure [69, 96] (as done in ADF and other molecular codes) or via updating the embedding potential at every SCF cycle as done in CP2K [78, 97] and Quantum-Espresso [81, 83, 98]. It is worth noting that FDE scales linearly with the number of subsystems provided that linear scaling methods for the solution of the electrostatic problem are employed [69].

The earliest example of diabatization by FDE was given in Ref. [48]. This is shown in Fig. 4.1, where the spin densities for a pair of guanines are calculated. KS-DFT of the supersystem carried out with semilocal XC functionals fails in the prediction of the spin density. This is because the self-interaction error makes the spin density spread on both guanines against the prediction given by more accurate theoretical work [35] and experimental studies [99–101]. On the contrary, FDE localizes the charge on a guanine of choice.

The fact that FDE was able to provide subsystem-localized electronic structures was known since its early application to systems with unpaired electrons [60–62, 102]. Later, this ability of FDE was explored for the computation of diabatic states for electron transfer [48, 49, 67, 103] and to compute hyperfine coupling constants [60, 61].

The question that one can raise is why FDE calculations yield charge localized states? We provide here four reasons [49, 103].

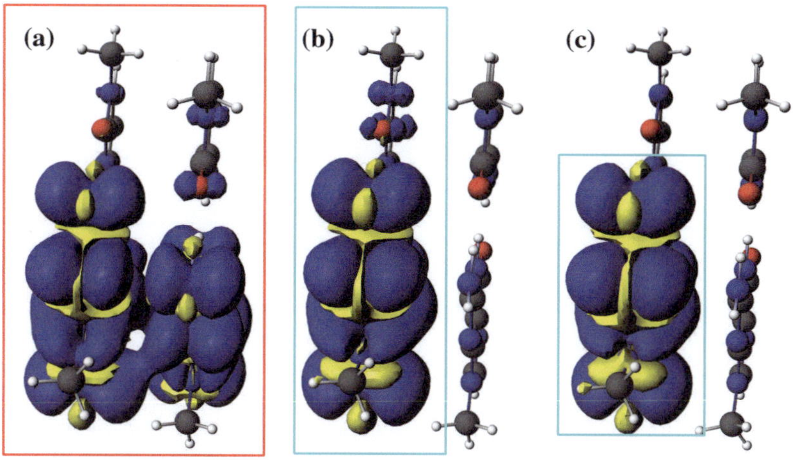

Fig. 4.1 Spin densities of a guanine-cytosine dimer radical cation, $(GC)_2^+$. **a** KS-DFT supramolecular calculation using PW91 functional, **b** FDE calculation considering two subsystems where the *left side* subsystems (*blue contour*) is positively charged and **c** FDE calculation for four subsystems with one subsystems (*blue contour*) is positively charged. The nucleobases structures and spin densities were taken from Ref. [48]

1. Orthogonality is not imposed between the molecular orbitals belonging to different subsystems.
2. FDE calculations can be executed in the monomer basis set. This is known as FDE(m) method [104].
3. FDE calculations are always initiated with a subsystem localized initial guess of the electron density.
4. Electrons of a subsystem, remain localized also because there are repulsive walls in the region of the surrounding (frozen) fragments.

The first reason, is important because it directly removes a bias towards delocalization which results due to orthonormalization of the molecular orbitals, as already noted by Dulak and Wesolowski [105]. The second and third reasons come together, the lack of basis functions on the surrounding subsystems, does not allow substantial charge transfer between the subsystems. As a consequence, the SCF is biased to converge to localized electronic structures.

The fourth reason makes reference to the approximate nature of the term $\frac{\delta T_s[\rho]}{\delta \rho(\mathbf{r})} - \frac{\delta T_s[\rho_I]}{\delta \rho_I(\mathbf{r})}$ (also known as nonadditive kinetic energy potential which is part of the embedding potential) in the region of the frozen fragments (e.g. in the region where ρ_J with $J \neq I$ is larger than any other subsystem electron density). Approximate nonadditive kinetic energy potentials fail in canceling out the attractive potential due to the nuclear charge in the vicinity of the nucleus of the surrounding frozen subsystems [104, 106], and they do not reproduce the exact potential at intermediate regions [107–109], especially in the vicinity of an atomic shell [106, 109]. In that region they cross the exact potential and large potential walls arise. A simplified depiction of this effect is devised in Fig. 4.2.

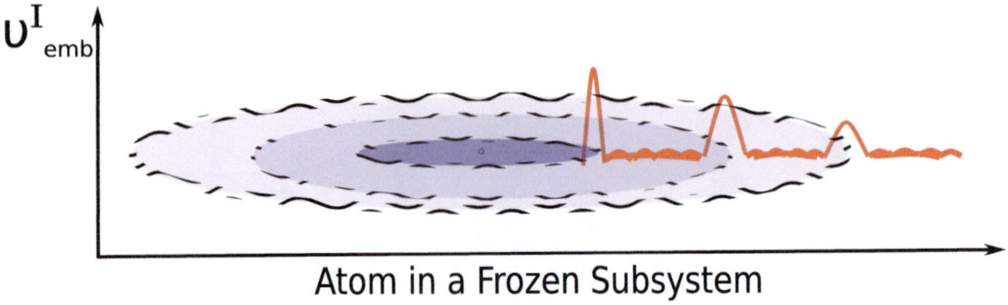

Fig. 4.2 Exemplification of the embedding potential at the atomic shells of the surrounding subsystems. Figure from Ref. [103]

In this scenario, diabatic states can be generated with FDE by performing at least two simulations, one featuring a hole/electron on the donor while the acceptor is neutral and one calculation in which the charge hole/electron is on the acceptor. The result is two charge localized states, whose, densities and Kohn–Sham orbitals are used in a later step in order to build the diabatic Hamiltonian and overlap matrices, needed to compute the diabatic coupling matrix element.

4.2.1.1 FDE-ET Method

FDE-ET is a methodology which computes Hamiltonian couplings from diabatic states generated by an FDE calculation. Electron transfer reaction are usually described in the basis of a two-state formalism [30], taking as basis set two broken-symmetry charge-localized states. This methodology can also approach models for the superexchange mechanism [7, 23, 110–113], where the transfer is still modelled by a Two-dimensional basis set but the coupling includes the effect of non-resonant bridges states. Figure 4.3, illustrates the difference between tunneling through the vacuum and through a set of bridge states. The bridge could be comprised of one or more molecules, a covalent bond or any other type o potential barrier as long as its height is lower than the one when vacuum separates donor and acceptor. As it is shown in Fig. 4.3, the higher the potential barrier the faster the coupling decays with respect to the donor–acceptor distance.

In FDE-ET we seek a method capable of computing the Hamiltonian matrix in the basis of charge-localized states generated with FDE. First, we have to define the needed matrix elements. As diabatic states are not the eigenfunctions of the molecular Hamiltonian, the off-diagonal elements of such Hamiltonian are not zero and can be approximated by the following formula [114, 115] if ψ_D and ψ_A are slater determinants representing the donor and acceptor diabats:

$$H_{DA} = \langle \psi_D | \hat{H} | \psi_A \rangle = S_{DA} E \left[\rho^{(DA)}(\mathbf{r}) \right] . \tag{4.5}$$

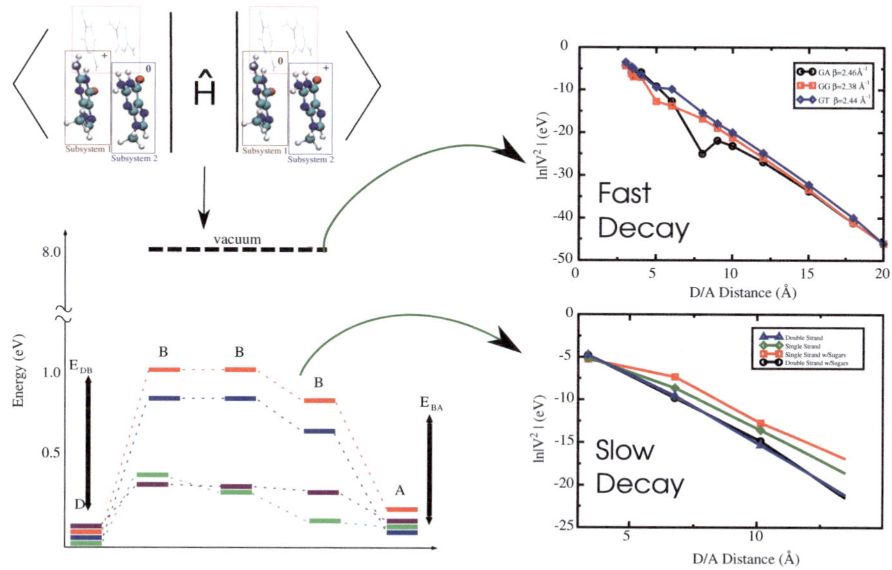

Fig. 4.3 Energy dependence of a charge transfer process. The off-diagonal element (Dirac notation) will decay as the potential well that the charge has to overcome increases. Two cases: for vacuum as a potential well we have a faster decay, and when molecules act as a bridge the transfer will decay slower

Here $\hat{H}$ is the molecular electronic Hamiltonian, and $\rho^{(DA)}(\mathbf{r})$ is the transition density defined as $\rho^{(DA)}(\mathbf{r}) = \langle \psi_D | \sum_{k=1}^{n_e} \delta(\mathbf{r}_k - \mathbf{r}) | \psi_A \rangle$, with n_e being the total number of electrons in the system (i.e. the sum of the electron number of all subsystems) and $E\left[\rho^{(DA)}(\mathbf{r})\right]$ is an energy density functional. The donor–acceptor overlap matrix elements are found by computing the following determinant:

$$S_{DA} = \det\left[\mathbf{S}^{(DA)}\right], \tag{4.6}$$

where $\mathbf{S}_{kl}^{DA} = \langle \phi_k^{(D)} | \phi_l^{(A)} \rangle$ is the transition overlap matrix in terms of the occupied orbitals ($\phi_{k/l}^{(D/A)}$) [114, 116]. Thus, the transition density is now written in the basis of all occupied orbitals which make up the diabatic states ψ_D and ψ_A.

$$\rho^{(DA)}(\mathbf{r}) = \sum_{kl}^{\text{occ}} \phi_k^{(D)}(\mathbf{r}) \left(\mathbf{S}^{(DA)}\right)_{kl}^{-1} \phi_l^{(A)}(\mathbf{r}). \tag{4.7}$$

The Hamiltonian coupling is not H_{DA}, but it is generally reported as the coupling between the Löwdin orthogonalized ψ_D and ψ_A. For only two states this takes the form,

$$V_{DA} = \frac{1}{1 - S_{DA}^2}\left(H_{DA} - S_{DA}\frac{H_{DD} + H_{AA}}{2}\right). \tag{4.8}$$

Turning to the superexchange picture, the effective coupling, is a summation of the contribution given by the interaction between D and A and the interaction of D and A with all bridges states, namely:

$$V_{DA}(E) = \tilde{V}_{DA} + \underbrace{\tilde{\mathbf{V}}^{T}_{DB}\mathbf{G}_{B}(E)\tilde{\mathbf{V}}_{BA}}_{V_{\text{bridge}}}, \tag{4.9}$$

where the superscript T stands for transpose, $\mathbf{G}_B(E)$ is the Green's operator, defined as

$$\mathbf{G}_B(E) = -(\tilde{\mathbf{V}}_B - E\,\tilde{\mathbf{I}}_B)^{-1}, \tag{4.10}$$

As shown in Eq. 4.9, $\tilde{V}_{DA}$ is the coupling for the donor–acceptor transfer, which in the absence of bridge states (CT through vacuum), would be the only contribution to $V_{DA}(E)$. On the other hand, if bridge states are present, the contribution to $V_{DA}(E)$ is given by the second addend in Eq. 4.9. Generally, E appearing above is the energy at which the tunneling event occurs (i.e. at the crossing seam of the Marcus parabolas). In our works [67], E was chosen to be in between E_A and E_B, and specifically to be $\frac{E_D+E_A}{2}$. This choice is invoked by several works in the literature [30, 117–119] where it is well known that there is a mild dependence of the coupling with the tunneling energy [118]. However, this equation holds when there is no resonance between D, A and the bridges states [23, 30, 120–123]. If near-degeneracies appear then the transport regime transitions to resonant tunneling or hopping.

4.2.1.2 Distance Dependence of the Electronic Coupling

In this section, we discuss calculations of the coupling matrix element (V_{DA}) of hole transfer from a donor to an acceptor molecule through the vacuum. This means that the initial state of hole transfer is the donor molecule (D), and the final state the acceptor molecule (A), and no intermediate bridge states are considered. Any reliable method for computing couplings should be able to reproduce high level calculations of CT coupling in small molecular dimers. For this purpose, we initially chose 23 biologically relevant π-stacks [67], in order to analyze the distance dependence of the coupling, separations of 3–20 Å were considered, as result a total of 276 coupling calculations were ran. Overall, our couplings show a good agreement with previous computations (e.g. we reproduced the decay factors, β, π-stacked dimers separated by vacuum).

When a test set for hole transfer couplings featuring high accuracy couplings became available [37] we could compare systematically the FDE-ET couplings with the benchmark values [103]. Benchmark calculations were ran on a set of 15 of π-stacked dimers. This study was rigorous, and tested the effect of the basis set size, nonadditive kinetic energy functionals (NAKE) and exchange-correlation functionals (XC) on the value of the computed couplings. The most important finding that resulted from the benchmark work resided in the fact that GGA functionals coupled

Table 4.1 Mean statistical values for the best XC-functional choices

Set	MUE (MeV)	MRUE (%)	MAX (meV)
PBE/PW91k/TZP	15.3	7.1	49.6
PW91/PW91k/TZP	15.2	7.1	49.1
B3LYP/PW91k/TZP	18.1	7.9	58.5
M06-2X/PW91k/TZP	18.0	8.2	54.9

Reproduced with permission from Ref. [103]

with a medium sized basis set and the PW91k NAKE functional allow the FDE-ET method to yield reliable electronic couplings as tested against high-level correlated wavefunction (MRCI+Q, NEVPT2 and SCS-CC2) methods applied to the array of dimers. The PBE and PW91 functionals are found to be a good choice in each case considered with a MAX error lower than 50 meV and an overall MRUE of a little over 7 % in both cases [103]. Statistically, we found that hole transfer couplings are relatively insensitive to the choice of NAKE functionals, while our analysis of the basis set dependence shows that QZ4P basis set is the most problematic, as it often biases the FDE convergence to nonphysical states at short intersubsystem separations—a problem already well documented in the FDE literature [124, 125]. Finally, Table 4.1 compares the performance of FDE-ET for different levels of theory. The results for GGAs are in good agreement with the benchmark values, and in some cases they showed to be superior to hybrid and meta-GGA functionals, particularly PBE and PW91. B3LYP also stands out as another valuable choice.

Generally, all functionals perform well in the FDE-ET coupling calculations making FDE-ET a method that is relatively insensitive to the XC and NAKE functional choice.

4.2.1.3 Hole Transfer in DNA Oligomers

In this section, we discuss an interesting application of FDE-ET to charge transfer in biosystems.The electronic coupling for hole transfer in a completely dry B-DNA structure of G(T)$_N$G and G(A)$_N$G was calculated. The structures considered lack water molecules, metal counterions and phosphate linker groups. The latter is because the applicability of FDE is restricted to non-covalently bound molecular fragments. Consequently, appropriate modifications to the B-DNA structure had to be made: we have removed the phosphate groups and capped the dangling bonds with hydrogen atoms at 1.09 Å from the bonding atom. The resulting structure of the modified G(T)$_N$G is depicted in Fig. 4.4. The largest system considered is the double strand with ribose groups and counts 308 atoms and 1322 electrons. In this study, the role of the environment on the CT in DNA is elucidated and analyzed on the basis of an all-electron computation.

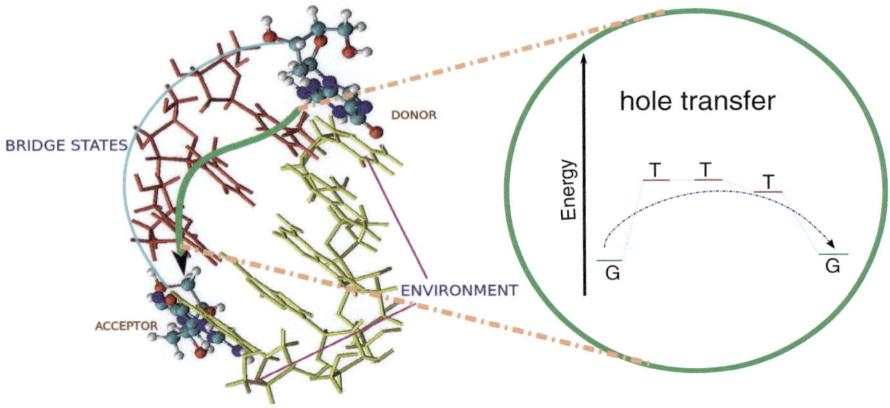

Fig. 4.4 The dephosphorylated $G(T)_NG$ B-DNA oligomer employed in the hole transfer coupling calculations. As the figure depicts, the hole tunnels from the *bottom* guanine (in *balls* and *sticks*) to the top guanine. The tunneling wall is provided by a series of three thymines (*red* branch, labeled as "bridge"). The counterstrand, $C(A)_NC$, acts as a solvating environment (in *yellow*, labeled as "spectators") and no hole is allowed to localize on it. Taken from Ref. [67]

Regarding the energetics (site energies), an uneven stabilization of the bridge states compared to donor/acceptor states occurs in both type of oligomers, being this effect more pronounced in the $G(T)_NG$ system than in the $G(A)_NG$ system. By inspection of the overall electrostatics of the interaction between G:C and T:A [126], we notice that T has a strong permanent dipole pointing towards A, similarly to C:G. Instead, A has a much weaker dipole compared to C or T and thus upon contact of the GTG strand with the CAC strand the cytosines will stabilize much more the holes on Gs than the adenines can stabilize the holes on Ts, hence the tunneling wall increases from single strand to double strand.

Regarding the couplings, when the magnitude of the through space and through bridge couplings are inspected, our calculations show that the effects of the ribose groups and the nucleobases in the counterstrand are opposite and different in magnitude depending on the oligomer size (see Table 4.2). We conclude, however, that the effect of the counterstrand on the computed superexchange couplings completely overpowers any effect due to the presence of the ribose groups.

4.2.2 Constrained Density Functional Theory Applied to Electron Transfer Simulations

Alternatively to FDE-ET methodology, constrained DFT (CDFT hereafter), a DFT-based procedure that was initially proposed by Dederichs et al. [127], and later introduced by Van Voorhis and Wu [128] with the aim of applying it to charge transfer reactions. CDFT is an effective method for calculating diabatic states for electron

Table 4.2 Through-space and through-bridge electronic couplings and tunneling energy gaps for single and double strand $G(T)_NG$ B-DNA, including the effects of the backbone (sugars)

	$\tilde{V}_{DA}$ (meV)	V_{bridge} (meV)	E_{DB} (eV)	E_{BA} (eV)
SINGLE STRAND NO RIBOSE				
GG	78.13			
GTG	0.76	12.46	0.71	0.50
$G(T)_2G$	0.01	1.13	0.79	0.66
$G(T)_3G$	–	0.09	0.79	0.77
DOUBLE STRAND NO RIBOSE				
GG	92.6			
GTG	0.65	7.66	0.93	0.96
$G(T)_2G$	0.01	0.47	1.11	0.94
$G(T)_3G$	–	0.02	0.99	1.16
SINGLE STRAND WITH RIBOSE				
GG	71.38			
GTG	0.18	25.01	0.43	0.37
$G(T)_2G$	0.02	1.70	0.58	0.37
$G(T)_3G$	–	0.21	0.41	0.41
DOUBLE STRAND WITH RIBOSE				
GG	91.07			
GTG	0.02	7.35	0.62	0.87
$G(T)_2G$	0.02	0.61	0.93	0.60
$G(T)_3G$	–	0.02	0.50	0.82

A–is shown for values below 0.01 meV. Reproduced with permission from Ref. [67]

transfer, it relies on the idea of seeking the ground state of a system subject to a constraint. This can be achieved by adding to the conventional KS Lagrangian an additional term that accounts for the constraining external potential, this reads as [40]:

$$\mathcal{L}_{CDFT}[\rho] = \underbrace{E_{HK}[\rho] + \int v_{\text{ext}}(\mathbf{r})\rho(\mathbf{r})d\mathbf{r} - \mu\left[\int \rho(\mathbf{r})d\mathbf{r} - N_e\right]}_{\text{same as regular KS-DFT}} + V_c\left[\int \omega_c(\mathbf{r})\rho(\mathbf{r})d\mathbf{r} - N_c\right] \tag{4.11}$$

where V_c is the Lagrange multiplier of the constraint, $\omega_c(\mathbf{r})$ acts as the weight function that defines the constraint, typically a population analysis based on a real-space [25] partitioning (such as Becke pop. analysis). N_c is the value of the constraint, and at self consistency it should satisfy the following tautology:

$$N_c = \int \omega_c(\mathbf{r})\rho(\mathbf{r})d\mathbf{r} \tag{4.12}$$

Having defined the constraint parameters, the energy of the system can be computed by solving the KS equation for the constrained system:

$$\left(-\frac{1}{2}\nabla^2 + \int \frac{\rho(\mathbf{r}')}{|\mathbf{r}-\mathbf{r}'|}d\mathbf{r}' + \upsilon_{xc}(\mathbf{r}) + V_c\omega_c(\mathbf{r})\right)\phi_i[V_c](\mathbf{r}) = \epsilon_i[V_c]\phi_i[V_c](\mathbf{r}) \tag{4.13}$$

where we have emphasized the functional dependence of the orbitals and orbital energies to the CDFT Lagrange multiplier. Clearly, the integral in (4.12) is only satisfied when an appropriate choice of V_c is employed. The term υ_{xc} is the exchange-correlation potential and ϕ_i are the KS-orbitals. Note that $\rho(\mathbf{r}) \equiv 2\sum_i |\phi_i(\mathbf{r})|^2$ for closed shell systems. Thus also the density is a functional of the CDFT Lagrange multiplier. To our knowledge, the CDFT algorithm can be found on NWChem [129], Q-Chem [73], CPMD [130], PSI [131], SIESTA [132], and ADF [133]. Computing the electronic coupling on a diabatic basis can be carried out similarly to (4.5–4.8) or using a CDFT-specific prescription [41].

An example is the long range charge transfer excited states of the zincbacteriochlorin-bacteriochlorin complex (ZnBC-BC), an important structure in photosynthetic process in bacteria, has been calculated on the basis of CDFT procedure [41, 128]. In Fig. 4.5, the excited states at different intersubsystem distances is depicted, where the last point of each curve represent the CT excitation energy of the linked complex. These energies are in good agreement with previous methodologies [134], and also demonstrates that by constraining CDFT ground state the excitations are more accurate than TDDFT energies (1.32–1.46 eV) [128].

Additionally, CDFT can generate states with partial charges [128], this is of particular importance, for example in metal-ligand CT processes, where the diabatic states can be generated by constraining the charge on the ligand and metal center.

Recently, the CDFT implementation of CPMD was tested against high-level wave function methods in the computation of electronic couplings for hole and excess electron transfer [37, 135]. CDFT was shown to be on average within 5.3 % of the benchmark calculations if 50 % HF exchange was introduced (the average deviation goes up to 38.7 % if HF exchange is not used).

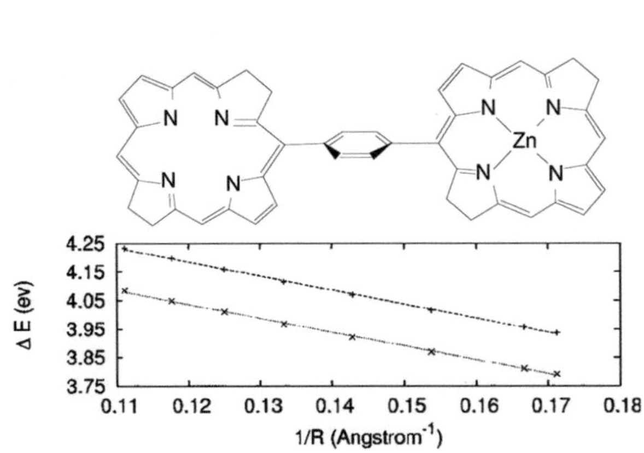

Fig. 4.5 Charge-transfer state energies of ZnBC-BC as compared to its ground-state energy at 5.84 Å separation. *Lower line* $Zn^{+}BC^{-}$; *upper line* $Zn^{-}BC^{+}$. Taken from Ref. [128]

4.2.3 Fragment Orbital DFT

The fragment orbital DFT or FODFT is a computationally low-cost method to calculate electronic couplings. This is because the wavefunctions of each diabatic state are approximated by the fronteer orbitals of the isolated donor/acceptor fragments [136–138]. The underlying approximations in FODFT are that (1) the interactions between donor and acceptor have not effect on the orbital shape, (2) the coupling component related to orbitals below the fronteer is neglected (e.g. frozen core). In FODFT, the wavefunctions can be described by a single determinant of $N-1$ spin-orbitals ϕ, where $N = N_A + N_D$ i.e. the sum of the number of electrons of the neutral donor and acceptor. These determinants are built from the KS orbitals of the noninteracting isolated donor and acceptor fragments.

$$\begin{aligned}\psi_a \approx \psi_a^{D^+A} &= \frac{1}{\sqrt{(N_D-1+N_A)!}} \det\left(\phi_D^1 \dots \phi_D^{N_D-1} \phi_A^1 \dots \phi_A^{N_A}\right) \\ \psi_b \approx \psi_b^{DA^+} &= \frac{1}{\sqrt{(N_D+N_A-1)!}} \det\left(\phi_D^1 \dots \phi_D^{N_D} \phi_A^1 \dots \phi_A^{N_A-1}\right)\end{aligned} \tag{4.14}$$

The Hamiltonian used to calculate the CT matrix elements is the KS-Hamiltonian. Namely,

$$\begin{aligned}H_a^{KS} &= \sum_{i=1}^{N_D+N_A-1} h_{a,i}^{KS} \\ H_b^{KS} &= \sum_{i=1}^{N_D+N_A-1} h_{b,i}^{KS}\end{aligned} \tag{4.15}$$

where $h_{a,i}^{KS}$ are the one-particle KS-Hamiltonians for either the “a” diabat or the “b” diabat. One feature of these Hamiltonians is that they are state dependent, thus, they are made of the combination of orbitals of donor and acceptor species at the given state. The transfer integral, or coupling between states, is calculated as:

$$\begin{aligned}H_{a,b} &= \langle \psi_a | H | \psi_b \rangle \\ &\approx \langle \psi_a^{D^+A} | H_a^{KS} | \psi_b^{DA^+} \rangle \\ &\approx \langle \phi_a^N | h_{a,i}^{KS} | \phi_b^N \rangle\end{aligned} \tag{4.16}$$

where N above is the fronteer orbital for D or A. Recently, Kubas et al. [37] have shown the differences of two FODFT flavors in the calculation of the hole transfer coupling for the HAB11 database. As we can see in Table 4.3, the implementation including $N_A + N_D$ orbitals in the KS Hamiltonian (indicated by $2N$ in the table) as done in ADF [136] is more accurate than the implementation using one of the Hamiltonians in (4.15) (which is indicated by $2N-1$ in the table).

Table 4.3 H_{DA} (meV) calculated with various FODFT approaches for HAB11 dimers at intermolecular separation of 3.5 Å

	FODFT(2N-1)	FODFT(2N)	ADF(2N)	FODFTB	REF
Ethylene	367.7	389.2	388.4	343.7	519.2
Acetylene	316.9	345.8	345.3	212.0	460.7
Cyclopropene	418.8	443.7	439.4	367.4	536.6
Cyclobutadiene	323.3	346.9	345.6	261.6	462.7
Cyclopentadiene	343.3	360.6	358.7	283.2	465.8
Furane	315.6	334.0	333.7	280.3	440.3
Pyrrole	328.7	347.8	347.7	286.2	456.3
Thiophene	341.2	357.8	356.1	264.8	449.0
Imidazole	310.7	328.9	328.2	277.5	411.6
Benzene	342.4	353.5	354.1	299.9	435.2
Phenol	190.5	211.3	279.5	231.4	375.0

Reproduced with permission from table XI of Kubas et al. [37]

FODFT has been successfully applied to models of CT in molecular semiconductors [139, 140] and also for modeling CT in biosystems. In the following, we provide applications of FODFT to biological CT: such as the determination of the hole rates on DNA hairpins linked by stilbenedicarboxamide, and the we also touch upon the electron transfer between two cofactors in the SO enzyme.

4.2.3.1 Hole Transfer Rates on DNA Hairpins

The absolute rates were determined by using Marcus theory, in (4.1), where the electronic coupling was calculated according to FODFT and the superexchange regime (see Sect. 4.2.1.1). Knowledge of the forward and backward rates enables one to determine the equilibrium constant $K = k_t/k_{-t}$ and the free energy change $-\Delta G = -k_B Tln(K)$. Comparable results with experiments [141] were obtained.

4.2.3.2 The Curious Case of Sulfite Oxidase

An interesting and elusive candidate for electron transfer studies is the Sulfite Oxidase protein [143]. For this protein, theory predicts an electron transfer rate between the cofactors (a heme and a molybdenum complex) about two orders of magnitude lower than what is measured experimentally [2]. To address this issue, Beratan et. al., using the Pathways model, suggested that the donor and the acceptor are joined together by a flexible tether [144]. As the tether allows the two cofactors to come sufficiently close to each other, electron transfer occurs at the rate shown by experiment. A recent simulation of this mechanism was carried out so that the protein was taken out of equilibrium and positioned in a new folded state featuring a much

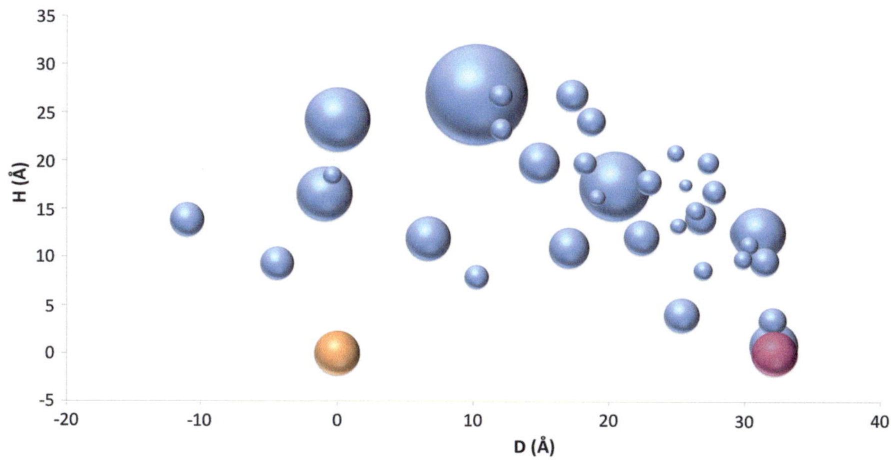

Fig. 4.6 Energy landscape for the hole transfer in Sulfite Oxidase: the *spheres* represent the position of the center of mass of each fragment with respect to the electron transfer vector coordinate(the distance between the center of mass of the HEME complex (*orange sphere*) and the MOCO complex (*pink sphere*)); the size of the sphere is $1/(x^2)$ in which x is the difference between the HOMO energy of the fragment and the Fe (that is HOMO of Fragment—HOMO of Fe). Fe and Mo were given a size of 1 for scale

decreased cofactor distance (about 10 Å). However, recent pulsed electron paramagnetic resonance measurements [143] indicated that the distance between the cofactors is unchanged on average from the one available in the crystal structure (32 Å). To approach this problem using FODFT, the crystal structure of the protein is obtained and only the chains of the protein between the two cofactors are considered. The chains of interest are broken into individual molecules and treated as separate bridges.

The FODFT computations that we present here will be part of a more in-depth study in a future publication [145]. Two ingredients are available from the simulations, the site energies, and the couplings between the sites. The energies of the hole transfer pathway for the electron transfer between the iron and the molybdenum is presented in Fig. 4.6. With the aid of Koopman's theorem, the HOMO energies computed with FODFT are taken here as a measure of the ionization potential of each site. The simulation was able to shed light on some very interesting aspects of the couplings and the energy landscapes. The landscape also shows the possibility of hopping stations-molecules that exist between the donor and acceptor and are close to them in energy. These two aspects of the landscape alone hint the possibility of a hole transfer occurring over the 32 Å. However, proteins are very complex structures with many variables such as size, dynamics and environment. Therefore, providing a quantitative analysis of the kinetic constant would require incorporating unbiased molecular dynamics and a more comprehensive structure to further characterize the role of these hopping stations.

4.2.4 Ultrafast Computations of the Electronic Couplings: The AOM Method

Recently, an ultrafast method to calculate electronic couplings was developed by Blumberger and coworkers [135]. The analytic overlap method or AOM is a useful method if CT simulations need to be coupled with molecular dynamics, like in proteins [146] or in organic semiconductors [147]. This quest requires hundreds or maybe thousands of H_{DA} and site energy calculations. AOM offers an interesting alternative for such simulations. As in FODFT, AOM assumes that CT is only mediated by two SOMO orbitals (fronteer orbitals, similarly to FODFT), which correspond to each fragment. Then, small Slater type orbital basis for the valence states is generated. Thus the overlap integral is evaluated as follows:

$$S_{DA} = \langle \Psi_D | \Psi_A \rangle = \langle \phi_D^N | \phi_A^N \rangle \approx$$
$$\approx \bar{S}_{DA} = \sum_{i \in D}^{atoms} \sum_{j \in A}^{atoms} c^*_{p\pi,i} c_{p\pi,j} \langle p_{\pi,i} | p_{\pi,j} \rangle$$

AOM further assumes contributions only from p-orbitals, particularly in organic compounds with π-conjugation, the p-orbital considered is that one perpendicular to the plane of π-conjugation.

Correlation of the overlap $\bar{S}_{DA}$ and the electronic coupling given by FODFT is shown in Fig. 4.7 for a set of dimers and their geometries. In this picture, a satisfactory linearity between these two parameters is witnessed, therefore reliable proportionality constants can be achieved. $H_{DA} = \bar{C}\bar{S}_{DA}$ is used in order to obtain the constant C that can be used to get couplings of similar compounds. The first version

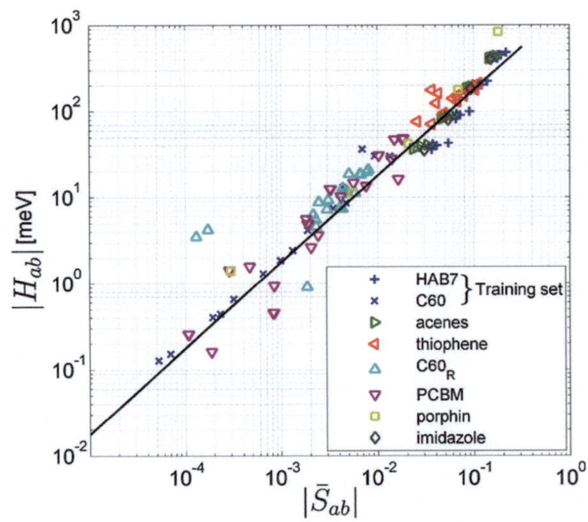

Fig. 4.7 Correlation between electronic coupling matrix element from sFODFT and overlap between SOMO orbitals of donor–acceptor fragments. Taken from Ref. [135]

of this method shows transferability for homo-dimers. However, transferability when non equivalent donor and acceptor systems are considered needs to be explored. Nevertheless, AOM's speed makes it a very valuable option.

4.2.5 Note on Orthogonality

When carrying out a large number of coupling calculations, one encounters all those low probability situations in which a method fails. In the case of FDE-ET, we probed a large number of so-called "difficult cases". Specifically, we faced two limitations of the FDE-ET method. If the diabats are orthogonal, or quasi orthogonal, numerical inaccuracies arise in the inversion of the transition overlap matrix in Eq. 4.7. This is not specific to FDE-ET, but is a problem shared by all those methods that assume the diabatic states to be nonorthogonal [30, 115, 148–150]. When they are orthogonal, some of the equations previously developed simply do not hold anymore. Yu et al. have applied equations similar to Eqs. (4.5–4.8) and obtained a picture of the behavior of the electronic couplings in the photosynthetic reaction center, see Fig. 4.8. If we concentrate on the left-side panels, we notice that in some cases the electronic coupling is proportional to the coupling, but in other cases (see lower left panel) the coupling seems to behave somewhat erratically as a function of the diabatic overlap. To understand this, let us consider two distinct limiting cases: (1) orthogonality by symmetry considerations, and (2) spatial separation of the orbitals. We found that the second case is the predominant, as the distance between donor–acceptor increases the diabatic overlap becomes increasingly small. In the asymptotic limit [44], there is a linear relationship between the coupling, the diabatic energy difference, and the diabatic overlap. If the overlap is small due to case (1), the asymptotic formula is not expected to hold. This explains the apparently contradictory results presented in Fig. 4.8.

Regarding FDE-ET, both cases can be circumvented computationally by performing a singular value decomposition of the overlap matrix and then invert only those values which are larger than a threshold (i.e. Penrose inversion). For DNA presented in Sect. 4.2.1.3, the default inversion threshold of 10^{-3} was appropriate in most cases [67]. However, three systems stood out: AG, GA and TT nucleobase pairs. All the systems above showed erratic behavior of the computed couplings for some specific donor–acceptor distances, specifically 4.0 Å for AG, 3.5 and 8.0 Å for GA and 9.0 Å for TT. We found that at those distances, the near singularity of the overlap matrix due to symmetry considerations (case 1 above) was the source of the erratic behavior. To circumvent these numerical issues, a threshold of 10^{-2} was adopted in these cases.

We thus conclude that although there is a formal relationship linking the diabatic overlap with the value of the coupling at large donor–acceptor distances [44], generally assuming linearity in the coupling versus overlap (as mentioned in the previous section) can lead to large errors in the magnitude of the computed couplings. In the future, inspired by a recent work by Evangelista et al. [131], a more stable FDE-ET

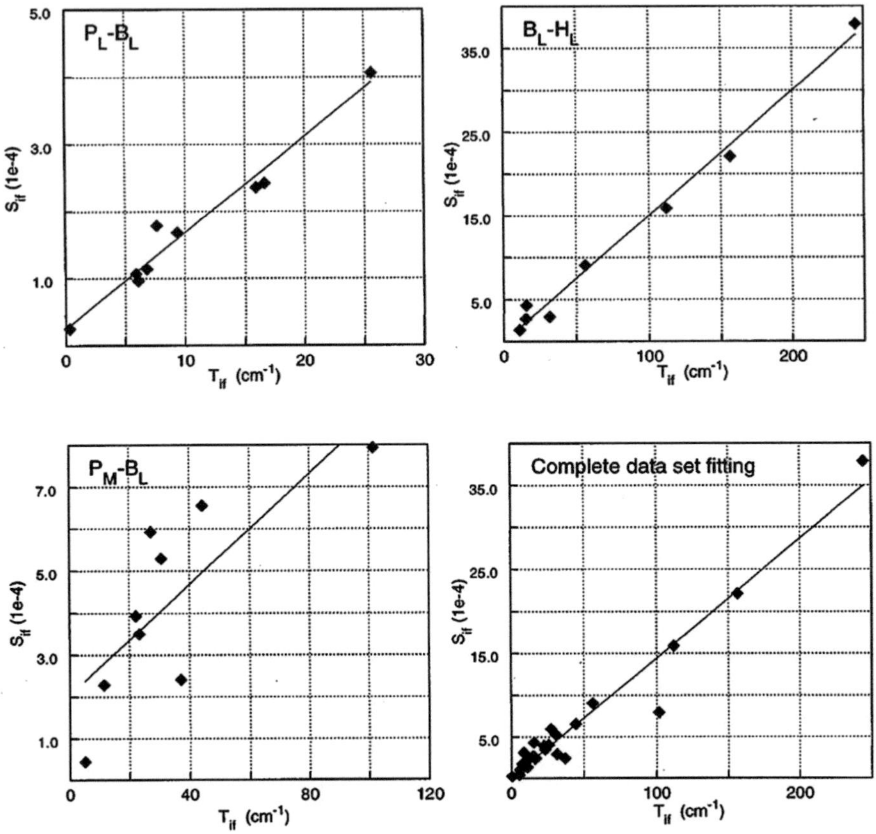

Fig. 4.8 Correlation between electronic coupling matrix element and overlap between diabatic states for electron transfer. Taken from Ref. [148]

algorithm that invokes an orthogonalization first, and then the computation of the couplings will be developed in our group.

4.2.6 A Fully Semiempirical Method: Pathways

Pathways [2, 50] is a semiempirical model which it is designed to reproduce electron transfer rates between cofactors in proteins [50, 151]. In essence, Pathways includes the contributions to the electronic tunneling from a stepwise path covering all non-bonded interactions, as well as the bonded ones at the nearest neighbor level. Namely:

$$|H_{DA}|^2 = A^2 \left(\prod_i \epsilon_i \right)^2 \tag{4.17}$$

where ϵ_i are the steps the charge need to make from donor to acceptor. For example, a hydrogen bond is one of such steps. The above product is maximized by searching all possible steps that contribute to the tunneling. The coupling is further split into three kinds of interactions:

$$|H_{DA}|^2 = A^2 \left(\prod_i \epsilon_{bond}(i) \right)^2 \left(\prod_j \epsilon_{space}(j) \right)^2 \left(\prod_k \epsilon_{H\text{-}bond}(k) \right)^2 \tag{4.18}$$

Pathways can yield reliable predictions of the electronic couplings, where the CT process in proteins are mediated by the interactions of a single or multiple configurations that the protein can adopt [50]. Pathways has been successfully applied to a number of CT processes in protein environment. For instance, the electron transfer between the proteins cytochrome c2 (cytc2) and the photosynthetic reaction center (RC) [152] in order to determine the protein structural dependence of this CT reaction, also, to look at the impact of structural and conformational variations on the electronic coupling between the proteins methylamine dehydrogenase and amicyanin from Paracoccus denitrificans [153].

4.3 High-Accuracy Electronic Couplings

This section is devoted to describing those methods which are able to predict the electronic couplings accurately given a certain definition of the corresponding diabatic states. These methods start with a mathematical definition of diabatic states (usually a definition that involves localization of the electronic structure) such that the resulting states resemble the donor and acceptor states in the electron transfer reaction. Once this is achieved, an adiabatic-to-diabatic transformation matrix is generated which can be applied to the adiabatic Hamiltonian to result in the diabatic Hamiltonian featuring the sought electronic couplings in the off-diagonal elements. Usually, an accurate wave function based level of theory is used for computing the adiabatic states and Hamiltonian [29]. Examples of such techniques are, the Generalized Mulliken–Hush method developed by Newton and Cave [29, 30, 154], Boys and Edmiston-Ruedenberg localizations of Subotnik et al. [38, 154, 155], and fragment charge difference proposed by Voityuk and Rösch [42, 156]. Their utility lies on the possibility of a very accurate computation of the corresponding adiabatic states, as was done for the hole transfer on π-stack DNA nucleobases at a CASPT2 and CASSCF level of theory accomplished by Voityuk et al. [35]. That computation has served as the benchmark reference for many recently developed methodologies [49, 99, 137, 157].

Taking this as a motivation, let us briefly introduce each of the above methods, followed by some examples in which these methods were employed.

4.3.1 GMH Method

In the two-state model, the charge localized diabatic states are related with the adiabatic states by the formula:

$$E_{2,1} = \frac{1}{2}\left(E_D + E_A \pm \sqrt{(E_D - E_A)^2 + 4|H_{DA}|^2}\right) \tag{4.19}$$

where $E_{D/A}$ are the energies of the donor and acceptor states respectively, E_1 and E_2 are the energies of the adiabatic states, this means the energy of the ground state (E_1) and the first excited state (E_2). E_1 and E_2 can be obtained with any quantum chemistry method, however when highly accurate wavefunctions methods are employed also the resulting couplings will be of high quality. We now distinguish two cases: a symmetric case, for instance homo-dimers, and the general asymmetric case. In the symmetric case, we have that $E_D = E_A$, and thus the electronic coupling does not depend on the diabatization procedure. Namely,

$$2|H_{DA}| = \Delta E_{12} \tag{4.20}$$

ΔE_{12} is the difference on energy between the adiabatic ground state E_1 and the first adiabatic excited state E_2.

For asymmetric cases the GMH method prescribes that the proper diabatic states are those that diagonalize the adiabatic dipole moment matrix. In the two-state problem this is calculated as follows:

$$|H_{DA}| = \frac{|\mu_{12}|\Delta E_{12}}{\sqrt{(\mu_{11} - \mu_{22})^2 + 4\mu_{12}^2}}, \tag{4.21}$$

where $\mu_{ij} = \langle \Psi_i | \mu_{ET} | \Psi_j \rangle$, with μ_{ET} being the dipole moment in the direction of the electron transfer. The power of GMH lies on the way one calculates the adibatic states. As we have seen through all sections is that the authors benchmark their own method by calculating proper diabatic couplings by using ab initio methods as multireference CI [37] and CASPT2 [35].

As an example of this method, let us discuss the very first example given by Cave and Newton on their paper [34]. The system $Zn_2H_2O^+$, the transfer of a hole is done over the Zn atoms. However, the water molecule, which is located at a fixed distance opposite to the Zn distance, causes an energy splitting of the Zn orbitals. Thus the electronic coupling is determined for the following diabatic states:

$$Zn(^1S) + Zn^+(^2S) \rightarrow Zn^+(^2S) + Zn(^1S)\ s \rightarrow s' \tag{4.22}$$

$$Zn(^1S) + Zn^+(^2S) \rightarrow Zn^+(^2S) + Zn(^1S)\ p \rightarrow p' \tag{4.23}$$

$$Zn(^1S) + Zn^+(^2S) \rightarrow Zn^+(^2S) + Zn(^1S)\ s \rightarrow p' \tag{4.24}$$

$$Zn(^1S) + Zn^+(^2S) \rightarrow Zn^+(^2S) + Zn(^1S)\ p \rightarrow s' \tag{4.25}$$

where the transition are between the orbitals of the diabatic states (prime correspond to the acceptor state) (s-s'), (p-p'), (s-p') and (p-s'). CASSCF wavefunction method was used in all calculations. In Table 4.4 we collect the values for the different couplings and βs. There the analysis of the distance dependence of the coupling is carried out for several R_{OZn} distances. Note that for an infinite R_{OZn} distance, the couplings for (s-p') and (p-s') are equal.

We refer the reader to other publications which have evaluated the GMH method in detail in regards to its suitability in modeling two-state as well as multi-state problems for both excitation energy transfer and electron transfer processes [39, 156, 158].

Table 4.4 (a) Electronic coupling elements versus distance (r_{ZnZn}) for $Zn_2H_2O^+$ with $r_{ZnO} =$ 2.05 Å. (b) Electronic coupling elements versus distance (r_{ZnZn}) for $Zn_2H_2O^+$ with $r_{ZnO} = 3.05$ Å. Results are in milihartree, the β values were calculated on the range of 5–9 Å. Taken from Ref. [29]

r_{ZnZn} (Å)	$H_{ss'}$	$H_{pp'}$	$H_{sp'}$	$H_{ps'}$
(a)				
4.0	28.3	23.6	50.4	42.7
5.0	10.5	13.0	51.7	22.3
6.0	3.73	7.55	41.1	10.1
7.0	1.09	4.23	21.8	4.08
8.0	0.340	2.57	13.8	1.62
9.0	0.0958	1.44	7.36	0.611
β	2.28	1.11	0.81	1.71
(b)				
4.0	29.7	34.4	59.3	41.7
5.0	7.95	14.7	38.5	22.1
6.0	2.34	7.83	19.1	9.44
7.0	0.698	4.25	9.56	3.84
8.0	0.203	2.27	4.78	1.51
9.0	0.0558	1.16	2.32	0.574
β	2.49	1.32	1.32	1.74

4.3.2 Other Adiabatic-to-Diabatic Transformation Methods

Inspired by GMH, the electronic coupling can generally be obtained by rotating the corresponding adiabtic states into a set of diabatic states. Thus, each diabatic state can be expressed as a linear combination of rotated adiabatic states as [154]:

$$|\Xi\rangle = \sum_{j=1}^{N_{states}} |\Phi_j\rangle U_{ji} \tag{4.26}$$

Under specific assumptions of the nature of the system-bath interaction (the following is valid for the condensed phase), the coupling can be estimated by constructing diabatic states based on Boys, Edminton–Ruedenberg (ER) or von Niessen–Edminton–Ruedenberg (VNER) localizations. In Boys diabatization, the bath exerts a linear electrostatic potential on the system, thus the rotation matrix can be found by minimizing the following localizing function [24, 39]:

$$f_{Boys}(U) = f_{Boys}(\Xi) = \sum_{i,j=1}^{N_{states}} = |\langle \Xi_i|\mu_{ET}|\Xi_i\rangle - \langle \Xi_j|\mu_{ET}|\Xi_j\rangle|^2. \tag{4.27}$$

Boys localization was shown to be equivalent to GMH for CT reactions [39].

ER diabatization, dictates that the bath exerts an electrostatic potential that responds linearly to the field generated by the molecular system (sum of donor and acceptor) system:

$$f_{ER}(U) = f_{ER}(\Xi) = \sum_{i=1}^{N_{states}} \int d\mathbf{r}_1 \int d\mathbf{r}_2 \frac{\langle \Xi_i|\hat{\rho}(\mathbf{r}_2)|\Xi_i\rangle - \langle \Xi_j|\hat{\rho}(\mathbf{r}_1)|\Xi_j\rangle}{|\mathbf{r}_1 - \mathbf{r}_2|}. \tag{4.28}$$

In VNER diabatization, the bath exerts an electrostatic potential that responds linearly to the field of the total system, but the interaction potential is a Dirac delta function:

$$f_{VNER}(U) = f_{VNER}(\Xi) = -\sum_{i=1}^{N_{states}} \int d\mathbf{r}(\langle \Xi_i|\hat{\rho}^2(\mathbf{r})|\Xi_i\rangle - \langle \Xi_i|\hat{\rho}(\mathbf{r})|\Xi_i\rangle^2). \tag{4.29}$$

Just like GMH, once diabats are generated, the electronic coupling readily arises from the off-diagonal element of the electronic Hamiltonian and is equal to:

$$H_{DA} = \langle \Xi_D|H^{el}|\Xi_A\rangle \tag{4.30}$$

For a series of bridge mediated excitation energy transfer experiments, where the donor is benzaldehyde and the acceptor naphthalene [159, 160], the transfer rates and couplings were calculated using Boys diabatization. In Fig. 4.9 it is shown how well

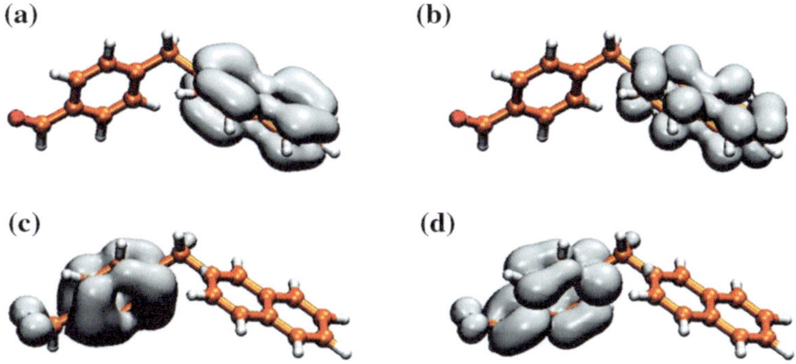

Fig. 4.9 Attachment/detachment plots for the occupiedvirtual separated Boys localized diabatic excited triplet states near the avoided crossing. The molecule here is donor–CH_2–acceptor. Taken from Ref. [154]. **a** Detachment density for the first localized diabet. **b** Attachment density for the first localized diabet. **c** Detachment density for the second localized diabet. **d** Attachment density for the second localized diabet

the orbitals are localized on either donor or acceptor edges in the various diabatic states. Although this computation does not concern a CT process, we want to stress the ability of this localization procedures in generating true diabatic states.

4.3.3 Fragment Charge Difference

Similarly to the GMH method, the fragment charge difference (FCD) method yields a donor to acceptor coupling [42]:

$$|H_{DA}| = \frac{|\Delta q_{12}|\Delta E_{12}}{\sqrt{(\Delta q_{11} - \Delta q_{22})^2 + 4\Delta q_{12}^2}} \tag{4.31}$$

where Δq_1 and Δq_2 are the donor–acceptor charges differences in the respective adiabatic states ψ_1 and ψ_2. Δq_{12} is the off diagonal term and is defined in a general form as $\Delta q_{ij} = q_{ij}(D) - q_{ij}(A)$, i.e. the difference of the populations of the transition charges.

Finally, when donor and acceptor are in resonance, i.e. when $E_D = E_A$ or $\Delta q = 0$, $H_{DA} = \frac{1}{2}\left(E_1 - E_2\right)$.

FCD method and its simplfied from (SFCD) are compared against GMH in the calculation of H_{DA} for two Watson–Crick pairs GC and AT. In Table 4.5, the couplings calculated from FCD are in good agreement with GMH, the SCFD is also quite reasonable. Because the energy gap between donor and acceptor are large, the charge is completely localized on purines (lowest IP). However, if an electric field F (water molecule for instance) is tuned on near the pairs, the energy gap is then

Table 4.5 Hole coupling matrix element H_{DA} of nucleobases within a Watson–Crick pairs

				GMH	SFCD	FCD
	Basis set	$E_2 - E_1$	$E_D - E_A$	H_{DA}	H_{DA}	H_{DA}
GC	6-31G*	2.163	2.159	0.0569	0.0663	0.0547
	6–311++G**	2.092	2.086	0.0679	0.0760	0.0705
AT	6-31G*	1.505	1.502	0.0421	0.0524	0.0363
	6-311++G**	1.462	1.459	0.0474	0.0528	0.0425

Energies in eV, dipole moment matrix elements in Debye, charges in a.u. Taken from Ref. [42]

reduced and the coupling strength is enhanced. Overall, FCD is another good alternative to compute accurate couplings. However, the computation of the charges and the transition charges is dependent on the specific population analysis chosen. To our knowledge, only the Mulliken population analysis was used so far (i.e. the transition charges are evaluated on the basis of the MO coefficients over the atomic orbital basis set).

4.4 Practical Aspects: A Protocol for Running FDE-ET Calculations

In order to obtain the electronic coupling for a CT reaction using FDE-ET, three different single point (SP) calculations have to be performed. FDE-ET is available in ADF [51]. In Figs. 4.10, 4.11 and 4.12, the input files corresponding to the FDE-ET methodology are described. First, a single point calculation for each isolated fragment present in the system is carried out. This gives the initial density and energy of each subsystem without any interaction between them. It is important to save the check point files (TAPE files in ADF), because they contain all fragment information needed in the subsequent calculations. Following all SP jobs for each isolated fragment, an FDE calculation is performed by taking into the account the whole supramolecular structure. So that, we create a diabatic state for each of the present subsystems. This is done by placing a charge different from neutrality in each subsystem, see Fig. 4.11. In this manner, two different directories are made: one in which an FDE calculation is carried out with subsystem 1 positively charged and one where subsystem 2 has the positive charge. In both cases, the SCF converges on the basis of subsystem DFT, thus, a series of three freeze-and-thaw procedure are done for each subsystem in each diabat.

Once both isolated and embedded densities are obtained from the FDE calculations, an electrontransfer job is run whose purpose is to compute (4.5–4.8). As in the FDE calculation, the information about the fragment is of paramount importance in this electrontransfer job. In Fig. 4.12, the input file that calculates the diabatic energies and the electronic coupling between them is showed, in pink we can see that the check point files (t21.emb.rho* in the figure) corresponding to the embedded

```
$ADFBIN/adf << eor      Bash options
Title          : Fragment no.       1; (isolated)    Job Title

EPRINT       prints the variables that you want in the outputfile
 SFO NOEIG NOOVL NOORBPOP
END

NOPRINT BAS FUNCTIONS

UNITS      units to work with
 length angstrom
angle degree
END

SYMMETRY NOSYM    no groups of symmetry

XC           XC functional
 GGA PBE
END

GEOMETRY    Type of calculation: Single Point
 SP
END

SCF            SCF cycles
 iterations 90
END

BASIS          basis sets to be use
 Type TZP
Core None
END

ATOMS        Cartesian coordinates
    1 C     0.000000  0.000000  0.000000
    2 C     1.332000  0.000000  0.000000
    3 H    -0.574301  0.000000 -0.928785
    4 H    -0.574301  0.000000  0.928785
    5 H     1.906301  0.000000  0.928785
    6 H     1.906301  0.000000 -0.928785
END

END INPUT
eor
                              Bash options
mv TAPE21 t21.iso.rho1
```

Fig. 4.10 Single point (SP) calculation input file for an isolated ethylene

```
$ADFBIN/adf << eor                    Bash options
Title          : Fragment no.     1; (polarized)          Job Title

EPRINT
 SFO NOEIG NOOVL NOORBPOP
END

NOPRINT BAS FUNCTIONS

UNITS                   units to work with
 length angstrom
 angle degree
END

SYMMETRY NOSYM          no groups of symmetry

XC       XC functional
 GGA PBE
END

GEOMETRY Type of calculation: Single Point
 SP
END

SCF     SCF cycles
 iterations 90
END

CHARGE 1 1 Charge and #of unpair electrons
UNRESTRICTED

FRAGMENTS        Subsystems involve in the calculation
 rho1 t21.iso.rho1
 rho2 t21.iso.rho2 type=fde Frozen subsystem
END

ATOMS     Cartesian coordinates
   1 C     0.000000    0.000000    0.000000   f=rho1
   2 C     1.332000    0.000000    0.000000   f=rho1
   3 H    -0.574301    0.000000   -0.928785   f=rho1
   4 H    -0.574301    0.000000    0.928785   f=rho1
   5 H     1.906301    0.000000    0.928785   f=rho1
   6 H     1.906301    0.000000   -0.928785   f=rho1
   7 C     0.000000    3.50000     0.0000000  f=rho2
   8 C     1.332000    3.50000     0.0000000  f=rho2
   9 H    -0.574301    3.500000    0.928785   f=rho2
  10 H    -0.574301    3.50000    -0.928785   f=rho2
  11 H     1.906301    3.50000    -0.928785   f=rho2
  12 H     1.906301    3.50000     0.928785   f=rho2
END

ALLOW PARTIALSUPERFRAGS

FDE       FDE options
 PW91k
XCNADD PBE
END


END INPUT
eor
                          Bash options
mv TAPE21 t21.emb.rho1
```

Fig. 4.11 Single point FDE calculation input file for ethylene dimer. Both fragments rho1 and rho2 come from two SP calculations for each isolated fragment

```
#!/bin/bash                                              Bash options

fetch="#fetch frag??.t21"
fetch2="scp ../FDE/diabatA/t21.iso.rho? ."
cp ../FDE/diabatA/t21.emb.rho1 fragA1.t21                Densities and energies
cp ../FDE/diabatA/t21.emb.rho2 fragA2.t21                  of both Diabats
cp ../FDE/diabatB/t21.emb.rho1 fragB1.t21
cp ../FDE/diabatB/t21.emb.rho2 fragB2.t21

$fetch
$fetch2
$ADFBIN/adf << eor
Title electrontransfer run  Job Title

EPRINT
 SFO NOEIG NOOVL NOORBPOP
END

NOPRINT BAS FUNCTIONS

UNITS                        units to work with
 length angstrom
 angle degree
END

SYMMETRY NOSYM               no groups of symmetry

XC
 GGA PBE  XC functional
END

GEOMETRY   Type of calculation: Single Point
 SP
END

SCF
 iterations 0  SCF cycles
END

INTEGRATION 6.0 6.0  Numerical integration

CHARGE 1 1
UNRESTRICTED   Charge and #of unpair electrons

FRAGMENTS     Subsystems involve in the calculation
 rho1 t21.iso.rho1
 rho2 t21.iso.rho2
END

ATOMS        Cartesian coordinates
  1 C     0.000000 0.000000  0.000000  f=rho1
  2 C     1.332000 0.000000  0.000000  f=rho1
  3 H    -0.574301 0.000000 -0.928785  f=rho1
  4 H    -0.574301 0.000000  0.928785  f=rho1
  5 H     1.906301 0.000000  0.928785  f=rho1
  6 H     1.906301 0.000000 -0.928785  f=rho1
  7 C     0.000000 3.500000  0.000000  f=rho2
  8 C     1.332000 3.500000  0.000000  f=rho2
  9 H    -0.574301 3.500000  0.928785  f=rho2
 10 H    -0.574301 3.500000 -0.928785  f=rho2
 11 H     1.906301 3.500000 -0.928785  f=rho2
 12 H     1.906301 3.500000  0.928785  f=rho2
END

ELECTRONTRANSFER    ET options
 numfrag 2
 InvThr 1.0e-3
END

ZLMFIT    Density fitting
END

END INPUT
eor
```

Fig. 4.12 Single point ET calculation input file for ethylene dimer. In *pink* there some bash-shell options in order to copy the information for each diabatic state calculated before with FDE fragments in each diabat is copied directly to the ET directory, where the electron-transfer calculation is done. These files are renamed as fragA*.t21 for those ones from the diabat A (positive charge on the donor fragment) and fragB*.t21 for those that come from diabat B (positive charge on the acceptor fragment). It is worth mentioning that special care has to be taken in the management of the file names. As it is illustrated in Fig. 4.12, the fragments are numerated as 1 and 2, that means that the charge, departs from fragA1 while fragA2 is neutral and arrive to fragB2 while fragB1 becomes neutral. This is very important when the system is comprised of more than two fragments, and the charge is moving throughout all of them.

4.5 Conclusions and Future Directions

To conclude, we have presented our (fairly subjective) view of what tools are available nowadays to compute electronic couplings for charge transfer processes. We have surveyed in detail the FDE-ET method simply because we are among the developers of this method. Other methods based on DFT, and those that are best suited for being coupled with wavefunction based methods have also been discussed. The discussion also touches on the strengths and limitations of the various methods.

When discussing the practical aspect of a coupling calculation, one must expose completely the methodology. We have done so for the FDE-ET method, and provided the reader with a step-by-step protocol on how to run such computations. This is important also for outsiders (such as experimentalists) as they can appreciate the kind of effort the theoreticians have to put in computing quantities relevant for the interpretation of the experiments.

We apologize in advance to those authors who have developed all those methods that we have omitted from this presentation. Admittedly, we provide here a subjective view of the field.

Acknowledgments This work was funded by a grant from the National Science Foundation, Grant CBET-1438493. We also acknowledge Dr. Eric Klein for bringing to our attention the problematics related to the electron transfer dynamics in the Sulfite Oxidase.

References

1. Gray HB, Winkler JR (1996) Annu Rev Biochem 65:537–561
2. Kawatsu T, Beratan DN (2006) Chem Phys 326:259–269
3. Larsson S (1983) J Chem Soc Faraday Trans 2(79):1375–1388
4. Farid RS, Moser CC, Dutton PL (1993) Curr Opin Struct Biol 3:225–233
5. Winkler JR, Gray HB, Prytkova TR, Kurnikov IV, Beratan DN (2005) Bioelectronics. Wiley-VCH Verlag GmbH & Co. KGaA, pp 15–33
6. Trumpower BL, Gennis RB (1994) Annu Rev Biochem 63:675–716
7. Bixon M, Giese B, Wessely S, Langenbacher T, Michel-Beyerle ME, Jortner J (1999) Proc Natl Acad Sci USA 96:11713–11716

8. Giese B (2004) In: Schuster G (ed) Longe-Range charge transfer in DNA I. Topics in current chemistry. vol 236. Springer, Berlin, pp 27–44
9. Giese B (2002) Annu Rev Biochem 71:51–70
10. Stowell MHB, McPhillips TM, Rees DC, Soltis SM, Abresch E, Feher G (1997) Science 276:812–816
11. Balabin IA, Onuchic JN (2000) Science 290:114–117
12. Troisi A (2011) Chem Soc Rev 40:2347–2358
13. McMahon DP, Troisi A (2011) Phys Chem Chem Phys 13:10241–10248
14. Marcus RA (1956) J Chem Phys 24:966–978
15. Marcus RA, Sutin N (1985) Biochim Biophys Acta 811:265–322
16. Landau LD (1932) Phys Soviet Union 2:46–51
17. Zener C (1932) Proc R Soc A 137:696–702
18. Condon E (1928) Phys Rev 32:858–872
19. Franck J, Dymond EG (1926) Trans Faraday Soc 21:536–542
20. Ladanyi BM, Skaf MS (1993) Annu Rev Phys Chem 44:335–368
21. Ingram JA, Moog RS, Ito N, Biswas R, Maroncelli M (2003) J Phys Chem B 107:5926–5932
22. Maroncelli M (1993) J Mol Liq 57:1
23. Nitzan A (2006) Chemical dynamics in condensed phases. Oxford University Press, Oxford
24. Subotnik JE, Cave RJ, Steele RP, Shenvi N (2009) J Chem Phys 130:234102
25. Van Voorhis T, Kowalczyk T, Kaduk B, Wang L-P, Cheng C-L, Wu Q (2010) Annu Rev Phys Chem 61:149–170
26. London F (1932) Z Phys 74:143
27. Mead CA, Truhlar DG (1982) J Chem Phys 77:6090–6098
28. Pavanello M, Neugebauer J (2011) J Chem Phys 135:134113
29. Cave RJ, Newton MD (1997) J Chem Phys 106:9213–9226
30. Newton MD (1991) Chem Rev 91:767–792
31. Warshel A, Weiss RM (1980) J Am Chem Soc 102:6218–6226
32. Mulliken RS (1952) J Am Chem Soc 64:811
33. Hush N (1968) Electrochim Acta 13:1005–1023
34. Cave RJ, Newton MD (1996) Chem Phys Lett 249:15–19
35. Blancafort L, Voityuk AA (2006) J Phys Chem A 110:6426–6432
36. Voityuk AA (2007) J Chem Phys 128:115101
37. Adam K, Hoffmann F, Heck A, Oberhofer H, Elstner M, Blumberger J (2014) J Chem Phys 140:104105
38. Fatehi S, Alguire E, Subotnik JE (2013) J Chem Phys 139
39. Subotnik JE, Yeganeh S, Cave RJ, Ratner MA (2008) J Chem Phys 129
40. Kaduk B, Kowalczyk T, Van Voorhis T (2012) Chem Rev 112:321–370
41. Wu Q, Van Voorhis T (2006) J Chem Phys 125:164105
42. Voityuk AA, Rösch N (2002) J Chem Phys 117:5607
43. Hsu C-P, You Z-Q, Chen H-C (2008) J Phys Chem C 112:1204–1212
44. Migliore A (2011) J Chem Theory Comput 7:1712–1725
45. Migliore A, Corni S, Di Felice R, Molinari E (2006) J Chem Phys 124
46. Migliore A, Sit PH-L, Klein ML (2009) J Chem Theory Comput 5:307–323
47. Cembran A, Song L, Mo Y, Gao J (2009) J Chem Theory Comput 5:2702–2716
48. Pavanello M, Neugebauer J (2011) J Chem Phys 135:234103
49. Pavanello M, Van Voorhis T, Visscher L, Neugebauer J (2013) J Chem Phys 138:054101
50. Beratan D, Betts J, Onuchic J (1991) Science 252:1285–1288
51. Amsterdam Density Functional program. Theoretical chemistry. Vrije Universiteit, Amsterdam. http://www.scm.com
52. Jacob CR, Neugebauer J (2014) WIREs: Comput Mol Sci 4:325–362
53. Wesolowski TA, Warshel A (1993) J Chem Phys 97:8050
54. Wesolowski TA (2006) In: Computational chemistry: reviews of current trends. vol 10. World Scientific, Singapore, pp 1–82
55. Neugebauer J, Louwerse MJ, Baerends EJ, Wesolowski TA (2005) J Chem Phys 122:094115

56. Neugebauer J (2010) Phys Rep 489:1–87
57. Humbert-Droz M, Zhou X, Shedge S, Wesolowski TA (2013) Theor Chem Acc 133
58. Jacob CR, Visscher L (2006) J Chem Phys 125:194104
59. Bulo RE, Jacob CR, Visscher L (2008) J Phys Chem A 112:2640–2647
60. Neugebauer J, Louwerse MJ, Belanzoni P, Wesolowski TA, Baerends EJ (2005) J Chem Phys 123:114101
61. Kevorkyants R, Wang X, Close DM, Pavanello M (2013) J Phys Chem B 117:13967–13974
62. Wesolowski TA (1999) Chem Phys Lett 311:87–92
63. Neugebauer J, Curutchet C, Munioz-Losa A, Mennucci B (2010) J Chem Theory Comput 6:1843–1851
64. Casida ME, Wesolowski TA (2004) Int J Quantum Chem 96:577–588
65. Pavanello M (2013) J Chem Phys 138:204118
66. García-Lastra JM, Wesolowski TA, Barriuso MT, Aramburu JA, Moreno M (2006) J Phys: Condens Matter 18:1519–1534
67. Ramos P, Pavanello M (2014) J Chem Theory Comput 10:2546–2556
68. Solovyeva A, Pavanello M, Neugebauer J (2014) J Chem Phys 140:164103
69. Jacob CR, Neugebauer J, Visscher L (2008) J Comput Chem 29:1011–1018
70. Helgaker T, Jensen HJA, Jørgensen P, Olsen J, Ruud K, Ågren H, Auer AA, Bak KL, Bakken V, Christiansen O, Coriani S, Dahle P, Dalskov EK, Enevoldsen T, Fernandez B, Hättig C, Hald K, Halkier A, Heiberg, H, Hettema H, Jonsson D, Kirpekar S, Kobayashi R, Koch H, Mikkelsen KV, Norman P, Packer MJ, Pedersen TB, Ruden TA, Sanchez A, Saue T, Sauer SPA, Schimmelpfennig B, Sylvester-Hvid KO, Taylor PR, Vahtras O (2001) DALTON, a molecular electronic structure program, Release 1.2. 2001
71. Aidas, K, Angeli C, Bak KL, Bakken V, Bast R, Boman L, Christiansen O, Cimiraglia R, Coriani S, Dahle P, Dalskov EK, Ekstrm U, Enevoldsen T, Eriksen JJ, Ettenhuber P, Fernndez B, Ferrighi L, Fliegl H, Frediani L, Hald K, Halkier A, Httig C, Heiberg H, Helgaker T, Hennum AC, Hettema H, Hjertens E, Hst S, Hyvik I-M, Iozzi MF, Jansk B, Jensen HJA, Jonsson D, Jrgensen P, Kauczor J, Kirpekar S, Kjrgaard T, Klopper W, Knecht S, Kobayashi R, Koch H, Kongsted J, Krapp A, Kristensen K, Ligabue A, Lutns OB, Melo JI, Mikkelsen KV, Myhre RH, Neiss C, Nielsen CB, Norman P, Olsen J, Olsen JMH, Osted A, Packer MJ, Pawlowski F, Pedersen TB, Provasi PF, Reine S, Rinkevicius Z, Ruden TA, Ruud K, Rybkin VV, Saek P, Samson CCM, de Mers AS, Saue T, Sauer SPA, Schimmelpfennig B, Sneskov K, Steindal AH, Sylvester-Hvid KO, Taylor PR, Teale AM, Tellgren EI, Tew DP, Thorvaldsen AJ, Thgersen L, Vahtras O, Watson MA, Wilson DJD, Ziolkowski M, Gren H (2014) WIREs: Comput Mol Sci 4:269–284
72. Goodpaster JD, Ananth N, Manby FR, Miller III TF (2010) J Chem Phys 133:084103
73. Shao Y, Fusti-Molnar L, Jung Y, Kussmann J, Ochsenfeld C, Brown ST, Gilbert ATB, Slipchenko LV, Levchenko SV, O'Neill DP, DiStasio Jr RA, Lochan RC, Wang T, Beran GJO, Besley NA, Herbert MJ, Lin CY, Voorhis TV, Chien SH, Sodt A, Steele RP, Rassolov VA, Maslen PE, Korambath PP, Adamson RD, Austin B, Baker J, Byrd EFC, Dachsel H, Doerksen RJ, Dreuw A, Dunietz BD, Dutoi AD, Furlani TR, Gwaltney SR, Heyden A, Hirata S, Hsu C-P, Kedziora G, Khalliulin RZ, Klunzinger P, Lee AM, Lee MS, Liang W, Lotan I, Nair N, Peters B, Proynov EI, Pieniazek PA, Rhee YM, Ritchie J, Rosta E, Sherrill CD, Simmonett AC, Subotnik JE, Woodcock III HL, Zhang W, Bell AT, Chakraborty AK, Chipman DM, Keil FJ, Warshel A, Hehre WJ, Schaefer III HF, Kong J, Krylov AI, Gill PMW, Head-Gordon M (2006) Phys Chem Chem Phys 8:3172–3191. http://www.q-chem.com
74. Ahlrichs R et al. http://www.cosmologic.de/turbomole.html
75. Ahlrichs R, Bär M, Häser M, Horn H, Kölmel C (1989) Chem Phys Lett 162:165–169
76. Laricchia S, Fabiano E, Sala FD (2010) J Chem Phys 133:164111
77. CP2k: A General Program to Perform Molecular Dynamics Simulations. http://www.cp2k.org
78. Iannuzzi M, Kirchner B, Hutter J (2006) Chem Phys Lett 421:16–20
79. Clark SJ, Segall MD, Pickard CJ, Hasnip PJ, Probert MIJ, Refson K, Payne MC (2005) Z Kristallogr 220:567570

80. Lahav D, Klner T (2007) J Phys: Cond Mat 19:226001
81. Genova A, Krishtal A, Ceresoli D, Pavanello M (2013) Frozen Density Embedding Project of Quantum Espresso. http://qe-forge.org/gf/project/fde
82. Giannozzi P, Baroni S, Bonini N, Calandra M, Car R, Cavazzoni C, Ceresoli D, Chiarotti GL, Cococcioni M, Dabo I, Dal Corso A, de Gironcoli S, Fabris S, Fratesi G, Gebauer R, Gerstmann U, Gougoussis C, Kokalj A, Lazzeri M, Martin-Samos L, Marzari N, Mauri F, Mazzarello R, Paolini S, Pasquarello A, Paulatto L, Sbraccia C, Scandolo S, Sclauzero G, Seitsonen AP, Smogunov A, Umari P, Wentzcovitch RM (2009) J Phys: Cond Mat 21:395502
83. Genova A, Ceresoli D, Pavanello M (2014) J Chem Phys 141:174101
84. Gonze X, Amadon B, Anglade P-M, Beuken J-M, Bottin F, Boulanger P, Bruneval F, Caliste D, Caracas R, Ct M, Deutsch T, Genovese L, Ghosez P, Giantomassi M, Goedecker S, Hamann D, Hermet P, Jollet F, Jomard G, Leroux S, Mancini M, Mazevet S, Oliveira M, Onida G, Pouillon Y, Rangel T, Rignanese G-M, Sangalli D, Shaltaf R, Torrent M, Verstraete M, Zerah G, Zwanziger J (2009) Comput Phys Commun 180:2582–2615
85. Govind N, Wang YA, da Silva AJR, Carter EA (1998) Chem Phys Lett 295:129–134
86. Senatore G, Subbaswamy KR (1986) Phys Rev B 34:5754–5757
87. Cortona P (1991) Phys Rev B 44:8454
88. Kolos W, Radzio E (1978) Int J Quantum Chem 13:627–634
89. Gordon RG, Kim YS (1972) J Chem Phys 56:3122
90. Götz A, Beyhan S, Visscher L (2009) J Chem Theory Comput 5:3161–3174
91. Wesolowski TA, Chermette H, Weber J (1996) J Chem Phys 105:9182
92. Ludeña EV, Karasiev VV, Nieto P (2003) Theor Chem Acc 110:395–402
93. Ludeñ EV, Karasiev VV (2002) Reviews of modern quantum chemistry, pp 612–665. Chapter 22
94. Fux S, Kiewisch K, Jacob CR, Neugebauer J, Reiher M (2008) Chem Phys Lett 461:353–359
95. Kiewisch K, Eickerling G, Reiher M, Neugebauer J (2008) J Chem Phys 128:044114
96. Wesolowski TA, Weber J (1996) Chem Phys Lett 248:71–76
97. Luber S (2014) J Chem Phys 141:234110
98. Krishtal A, Sinha D, Genova A, Pavanello M (2015) J Phys Condens Matter. accepted
99. Kumar A, Sevilla MD (2011) J Phys Chem B 115:4990–5000
100. Mantz YA, Gervasio FL, Laino T, Parrinello M (2007) Phys Rev Lett 99:058104
101. Burin AL, Uskov DB (2008) J Chem Phys 129:025101
102. Solovyeva A, Pavanello M, Neugebauer J (2012) J Chem Phys 136:194104
103. Ramos P, Papadakis M, Pavanello MJ (2015) Phys Chem B. Vol 119, pp 7541–7557
104. Jacob CR, Beyhan SM, Visscher L (2007) J Chem Phys 126:234116
105. Dulak M, Wesolowski TA (2006) J Chem Phys 124:164101
106. Wang YA, Carter EA (2000) In: Schwartz SD (ed) Theoretical methods in condensed phase chemistry Kluwer. Dordrecht, pp 117–184
107. Thakkar AJ (1992) Phys Rev A 46:6920–6924
108. Perdew JP (1992) Phys Lett A 165:79–82
109. Fux S, Jacob CR, Neugebauer J, Visscher L, Reiher M (2010) J Chem Phys 132:164101
110. Lewis FD, Wu T, Zhang Y, Letsinger RL, Greenfield SR, Wasielewski MR (1997) Science 277:673–676
111. Jortner J, Bixon M, Langenbacher T, Michel-Beyerle ME (1998) Proc Natl Acad Sci USA 95:12759–12765
112. Renaud N, Berlin YA, Lewis FD, Ratner MA (2013) J Am Chem Soc 135:3953–3963
113. Giese B (2000) Acc Chem Res 33:631–636
114. Thom AJW, Head-Gordon M (2009) J Chem Phys 131:124113
115. Farazdel A, Dupuis M, Clementi E, Aviram A (1990) J Am Chem Soc 112:4206–4214
116. Mayer I (2002) Int J Quantum Chem 90:63–65
117. Hatcher E, Balaeff A, Keinan S, Venkatramani R, Beratan DN (2008) J Am Chem Soc 130:11752–11761
118. Marcus R (1987) Chem Phys Lett 133:471–477
119. Voityuk AA (2012) Phys Chem Chem Phys 14:13789–13793

120. Evenson JW, Karplus M (1992) J Chem Phys 96:5272–5278
121. Löwdin P-O (1963) J Mol Spectrosc 10:12–33
122. Larsson S (1981) J Am Chem Soc 103:4034–4040
123. Priyadarshy S, Skourtis SS, Risser SM, Beratan DN (1996) J Chem Phys 104:9473–9481
124. Fradelos G, Wesoowski TA (2011) J Phys Chem A 115:10018–10026
125. Fradelos G, Wesolowski TA (2011) J Chem Theory Comput 7:213–222
126. Miller KJ (1990) J Am Chem Soc 112:8533–8542
127. Dederichs PH, Blügel S, Zeller R, Akai H (1984) Phys Rev Lett 53:2512–2515
128. Wu Q, Van Voorhis T (2005) Phys Rev A 72:024502
129. Valiev M, Bylaska E, Govind N, Kowalski K, Straatsma T, Dam HV, Wang D, Nieplocha J, Apra E, Windus T, de Jong W (2010) Comput Phys Commun 181:1477–1489
130. Oberhofer H, Blumberger J (2010) J Chem Phys 133:244105
131. Evangelista FA, Shushkov P, Tully JC (2013) J Phys Chem A 117:7378–7392
132. Souza AM, Rungger I, Pemmaraju CD, Schwingenschloegl U, Sanvito S (2013) Phys Rev B 88:165112
133. Ramos P, Pavanello M (2016) Phys Chem Chem Phys. submitted
134. Dreuw A, Head-Gordon M (2004) J Am Chem Soc 126:4007–4016
135. Gajdos F, Valner S, Hoffmann F, Spencer J, Breuer M, Kubas A, Dupuis M, Blumberger J (2014) J Chem Theory Comput 10:4653–4660
136. Senthilkumar K, Grozema FC, Bickelhaupt FM, Siebbeles LDA (2003) J Chem Phys 119:9809–9817
137. Kubar T, Elstner M (2013) Phys Chem Chem Phys 15:5794–5813
138. Oberhofer H, Blumberger J (2012) Phys Chem Chem Phys 14:13846–13852
139. Coropceanu V, Cornil J, da Silva Filho DA, Olivier Y, Silbey R, Brédas J-L (2007) Chem Rev 107:926–952
140. Gajdos F, Oberhofer H, Dupuis M, Blumberger J (2013) J Phys Chem Lett 4:1012–1017
141. Lewis FD, Liu J, Zuo X, Hayes RT, Wasielewski MR (2003) J Am Chem Soc 125:4850–4861
142. Senthilkumar K, Grozema F, Guerra C, Bickelhaupt F, Lewis F, Berlin Y, Ratner M, Siebbeles L (2005) J Am Chem Soc 127:14894–14903
143. Astashkin AV, Rajapakshe A, Cornelison MJ, Johnson-Winters K, Enemark JH (2012) J Phys Chem B 116:1942–1950
144. Utesch T, Mroginski MA (2010) J Phys Chem Lett 2159–2164
145. Ramos P, Mankarious M, Klein E, Pavanello M (2015), in preparation
146. Balabin IA, Beratan DN, Skourtis SS (2008) Phys Rev Lett 101:158102
147. Troisi A, Orlandi G (2006) Phys Rev Lett 96:086601
148. Zhang LY, Friesner RA, Murphy RB (1997) J Chem Phys 107:450–459
149. Cave RJ, Baxter DV, Goddard WA, Baldeschwieler JD (1987) J Chem Phys 87:926–935
150. King HF, Stanton RE, Kim H, Wyatt RE, Parr RG (1967) J Chem Phys 47:1936–1941
151. Jones ML, Kurnikov IV, Beratan DN (2002) J Phys Chem A 106:2002–2006
152. Aquino A, Beroza P, Beratan D, Onuchic J (1995) Chem Phys 197:277–288
153. de la Lande A, Babcock NS, Rezac J, Sanders BC, Salahub DR (2010) Proc Natl Acad Sci USA 107:11799–11804
154. Subotnik JE, Vura-Weis J, Sodt AJ, Ratner MA (2010) J Phys Chem A 114:8665–8675
155. Vura-Weis J, Newton MD, Wasielewski MR, Subotnik JE (2010) J Phys Chem C 114:20449–20460
156. Hsu C-P (2009) Acc Chem Res 42:509–518
157. Migliore A (2009) J Chem Phys 131:114113
158. Yang C-H, Hsu C-P (2013) J Chem Phys 139:154104
159. Closs GL, Piotrowiak P, MacInnis JM, Fleming GR (1988) J Am Chem Soc 110:2652–2653
160. Closs GL, Johnson MD, Miller JR, Piotrowiak P (1989) J Am Chem Soc 111:3751–3753

Chapter 5
Methods for Computing Ro-vibrational Energy Levels

Tucker Carrington

Abstract In this article I review methods for computing ro-vibrational energy levels of small polyatomic molecules. The principal impediment to the calculation of energy levels is the size of the required basis set. If one uses a product basis the Hamiltonian matrix for a four-atom molecule is too large to store in core memory. Iterative methods enable one to use a product basis to compute energy levels (and spectra) without storing a Hamiltonian matrix. Despite the advantages of iterative methods it is not possible, using product basis functions, to calculate ro-vibrational spectra of molecules with more than four atoms. A recent method combining contracted basis functions and the Lanczos algorithm is described.

5.1 Introduction

Most methods for solving the Schroedinger equation represent wavefunctions as linear combinations of basis functions and solve a matrix eigenvalue problem. In this article, I shall focus on solving the Schroedinger equation describing the motion of nuclei on a potential energy surface (generated using the Born-Oppenheimer approximation). For a molecule with more than four atoms, computing ro-vibrational energy levels requires using a huge number of basis functions. It is critical to devise theoretical/computational methods to either reduce the size of the basis and/or cope with large matrices. It is straightforward to use simple product basis function and exploit the structure of the product basis set to calculate the energy levels of interest. This can be done using the Lanczos algorithm and evaluating matrix-vector products by doing sums sequentially. Another option is to use more complicated basis functions that account of coupling and therefore reduce the size of the basis. These ideas can be used together.

To use any basis set method one must first determine the Hamiltonian operator. If all vibrations are of small amplitude it is sufficient to use normal coordinates (to describe the shape of the molecule), Eckart axes (to describe its orientation) and

T. Carrington (✉)
Chemistry Department, Queen's University, Kingston, ON K7L 3N6, Canada
e-mail: Tucker.Carrington@queensu.ca

J. Leszczynski and M.K. Shukla (eds.), *Practical Aspects of Computational Chemistry IV*, DOI 10.1007/978-1-4899-7699-4_5

perturbation theory (to calculate energy levels) [1]. Normal coordinates are,however, not the best coordinates if vibrations are of large amplitude. To compute spectra of molecules for which large amplitude motion is important it is best to choose geometrically defined curvilinear coordinates to describe the shape of the molecule, derive a kinetic energy operator (KEO) in terms of the curvilinear coordinates, and calculate transition energies and intensities from eigenvalues and eigenvectors of a matrix that represents the Hamiltonian operator in a basis [2–7]. Making a potential energy surface (PES) is also a difficult problem. To do this one must calculate electronic energies at a large number of points, and either fit a functional form to them or employ an interpolation scheme. Methods for the calculation of energy levels and wavefunctions have evolved with the ability of quantum chemists to calculate reasonably accurate potential energy surfaces and the ability of experimentalists to measure highly resolved ro-vibrational spectra of small polyatomic molecules. Although most methods for computing energy levels do require a potential, in this article I focus on KEOs and the computation of the energy levels.

5.2 Deriving the Kinetic Energy Operator

To compute energy levels of a semi-rigid molecule one frequently uses a KEO in rectilinear normal coordinates. A general (i.e. for a molecule with N atoms) normal coordinate KEO, for a molecule with N atoms, has been known for many years [8]. The Watson KEO (atomic units are used in this article) is

$$\hat{K}_W = \frac{1}{2}\sum_{\alpha,\beta}\left(\hat{J}_\alpha - \hat{\pi}_\alpha\right)\mu_{\alpha\beta}\left(\hat{J}_\beta - \hat{\pi}_\beta\right) - \frac{1}{2}\sum_k \frac{\partial^2}{\partial {Q_k}^2} - \frac{1}{8}\sum_\alpha \mu_{\alpha\alpha}, \tag{5.1}$$

where

$$\mu_{\alpha\beta} = \left(\mathbf{I}'^{-1}\right)_{\alpha\beta}; \quad \mathbf{I}'_{\alpha\beta} = \mathbf{I}_{\alpha\beta} + \sum_{k,l,m}\zeta^\alpha_{km}\zeta^\beta_{lm}Q_kQ_l \tag{5.2}$$

and where $I_{\alpha,\beta}$ is the inertia tensor and ζ^α_{km} are Coriolis parameters defined for example in Ref. [8]. The vibrational angular momentum terms, π_α, are given by

$$\pi_\alpha = -i\sum_{k,l}\zeta^\alpha_{kl}Q_k\frac{\partial}{\partial Q_l}. \tag{5.3}$$

Many programs for computing energy levels use normal coordinates [9–14]. An important disadvantage of the Watson normal coordinate KEO is its complexity. The vibrational KEO is simple, only if one discards all the $\pi\pi$ cross terms. If the $\pi\pi$ cross terms are retained, the KEO is complicated because $\boldsymbol{\mu}$ depends on all the coordinates. If the mass of the molecule is large the $\boldsymbol{\pi}^t\boldsymbol{\mu}\boldsymbol{\pi}$ terms are, in general, small.

If motions of large amplitude have an important effect on the spectrum, it is best not to use normal coordinates, but instead to use geometrically defined internal coordinates describing the shape of the molecule. Using the polar coordinates associated with any set of $N-1$ vectors that specifies the shape and orientation of the molecule, a simple and general KEO can be derived. Chapuisat and Iung [15] were the first to adopt this approach, which was also used recently in Refs. [16–22]. The KEO is simpler if one uses "orthogonal" vectors. "Orthogonal" in this context means that the mass-weighted vectors are related by an *orthogonal* transformation to the mass-weighted nuclear position vectors.

Having defined coordinates, a KEO can be derived by using the chain rule. Sutcliffe and Tennyson [24–28] and Handy [29] used the chain rule. Mladenovic [21, 22] has also implemented this approach. Written in terms of angular momenta operators for $N-1$ vectors which specify the shape and orientation of a molecule in the center of mass frame, the ro-vibrational KEO is compact and general. One begins with the KEO for N nuclei in space-fixed Cartesian coordinates,

$$\begin{aligned} \hat{T}_{\mathrm{N}} &= -\frac{1}{2}\sum_{i=0}^{N-1}\frac{1}{M_i}\left(\frac{\partial^2}{\partial X_i^2}+\frac{\partial^2}{\partial Y_i^2}+\frac{\partial^2}{\partial Z_i^2}\right) \\ &\equiv -\frac{1}{2}\left(\hat{T}_X+\hat{T}_Y+\hat{T}_Z\right) \end{aligned} \tag{5.4}$$

where i is a nuclear label and M_i a nuclear mass. For $\hat{T}_X$ one then transforms: (i) to mass-weighted coordinates $\{\bar{X}_i = M_i^{1/2}X_i\}$; (ii) to coordinates $\{\bar{P}_\alpha\}$ linearly related to $\{\bar{X}_i\}$ by an $N \times N$ orthonormal transformation; and (iii) to mass-unweighted coordinates $\{P_\alpha = \mu_\alpha^{-1/2}\bar{P}_\alpha\}$. The third step introduces arbitrary masses $\{\mu_\alpha\}$:

$$\hat{T}_X = -\frac{1}{2}\sum_{\alpha=0}^{N-1}\frac{1}{\mu_\alpha}\left(\frac{\partial^2}{\partial P_\alpha^2}\right) \tag{5.5}$$

Applying the same transformation to $\hat{T}_Y$ and $\hat{T}_Z$ (introducing coordinates $\{Q_\alpha\}$ and $\{R_\alpha\}$), one obtains N vectors $\{\mathbf{r}_\alpha\}$ (with space-fixed Cartesian components $(P_\alpha, Q_\alpha, R_\alpha)$) that are linear combinations of the space-fixed Cartesian nuclear position vectors with coefficients that are elements of a matrix

$$\mathbf{J} = \mu^{-1/2}\mathbf{U}\mathbf{M}^{1/2} \tag{5.6}$$

where $\mathbf{M}$ and μ are diagonal matrices of masses, and $\mathbf{U}$ is orthonormal [20]. $\mathbf{r}_{N-1}$ is often chosen as the position of the nuclear centre of mass. Obvious coordinates for describing the shape and orientation of the molecule are: the lengths of the remaining $N-1$ vectors $r_0, r_1, \ldots, r_{N-2}$; $N-2$ polar angles, θ_α $(\alpha = 1 \cdots N-2)$ between r_0 and r_α; $N-3$ angles ϕ_β $(\beta = 2 \cdots N-2)$ between the plane that contains r_0 and r_1 and the planes that contain r_0 and r_β ; and three Euler angles that specify the orientation of the molecule-fixed axis system with respect to the space-fixed axes.

The molecule-fixed z axis is parallel to r_0; the molecule-fixed y axis is along $\mathbf{r_0} \times \mathbf{r_1}$. The KEO in these coordinates is given in many papers. [18, 30, 31]. One convenient form [32] is,

$$T = T_{\rm s} + T_{\rm br} + T_{\rm cor} \tag{5.7}$$

with

$$T_{\rm br} = T_{\rm br,diag} + T_{\rm br,off} \; . \tag{5.8}$$

and

$$\begin{aligned}
T_{\rm s} &= -\sum_{k=0}^{N-2} \frac{1}{2\mu_k} \frac{\partial^2}{\partial r_k^2} \\
T_{\rm br,diag} &= \left[B_0(r_0) + B_1(r_1)\right] \left[-\frac{1}{\sin\theta_1} \frac{\partial}{\partial\theta_1} \sin\theta_1 \frac{\partial}{\partial\theta_1} + \frac{1}{\sin^2\theta_1}(J_z - L_z)^2 \right] \\
&+ \sum_{k=2}^{N-2} \left[B_0(r_0) + B_k(r_k)\right] l_k^2 \\
&+ B_0(r_0) \left[J^2 - 2(J_z - L_z)^2 - 2J_z(L_z) + 2 \sum_{k \neq k'=2}^{N-2} l_{kz} l_{k'z} \right] \\
T_{\rm br,off} &= B_0(r_0) \left[(L_+)a_1^- + (L_-)a_1^+ + \sum_{k \neq k'=2}^{N-2} (l_{k+} l_{k'-} + l_{k-} l_{k'+}) \right] \\
T_{\rm cor} &= -B_0(r_0) \left[J_-(a_1^+ + L_+) + J_+(a_1^- + L_-) \right]
\end{aligned} \tag{5.9}$$

where

$$B_i(r_i) = \frac{1}{2\mu_i r_i^2} \tag{5.10}$$

$$L_z = \sum_{k=2}^{N-2} l_{kz} \tag{5.11}$$

$$L_- = \sum_{k=2}^{N-2} l_{k-} \tag{5.12}$$

$$L_+ = \sum_{k=2}^{N-2} l_{k+} \tag{5.13}$$

$$l_{i\pm} = l_{ix} \pm \imath l_{iy} \quad (i = 2, \ldots, N-2) \tag{5.14}$$

$$J_\pm = J_x \pm \imath J_y \tag{5.15}$$

$$a_1^{\pm} = \pm\frac{\partial}{\partial\theta_1} - \cot\theta_1(J_z - L_z)\,. \quad (5.16)$$

and $l_{kx}, l_{ky}, l_{kz}, l_k^2$ are the usual angular momentum operators [33]. a_1^+ and a_1^- are raising and lowering operators [32]. This KEO is associated with the volume element $\sin\theta_1 \sin\theta_2 \cdots, \sin\theta_{N-2}\, dr_o dr_1 \cdots dr_{N-2} d\theta_1 \cdots d\theta_{N-2} d\phi_2 \cdots d\phi_{N-2}$. It is valid for any choice of the orthogonal vectors. Several four-atom choices are given explicitly in Ref. 20. There are momentum cross terms between angles, but none between angles and lengths. This general "polyspherical" KEO is compact only if it is written in terms of angular momentum operators.

Although compact and general the utility of this KEO is sometimes limited by coupling. Coupling can be reduced by making a judicious choice of the z-axis which is used to define the θ and ϕ coordinates. It is best to put the z axis along the vector for which $B_i(r_i) = 1/(2\mu_i r_i^2)$, where μ_i and r_i are the mass for and the length of the vector $\mathbf{r_i}$ in some representative reference configuration, is the smallest. A disadvantage of the above KEO is its lack of flexibility: one must place the molecule-fixed z axis along one of the $\mathbf{r_i}$ vectors. If $r_j (j = 2, \ldots, N-2)$ are orthogonal to r_0 and r_1 a general and flexible molecule-fixed KEO can be derived, even if r_0 and r_1 are themselves not orthogonal, by attaching the molecule-fixed axis system to the plane spanned by r_0 and r_1 . This is explained in Ref. [34].

It is not necessary to have an equation for the KEO. One must only be able to calculate matrix elements of the KEO which can be done (with quadrature) as long as it is possible to evaluate (coordinate dependent) coefficients of differential operators in the KEO at the quadrature points. This numerical approach is very useful. It has a long history [35–39], but its utility has been greatly enhanced by the TNUM program of Lauvergnant [40]. One way to implement a numerical approach is to determine $\frac{\partial x_{\gamma i}}{\partial Q_k}$, where $x_{\gamma i}$ are Cartesian coordinates in a molecule-fixed frame and Q_k are the internal coordinates in terms of which one wishes the KEO. To determine at points coefficients of differential operators in the KEO, one needs to invert the matrix whose elements are $\frac{\partial x_{\gamma i}}{\partial Q_k}$. Another numerical approach uses finite difference derivatives to directly compute $\frac{\partial Q_k}{\partial x_{\alpha i}}$.

5.3 Basis Functions

5.3.1 Vibrational Basis Functions

5.3.1.1 Product Basis Functions

A product basis is a basis each of whose functions is a product of functions of a single coordinate. Often, the single coordinate functions are labelled with one index and the basis is a direct product basis, may be written,

$$\Phi_{i_1,i_2,\ldots,i_f} = \phi_{i_1}(q_1)\phi_{i_2}(q_2)\cdots\phi_{i_f}(q_f), \tag{5.17}$$

If the Hamiltonian is a sum of products (SOP), i.e.,

$$\hat{H} = \sum_{l=1}^{g}\prod_{k=1}^{f}\hat{h}^{(k,l)}(q_k), \tag{5.18}$$

and a direct product basis set is used then matrix elements of a term can be written as products of matrix elements of the factors. As a rule of thumb one needs about ten basis functions per degree of freedom and hence about 10^f product basis functions to calculate vibrational energy levels and wavefunctions of a molecule with f vibrational degrees of freedom. For a triatomic molecule the product basis Hamiltonian matrix is only about 1000×1000, but for a four-atom molecule the product-basis Hamiltonian matrix is about 1000000×1000000. Storing this matrix in the main memory of a computer would require about 8000 GB.

5.3.1.2 Contracted Basis Functions

Whenever coupling is important the number of product basis functions required to converge even a small number of energy levels of a molecule with four or more atoms large. Contracted basis sets are smaller. They are effective tools for computing vibrational energy levels [41, 42] and are usually obtained by diagonalizing reduced-dimension Hamiltonian matrices. There are two popular ways to define effective reduced-dimension Hamiltonian matrices: 1) one diagonalizes blocks of the full Hamiltonian matrix in a product basis; 2) one diagonalizes matrices representing the Hamiltonian with one or more coordinates fixed. The first route yields nondirect product basis functions [42–49]. The second route yields basis functions that are direct products of functions of different coordinates or groups of coordinates [41, 50]. A triatomic example will make this clear. A type 1 contraction scheme is obtained by diagonalizing (stretch) blocks $< \alpha(\theta)|\hat{H}(\theta, r_1, r_2)|\alpha(\theta) >$ to get eigenfunctions $\phi_n^\alpha(r_1, r_2)$ where $\alpha(\theta)$ is a discrete variable representation (DVR) [51] bend function, $\hat{H}(\theta, r_1, r_2)$ is the full Hamiltonian operator, and θ, r_1, r_2 are the bend and stretch coordinates. The contracted basis functions are $\alpha(\theta)\phi_n^\alpha(r_1, r_2)$. A type 2 contraction scheme is obtained by making products of eigenfunctions of $H(\theta_e, r_1, r_2)$, denoted $\psi(r_1, r_2)$ and eigenfunctions of $H(\theta, r_1^e, r_2^e)$, denoted $\chi(\theta)$. The contracted basis functions are $\psi(r_1, r_2)\chi(\theta)$. r_1^e, r_2^e, and θ_e are specific values of r_1, r_2, and θ. Equilibrium values are often a good choice. Regardless of the route one takes, only a small fraction of the eigenvectors of the reduced-dimension Hamiltonian matrices are retained. These methods are sometimes called diagonalization-truncation-recoupling methods [51, 52].

The type 1 approach has been extensively used by Bacic, Bowman, Light, Tennyson and their coworkers [42, 43, 45, 46, 51, 52]. For 6D calculations, they have been used by Luckhaus [49] and Mladenović [47, 48]. The type 2 approach has been used to compute rovibrational energy levels of four-atom molecules, e.g. C_2H_2 [53], H_2O_2 [54], NH_3 (vibration only) [55] and was favoured by Carter and Handy [41, 50].

5.3.2 Ro-vibrational Basis Functions

5.3.2.1 Product Basis Functions

When, as explained in Sect. 5.2, the molecule-fixed frame is attached to two vectors, the ro-vibrational basis functions for a molecule with 5 atoms are

$$|l_1(m_1)l_2m_2l_3m_3; JKM\rangle = \Theta_{l_1}^{m_1}(\theta_1)Y_{l_2}^{m_2}(\theta_2, \phi_2)Y_{l_3}^{m_3}(\theta_3, \phi_3)\bar{D}_{MK}^{J}(\alpha, \beta, \gamma)^* , \quad (5.19)$$

with the constraint $K = m_1 + m_2 + m_3$. This constraint removes the $1/\sin^2\theta_1$ singularity in the KEO. In this equation, $\bar{D}_{mk}^{j}(\alpha, \beta, \gamma) = \sqrt{\frac{2j_A+1}{8\pi^2}} D_{mk}^{j}(\alpha, \beta, \gamma)$ is a normalized Wigner function.

5.3.2.2 Contracted Basis Functions

The size of the basis required to achieve converged energy levels can be significantly reduced by contracting. Contracted bases for the $J > 0$ problem have been used for years. Tennyson and Sutcliffe's two-step method is similar to a type 1 vibrational contraction: for each K (quantum number for the molecule-fixed (MF) z component of the angular momentum) eigenfunctions of a Hamiltonian depending only on vibrational coordinates are computed [56]. For triatomic molecules this works well. This sort of idea was used in 2004, with MF axes (vector-z frame) attached to two Radau vectors, to compute accurate $J = 1$ energy levels of methane [57]. Another possible contracted ro-vibrational basis is composed of products of vibrational wavefunctions, $|v\rangle$ and Wigner functions, $|JKM\rangle$ [58, 59]. As is the case for the type 2 contracted vibrational functions used for the $J = 0$ problem, [60] the $|v\rangle$ and $|JKM\rangle$ factors do not have shared labels. That is, the full-dimensional basis is a direct product and not an indirect product basis. When the $|v\rangle$ factors of the $|v\rangle$ $|JKM\rangle$ basis are also obtained from a basis of products of contracted bend $|b\rangle$ and stretch $|s\rangle$ functions, one is using nested contractions.

The meaning of the $|v\rangle$ $|JKM\rangle$ functions is different for different choices of the molecule-fixed (MF) frame and the size of the required basis depends on the definition of the MF frame. For many molecules an Eckart frame, designed to minimize Coriolis coupling close to a reference configuration, is a good choice. When normal coordinates are used it is easy to use an Eckart frame [1, 61], however, the Eckart

KEO in internal coordinates is very complicated. Eckart KEOs have been derived for triatomic molecules, but never employed to compute spectra [62–65]. This problem is resolved by computing G matrix elements numerically [35–39].

5.4 Eigensolvers

5.4.1 Direct Methods

Eigenvalues and eigenvectors of matrices not larger than about 50000×50000 can be easily computed using a standard implementation of Householder's algorithm [66]. This requires storing the matrix, i.e., N^2 numbers, where N is the size of the matrix and the cost of the calculation scales as N^3. These disadvantages of Householder's algorithm are debilitating if one wishes to calculate ro-vibrational energy levels of a triatomic molecule or vibrational energy levels of a molecule with more than three atoms.

5.4.2 Iterative Methods

Iterative methods are often used for calculating spectra [67–80]. Their most obvious advantage is that they require only the computation of matrix-vector products and therefore can be used without modifying or storing the matrix [81]. It is now widely recognized that Hamiltonian matrix-vector products can be computed without storing the Hamiltonian matrix and without even calculating its matrix elements [71, 82, 83]. The Lanczos algorithm is one popular iterative methods used to calculate spectra [81, 90]. It generates an $M \times M$ tridiagonal matrix T_M recursively by evaluating M matrix-vector products. When M is large enough, among the eigenvalues of T_M are eigenvalues of the original matrix. Refined related algorithm, designed to accelerate the convergence of selected eigenvalues, exist [91–93]. These methods require considerably more storage than the straightforward Lanczos method of Cullum and Willoughby (C&W). It is simple to use the C&W Lanczos method to calculate [90] eigenvalues. Computing eigenvectors is harder. One must either re-calculate or store the Lanczos vectors. The C&W Lanczos method is simpler to implement and efficient [94].

5.5 Using Iterative Methods with a Product Basis Set

A product basis has structure that can be exploited to reduce the cost of evaluating matrix-vector products so that their cost scales as n^{f+1}, where n is a representative number of basis functions for a single degree of freedom, and f is the number of degrees of freedom. If H is factorizable (KEOs are usually factorizable and potentials are sometimes factorizable) then it has the form

$$H = \sum_{l=1}^{g} \prod_{k=1}^{f} \hat{h}^{(k,l)}(q_k), \tag{5.20}$$

and a matrix-vector product $\mathbf{Hu} = \mathbf{u}'$ is efficiently evaluated by doing sums sequentially,

$$\sum_{l=1}^{g} \sum_{i_1} h^{(1,l)}_{i'_1,i_1} \sum_{i_2} h^{(2,l)}_{i'_2,i_2} \cdots \sum_{i_f} h^{(f,l)}_{i'_f,i_f} u_{i_1,i_2,\ldots,i_f} = u'_{i'_1,i'_2,\ldots,i'_f} \tag{5.21}$$

where $h^{(k,l)}_{i'_k,i_k}$ is an element of the $n \times n$ matrix representation of the factor $\hat{h}^{(k,l)}(q_k)$ [71, 82, 83] There are several established methods for making PESs with the sum-of-products form of Eq. (5.20) [84–89].

Perhaps surprisingly, even if the potential is not factorizable it is still possible to do matrix-vector products at a cost that scales as n^{f+1}, *if a direct product quadrature grid is used*. For example, the quadrature approximation for the integral

$$V_{n'm',nm} = \int dq_1 dq_2 f_{n'}(q_1) g_{m'}(q_2) V(q_1, q_2) f_n(q_1) g_m(q_2) \tag{5.22}$$

is

$$\sum_{\alpha} \sum_{\beta} \sqrt{\frac{\omega_\alpha}{w((q_1)_\alpha)}} \sqrt{\frac{\omega_\beta}{w((q_2)_\beta)}} f_{n'}((q_1)_\alpha) g_{m'}((q_2)_\beta)$$
$$\times V((q_1)_\alpha, (q_2)_\beta) f_n((q_1)_\alpha) g_m((q_2)_\beta) \sqrt{\frac{\omega_\alpha}{w((q_1)_\alpha)}} \sqrt{\frac{\omega_\beta}{w((q_2)_\beta)}} \tag{5.23}$$

where $(q_1)_\alpha$ and $(q_2)_\beta$ are quadrature points and ω_α and ω_β are the corresponding quadrature weights and $w(q_1)$ and $w(q_2)$ are the corresponding weight functions. Re-writing this equation by defining,

$$({}^{[q_1]}T)_{n',\alpha} = f_n((q_1)_\alpha) \sqrt{\frac{\omega_\alpha}{w((q_1)_\alpha)}} \tag{5.24}$$

$$({}^{[q_2]}T)_{n',\beta} = f_n((q_2)_\beta) \sqrt{\frac{\omega_\beta}{w((q_2)_\beta)}} \tag{5.25}$$

one obtains

$$V_{n'm',nm} \approx \sum_{\alpha} \sum_{\beta} (({}^{[q_1]}T)^\dagger)_{n',\alpha} ({}^{[q_2]}T^\dagger)_{m',\beta} V((q_1)_\alpha, (q_2)_\beta) ({}^{[q_1]}T)_{\alpha,n} ({}^{[q_2]}T)_{\beta,m} \,. \tag{5.26}$$

A matrix-vector product can then be written,

$$\sum_{\alpha}(({}^{[q_1]}T)^{\dagger})_{n',\alpha}\sum_{\beta}({}^{[q_2]}T^{\dagger})_{m',\beta}V((q_1)_{\alpha},(q_2)_{\beta})\sum_{m}({}^{[q_2]}T)_{\beta,m}\sum_{n}({}^{[q_1]}T)_{\alpha,n}u_{nm}=u'_{n'm'} \tag{5.27}$$

and the n^{f+1} scaling is evident.

The n^{f+1} scaling is due to the structure of the product basis. It is not a consequence of the sparsity of the Hamiltonian matrix or the sparsity of the $\mathbf{h}^{(k,l)}$ matrices (although it is sometimes advantageous to choose the single coordinate functions so that the $\mathbf{h}^{(k,l)}$ matrices are sparse to further accelerate the matrix-vector product [95]). Evaluating matrix-vector products by doing the sums sequentially means that one never calculates matrix elements of the Hamiltonian (not even "on the fly"). It is important that it is possible to evaluate Hamiltonian matrix-vector products without first building the Hamiltonian matrix.

5.6 Using Contracted Bases with the Lanczos Method

Iterative methods for solving the Schroedinger equation are efficient only if matrix-vector products can be computed cheaply. As explained in the previous section, in the product basis case, matrix-vector products can be evaluated at a cost that scales as n^{f+1}. Contracted basis sets, defined in Sect. 5.3 A have many advantages, but if they are to be used with an iterative method it is essential that it be possible to efficiently evaluate the corresponding matrix-vector products. Contracted basis functions (cf. Sect. 5.3) of type 1 are excellent but have important disadvantages: The matrix-vector products are costly; many reduced-dimension eigenvalue problems must be solved, and many sets of eigenvectors must be stored. For molecules with more than three atoms, contracted basis functions of the type 2 are certainly better [101, 102]. In orthogonal polyspherical coordinates the $J = 0$ KEO is [21, 103]

$$H = T_{ben}(\theta, r) + T_{str}(r) + V(\theta, r) \tag{5.28}$$

with

$$T_{ben}(\theta, r) = \sum_i G_b^{(i)}(r)T_b^{(i)}(\theta)$$
$$T_{str}(r) = \sum_i G_s^{(i)}\frac{\partial^2}{\partial r_i^2}\,. \tag{5.29}$$

θ represents all the bend coordinates and r represents all the stretch coordinates. The functions $G_b^{(i)}(r)$, $G_s^{(i)}$, and the operators $T_b^{(i)}(\theta)$ are well-known [21].

We make contracted bend functions from a Hamiltonian obtained by fixing all stretch coordinates at the values of a reference configuration (often the equilibrium configuration) and contracted stretch functions from a Hamiltonian obtained by fixing all bend coordinates at reference configuration values. The final basis functions are products of stretch contracted and bend contracted functions.

The reduced-dimension bend Hamiltonian is,

$$H^{(b)} = T_{ben}(\theta, r_e) + V(\theta, r_e). \tag{5.30}$$

Its wavefunctions are

$$X_b(\theta) = \sum_l C_{lb} f_l(\theta) \tag{5.31}$$

and the energies are denoted E_b. The f_l are primitive FBR (finite basis representation) or DVR bend basis functions (when there is more than one bend coordinate l is a composite index) and the number of retained bend wavefunctions is n_b. The stretch reduced-dimension Hamiltonian is,

$$H^{(s)} = T_{str}(r) + V(\theta_e, r). \tag{5.32}$$

and its wavefunctions are,

$$Y_s(r) = \sum_\alpha D_{\alpha s} g_\alpha(r) \tag{5.33}$$

and its energies are E_s. The g_α are primitive DVR stretch basis functions (if there is more than one stretch coordinate α is a composite index) and the number of retained stretch wavefunctions is denoted by n_s. θ_e and r_e represent reference values of the bend coordinates and the stretch coordinates. The final basis functions are products

$$|bs\rangle = |X_b\rangle |Y_s\rangle \tag{5.34}$$

The full Hamiltonian is written

$$H = H^{(b)} + H^{(s)} + \Delta T + \Delta V \tag{5.35}$$

where

$$\Delta V(r, \theta) = V(r, \theta) - V(r_e, \theta) - V(r, \theta_e) \tag{5.36}$$

and

$$\Delta T = \sum_i \Delta G_b^{(i)}(r) T_b^{(i)}(\theta) \tag{5.37}$$

with

$$\Delta G^{(i)}(r) = G_b^{(i)}(r) - G_b^{(i)}(r_e)\,. \tag{5.38}$$

Matrix elements of ΔT and ΔV in the product contracted basis must be computed. If the contracted basis is small enough, the Hamiltonian matrix in the the product contracted basis can be diagonalized with a direct linear algebra method. When this is not possible one uses an iterative method and evaluates matrix-vector products. Owing to the product structure of Eq. (5.37), matrix-vector products for ΔT are straightforward. When computing matrix-vector products for ΔV it is critical to avoid storing the potential on a stretch-bend grid. This is the subject of the next section.

5.6.1 Matrix-Vector Products for ΔV

When using an FBR primitive bend basis, a matrix element of ΔV in the product contracted basis [102] is,

$$\begin{aligned} \langle b's'|\Delta V(\theta, r)|bs\rangle &= \sum_{\substack{l'l \\ \alpha}} C_{l'b'} C_{lb} D_{\alpha s'} D_{\alpha s} \langle l'\alpha|\Delta V(\theta, r)|l\alpha\rangle \\ &= \sum_{\substack{l'l \\ \alpha\beta}} C_{l'b'} C_{lb} D_{\alpha s'} D_{\alpha s} T_{l'\beta} T_{l\beta} \Delta V_{\beta\alpha} , \end{aligned} \tag{5.39}$$

where

$$\begin{aligned} T_{l\beta} &= \sqrt{w_\beta} f_l(\theta_\beta) \\ \Delta V_{\beta\alpha} &= \Delta V(\theta_\beta, r_\alpha) . \end{aligned} \tag{5.40}$$

(θ_β, w_β) are points and weights for the quadrature and l, β, and α are composite indices. The columns of $\mathbf{C}$ are the eigenvectors of $H^{(b)}$ in the FBR basis.

It is best to evaluate ΔV matrix-vector product by doing sums sequentially [101, 104–106], A potential matrix element in the contracted basis is rewritten

$$\begin{aligned} \langle b's'|\Delta V(\theta, r)|bs\rangle &= \sum_{\alpha\beta} \tilde{C}_{\beta' b'} \tilde{C}_{\beta b} D_{\alpha s'} D_{\alpha s} \Delta V_{\beta\alpha} \\ &= \sum_{\alpha} F_{b'b,\alpha} D_{\alpha s'} D_{\alpha s} \end{aligned} \tag{5.41}$$

where $\mathbf{F}$ is defined by,

$$F_{b'b,\alpha} = \sum_{\beta} \tilde{C}_{\beta b'} \tilde{C}_{\beta b} \Delta V_{\beta\alpha} . \tag{5.42}$$

All the $F_{b'b,\alpha}$ are calculated and stored and the ΔV matrix-vector product,

$$u'_{b's'} = \sum_{bs} \langle b's'|\Delta V|bs\rangle u_{bs} . \tag{5.43}$$

is most efficiently done in three steps:

$$
\begin{aligned}
u^{(1)}_{b\alpha} &= \sum_s D_{\alpha s} u_{bs} \\
u^{(2)}_{b'\alpha} &= \sum_b F_{b'b\alpha} u^{(1)}_{b\alpha} \\
u'_{b's'} &= \sum_\alpha D_{\alpha s'} u^{(2)}_{b'\alpha}
\end{aligned} \tag{5.44}
$$

The CPU cost of this matrix-vector product is $n_\alpha(n_b^2 + 2n_b n_s)$. One can reduce the total time by parallelizing the calculation of **F**.

5.7 Conclusion

Using iterative eigensolvers has made it possible to compute vibrational spectra with product basis sets for molecules with 4 atoms. The structure of the product basis set is exploited to evaluate the required matrix-vector products. To compute vibrational spectra of larger molecules it is helpful to use contracted basis sets. Contraction necessarily complicates the evaluation of matrix-vector products. Simply contracted basis functions (type 2) and iterative methods can be efficiently combined. It is imperative to avoid storing the potential on a large direct-product quadrature grid. This can be done by using the F matrix idea. It obviates the need to transform from the contracted basis to the primitive basis to do matrix-vector products. Contracted basis functions are also advantageous for computing ro-vibrational spectra, even for molecules with only four atoms. Products of symmetric top functions and vibrational wavefunctions are good contracted ro-vibrational functions.

Acknowledgments This work has been supported by the Natural Sciences and Engineering Research Council of Canada

References

1. Aliev MR, Watson JKG (1985) Molecular spectroscopy: modern research, vol 3
2. Carter S, Handy NC (1986) Comput Phys Rep 5:115
3. Tennyson J (1986) Comput Phys Rep 4:1
4. Bačić Z, Light JC (1989) Annu Rev Phys Chem 40:469
5. Sibert EL (1990) Int Rev Phys Chem 9:1
6. Carrington T Jr (1998) Encyclopedia of computational chemistry, von Ragué Schleyer P (ed), vol 5. Wiley, New York
7. Spectrochim (2002) Acta Part A 58 A special issue on First principles rovibrational spectroscopy
8. Watson JKG (1968) Mol Phys 15:479–490
9. Romanowski H, Bowman JM, Harding L (1985) J Chem Phys 82:4155

10. Carter S, Culik SJ, Bowman JM (1997) J Chem Phys 107:10458
11. Carter S, Bowman JM, Handy NC (1998) Theor Chem Acta 100:191
12. Bowman JM, Carter S, Huang X (2003) Int Rev Phys Chem 22:533–549
13. Bowman JM, Carter S, Handy NC(2005) Theory and applications of computational chemistry: the first forty years, Dykstra C et al (ed), Chapter 11. Elsevier, New York
14. Oyanagi C, Yagi K, Taketsugu T, Hirao K (2006) J Chem Phys 124:064311
15. Chapuisat X, Iung C (1992) Phys Rev A 45:6217–6235
16. Gatti F, Iung C, Menou M, Justum Y, Nauts A, Chapuisat X (1998) J Chem Phys 108:8804
17. Iung C, Gatti F, Viel A, Chapuisat X (1999) Phys Chem Chem Phys 1:3377
18. Mladenović M (2000) J Chem Phys 112:1070–1081
19. Mladenović M (2000) J Chem Phys 112:1082–1095
20. Bramley MJ, Carrington T Jr (1993) J Chem Phys 99:8519
21. Mladenović M (2000) J Chem Phys 112:1070
22. Mladenović M (2000) J Chem Phys 112:1082
23. Chapuisat X, Iung C (1992) Phys Rev A 45:6217
24. Sutcliffe BT (1982) Current aspects of quantum chemistry. In: Carbo R (ed) Studies in theoretical and physical chemistry, vol 21. Elsevier, Amsterdam, pp 99–125
25. Sutcliffe BT, Tennyson J (1991) Int J Quant Chem 39:183
26. Tennyson J, Sutcliffe BT (1982) J Chem Phys 77:4061
27. Brocks G, Van Der Avoird A, Sutcliffe BT, Tennyson J (1983) Mol Phys 50:1025
28. Xantheas SS, Sutcliffe BT (1995) J Chem Phys 103:8022
29. Handy NC (1987) Mol Phys 61:207
30. Gatti F, Iung C, Leforestier C, Chapuisat X (1999) J Chem Phys 111:7236
31. Gatti F, Munoz C, Iung C (2001) J Chem Phys 114:8275
32. Wang X-G, Carrington T Jr (2004) J Chem Phys 121:2937
33. Zare RN (1988) Angular momentum. Wiley, New York
34. Wang X-G, Carrington T Jr (2000) J Chem Phys 113:7097–7101
35. Laane J, Harthcock MA, Killough PM, Bauman LE, Cooke JM (1982) J Mol Spec 91:286; Harthcock MA, Laane J (1982) J Mol Spectrosc 91:300
36. Meyer R (1979) J Mol Spectrosc 76:266
37. McCoy AB, Burleigh DC, Sibert EL (1991) J Chem Phys 95:7449
38. Lauvergnat D, Nauts A (2002) J Chem Phys 116:8560
39. Mátyus E, Czakó G, Császár AG (2009) J Chem Phys 130:134112
40. Lauvergnat David (2002) J Chem Phys 116:8560
41. Carter S, Handy NC (1988) Comput Phys Commun 51:49
42. Bačić Z, Light JC (1989) Annu Rev Phys Chem 40:469
43. Henderson JR, Tennyson J (1990) Chem Phys Lett 173:133
44. Wu XT, McCoy AB, Hayes EF (1999) J Chem Phys 110:2354
45. Qiu Y, Zhang JZH, Bačić Z (1998) J Chem Phys 108:4804
46. Bahel A, Bačić Z (1999) J Chem Phys 111:11164
47. Mladenović M (2002) Spectrochim Acta Part A 58:809
48. Mladenović M (2002) Spectrochim Acta Part A 58:795
49. Luckhaus D (2000) J Chem Phys 113:1329
50. Bramley MJ, Handy NC (1993) J Chem Phys 98:1378
51. Light JC, Carrington T Jr (2000) Adv Chem Phys 114:263
52. Bowman JM, Gazdy B (1991) J Chem Phys 94:454
53. Carter S, Handy NC (2002) Mol Phys 100:681
54. Koput J, Carter S, Handy N (2001) J Chem Phys 115:8345
55. Handy NC, Carter S, Colwell SM (1999) Mol Phys 96:477
56. Tennyson J, Sutcliffe BT (1986) Mol Phys 58:1067; Sutcliffe BT, Tennyson J (1986) Mol Phys 58:1053
57. Wang X-G, Carrington T Jr (2004) J Chem Phys 121:2937
58. Carter S, Bowman JM, Handy NC (1998) Theor Chem Acc 100:191
59. Fábri C, Mátyus E, Császár AG (2011) J Chem Phys 134:074105

60. Wang X-G, Carrington T Jr (2003) J Chem Phys 119:101
61. Papousek D, Aliev MR (1982) Molecular vibrational-rotational spectra. Elsevier, Amsterdam
62. Wei H, Carrington T (1997) J Chem Phys 107:2813–2818
63. Wei H, Carrington T Jr (1997) J Chem Phys 107:9493
64. Wei H, Carrington T Jr (1998) Chem Phys Lett 287:289
65. Wei H (2003) J Chem Phys 118:7208
66. Press WH, Flannery BP, Teukolsky SA, Vetterling WT (1989) Numerical recipes. Cambridge University Press, Cambridge
67. Wyatt RE (1989) Adv Chem Phys 73:231
68. Koeppel H, Domcke W, Cederbaum LS (1984) Adv Chem Phys 57:59
69. Iung C, Leforestier C (1989) J Chem Phys 90:3198
70. Bentley JA, Brunet J-P, Wyatt RE, Friesner RA, Leforestier C (1989) Chem Phys Lett 161:393
71. Bramley MJ, Carrington T Jr (1993) J Chem Phys 99:8519
72. Bramley MJ, Carrington T Jr (1994) J Chem Phys 101:8494
73. Neuhauser D (1990) J Chem Phys 93:2611
74. Lehoucq RB, Gray SK, Zhang D-H, Light JC (1998) Comput Phys Commun 109:15
75. Mussa HY, Tennyson J (1998) J Chem Phys 109:10885
76. Wall NR, Neuhauser D (1995) J Chem Phys 102:8011
77. Mandelshtam VA, Taylor HS (1995) J Chem Phys 102:7390
78. Mandelshtam VA, Taylor HS (1997) J Chem Phys 106:5085
79. Yu H-G, Smith SC (1997) Ber Bunsenges Phys Chem 101:400
80. Chen R, Guo H (1998) J Chem Phys 108:6068
81. Golub GH, Van Loan CF (1989) Matrix computations. Johns Hopkins University Press, Baltimore
82. Friesner RA, Wyatt RE, Hempel C, Criner B (1986) J Comput Phys 64:220
83. Manthe U, Koeppel H (1990) J Chem Phys 93:345
84. Beck MH, Jaeckle A, Worth GA, Meyer H-D (2000) Phys Rep 324:1
85. Pelaez D, Meyer H-D (2013) J Chem Phys 138:014108
86. Manzhos S, Carrington T (2006) J Chem Phys 125:194105
87. Manzhos S, Carrington T (2007) J Chem Phys 127:014103
88. Manzhos S, Carrington T (2008) J Chem Phys 129:224104
89. Pradhan JCMSE, Carreon-Macedo J-L, Brown A (2013) J Phys Chem A 117:6925
90. Cullum JK, Willoughby RA (1985) Lanczos algorithms for large symmetric eigenvalue computations. Birkhäuser, Boston
91. Sorensen DC (1992) SIAM. J Math Anal Appl 13:357
92. Lehoucq RB, Sorensen DC, Yang C (1998) ARPACK USERS GUIDE: solution of large scale eigenvalue problems by implicitly restarted Arnoldi methods. SIAM, Philadelphia
93. Baglama J, Calvetti D, Reichel L (1996) BIT 36:400
94. Huang S, Carrington T Jr (1999) Chem Phys Lett 312:311–318
95. Roy P-N, Carrington T Jr (1996) Chem Phys Lett 257:98–104
96. Bunker PR (1998) Symmetry Per Jensen molecular, spectroscopy, 2nd edn. NRC Research Press, Ottawa
97. Wang X-G, Carrington T Jr (2001) J Chem Phys 114:1473
98. Chen R, Guo H (2001) J Chem Phys 114:1467
99. Wang X-G, Carrington T Jr (2001) J Chem Phys 115:9781
100. Wang X-G, Carrington T Jr (2003) J Chem Phys 118:6946–6956
101. Bramley MJ, Carrington T Jr (1994) J Chem Phys 101:8494
102. Wang X-G, Carrington T Jr (2002) J Chem Phys 117:6923
103. Chapuisat X, Belfhal A, Nauts A (1991) J Mol Spectrosc 149:274
104. Wu XT, Hayes EF (1997) J Chem Phys 107:2705
105. Friesner RA, Bentley JA, Menou M, Leforestier C (1993) J Chem Phys 99:324
106. Viel A, Leforestier C (2000) J Chem Phys 112:1212

Chapter 6
Effectively Unpaired Electrons for Singlet States: From Diatomics to Graphene Nanoclusters

Anatoliy V. Luzanov

Abstract Formal and computational models within the effectively unpaired electron (EUE) theory are reviewed and extended. In the first part, we analyze open-ended aspects of the existing EUE measures and find additional advantages of the Head-Gordon index (2003) over the very first (Yamaguchi et al. 1978) index. In particular, for ground states the Head-Gordon index estimates an average occupation of virtual holes and particles, which occur due to electron correlation. Additional hole-particle indices for describing EUE are proposed and analyzed. The second part of the paper is focuses on practical aspects and EUE computational schemes in small molecules (at the ab initio level) and large-scale polyaromatic and graphene-like structures (at the semi-empirical level). Here the unrestricted Hartree-Fock (UHF) schemes and their recently proposed simplistic versions turn out to be a suitable tool producing meaningful EUE characteristics for the extended π-electron systems (with number of carbon atoms $\sim 10^3$ and more) in a fast and simple way. We emphasize that UHF solutions should be regarded not as invalid spin-contaminated states but as precursors of the appropriate spin-projected states of the Lowdin's extended Hartree-Fock type. The influence of the static and variable electric fields on π-electron systems is also studied. It is shown that strong perturbations drastically increase the electron unpairing in aromatic hydrocarbons, especially those with the initially stable Clar-type structure.

Dedicated to Late Prof. O.V. Shishkin.

A.V. Luzanov (✉)
STC "Institute for Single Crystals" of National Academy of Sciences of Ukraine, Kharkiv, Ukraine
e-mail: avluzanov@gmail.com

J. Leszczynski and M.K. Shukla (eds.), *Practical Aspects of Computational Chemistry IV*, DOI 10.1007/978-1-4899-7699-4_6

6.1 Introduction

The notion of effectively unpaired electrons (EUE) in molecules originates from the seminal paper of Yamaguchi and colleague [1]. The EUE analysis of wave functions has become a very useful tool for understanding electronic structure of complex in particular, conjugated molecules. In case of singlet states all the approaches to EUE are inevitably rooted in high-level many-electron theories because no unpaired electrons are possible in a one-electron picture of closed-shell systems. Indeed, by construction, each electron pair (with opposite spins) occupies exactly one suitable MO, as prescribed by any correct independent-particle model. Hence, only bona fide wave functions accounting for electron correlation should be employed for treating EUE. This makes the problem interesting and difficult simultaneously.

It should be also emphasized that there exist no spin density effects in spin-singlet (diamagnetic) molecules if relativistic effects are neglected (see Eqs. (5.2) and (5.3) in Ref. [2]). Hence, "effectively unpaired electrons" (as in the dissociated hydrogen molecule or in radical pairs) remain, as a whole, in the singlet state due to the total spin conservation law. This fact indicates some type of spin correlation between particles, particularly between spatially separated ones, as it is in the classical Einstein-Podolsky-Rosen pair [3, 4]. At the same time, systems with large unpairing effects behave as singlet diradicaloids or even polyradicals which feature many unusual properties. The problem is significant because a considerable occurrence of the effectively unpaired electrons is an instability factor of the system of interest. In particular, an EUE analysis of wave functions can easily point to a diradical or, generally, polyradical character of electronic states. In this respect, the EUE problem is also important for designing new molecular-based materials containing giant molecules. A separate issue is electron unpairing in excited states, which has attracted attention only recently.

There exist various quantum chemical approaches to define and quantify "odd" electrons (this very term is used in [1] for effectively unpaired electrons). Obviously, it is important to rightly choose the scheme describing EUE. Two key approaches are now popular in this field. The first employs the so-called Yamaguchi index from [1]; the other uses the Head-Gordon indices from [5]. A noteworthy progress was recently made in ab initio applications of the EUE theory [6–11]. And yet, high-level electron correlation methods are very computationally demanding or, more frequently, not available for large-scale and super-large-scale systems. Therefore, it was important to develop a simplified semi-empirical, but at the same time physically meaningful EUE theory for huge systems such as π-electron-containing graphene-like structures and finite-sized carbon nanotubes. Preliminary attempts in this direction were made in [12, 13]. It is worth mentioning some interesting results produced for giant graphene molecules obtained by the density functional theory (DFT) [14]. At the same time, in practice standard DFT approaches typically fail to produce correctly formed density matrices, which presents a stumbling block for the consistent analysis of molecular electronic structures.

In light of the above, the goal of the present contribution is to describe main trends in this field and to give a comparable analysis of different approaches as well as to demonstrate the utility of the EUE indices for interpreting complex structures—up to covalent polymeric networks. In our applications of the EUE theory the stress will be on strongly correlated molecular states, in particular large π-conjugated systems. Here, we revisit the EUE problem in the context of previous studies [12, 15, 16], and find the Head-Gordon index to be more appropriate and more consistent from a physical viewpoint as well. We also describe new applications of EUE indices for molecules in strong static and nonstationary electric fields. We aim to provide a self-contained introduction to and a concise overview of the EUE theory.

6.2 General Definitions and Yamaguchi's Index

We start with paper [1]. This work had put forward a first possible definition of the EUE density for an arbitrary wave function with any permitted spin value $s \geq 0$. As mentioned in the introduction, our main interest is the case of singlet states, and for them the EUE effects are really important and interesting. Indeed, for nonzero spin states (doublet-state radicals, triplet-state diradicals etc.), the manifestations of unpaired electrons can be described even within the restricted open-shell Hartree-Fock (ROHF) theory. The latter characterizes the unpaired spins by standard spin density matrices. In the singlet state, the spin density matrix disappears [2], and yet, electron correlation enforces electrons to be unpaired if physical and chemical circumstances require it (e.g., in bond breaking processes).

First, we provide the main EUE definitions using the conventional reduced density matrix (RDM) methods. In singlet states, the first-order RDM (1-RDM) can be defined as a spin-free matrix which is also termed the charge density matrix. Throughout the paper, the capital letter D will be denoted the charge density matrix. In Dirac's bra-ket notation, D conveniently takes a compact form of a spectral resolution, that is the following diagonal form:

$$D = \sum_k \lambda_k |\varphi_k\rangle\langle\varphi_k|, \tag{6.1}$$

with $|\varphi_k\rangle$ being the eigenvectors (natural orbitals), and $\lambda_k(\geq 0)$ being the respective eigenvalues. The quantities λ_k are usually called natural orbital occupancy numbers (NOON). Due to Pauli's principle, $0 \leq \lambda_k \leq 2$. Furthermore, NOON are usually normalized to a total (always even in our case) number of electrons, $N = 2n$, where n is a number of electron pairs in the given spin-singlet molecule. Thus,

$$\mathrm{Tr}\, D = \sum_k \lambda_k = N. \tag{6.2}$$

If unpaired electrons are absent then all molecular orbitals are doubly occupied. Obviously, for the fully closed-shell system the all nonzero NOON are equal to 2:

$$\lambda_i = 2, \quad 1 \le i \le n \tag{6.3}$$

(pair orbital occupancies), other $\lambda_{k>n} = 0$. In this case one deals actually with the customary independent-particle model, more exactly, the restricted Hartree-Fock (RHF) approach. The RHF density matrix is denoted as D_0. Explicitly,

$$D_0 = 2 \sum\nolimits_{1 \le i \le n} |\varphi_i\rangle\langle\varphi_i|. \tag{6.4}$$

The structure (6.4) is certainly changed under the influence of electron correlation. So, expression (6.1) with a more involved NOON spectrum $\{\lambda_k\}$ is generally valid, and a deviation of a realistic NOON distribution from Eq. (6.3) properly characterizes EUE. In [1] this simple reasoning was the underlying rationale for introducing the EUE notion and the corresponding deviation measure.

Proceeding in a more formal fashion, we introduce the EUE density matrix, D^{eff}, as a function of density matrix D (in the operatorial sense). It means that

$$D^{\text{eff}} = \sum_k \lambda_k^{\text{eff}} |\varphi_k\rangle\langle\varphi_k|, \tag{6.5}$$

and the new, also nonnegative, 'occupation numbers' λ_k^{eff} are generated by a certain function of the initial NOON spectrum, that is

$$\lambda_k^{\text{eff}} = f(\lambda_k) \tag{6.5$'$}$$

(see [5]). The total EUE number, or better still, the EUE index, N_{eff}, can be identified with a sum of these λ_k^{eff}:

$$N_{\text{eff}} = \text{Tr}\, D^{\text{eff}} = \sum_k \lambda_k^{\text{eff}}. \tag{6.6}$$

For singlet states, the natural requirement is $f(\lambda_k) \equiv 0$ for any one-determinant wave function satisfying Eq. (6.3). In this case $\lambda_k^2 = 2\lambda_k$, and this can be condensed into the matrix identity

$$2D_0 - (D_0)^2 = 0 \tag{6.7}$$

(the duodempotency relation). Thus, function $f(\lambda) = 2\lambda - \lambda^2$ provides an admissible choice for a function which vanishes in the case of singlet state determinants. This leads to the simplest solution of the EUE problem: $\lambda_k^{\text{eff}} = \lambda_k^{\text{odd}}$ where

$$\lambda_k^{\text{odd}} = 2\lambda_k - \lambda_k^2. \tag{6.7$'$}$$

Here and elsewhere superscript and subscript 'odd' denote that the Yamaguchi index and related quantities are considered. As a result, the matrix representation of Eq. (6.5) takesthe form

$$D^{\text{odd}} = 2D - D^2, \tag{6.8}$$

and we can easily specify Eq. (6.6) as

$$N_{\text{odd}} = 2N - \text{Tr } D^2 = 4n - \sum_k \lambda_k^2. \tag{6.9}$$

The above approach gained more attention after this method was restated in [17] (see also [18] about the history of N_{odd} and related measures in earlier literature on valency). We now understand that for singlet states, Yamaguchi's index N_{odd} is merely a possible measure of the wave function departure from a single determinant. In a different context, a closely related nonidempotency measure of 1-RDM was independently introduced in Mestechkin's book [19]. Furthermore, N_{odd} was examined, carefully analyzed and extended in many later works, such as [6, 20–24].

Let us now give a simple example of using N_{odd}. Following mainly [1, 5, 12], we consider the unrestricted Hartree-Fock (UHF) method for singlet states. In this case Eqs. (6.8) and (6.9) can be easily rewritten, based on the known UHF relation

$$D = \rho_\alpha + \rho_\beta, \tag{6.10}$$

where ρ_α and ρ_β are usual one-electron density matrices for α - и β-shells of the UHF determinant. Then, the working equation is

$$N_{\text{odd}}[\text{UHF}] = N - 2\text{Tr}\,(\rho_\alpha \rho_\beta), \tag{6.11}$$

that is equivalent to

$$N_{\text{odd}}[\text{UHF}] = N - 2\text{Tr}\,(\rho_\beta \rho_\alpha \rho_\beta), \tag{6.12}$$

as a result of idempotency relations $\rho_\alpha = (\rho_\alpha)^2$ and $\rho_\beta = (\rho_\beta)^2$. The equivalent representation is

$$N_{\text{odd}}[\text{UHF}] = ||\rho_\alpha - \rho_\beta||^2,$$

where $||Z|| = (\text{Tr } Z^+ Z)^{1/2}$ is the familiar Euclidean (Frobenius) matrix norm for an arbitrary matrix Z. When $\rho_\beta = \rho_\alpha$ (no spin polarization) we return to RHF, and $N_{\text{odd}} = 0$, as it should be. Another form of Eq. (6.11) is

$$N_{\mathrm{odd}}[\mathrm{UHF}] = 2\langle \mathbf{S}^2 \rangle_{\mathrm{UHF}}, \tag{6.13}$$

with $\langle \mathbf{S}^2 \rangle$ being an average value of the squared spin [18].

6.3 Head-Gordon's Index

In spite of many useful applications of Yamaguchi's index, it fails in many cases. It was first shown in [5] where one interesting example (dissociation of the triplet oxygen molecule) was considered, and an incorrect behavior of N_{odd} was observed for the dissociation limit. In this work a new approach was formulated in such a way that could handle difficult cases as well. We will further refer to the EUE index from [5] as the Head-Gordon index, and use the more compact term "H-G index", denoting it simply by N_{eff}. This index is based on the following choice:

$$\lambda_k^{\mathrm{eff}} = 1 - |\lambda_k - 1| \equiv \mathrm{Min}[\lambda_k, 2 - \lambda_k], \tag{6.14}$$

so that

$$N_{\mathrm{eff}} = \sum_k \mathrm{Min}[\lambda_k, 2 - \lambda_k]. \tag{6.15}$$

By construction, the index satisfies the inequality

$$N_{\mathrm{eff}} \leq N_{\mathrm{odd}} \tag{6.16}$$

(see Fig. 1 in [5]). The inequality is rather easy to demonstrate by considering the UHF model. Using the corresponding spectrum $\{\lambda_k\}$ from [25] an explicit expression is derived to be

$$N_{\mathrm{eff}}[\mathrm{UHF}] = N - 2\sum_{j=1}^{n} \sqrt{\lambda_j^{\alpha\beta}}, \tag{6.17}$$

where $\lambda_j^{\alpha\beta}$ are eigenvalues of $\rho_\beta \rho_\alpha \rho_\beta$. We see that indeed $N_{\mathrm{eff}}[\mathrm{UHF}] \leq N_{\mathrm{odd}}[\mathrm{UHF}]$ because eigenvalues of the product of idempotent matrices are less than 1, and moreover, in the same notation we have from Eq. (6.12)

$$N_{\mathrm{odd}}[\mathrm{UHF}] = N - 2\sum_{j=1}^{n} \lambda_j^{\alpha\beta}.$$

There are many researchers who exploit H-G index (e.g., see [6–12, 26, 27]). In several papers, indices N_{odd} and N_{eff} are considered concurrently, and only few authors advocate a preference of N_{odd}. Notice the polemic papers [28, 29] which present conficting viewpoints on the EUE problem. We will discuss a difference between N_{odd} and N_{eff} in Sects. 6.5, 6.14, and Appendix C. Based on this consideration it appears that N_{eff} provides a more consistent way to quantify the diradical (or polyradical) character in terms of traditional NOON. It is worth mentioning the earlier works [30] where NOON have been applied for a qualitative identification of diradical states and diagnostics of multiconfigurational character.

Let us review some common properties of the N_{odd} and N_{eff} measures and the corresponding EUE occupancies, that is numbers λ_k^{odd} in Eq. (6.7′) and λ_k^{eff} in Eq. (6.14). Returning to Eq. (6.5′) we write understandable requirements of the nonnegative EUE function $f(\lambda)$ to be defined only in the closed interval [0, 2], so that

$$f(0)=f(2)=0,\ f(1)=1.$$

Of course, N_{odd} and N_{eff} obey these equations. Less trivial is that $f(\lambda)$ is symmetric with respect to axis $\lambda=1$, that is

$$f(\lambda)=f(2-\lambda). \tag{6.18}$$

In other words, Eq. (6.18) is satisfied by an appropriate function $f(\lambda)=F(|\lambda-1|)$. Obviously, λ_k^{odd} and λ_k^{eff} obey the above relations. The reason for symmetry (6.18) will be explained in Appendix A in terms of a duality transformation well-known in the multilinear algebra literature. In Appendix A one can also find a possible generalization of indices N_{odd} and N_{eff}. Various examples of N_{eff} and related indices will be given throughout the rest of this chapter.

6.4 Unpairing Indices from Collectivity and Entropy Numbers

Another EUE quantification scheme appeared in [15, 31]. This scheme is based on the so-called collectivity numbers first introduced in [32] for describing electronic excitations within the single configuration interaction (CIS) method. More general collectivity numbers were subsequently given in [33] for the full configuration interaction (FCI). The related logarithmic measures are considered in [15, 34, 35]. Below we follow [15] from which a few illustrations (with a slight modification) are taken as well.

The collectivity number can be treated as a statistical measure. A similar statistical measure has been defined before in [36] for analyzing localization of vibrational modes. It was named the participation ratio. We define it as an average rank, κ, of the given normalized probability distribution $\{w_k\}$:

$$\kappa = 1/\sum_k (w_k)^2, \tag{6.19}$$

(for more detail see [15, 32, 37]). The squared expansion coefficients can be used as a possible $\{w_k\}$ set. Generally, the resulting κ will be crucially dependent of the representation chosen (the AO or MO representation, for instance). To make Eq. (6.19) basis-independent, a matrix construction is required. The result is automatically attained within the FCI matrix theory [38] based on the conventional determinant FCI method [39].

For singlet states the FCI wave function $|\Psi\rangle$ can be cast as follows:

$$|\Psi\rangle = \sum\nolimits_{I,J} X_{IJ} |I; J\rangle. \tag{6.20}$$

Here real-valued (for simplicity) expansion coefficients X_{IJ} are normalized, and $|I; J\rangle$ symbolizes the N-electron Slater determinants comprised of spin-up orbital subset $\{\overset{+}{\chi}_{i_1}, \ldots, \overset{+}{\chi}_{i_n}\}$ and spin-down orbital subset $\{\bar{\chi}_{p_1}, \ldots, \bar{\chi}_{p_n}\}$. The orbitals are all taken from the chosen "full" basis set

$$\{|\chi_\mu\rangle\}.$$

Furthermore, in Eq. (6.20) indices I and J are in fact ordered multi-indices (strings in [39]) of the form

$$I \equiv \{i_1, i_2 \ldots, i_n\}, \; J \equiv \{j_1, j_2, \ldots, j_n\}.$$

As a consequence of the determinantal nature of the basis set $\{|I; J\rangle\}$ used in Eq. (6.20), the matrix

$$X = ||X_{IJ}||$$

has regular transformation properties, and thus all matrix invariants of X are basis-independent. Moreover, for singlet states X should be a symmetric $(X_{IJ} = X_{JI})$ and normalized matrix (due to $\langle \Psi \mid \Psi \rangle = 1$), so that $\sum_{I,J} (X_{IJ})^2 = \mathrm{Tr}(X^2) = 1$. The eigenvalues $\{x_k\}$ of X produce a normalized probability distribution, that is $\{w_k\} = \{x_k^2\}$. Thus, the counterpart of Eq. (6.19) for FCI is

$$\kappa = 1/\sum_k x_k^4 = 1/\mathrm{Tr}(X^4). \tag{6.21}$$

This is just the collectivity number which was introduced in [33] and systematically studied in [15]. For single-determinant models we have $X^2 = X$. In this case, $|\Psi\rangle$ in Eq. (6.19) can be reduced to one term, that is to a single determinant, so $\kappa = 1$. More preferable is a logarithmic quantity which we define by the expression:

$$N_\kappa = -2\ \ln \mathrm{Tr}\,(X^4)/\ln 2. \tag{6.22}$$

We term this quantity as the EUE κ-index, or simply κ-index. As shown in [15, 31], this index provides the needed properties (nonnegativity, additive separability, and clearly interpreted results for simple chemical systems in extreme cases).

Consider two examples. The first is the two-electron hydrogen molecule treated in a minimal basis of two atom-centred orthonormal AOs, $\{|\chi_1\rangle, |\chi_2\rangle\}$. From the symmetry and normalization we directly obtain matrix X, as follows

$$X[H_2] = \frac{1}{\sqrt{2}} \begin{Vmatrix} \sqrt{1-z} & \sqrt{z} \\ \sqrt{z} & \sqrt{1-z} \end{Vmatrix}, \tag{6.23}$$

with z being a variational parameter. This z has a meaning of a covalency parameter (a weight of the Heitler-London geminal $|\chi_1\chi_2 + \chi_2\chi_1\rangle/\sqrt{2}$ in the total wave function). Elementary computations on Eq. (6.22) give

$$N_\kappa = 2 - (2/\ln 2)\ \ln(1 + 4z - 4z^2). \tag{6.24}$$

It can be compared with the respective values of N_{odd} and N_{eff} indices:

$$N_{\mathrm{odd}}[\mathrm{H}_2] = 2(1-2z)^2, \quad N_{\mathrm{eff}}[\mathrm{H}_2] = 2\left(1 - 2\sqrt{(1-z)z}\right) \tag{6.25}$$

(they follow from the respective density matrix $D[\mathrm{H}_2] = 2(X[\mathrm{H}_2])^2$. Notice that there is a misprint for $\kappa[\mathrm{H}_2]$ in [15, 31]. The value $z = 1/2$ describes the Hartree-Fock ground state. All indices, Eqs. (6.24) and (6.25), go to 2 when $z \to 1$, that corresponds to the dissociation limit of the hydrogen molecule (see Fig. 6.1).

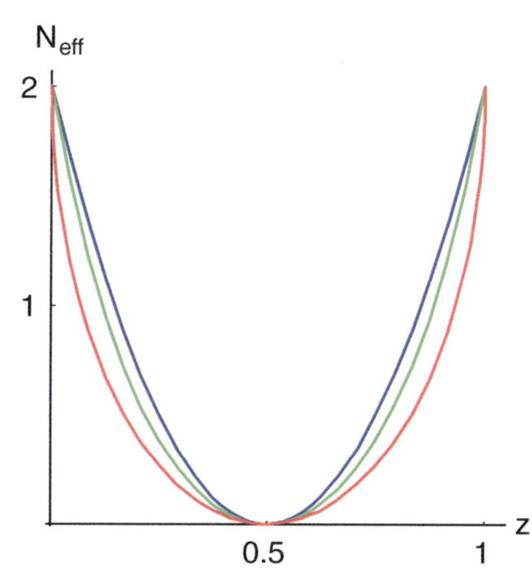

Fig. 6.1 Dependence of EUE indices from the covalency parameter z in the H_2 molecule treated in minimal basis: Yamaguchi's index N_{odd} in *blue*, κ-index N_κ in *green*, and Head-Gordon's index N_{eff} in *red*

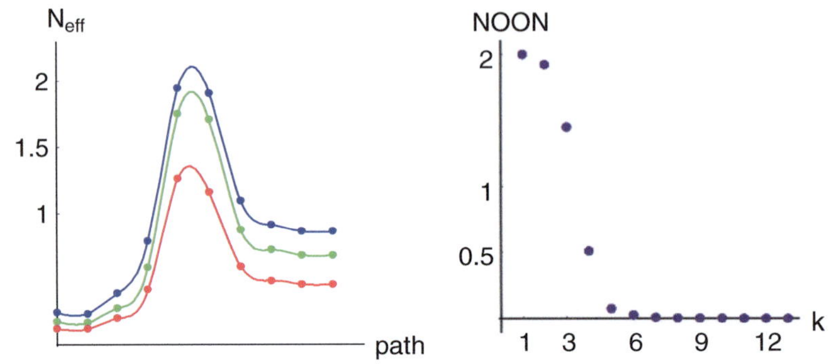

Fig. 6.2 *Left panel* N_{odd} (*in blue*), N_{κ} (*in green*), and N_{eff} (*in red*) for insertion reaction (6.26) within FCI/6-31G. *Right panel* the NOON spectrum corresponding to the region near the transition state (for point 5 on the reaction path)

We see from Fig. 6.1 that the κ-index is intermediate between Yamaguchi's and Head-Gordon's indices: $N_{eff} \leq N_{\kappa} \leq N_{odd}$. As our experience testifies, this is, in fact, the typical result.

As a second more complex example, consider the insertion reaction:

$$\mathrm{Be} + \mathrm{H}_2 \rightarrow \mathrm{Be\,H}_2 \tag{6.26}$$

(see [40] for the molecular geometry in the selected 10 points on the reaction path). For reaction (6.26) the results (partially taken from [15]) are presented in Fig. 6.2. Evidently, each of the approaches gives a similar picture with a maximum near a transition region (the fifth and sixth points in Fig. 6.2). A more detailed analysis reveals that the transition state (TS) corresponds, only very approximately, to a diradicaloid state in which unpaired electrons should be significantly localized on the H-atoms. Really, from the NOON spectrum (the right panel in Fig. 6.2) it is clear that there are no NOON near 1. Only the values 1.45 and 0.51 in this spectrum appear to give a quasi open-shell TS structure. The value $N_{eff} = 1.26$ for the fifth point is quite compatible with this situation. At the same time, the diradical character of this TS appears to be overestimated by the corresponding values $N_{odd} = 1.94$ and $N_{\kappa} = 1.75$. The fact that N_{odd} and N_{κ} overestimate the radicaloid character is typical.

6.5 Hole-Particle Densities and Head-Gordon's Index

We now look at the EUE problem from the viewpoint of the general theory of hole-particle distributions given in [16]. The related indices (in a different disguise) appeared in [15, 31]. In these works, Kutzelnigg's original idea [41] about an openness measure of electronic shells was discussed as well. A suitable description of EUE follows from a direct analysis of the RDM hole-particle components [16]. We introduce the latter by considering the general type of wave functions in terms of the so-called excitation operators [39, 42]. They create the CI (configuration interaction) wave-function of arbitrary order, up to FCI. Expansions of this type are well known long ago [43, 44]. But only in [45] the one-electron and two-electron RDMs were presented explicitly in terms of excitation operator matrices, more exactly, elementary transition matrices (see also [16, 46]).

As usual, we must choose an appropriate reference determinant $|\Phi\rangle$ from which one can generate singly excited $\{|\Phi_i^a\rangle\}$, doubly excited $\{|\Phi_{ij}^{ab}\rangle\}$, and so on configurations (as usually, indices *i, j,...* refer to occupied orbitals of the reference, and indices *a, b,...* refer to virtual orbitals). Thus, the *k*-excited configurations are taken from the set $\{|\Phi_{i_1...i_k}^{a_1...a_k}\rangle\}$, and all possible *k* must be taken into account in an exact (FCI) consideration. Each configuration contributes, to the considered FCI function, with a respective configuration coefficient, $C_{a_1...a_k;\,i_1...i_k}$, or explicitly

$$|\Psi\rangle = \sum_{0\leq k\leq N} \sum_{\substack{i_1...i_k \\ a_1...a_k}} C_{a_1...a_k;\,i_1...i_k} |\Phi_{i_1...i_k}^{a_1...a_k}\rangle. \tag{6.27}$$

For our formal consideration, the full set

$$\{C_{a_1...a_k;\,i_1...i_k}\}, \quad 0\leq k\leq N, \tag{6.28}$$

is assumed to be fully known for a while. As usual, it is normalized to 1. Coefficients (6.28) can be packed into the corresponding multi-index matrices

$$C_k = [C_{a_1...a_k;\,i_1...i_k}], \quad 0\leq k\leq N. \tag{6.29}$$

The same quantities (6.28) are identified with hole-particle amplitudes which are just equal to matrix elements of hole-particle excitation operators $\hat{C}_k$. By definition, $\hat{C}_k$ generates the superposition of *k*-excited configurations of the corresponding order *k* (for more detail see [39, 42]). Within the customary hole-particle formalism, the first *k* indices $a_1 \ldots a_k$ in $C_{a_1...a_k;\,i_1...i_k}$ are related to states of 'particles' which are excited above a 'sea' of occupied states, whereas the second *k* indices $i_1 \ldots i_k$ (occupied orbitals) are related to the possible hole states in the same sea. This well-known interpretation is also suitable for designing correlation indices. To this end, let us consider the normalization condition which is, evidently,

$$\sum_{k=0}^{N} ||C_k||^2 = 1 \tag{6.30}$$

where obviously

$$||C_k||^2 = \sum_{a_1 \ldots a_k;\, i_1 \ldots i_k} |C_{a_1 \ldots a_k;\, i_1 \ldots i_k}|^2. \tag{6.31}$$

The squared norm $||C_k||^2$ can be presented in two equivalent forms: as $||C_k||^2 = \mathrm{Tr}\ C_k(C_k)^+$, and as $||C_k^+||^2 = \mathrm{Tr}\ (C_k)^+ C_k$. Evidently, the first form is relevant to the particles, whereas the second to the holes.

Now look at an average number of holes, that is index N^{h} (in notations from [16]):

$$N^{\mathrm{h}} = \sum_{k=0}^{N} k||C_k^+||^2.$$

But the same expression is valid for the average number of particles:

$$N^{\mathrm{p}} = \sum_{k=0}^{N} k\, ||C_k||^2.$$

so

$$N^{\mathrm{h}} = N^{\mathrm{p}} = \sum_{k=0}^{N} k\, \mathrm{Tr}\ C_k(C_k)^+. \tag{6.32}$$

Thus, we find the sum

$$N_{\mathrm{h-p}} = N^{\mathrm{p}} + N^{\mathrm{h}} \equiv 2N^{\mathrm{p}} \tag{6.33}$$

as an admissible hole-particle EUE measure [16, 31]. It remains to add that the reference determinant $|\Phi\rangle$ in expansion (6.27) should be built up from natural orbitals of the state in question. But sometimes another choice can be also informative.

No practical difficulties exist in calculating $N_{\mathrm{h-p}}$ because in terms of spin-free RDMs we have the explicit relations

$$N^{\mathrm{h}} = \mathrm{Tr}\, D^{\mathrm{h}},\ \ N^{\mathrm{p}} = \mathrm{Tr}\, D^{\mathrm{p}}, \tag{6.34}$$

and D^{h} and D^{p} are the hole, and, respectively, particle components of D. These components, as defined in [16], are

$$D^{\mathrm{h}} = 2\rho - \rho D \rho,\ \ D^{\mathrm{p}} = (I - \rho)\, D(I - \rho), \tag{6.35}$$

Here and elsewhere, $\rho = D_0/2$, that is the spin-free projector on n maximally occupied natural orbitals of the reference determinant:

$$\rho = \sum_{i=1}^{n} |\varphi_i\rangle\langle\varphi_i|. \tag{6.36}$$

(the Dirac-Fock density matrix, in other terms). In the same notations the total hole-particle density matrix is $D^{\text{hp}} = D_{\text{h}} + D_{\text{p}}$, or

$$D^{\text{h}-\text{p}} = 2\rho + D - \rho D - D\rho. \tag{6.37}$$

Within the given hole-particle approach, D^{hp} is a counterpart of the corresponding EUE density matrix (6.5). Technical details for computing FCI and closed-shell CCSD (singles and doubles coupled cluster) approaches are given in [16, 47]. We write here only the simplest relation

$$N_{\text{h}-\text{p}} = 2\,(N - \text{Tr}\,D\rho) \tag{6.38}$$

following from Eqs. (6.37). The corresponding spectral sum is

$$N_{\text{h}-\text{p}} = 2N - 2\sum_{i \le n} \lambda_i = 2\sum_{a > n} \lambda_a. \tag{6.39}$$

It is this quantity from in [15] which was derived based on [41]. More than that, the numerical experience revealed that our hole-particle index (6.39) actually provides the same characterization of EUE, as H-G index does. This fact was recently subject to closer scrutiny [12]. The main inference from the analysis [12] is that the identity

$$N_{\text{eff}} = N_{\text{h}-\text{p}} \tag{6.40}$$

is true for 'normal' ground states, which have no pathology in the NOON spectrum $\{\lambda_k\}$ (see below). This becomes transparent if one considers the spectral representation

$$D^{\text{h}-\text{p}} = \sum_{i \le n} (2 - \lambda_i)|\varphi_i\rangle\langle\varphi_i| + \sum_{a > n} \lambda_a|\varphi_a\rangle\langle\varphi_a|. \tag{6.41}$$

The latter follows from Eqs. (6.1) and (6.37), as moment's inspection of definitions shows it. In the same fashion one can rephrase the matrix D^{eff}, Eq. (6.5), which is associated with H-G index by Eq. (6.14):

$$D^{\text{eff}} = \sum_{i \le n'} (2 - \lambda_i)|\varphi_i\rangle\langle\varphi_i| + \sum_{a > n'} \lambda_a|\varphi_a\rangle\langle\varphi_a|. \tag{6.42}$$

Here n' is a number of NOONs greater than 1. If $n' = n$ we have, by definition, the normal state, and then

$$D^{\text{eff}} = D^{\text{h-p}}. \tag{6.43}$$

Otherwise the state in question falls into the category of 'pathological' states. In practice, the excited states can be such ones, and in Sect. 6.10 (Table 6.4) we will provide an example of the pathological state.

As usual, the ground state is normal in this categorization. It means that in reality, identities (6.40) and (6.43) are valid even for highly correlated ground states. This fact serves as additional argument in favor of N_{eff} since a clear physical meaning can be ascribed to this index within the conventional hole-particle picture. Namely, for the normal ground states the N_{eff} index is the average number of holes and particles which are excited in the reference one-determinant state due to electron correlation. As for molecular excited states themselves, the situation is generally more involved, and will be addressed in Sect. 6.10. Incidentally, from Eq. (6.41) it follows that the hole and, respectively, particle occupancy spectra are of the form

$$\{2-\lambda_i\},\ \ \{\lambda_a\}, \tag{6.41$'$}$$

where λ_i are related to 'occupied' natural orbitals ($\lambda_i > 1$), and λ_a to 'vacant' natural orbitals ($\lambda_a \leq 1$). A possible generalization of hole-particle EUE measure (6.39) is postponed to Sect. 6.14.

6.6 Using the High-Order Density Matrices

The fact that the EUE theory [1, 5, 15] can be chiefly founded on the one-electron RDM is remarkable per se. However, electron correlation effects are at least two-electron in nature, and it is no wonder that the second-order RDM was applied for quantifying EUE and related electron-correlation properties. Seemingly, the first investigation in this direction was presented in book [19] where in Sect. 6.5 a special operator named 'correlation operator' was introduced. Actually, in [19] the two-electron counterpart of D^{eff} was examined. In this section we will denote RDMs of order k by D_k^{so}. The superscript 'so' shows that the full RDM (in spin-orbital basis) is considered. For instance, D_1^{so} and D_2^{so} are the conventional one-electron and two-electron RDMs.

As well known, for the single Slater determinant (SD), that is for independent-particle models, the two-electron RDM is the antisymmetrized product of one-electron RDMs [48, 49]:

$$D_2^{\text{so}} \xrightarrow{\text{SD}} \frac{1}{2}(I - P_{12}) D_1^{\text{so}}(1)\, D_1^{\text{so}}(2), \tag{6.44}$$

where P_{12} represents the full (with spin variables) transposition operator. Following [19], we introduce the correlation operator Δ_2^{so} as a difference between the exact two-electron RDM and the SD approximation (6.44):

$$\Delta_2^{so} = D_2^{so} - \frac{1}{2}(I - P_{12})D_1^{so}(1)D_1^{so}(2). \tag{6.45}$$

By contracting Δ_2^{so} over variables of the second electron, we find

$$-2\mathrm{Tr}_{(2)}\Delta_2^{so} = D_1^{so} - (D_1^{so})^2, \tag{6.46}$$

that is but a 'nonidempotency matrix'. This fact was independently discovered later in [21] (the first paper in this reference entry). Really, contracting Eq. (6.46) over spin variables just produces the EUE density matrix D^{odd} in Eq. (6.8). We see that using Eq. (6.46) does not provide us a new quantification scheme, not to mentioning that Eq. (6.8) gives not very good approach, as argued previously.

A significant advance has been made in [50] where the completely two-electron measure was introduced. In this work the squared norm of Δ_2^{so}, that the quantity $||\Delta_2^{so}||^2$, was proposed as a new correlation and entanglement measure. Admittedly, the EUE aspect was not within the scope of [50]. This aspect is discussed in [35] (among other approaches). An appropriate rescaling, by constant factor 8/7, guarantees a correct number of unpaired electrons in the dissociated H_2 molecule and in arbitrary cluster of dissociated two-electron systems. Therefore, it is simply to modify the above-mentioned measure, as follows:

$$N_{cum} = \frac{8}{7}||\Delta_2^{so}||^2. \tag{6.47}$$

Here subscript 'cum' in N_{cum} means that this EUE index is produced by the so-called cumulant density matrix (6.45), as such RDM constructions are termed in the current RDM theory [51]. For practical computations, within FCI or RAS-CI (restricted active space CI), more suitable is a spin-free expression from [35].

Some results (the data partially from [35]) are presented in Table 6.1. We see that N_{cum} gives the values which are somewhat close to N_{eff} than other indices. We also observe that all the indices provide a similar qualitative picture. For instance, in a case of the fully dissociated BeH_2, we must obtain $N_{eff}[\mathrm{Be}+\mathrm{H}+\mathrm{H}] = N_{eff}[\mathrm{Be}] + 2$ where $N_{eff}[\mathrm{Be}]$ is a non-zero value which results from the effect of intra-atomic electron correlation in a free Be atom. At the FCI/6-31G level, we obtain $N_{odd}[\mathrm{Be}] = 0.74$ and $N_{eff}[\mathrm{Be}] = 0.39$. Thus, it is expected that for the full dissociation $N_{odd} = 2.74$ and $N_{eff} = 2.39$, as it is the case, judging from Table 6.1. We note also that at present using N_{cum} is rather restricted because a direct handling with 2-RDM is avoided as a rule when treating large scale problems.

Table 6.1 Comparison of Yamaguchi's index N_{odd}, κ-index N_κ, cumulant index N_{cum}, and Head-Gordon's index N_{eff}, for small molecules at the FCI/6-31G level

System	Geometry	N_{odd}	N_κ	N_{cum}	N_{eff}
BeH_2	R_e	0.24	0.17	0.14	0.12
BeH_2	$3R_e$	2.74	2.59	2.47	2.39
BH	R_e	0.62	0.48	0.39	0.33
BH	$3R_e$	2.38	2.24	2.15	1.86
CH_2	R_e	0.62	0.44	0.37	0.32
CH_2	$3R_e$	4.17	4.09	4.04	3.62
H_2O	R_e	0.36	0.22	0.21	0.18
H_2O	$3R_e$	4.07	3.98	3.93	3.57
HF	R_e	0.27	0.16	0.16	0.14
HF	$3R_e$	2.09	2.01	1.98	1.73

R_e is the equilibrium bond length, and $3R_e$ corresponds to the triply stretched equilibrium length

6.7 Algorithmic Aspects

Several schemes are possible for practical calculations of the main EUE indices. Frequently, all the elements of D are needed, e.g. for computing N_{odd} by Eq. (6.9) and N_{eff} by Eq. (6.15). In a number of cases we can simply exploit the explicit expressions, as in the case of the rather easily performed UHF-like models. When sophisticated multiconfigurational models are used, it is necessary to employ the technique which is elaborated for obtaining D within the restricted active space CI (RAS-CI) and coupled cluster schemes [39, 42]. However, the direct way is too demanding when large-scale systems need to be addressed. Sometimes, one can employ the RDM-free scheme from [52] that avoids the tedious computations of all matrix elements of D. This scheme (see Eqs. (107) and (111) in [52]) can be applied to the hole-particle quantification scheme described above. The respective technique is based on reverting the obvious relationship which connects expectation values to RDM. Namely, for the given spinless one-particle operator Z we have

$$\mathrm{Tr}\, ZD = \langle\Psi| \sum_{1\le k\le N} Z(k)|\Psi\rangle. \tag{6.48}$$

Then the N_{eff} index is simply computed in the equivalent form of N_{h-p}, Eq. (6.38), that is as usual one-electron average (6.48) with

$$Z = 2(I - \rho).$$

In this case we imply that ρ is known. This is a case when the Hartree-Fock reference determinant can be approximately used in the EUE analysis. Another case is the Brueckner coupled-cluster method [42] producing the reference molecular orbitals, almost the same as natural orbitals.

Additional indices which can by obtained by using Eq. (6.48) are the hole-particle atomic localization indices $\{D_A^{\mathrm{eff}}\}$ which are related to N_{h-p}

(in practice, N_{eff}). They can be defined by the customary partition procedure well known for other atomic indices [53, 54]. Let us introduce projector I_A on atomic orbitals belonging to the given atom (or fragment) A:

$$I_A = \sum_{\mu \in A} |\chi_\mu\rangle\langle\chi_\mu|. \tag{6.49}$$

Then

$$D_A^{\text{eff}} = \operatorname{Tr} I_A D^{\text{eff}} = \sum_{\mu \in A} (D^{\text{eff}})_{\mu\mu}, \tag{6.50}$$

where index μ numbers the standard orthonormalized AOs. Here the full orthonormalized AO basis is

$$\{|\chi_\mu\rangle\}_{1 \le \mu \le \dim}, \tag{6.51}$$

(dim is a size of the basis set). Hence, performing calculations using Eq. (6.48) with $Z = I_A + \frac{1}{n}(\operatorname{Tr} I_A\rho)\, I - I_A\,\rho - \rho\, I_A$ for each atom A, we find the full atomic distribution of the unpaired electrons in molecule. Obviously, the identity

$$\sum_A D_A^{\text{eff}} = N_{\text{eff}} \tag{6.52}$$

is guaranteed. For computing D and $\{D_A^{\text{eff}}\}$ in case of the CCSD model one can apply a suitable algorithm which resembles that of the CISD (CI singles and doubles) method (see Appendix in [47]).

6.8 Spin Correlations

In the introduction, we mentioned that the presence of unpaired electrons in singlet states gives indirect evidence in favor of the essential spin correlations between the electrons, especially when they are strongly localized. The following discussion highlights this issue. It is well known from the quantum theory of magnetism that spin correlations can be interpreted consistently by invoking the spin correlator formalism. In quantum chemistry, spin correlators had been introduced by Penny [55]. In the last two decades the interest to them revived (see [35, 56–60] and many others). We follow the notations and techniques from [35, 59].

For the given atoms or molecular subunits A and B, spin correlator $\langle \mathbf{S}_A \cdot \mathbf{S}_B \rangle$ is an average of the form

$$\langle \mathbf{S}_A \cdot \mathbf{S}_B \rangle \equiv \langle \Psi | \mathbf{S}_A \cdot \mathbf{S}_B | \Psi \rangle \tag{6.53}$$

where the local spin operator $\mathbf{S}_A$ can be taken as follows: $\mathbf{S}_A = \sum_{1 \le i \le N} \mathbf{s}(i) I_A(i)$. Here $\mathbf{s}(i)$ is the spin operator for the ith electron, and $I_A(i)$ is a local projector (6.49) for the ith electron. The diagonal correlators $\langle \mathbf{S}_A \cdot \mathbf{S}_A \rangle \equiv \langle \mathbf{S}_A^2 \rangle$ are usually named the (squared) local spins. The useful identity is

$$\sum_B \langle \mathbf{S}_A \cdot \mathbf{S}_B \rangle = (s+1) Q_A/2, \tag{6.54}$$

where Q_A is a spin density localized on A, and s is the total spin value for the state in study [35, 59]. For singlet states, spin densities identically dissappear, so

$$\sum_B \langle \mathbf{S}_A \cdot \mathbf{S}_B \rangle = 0. \tag{6.55}$$

Obviously, the full sum rule is

$$\sum_{A,B} \langle \mathbf{S}_A \cdot \mathbf{S}_B \rangle = s(s+1). \tag{6.56}$$

When analyzing spin correlators it is also suitable to pack the correlators into the matrix

$$\mathbf{K}_{\text{spin}} = \| \langle \mathbf{S}_A \cdot \mathbf{S}_B \rangle \|, \tag{6.57}$$

where subscripts A and B run over all atoms in molecule.

As a simple application of these rules, consider the system divided into two parts (subsystems) A and B. From Eq. (6.55) it follows

$$\langle \mathbf{S}_A^2 \rangle + \langle \mathbf{S}_A \cdot \mathbf{S}_B \rangle = 0, \quad \langle \mathbf{S}_B^2 \rangle + \langle \mathbf{S}_A \cdot \mathbf{S}_B \rangle = 0.$$

Thus, the spin-correlator matrix is of the form

$$\mathbf{K}_{\text{spin}} = \langle \mathbf{S}_A^2 \rangle \begin{Vmatrix} 1 & -1 \\ -1 & 1 \end{Vmatrix}, \tag{6.58}$$

where local correlator $\langle \mathbf{S}_A^2 \rangle$ takes the specific values. If the subsystems A and B are in a singlet state (as in the case of a van der Waals (vdW) dimer of singlet molecules) then the spin-correlator matrix is evidently

$$\mathbf{K}_{\text{spin}} = \begin{Vmatrix} 0 & 0 \\ 0 & 0 \end{Vmatrix}.$$

Now let A and B be subsystems which we assume to be in a triplet state. Then for a resulting singlet state of the entire system, matrix (6.57) is

$$\mathbf{K}_{\text{spin}} = \begin{Vmatrix} 2 & -2 \\ -2 & 2 \end{Vmatrix}. \tag{6.59}$$

This case occurs when we treat the singlet excited states of the vdW dimers and complexes (the so-called triplet-triplet (TT) excitations). The spin-correlator analysis for the vdW dimers was shortly mentioned in [61] where the singlet fission models are discussed. In context of the EUE problems we can connect spin correlators with the N_{eff} measure of the TT-type excited states. For instance, when no charge transfer effects are involved, the local spins $\langle \mathbf{S}_A^2 \rangle \equiv \langle \mathbf{S}_B^2 \rangle$ provide an estimate of a weight of double excitations $|A^*_{s=1}B^*_{s=1}\rangle$ in the total excited state of dimer AB (see Appendix C in [61]). Then $N_{\text{eff}} \approx 2\langle \mathbf{S}_A^2 \rangle$ because the singlet excitation $|A^*_{s=1}B^*_{s=1}\rangle$ has four unpaired electrons. In more general situation we must take into account interfragment charge-transfer states. It leads to a more complicated analysis which will be given in a forthcoming paper in collaboration with D. Casanova and A. Krylov. Additional aspects of the EUE analysis for excited states are considered in Sect. 6.10.

6.9 Spin-Polarization Indices and Antiferromagnetic Image of Molecule

We briefly considered in Sects. 6.2 and 6.3 how to treat EUE within the UHF approximation which admits to different orbitals for different spin (DODS). For singlet states the UHF scheme is usually called the spin-polarized HF method (then $\rho_\alpha \neq \rho_\beta$, unlike RHF where $\rho_\alpha = \rho_\beta$). Here we look at the problem from the more general viewpoint which allow us to introduce relevant spin-polarization indices for any singlet many-electron states [62].

It is well known that for singlet states, the UHF solutions with $\rho_\alpha \neq \rho_\beta$ are really possible when electron correlations become sufficiently strong. More exactly, the spin-polarized HF determinant $|\Phi\rangle$ appear only under the non-singlet (triplet) instability which was defined by Cizek and Paldus in [63]. At the same time, solutions of the spin-projected variational HF method (the *Löwdin*'s extended HF scheme) always exist [19]. The wave functions of this type will be signified by $|\Phi^{\text{ext}}\rangle$. This is usually defined by (apart from a normalization factor)

$$|\Phi^{\text{ext}}\rangle = O_s|\Phi\rangle, \tag{6.60}$$

with O_s being a projection operator onto a spin-pure N-electron state with the spin z-projection $s_z = s$ and the total spin value s.

In this context, it is pertinent to recall that in many cases one can obtain the so-called best overlap orbitals [64] of DODS type which are produced by the given many-electron wave function. These orbitals were considered in [65] where they were identified with spin-polarized Brueckner orbitals. However, they exist if and only if the so-called nonsinglet Brueckner instability conditions are satisfied. At last, if the correct spin-projected determinant $|\Phi^{\text{ext}}\rangle$ is involved in the consideration, then it is always possible to construct the best overlap orbitals of DODS type for the given exact or approximate state vector $|\Psi\rangle$. These orbitals were recently introduced [62] and named the spin-polarized extended Brueckner (SPEB) orbitals. By construction, they maximize $\langle \Phi^{\text{ext}} \mid \Psi\rangle$.

Typical overlap integrals between $|\Phi^{\text{ext}}\rangle$ and $|\Psi\rangle$ are found to be around 0.98 even for dissociated covalent molecules [62]. The corresponding EUE measures (6.1) and (6.5) were also studied in [62] along with appropriate spin-polarization indices. The latter are computed for $|\Phi^{\text{ext}}\rangle$ from matrices ρ_α and ρ_β in another way than in Eqs. (6.11) or (6.17). Following the cited work, let us introduce the intermediate matrices

$$\upsilon_0 = (I-\rho^\beta)\rho^\alpha, \quad \pi_0 = (I-\rho^\alpha)\rho^\beta, \tag{6.61}$$

and define for SPEB the special spin-polarization matrices

$$D^\alpha_{\text{pol}} = (\upsilon_0^+ \upsilon_0)^2, \; D^\beta_{\text{pol}} = (\pi_0^+ \pi_0)^2. \tag{6.62}$$

Then the indices

$$N^\alpha_{\text{pol}} = \text{Tr}\, D^\alpha_{\text{pol}}, \; N^\beta_{\text{pol}} = \text{Tr}\, D^\beta_{\text{pol}},$$

and the total spin-polarization index

$$N_{\text{pol}} = N^\alpha_{\text{pol}} + N^\beta_{\text{pol}} = 2N^\beta_{\text{pol}} \tag{6.63}$$

serve for the spin-up and spin-down characterization of EUE in the singlet states. Additionally, we can introduce the associated EUE α-and β-distributions which are composed of the atomic contributions, viz.,

$$\Pi^\alpha_A = \sum_{\mu\in A} (D^\alpha_{\text{pol}})_{\mu\mu}, \; \Pi^\beta_A = \sum_{\mu\in A} (D^\beta_{\text{pol}})_{\mu\mu}. \tag{6.64}$$

They are also helpful for the visual interpretation of the ESPB computations. Here, A is a selected atom in molecule, and subscript μμ, as previously in Eq. (6.50), indicates a diagonal element of the matrix in the orthonormal AO basis. The total spin polarization index assigned to atom A is evidently equal to

$$\Pi_A = \Pi_A^{\alpha} + \Pi_A^{\beta}, \tag{6.65}$$

so that summing Π_A over all A reproduces N_{pol}.

Let us look at Table 6.2 to understand what one can gain from this analysis. In the table, along with the above indices and distributions, we also give coefficient $C_{\mathrm{nat}} \equiv \langle \Phi^0 \mid \Psi \rangle$ at the reference determinant $|\Phi^0\rangle$ (the latter is taken as is the closed-shell natural orbital determinant). From Table 6.2 we see that $\langle \Phi^{\mathrm{ext}} \mid \Psi \rangle \cong C_{\mathrm{nat}}$ (so that $C_{\mathrm{nat}} \cong 1$) only for normal molecules which are far from quasi-degeneracy (H_6, CH_2 etc. in the table). In the H_8 cluster, due to the frontier orbital degeneracy, the ground state allows no symmetrical closed-shell structure, and it leads to the fact that even C_{nat} is not large. At the same time, the SPEB orbitals generate the spin-projected determinant which provides a sufficiently high overlap with the exact wave function. The same is true for dissociative states in Table 6.2. Interestingly, in this table the N_{pol} indices turn out, as a rule, to be more close to the N_{eff} values than to the N_{odd} ones. The spin–polarization diagrams (two columns in Table 6.2) deserve attention too. From them we see that the spin-up EUE distributions are preferably localized in those parts of molecules where the spin-down EUE distributions are localized poorly, and vice versa. This behaviour outwardly resembles features of the alternant MOs introduced by Löwdin (e.g., see Fig. 3 in [66]). The distinction between the two descriptions is in the fact that the $\{\Pi_A^{\alpha}\}$ and $\{\Pi_A^{\beta}\}$ describe the "spin" localization of EUE, that is purely correlation effects, whereas the spin-up and spin-down orbitals in UHF, EHF etc. correspond to individual one-electron states without specifying correlations per se. In particular, UHF orbitals are always nonzero whereas the spin-polarization indices can disappear (e.g. in 'one-electron' limit). It is a matter of no small importance that the SPEB orbitals and corresponding distributions $\{\Pi_A^{\alpha}\}$, $\{\Pi_A^{\beta}\}$ are generated by exact (FCI or RAS-CI) wave functions or high-level many-electron approximations. Opposite to the latter, UHF and EHF frequently provide only a small part of correlation effects for molecules in equilibrium or not too far from it (e.g., see [67]).

Alternatively, the EUE structure (within the SPEB) can be depicted by the special spin-arrow diagrams representing together the $\{\Pi_A^{\alpha}\}$ and $\{\Pi_A^{\beta}\}$ distributions. One may think of such diagrams as giving the antiferromagnetic EUE images of molecules. Some examples are given in Fig. 6.3. Notice that the interpretation of molecular structures in terms of antiferromagnetic coupling has a long history. Implicitly, it was used in the Hartmann work [68]. Usually this terminology is invoked when analyzing π-conjugated polymers and atomic clusters [69–75]. However, the nature of antiferromagnetism for the overall singlet state in molecules is not so simple as in the case of the solid state ferromagnetism [69, 70]. We return to this issue in Sect. 6.11.

Table 6.2 The reference coefficient C_{nat}, overlap integral $\langle \Phi^{ext} \mid \Psi \rangle$, EUE indices N_{odd} and N_{eff}, total polarization index N_{pol}, and atomic distributions $\{\Pi_A^{\alpha}\}$, $\{\Pi_A^{\beta}\}$ for small molecules in the basis set of 6-31G quality

System	C_{nat}	$\langle \Phi^{ext} \mid \Psi \rangle$	N_{odd}	N_{eff}	N_{pol}	$\{\Pi_A^{\alpha}\}$	$\{\Pi_A^{\beta}\}$
H_6	0.97	0.99	0.48	0.24	0.44	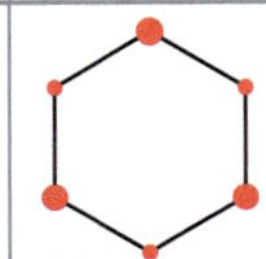	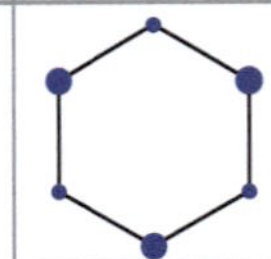
H_8	0.65	0.98	2.71	2.36	2.24	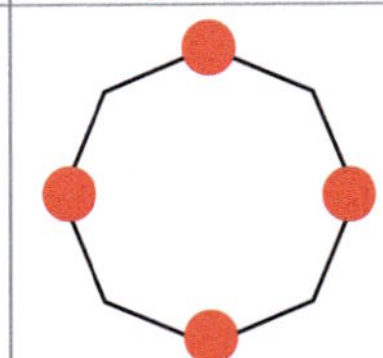	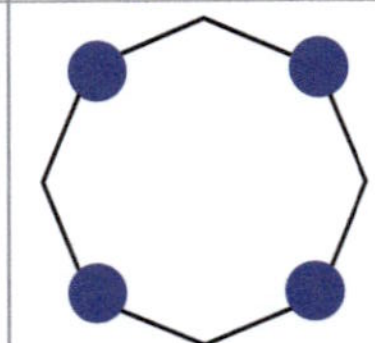
CH_2 (1Re)	0.96	0.98	0.62	0.32	0.65	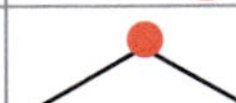	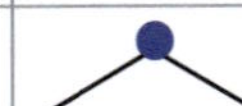
CH_2 (3Re)	0.50	0.98	4.17	3.62	3.94	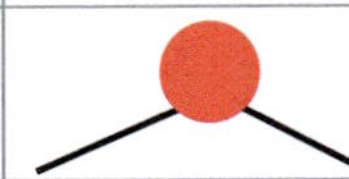	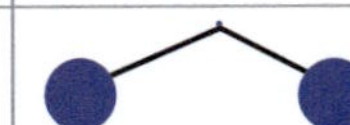
vinylidene	0.97	0.97	0.80	0.41	0.41	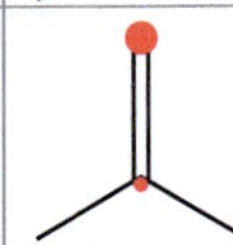	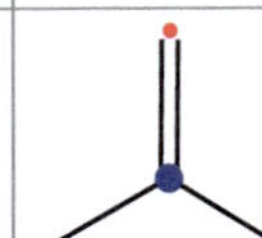

(continued)

Table 6.2 (continued)

System	C_{nat}	$\langle \Phi^{ext} \mid \Psi \rangle$	N_{odd}	N_{eff}	N_{pol}	$\{\Pi_A^\alpha\}$	$\{\Pi_A^\beta\}$
Cyclopropenylidene	0.97	0.97	0.61	0.31	0.27	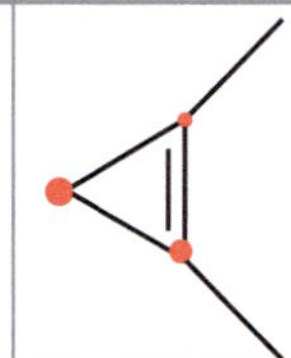	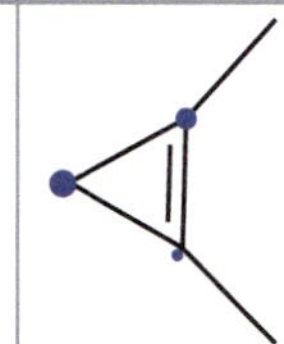
Allene	0.96	0.98	0.29	0.57	0.47		
Allene (TS)	0	1.00	2.20	2.39	2.28	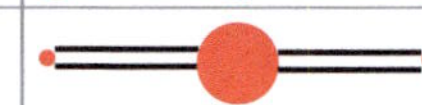	

For the CH_2 molecule, FCI with frozen core is used; for other carbon-containing systems, a CASSCF-like scheme is exploited

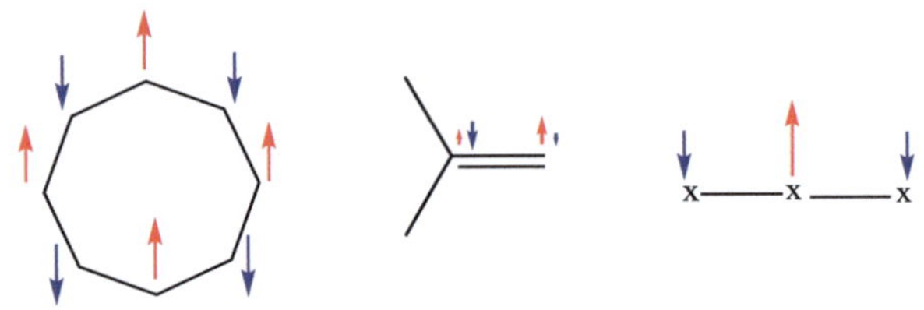

Fig. 6.3 Antiferromagnetic EUE images for the H_8 cluster, vinylidene and twisted allene molecules

6.10 Unpairing in Excited States

The molecular excited states are just those in which the electron unpairing is one of the key points in understanding the nature and properties of the electronic transitions. But only in few works, such as [15, 62, 76–78], the EUE characteristics were explicitly invoked for analyzing excited states. Recall that the simplified approximations are typically based on CIS (CI singles) and TDDFT (time-dependent DFT) models. For singlet excitations the CIS density matrices were first derived by McWeeny [79]. Then they were generalized [80] and extended to RPA (random phase approximation) and TDDFT [81, 82].

The CIS wave function, as a particular case of Eq. (6.27), can be written as follows:

$$|\Psi^{\mathrm{CIS}}\rangle = \sum_i \sum_a C_{ai}\,|\Phi_i^a\rangle, \tag{6.66}$$

with $|\Phi_i^a\rangle$ being the singly excited configurations, and C_{ai} the normalized amplitudes (configurational coefficients). For our purposes we will use the equivalent form of Eq. (6.66) which is based on spinless amplitudes τ_{ai}, so that

$$|\Psi^{\mathrm{CIS}}\rangle = \sum_{i=1}^{n} \sum_{a=n+1}^{r} \tau_{ai}\,|\Phi_{i\to a}\rangle, \tag{6.67}$$

where

$$|\Phi_{i\to a}\rangle = \left(|\Phi_{\overset{+}{i}}^{\overset{+}{a}}\rangle + |\Phi_{\overset{-}{i}}^{\overset{-}{a}}\rangle\right)/\sqrt{2} \tag{6.68}$$

are the standard spin-singlet configurations [83]. The charge density matrix for this $|\Psi^{\mathrm{CIS}}\rangle$ is

$$D^{\mathrm{CIS}} = 2\rho + \tau\tau^{+} - \tau^{+}\tau, \tag{6.69}$$

where the spin-free transition matrix τ is defined by the formula

$$\tau = \sum_{i=1}^{n} \sum_{a=n+1}^{r} \tau_{ai} \, |\varphi_a\rangle\langle\varphi_i|, \tag{6.70}$$

and ρ is the previously defined projector (6.36) on occupied spin-free MOs of the reference determinant (e.g., the RHF determinant). To guarantee the normalization condition $\langle \Psi^{\mathrm{CIS}} \mid \Psi^{\mathrm{CIS}} \rangle = 1$ we impose the condition $\mathrm{Tr}\tau^{+}\tau = 1$.

Having at disposal density matrix (6.69) it is easy to perform the hole-particle analysis of the CIS method. In this case, Eqs. (6.40) and (6.43) are valid because the CIS states have no anomalies in the density matrix spectrum. Simple manipulations on Eq. (6.37) lead to

$$D^{\mathrm{eff}}[\mathrm{CIS}] = \tau\tau^{+} + \tau^{+}\tau, \tag{6.71}$$

so

$$N_{\mathrm{eff}}[\mathrm{CIS}\ ] = N_{\mathrm{h-p}}[\mathrm{CIS}\ ] = 2 \tag{6.72}$$

for any CIS state [15]. The result is quite natural, and it is in agreement with the standard spin structure of each singlet-spin configuration $|\Phi_{i\to a}\rangle$, Eq. (6.68). The same result is obtained for κ-index (6.22): $N_{eff}^{\kappa}[\mathrm{CIS}] = 2$. On the other hand, computations on Eq. (6.9) give

$$N_{\mathrm{odd}}[\mathrm{CIS}] = 2\{\, 2 - \mathrm{Tr}(\tau^{+}\tau)^2 \,\}. \tag{6.73}$$

From this it follows that $2 \le N_{\mathrm{odd}}[\mathrm{CIS}] < 4$. Thus, we see that again Yamaguchi's index overestimates the EUE measure even for the discussed (very restricted) CI wave function. More important is the result (6.71). This EUE density matrix exactly coincides with the excitation localization operator, which was first introduced in [84] (see also Eq. (6.5) in [82]). In the notation adopted here this is of the form

$$\hat{L}^{*}[\mathrm{CIS}] = (\tau\tau^{+} + \tau^{+}\tau)/2, \tag{6.74}$$

Thus, in terms of Eq. (6.5),

$$\hat{L}^{*}[\mathrm{CIS}] = D^{\mathrm{eff}}[\mathrm{CIS}]/2. \tag{6.75}$$

We see that within the CIS approximation the excitation localization indices can be additionally treated as the localization indices (6.50) of the unpaired electrons occurring under excitation:

$$L_A^*[\mathrm{CIS}] = D_A^{\mathrm{eff}}[\mathrm{CIS}]/2. \tag{6.76}$$

This aspect of the EUE theory for CIS and CIS-like models was briefly outlined in the recent review [76] (in Sect. 14.4). It would be interesting to understand to what extend this holds true for more general models. We provide here only preliminary insight on this rather difficult question.

For the lowest transitions, the CIS and the related TDDFT excitation energies are often found in satisfactory or even good agreement with more refined theoretical estimates. Nevertheless, there are low-lying transitions that cannot be usefully studied in these popular approximations. Among such are the so-called double-excitation transitions for which the excited-state wave function has a significant proportion of a doubly excited configurations. Even sometimes, more efficient methods may fail as in the case presented for the methylene CH_2 in [85]. The methylene singlet excited states were also examined in terms of spin-polarization diagrams [62]. Here relevant supplementary results are added (Table 6.3). In the table, along with excitation energies and EUE indices, we present the most important squared norms $||C_k||^2$ defined by Eq. (6.31). These norms are computed in the basis of the ground-state natural orbitals.

From the table we observe that $1B_1$ and $1A_2$ terms are CIS-like states ($||C_1||^2 \cong 1$, all $N_{\mathrm{eff}} \cong 2$). The $2A_1$, excited state is the doubly excited state ($||C_2||^2 \cong 1$), which, however, has almost the same small N_{eff} value as that of the ground state (see Tables 6.2 and 6.3). Thus, the EUE indices may not reflect the multiconfigurational character of excited states. To elucidate this issue, let us consider the main part of the NOON spectrum for the ground and excited states of CH_2 (Table 6.4). In the table we omitted the maximal NOON value 2, which is due to the $(1s)_2$ frozen core of the carbon atom. Additionally, we included in Table 6.4 the main hole-particle index $N_{\mathrm{h-p}}$, Eq. (6.39), and the related index $N_{\mathrm{h-p}}^{(\mathrm{ref})}$. The latter was computed by Eq. (6.38) with ρ taken as the projector on the occupied MO of the reference determinant used in the CI expansion (6.27). Recall that in agreement with definitions (6.32) and (6.33), hole-particle index $N_{\mathrm{h-p}}$ can serve as a suitable measure for multiconfigurational character in the corresponding CI expansion (6.27). From Table 6.4 it is clear that the $1A_1$ and $2A_1$ states are very similar in their NOON spectrum. Only in the case of the $2A_1$ state its closed-shell

Table 6.3 Electronic characteristics of the CH_2 lowest excited states at the FCI/6-31G level

State	λ	$(C_0)^2$	$\|\|C_1\|\|^2$	$\|\|C_2\|\|^2$	N_{odd}	N_κ	N_{eff}
$1B_1$	1.74 (0.005)	0.	0.957	0.019	2.27	2.21	2.12
$2A_1$	4.48 (0.006)	0.028	0.002	0.931	0.68	0.49	0.35
$1A_2$	6.00	0.	0.926	0.051	2.33	2.27	2.16

Vertical excitation energy λ in eV, oscillator strength in parenthesis

Table 6.4 Hole-particle EUE indices $N_{\mathrm{h-p}}$ and $N_{\mathrm{h-p}}^{(\mathrm{ref})}$, and NOON spectra for methylene's lowest states at the FCI/6-31G level

State	$N_{\mathrm{h-p}}$	$N_{\mathrm{h-p}}^{(\mathrm{ref})}$	NOON
$1A_1$	0.32	0.32	1.9754, 1.9676, 1.8974, 0.0856, 0.0278 …
$1B_1$	2.13	2.14	1.9735, 1.9640, 0.9954, 0.9919, 0.0255 …
$2A_1$	0.35	4.00	1.9600, 1.9476, 1.9155, 0.0979, 0.0276 …
$1A_2$	2.16	2.19	1.9617, 1.9562, 1.0025, 0.9932, 0.0292 …

nature is combined with the high multiconfigurational character when $\|C_2\|^2 \cong 1$ (see Table 6.3). This picture is in accordance with $N_{\mathrm{h-p}}^{(\mathrm{ref})}$ $[2A_1]$ = 3.997. The example demonstrates a usefulness of the hole-particle indices as supplementary characteristics of excited states. Incidentally, one can observe from Table 6.4 that the $1B_1$ state of CH_2 provides an example of the pathological state (for definition, see Sect. 6.4).

The above outlined peculiarities in using EUE indices for the multiconfigurational states return us to the problem of constructing excitation localization indices for arbitrary excitations. We can proceed in many ways. In the scheme [82, 84] the operator modulus of density matrix difference, ΔD, is used. Namely, the normalized excitation operator, $\hat{L}^*$, can be naturally introduced as follows:

$$\hat{L}^* = (\mathrm{Tr}|\Delta D|)^{-1} \Delta D, \tag{6.77}$$

$$\Delta D = D^* - D,$$

where D^* is the charge density matrix for the excited state of interest, and by definition, $|\Delta D| = [(\Delta D)^2]^{1/2}$. Then, as usual, the atomic indices

$$L_A^* = \sum_{\mu \in A} (\hat{L}^*)_{\mu\mu} \tag{6.78}$$

furnish the excitation localization measure assigned to each atom of the excited molecule. Doing so for CIS-type states (6.67) we automatically produce indices (6.76).

To solve the same problem by another way, take atomic EUE distributions (6.50) and compute the corresponding normalized indices

$$L_A^{*(\mathrm{eff})} = |(D^*)_A^{\mathrm{eff}} - D_A^{\mathrm{eff}}| / \textstyle\sum_B |(D^*)_B^{\mathrm{eff}} - D_B^{\mathrm{eff}}|. \tag{6.79}$$

These indices characterize the excitation localization in its own manner. For CIS-like states the distributions $\{L_A^*\}$ and $\{L_A^{*(\mathrm{eff})}\}$ are sufficiently close, as the methylene molecule example shows. In particular, we have

$$L_{\mathrm{C}}^*[1\mathrm{B}_1] = 0.921,\ L_{\mathrm{H}}^*[1\mathrm{B}_1] = 0.039,$$

$$L_{\mathrm{C}}^{*(\mathrm{eff})}[1\mathrm{B}_1] = 0.971,\ L_{\mathrm{H}}^{*(\mathrm{eff})}\ [1\mathrm{B}_1] = 0.015.$$

However, for the double excited state the compared results are markedly different:

$$L_{\mathrm{C}}^*[2\mathrm{A}_1] = 0.933,\ L_{\mathrm{H}}^*[2\mathrm{A}_1] = 0.034,$$

$$L_{\mathrm{C}}^{*(\mathrm{eff})}[2\mathrm{A}_1] = 0.760,\ L_{\mathrm{H}}^{*(\mathrm{eff})}\ [2\mathrm{A}_1] = 0.120.$$

We conclude that for the depiction of excitation localization, the EUE indices should be used with a certain care.

6.11 Conjugated Hydrocarbons in π-Electron Schemes

In this section we consider computations of the EUE indices for moderate-sized systems within the easily implementable semi-empirical methods. Before doing so, we briefly touch on simplified ab initio approaches to polyaromatic hydrocarbons (PAHs). Many of the ab initio studies are based on various UHF and unrestricted DFT schemes [86–88]. By these schemes, crude estimates of EUE effects can be made even from the $\langle \mathbf{S}^2 \rangle_{\mathrm{UHF}}$ values. Really, for slightly correlated systems the semi-quantitative relation

$$N_{\mathrm{eff}} \approx N_{\mathrm{odd}}/2 \tag{6.80}$$

holds (e.g., compare with the results of Table 6.2 for equilibrium geometries). As an additional example, take the benzene molecule for which we have (in 6-31G basis set): $N_{\mathrm{odd}}[\mathrm{UHF}] = 1.101$, and $N_{\mathrm{eff}}[\mathrm{UHF}] = 0.584$. Recalling Eq. (6.13), we can expect that the rough estimate

$$N_{\mathrm{eff}} \approx \langle \mathbf{S}^2 \rangle^{\mathrm{UHF}} \tag{6.81}$$

can be utilized for a simplistic description in other moderately correlated systems. Such estimates are easy because the needed data can be routinely obtained by most quantum chemical programs. Besides, in the current literature, the $\langle \mathbf{S}^2 \rangle^{\mathrm{UHF}}$ data are available for many PAHs (e.g., see [87, 88]).

To be more specific, consider linear polyacenes for which a model geometry will be used here and throughout the paper: the carbon backbone is formed by regular hexagons with the C-C bond length of 1.4 Å and the C-H bond length of 1.09 Å. Using the Gaussian program package [89], we computed $\langle \mathbf{S}^2 \rangle^{\mathrm{UHF}}$ for the first ten linear polyacenes $C_{4n+2}H_{2n+4}$ at the UHF/6-31G level. The results are conveniently expressed via the linear regression

$$\langle \mathbf{S}^2 \rangle^{\mathrm{UHF}}(n) = 0.6462 \cdot n, \tag{6.82}$$

with residual variance 10^{-3}, thus reflecting a size-consistent behavior of the index.

In the case of large carbon-containing systems it is suitable to compare the EUE index value per carbon atom. For instance, we introduce

$$\overline{\langle \mathbf{S}^2 \rangle}^{\mathrm{UHF}} \equiv \langle \mathbf{S}^2 \rangle^{\mathrm{UHF}} / N_{\mathrm{C}}, \tag{6.83}$$

where N_{c} is a number of carbon atoms ($N_{\mathrm{c}} = 4n + 2$ for Eq. (6.82)). For large linear acenes we have from Eq. (6.82)

$$\overline{\langle \mathbf{S}^2 \rangle}^{\mathrm{UHF}} \cong 0.16.$$

Likewise, other EUE indices per carbon atom are defined:

$$\bar{N}_{\mathrm{odd}} \equiv N_{\mathrm{odd}} / N_{\mathrm{c}}, \tag{6.84}$$

$$\bar{N}_{\mathrm{eff}} \equiv N_{\mathrm{eff}} / N_{\mathrm{c}}, \tag{6.85}$$

Returning to Eqs. (6.81) and (6.83) we suggest a rough estimate,

$$\bar{N}_{\mathrm{eff}} \cong 0.16, \tag{6.86}$$

is reasonable for sufficiently large linear polyacenes treated with the 6-31G basis set.

An interesting point is a measure of the participation of π-electrons in the total unpairing. Our experience with small conjugated systems tells that usually $\approx 2/3$ of the average $D_{\mathrm{C}}^{\mathrm{eff}}$ value is from π-electrons. Together with Eq. (6.85) it gives a crude estimate

$$\bar{N}_{\mathrm{eff}}[\pi] \cong 0.1. \tag{6.87}$$

This approach is in concordance with the fact that in large conjugated systems, electron correlation largely influences the outer π- electron shells. That is why most physical models of conjugated polymers are based on one or another version of the π-electron approximation (most often in the form of a many-electron Hubbard model, see reviews [69, 90]). Furthermore, previous non-empirical studies [8, 9, 11] in large conjugated molecules (by using DMRG and MR-AQCC) have considered only π-electron contribution to EUE. All this has motivated us to undertake a detailed study [12] of the EUE effects within the conventional π-electron theory. Below we shortly outline the main results of this study.

For the selected PAHs, the results are presented in Table 6.5 where in the structural formulas the EUE atomic distributions are displayed in a qualitative manner. All EUE indices (except for $\overline{\langle \mathbf{S}^2 \rangle}^{UHF}$) are computed within the hole-particle approach, Eqs. (6.37) and (6.38), which, for ground states, is equivalent to the Head-Gordon approach. Here, the Parizer-Parr-Pople (PPP) π-electron approximation is employed. We see that again the UHF scheme based on Eqs. (6.10) and (6.17) works well (in respect to the CCD results), and this fact was emphasized in [12]. The π-electron UHF scheme (π-UHF) is favorable because of its simplicity of computation, and ease of interpretation. However, this method is not recommended for systems with a relatively small π-electron correlation effects, e.g. in the peropyrene molecule (the third entry in Table 6.5). In the case of too little electron correlation, the half-projected Hartree-Fock (HPHP) [91] and EHF schemes are

Table 6.5 Specific EUE indices $\overline{N}_{eff}^{UHF}$, $\overline{N}_{eff}^{EHF}$, $\overline{N}_{eff}^{CCD}$, and $\overline{\langle \mathbf{S}^2 \rangle}^{UHF}$ for π-systems in pentacene bisanthene, peropyrene, and pyranthrene within π-UHF

π-Structure	$\overline{\langle \mathbf{S}^2 \rangle}^{UHF}$	$\overline{N}_{eff}^{UHF}$	$\overline{N}_{eff}^{EHF}$	$\overline{N}_{eff}^{CCD}$
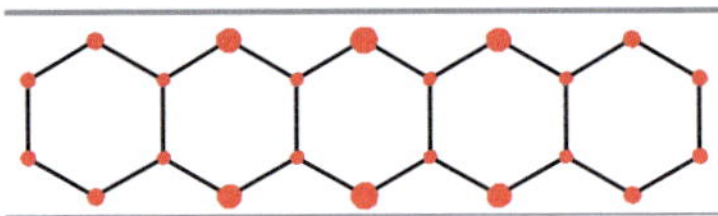	0.056	0.061	0.059	0.061
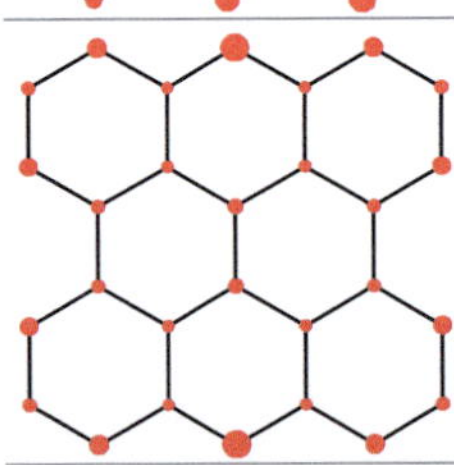	0.045	0.056	0.055	0.057
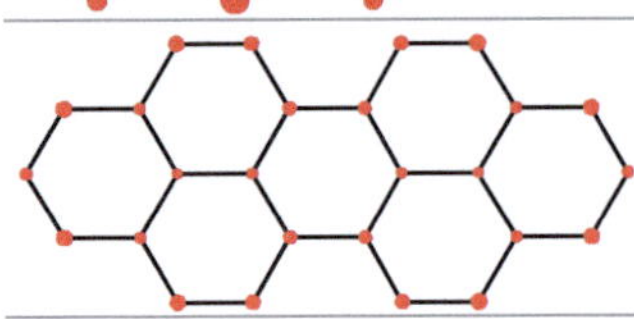	0.025	0.027	0.041	0.047
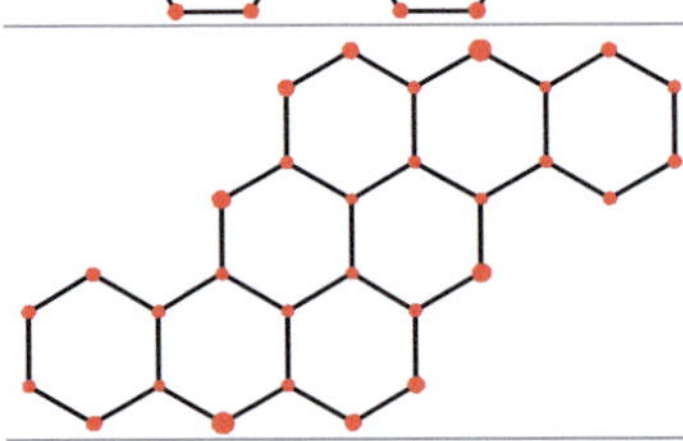	0.032	0.036	0.042	0.049

applicable. We remark that the EHF results in Table 6.5 are sufficiently close to the CCD ones even for the peropyrene and pyranthrene molecules which have modest electron correlation effects. These observations are the basis on which the UHF method, with obligatory use of Eq. (6.17) or more refined indices, can be recommended for studying EUE effects in large graphene-like molecules [12]. The usefulness of this approach shows Table 6.6 containing two examples taken from the cited paper. To the previously defined quantities, Eqs. (6.83) and (6.85), we have included in the table one additional characteristic index, $\aleph^{\text{loc}}$. The index $\aleph^{\text{loc}}$ gives a mean number of atoms (sites) on which the unpaired electrons are preferentially localized. Explicitly,

$$\aleph^{\text{loc}} = [\sum_A (D_A^{\text{eff}})^2]^2 / \sum_A (D_A^{\text{eff}})^4. \tag{6.88}$$

This equation is a generalization of the participation ratio (6.19) and gives a more sharp estimate for a number of strongly localized atomic centers (sites). The related index was employed in [92] where it is shown that the index can well distinguish between localized and extended states. From Table 6.6 we see that indeed $\aleph^{\text{loc}}$ gives an acceptable average number of the essentially localized unpaired electrons. When using $\aleph^{\text{loc}}$ one must keep in mind that this index is informative if $\aleph^{\text{loc}} \ll N_c$, that is in the case of a sufficiently sharp EUE localization.

Now we remark on the NOON spectrum $\{\lambda_k\}$ given in the fourth column of Table 6.6. Similar plots are frequently displayed when considering the nature of EUE in large molecules [8, 9, 11–13, 26]. The first system in the table belongs to the so-called periacene family. The earlier theoretical study of this family was given in [93] where a simplified Hubbard-like π-UHF method was applied. In the recent papers [8, 9, 11] the EUE analysis for the periacenes was given at the high-level ab initio level. Here, we can directly compare these reliable ab initio results and ours, thanking to the fact that the needed ab initio data were kindly provided by the authors of [9, 11]. The results are displayed in Fig. 6.4, where the same nomenclature of periacenes, as in [9, 11], is used. Comparing the corresponding plots, we observe their really close similarity. More specifically, the same localization of few NOON in the vicinity of 1 is found in the ab initio as well as the π-electron calculations, and this localization corresponds to a genuine open-shell (polyradical) singlet structure. A general view of the plots is also similar. Moreover, in the (5a, 6z) periacene the EUE atomic localization is comparable (see [12] for detail). On this account, we suggest that the π-electron EUE model, which is based on the simple UHF expression (6.17), should be useful for other large-scale conjugated systems, at least at a qualitative level. The symmetry (exact or approximate) of the NOON in respect to point $\lambda = 1$ is also deserve attention. For alternant hydrocarbons within any correct PPP scheme this symmetry is exact, and it is easy to prove (see the last paragraph in Appendix B). The ab initio results [9, 11] approximately fulfil this symmetry which implicitly reflects the physical equivalence of the holes and the particles in alternant π-systems [94] (see again plots in Fig. 6.4).

Table 6.6 EUE indices $\overline{N}_{\mathrm{eff}}^{\mathrm{UHF}}$ and $\overline{\langle \mathbf{S}^2 \rangle}_{\mathrm{UHF}}$, NOON spectrum $\{\lambda_k\}$, and average number of localized EUE sites, $\aleph^{\mathrm{loc}}$, within π-UHF for graphene megamolecules with armchair-zigzag edges

Megamolecule	$\overline{\langle \mathbf{S}^2 \rangle}^{\mathrm{UHF}}$	$\overline{N}_{\mathrm{eff}}^{\mathrm{UHF}}$	Spectrum	$\aleph^{\mathrm{loc}}$
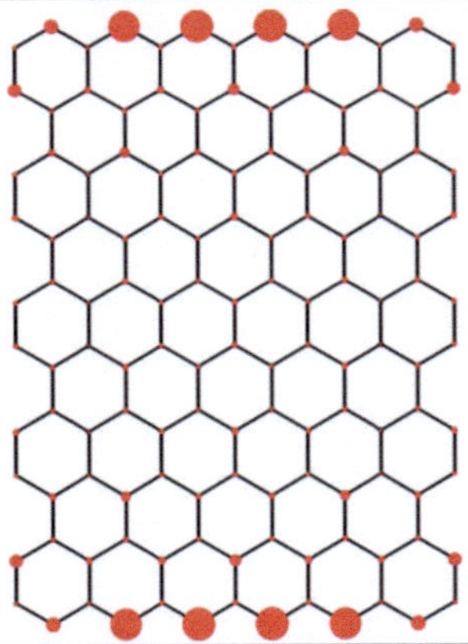	0.033	0.047	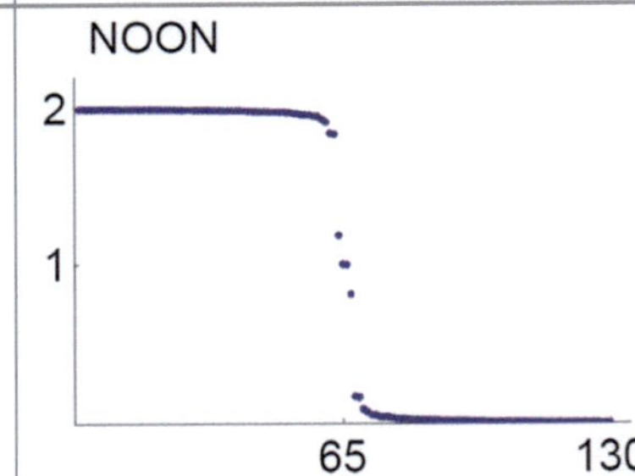	12.5
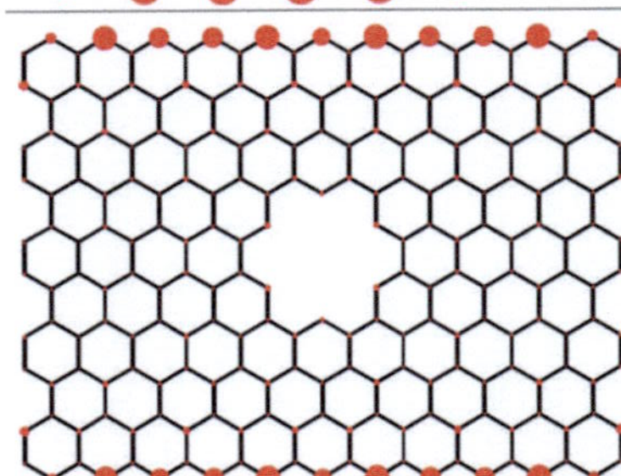	0.032	0.049	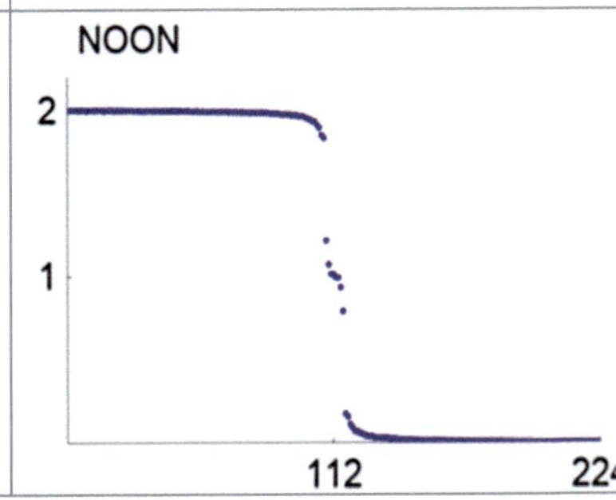	22.6

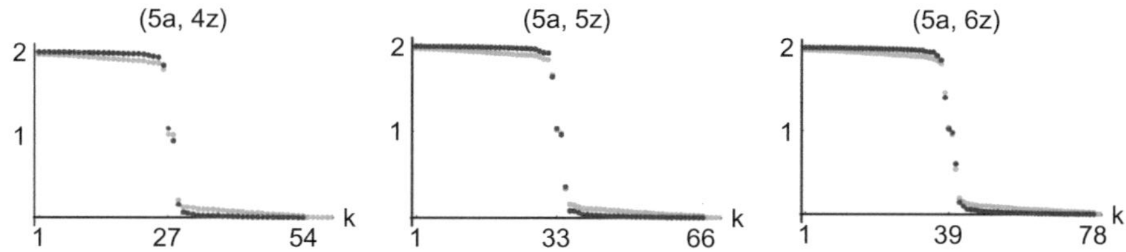

Fig. 6.4 Comparison of the ab initio [9] (*violet color*), and semi-empirical (*green color*) NOON spectra for (5a, nz) periacenes

Notice that the above cited EUE ab initio study was performed for active π-orbitals. Thus, the EUE indices obtained from corresponding NOON spectra, are related to respective π-electron contributions. In case of the periacenes (5a, 4z), (5a, 5z), and (5a, 6z), we find, from the respective NOON spectra, the following ab initio $\bar{N}_{\text{eff}}[\pi]$ values: 0.111, 0.108, 0.108, in agreement with rough estimation (6.87). The analogous π-electron PPP data will be given in Sect. 6.14, Table 6.11. They are approximately twice less than the ab initio values. At the same time, if we exploit the so-called Mataga formula for two-electron two-center integrals $\gamma_{\mu\nu}$ (with a Coulomb-like distance dependence), then we obtain the results closer to the ab initio ones. In particular, for the (5a, 6z) periacene the π-UHF scheme with Mataga's $\gamma_{\mu\nu}$ gives $\bar{N}_{\text{eff}} = 0.129$. Nevertheless, the standard π-parametrization we use ($\gamma_{\mu\nu}$ by Ohno's formula) is more appropriate for π-electron correlation effects, as was established long ago. We also computed index $\overline{\langle \mathbf{S}^2 \rangle}^{\text{UHF}}$ (by using the program package [89]) and get a crude ab initio estimate via Eqs. (6.81) and (6.83). For the (5a, 6z) periacene at the 6-31G level we thus obtained $\bar{N}_{\text{eff}}[\pi] \cong 0.09$ which seems quite reasonable in comparison with the above non-empirical value $\bar{N}_{\text{eff}}[\pi] \cong 0.108$ from [9].

As mentioned in Sect. 6.9, the EUE structure can be interpreted in somewhat notional terms of antiferromagnetism [95, 96]. Indeed, a local spin density is absent in any correctly defined singlet state, and, strictly speaking, the Néel-like spin structure is not possible for the single spin-singlet molecule. Thence, we cannot introduce, as usual, the antiferromagnetic order parameter (such as average difference of spin density between neighboring atoms). For the correlated singlet states, spin density matrix can be substituted with EUE density matrix (6.5). Consequently, index $\bar{N}_{\text{eff}}$ might serve as an appropriate order parameter for polymer structures,. This index satisfies inequality: $0 \le \bar{N}_{\text{eff}} \le 1$, that is natural to expect from the order parameter. In our case, $\bar{N}_{\text{eff}} = 1$ corresponds to the ordered Néel state with the maximal 'spin' value in each sublattice of the bipartite structure. The given interpretation introduces an obviousness in understanding EUE for bipartite network structures. By adopting this reasoning, one can, moreover, invoke the best spin-polarized orbitals, that is the SPEB solutions discussed in Sect. 6.9. It allows us to reinterpret $\bar{N}_{\text{eff}}$ as a "spin" order parameter for exact or almost exact wave functions too.

Before closing this section let us comment on the UHF calculations presented above. From the formal viewpoint, UHF is the one-electron model which deals with a single determinant wave function $|\Phi\rangle$. However, for strongly correlated systems the UHF wave function well mimics many properties of the spin-projected determinant $|\Phi^{\text{ext}}\rangle$ which is, of course, many-determinant state and which takes into account electron correlation. The closeness between $|\Phi\rangle$ and $|\Phi^{\text{ext}}\rangle$ had been demonstrated long ago [70] with the infinite polyene chain treated analytically within the 'diagonal' Hubbard Hamiltonian approximation. The authors had suggested that it is a general feature of UHF solutions in polymeric π-problems. Our experience with EHF computations on large π-systems confirms these expectations. In particular, for the large systems the UHF charge RDM, D^{uhf}, as in Eq. (1.10), is a good approximation to the EHF charge RDM, D^{ext}, which is provided by the variational $|\Phi^{\text{ext}}\rangle$ state. Nevertheless, the UHF spin density matrix does not vanish for the UHF (spin-polarized) singlet ground state. Therefore, upon obtaining the UHF solution, the spin density matrix should be ignored (fixed to zero) what corresponds to an implicit purification of the spin-contaminated singlet state. At the same time, charge density D^{uhf} is well defined, and indeed very close to the EHF counterpart. For instance, we find the following squared norms, $||D^{\text{ext}} - D^{\text{uhf}}||^2/N$ (deviation of D^{uhf} from D^{ext} per π-electron): 0.0007 for decacene $C_{42}H_{c4}$, and 0. 0002 for acene $C_{102}H_{54}$, respectively. These and many other examples (recall also Table 6.5) allow us to consider, for large systems, the usual spin-contaminated UHF solutions as a good approximation to main properties of the spin-adapted EHF solutions.

6.12 Giant Hydrocarbons and Nanographenes in a Spin-Polarized Hückel-like Scheme

In case of huge conjugated systems with several thousands of atoms, even the π-electron UHF method, in its full version, necessitates using high-performance computer clusters. Meanwhile, many important problems of nanoelectronics require studying novel molecular materials, including graphene nanoribbons, nanoislands, nanowiggles and other unusual giant honeycomb structures [74, 97–100]. Most of these structures are based on the so-called bipartite lattices. By definition, the bipartite lattice is formed by two interpenetrating sublattices, and each of these sublattices contains only one kind of atoms. Following [101], we will use the term "lattice" in an extended meaning, allowing the term for finite lattices and even for any finite-size atomic structures. In the theory of π-conjugated molecules, the standard term "alternant system" is a full counterpart of the term "bipartite lattice".

There are many remarkable theorems dealing with abstract and realistic models of bipartite lattices [94, 101–104]. The well-known Coulson-Rushbrooke pairing theorem [102] is one of them. Additionally, the pairing theorem has a nice and useful matrix representation due to Hall [105]. The Hall formula (see below Eq. (6.90)) is valid within the Hückel method, and there is its analogue within the PPP one-electron approximation. In solid state physics, the counterpart of the Hückel approach is known as the tight-binding (TB) model. TB schemes, now more refined than before in the old solid-state physics days, are very popular because they have advantages to handle atomic cluster with thousands of atoms, reaching experimental sizes [106, 107]. Unfortunately, all these methods ignore electron correlation. In [13] we modified the TB model for bipartite lattices in such a way that it can handle strongly correlated bipartite lattices, and describe in them the relevant EUE effects. Below we sketch the main results of this work, and leave most formal details to Appendix B.

We recall few simple facts from the TB (or Hückel) theory of bipartite lattices. For the carbon-containing conjugated systems, the usual basis set $\{|\chi_\mu\rangle\}$ of the orthonormalized $2p_z$-orbitals is employed. The corresponding one-electron Hamiltonian can be represented by the 2×2 block-structure matrix

$$h^0 = -\begin{pmatrix} 0 & B \\ B^+ & 0 \end{pmatrix}, \tag{6.89}$$

where all entries are expressed in units of $|\beta_0|$ with β_0 being the standard hopping (resonance) integral between nearest-neighbor sites (π-centers). The block B in Eq. (6.89) is the biadjacency matrix, that is $B_{\mu\nu} = 1$, if μ and ν are nearest-neighbor sites, otherwise $B_{\mu\nu} = 0$. Obviously, due to a bipartite structure of the considered lattices we can always renumber lattice sites in such a way that Eq. (6.89) holds true. From Eq. (6.89) it is not difficult to deduce the Hall formula [105] for the charge density matrix (or Coulson's bond-order matrix):

$$D_0 = \begin{pmatrix} I & B(B^+B)^{-1/2} \\ (B^+B)^{-1/2}B^+ & I \end{pmatrix}. \tag{6.90}$$

This and somewhat more general relations are rederived in Appendix B.

Certainly, Eq. (6.89) is only a specific case of Eq. (6.4), and no EUE effects are possible at this level of description. It would be important to extend the Hückel model in order to somehow account for electron correlation effects without over-simplifying the model. The approximation of this kind was given in [108] and applied to EUE problems in [13]. The most important expressions of this work are reproduced here (see cit. loc. for the argumentation and precursors of the model). The model was referred as to the quasi-correlated tight-binding (QCTB) method. Within QCTB, we construct the effective Hamiltonians matrices

$$h^\alpha = -\begin{pmatrix} \delta I & B \\ B^+ & -\delta I \end{pmatrix}, \quad h^\beta = -\begin{pmatrix} \delta I & B \\ B^+ & -\delta I \end{pmatrix} \tag{6.91}$$

where δ is treated as a fixed auxiliary parameter. The h^α and h^β matrices are the counterparts of common Fock matrices for spin-up and spin-down electrons, respectively. Unlike UHF, no self-consistency procedure is needed for obtaining the corresponding density matrices ρ_α and ρ_β. The approach used is the most similar to the earlier approximate one-parameter UHF theory (e.g., see [101], the second citation). However, we can always obtain nonzero correlation effects by a suitable choice of the fitting parameter δ, and it allows to extend the applicability of the whole approach. Only for very strongly correlated systems, QCTB and the one-parameter UHF theory scheme are virtually equivalent.

Now turn to computational aspects. For matrices ρ_α and ρ_β, a block representation is easy to find by simple algebra (see Appendix B). As a result, we get charge density matrix of the QCTB model, Eq. (B4), and the respective NOON spectrum, Eq. (B5). It comes to a suitable working formula for the main EUE index:

$$N_{\text{eff}} = N_{\text{h}-\text{p}} = N - 2\sum_{i=1}^{n} \varepsilon_i / \sqrt{\delta^2 + \varepsilon_i^2}. \tag{6.92}$$

Here $\varepsilon_i \equiv |\varepsilon_i|$ are eigenvalues of $(B^+B)^{1/2}$, that is $\{\varepsilon_i\}$ is precisely the Hückel energy spectrum (in modulus) of the respective alternant system (the bipartite graph spectrum). In specific computations we will use value $\delta = 7/24$ which was found by fitting. Incidentally, remark that for small δ it is easy to check that with second-order accuracy in δ, $N_{\text{odd}} = 2N_{\text{eff}}$, as suggested before from a numerical experience (see Eq. (6.80)).

The above quasi-Hückel approach to EUE turns out to be reasonable and sufficiently close to the UHF and even CCD schemes (see Table 1 in [12]). Here we extract from this reference two kind of representative examples. One kind of them is related to the conjugated polymer structures (Table 6.7), the other to the finite-size graphene nanoflakes (Table 6.8). Before considering Table 6.7, let us make brief preliminary remarks. For many π-electron structure, particularly, with translation symmetry the analytical solution of the Hückel band spectrum is well known. For instance, consider a long polyene chain [–(CH=CH)–]$_n$ (polyacetylene) as a paradigmatic example of strong correlation in the physics of conjugated polymers [69, 109]. In case of the finite polyene chain the Hückel spectrum is $\varepsilon_k = 2\cos[\pi k/(2n+1)]$(see any quantum chemistry textbook). For the asymptotic case, $n \to \infty$, straightforward computations on Eq. (6.92) (with approximating a sum by integration method) lead to

$$N_{\text{eff}} = N\left(1 - \frac{2}{\pi}\arcsin[1/\sqrt{1+\delta^2/4}]\right). \tag{6.93}$$

We see from this equation that in the limit of large δ (very strong correlation effects) the EUE index $N_{\text{eff}} \to N$, as it should be. Evidently, the value $N_{\text{eff}} = N$ corresponds to breaking each of π-bonds, when all π-electrons are unpaired. Remark that for infinite polymer chains the NOON spectrum $\{\lambda_k\}$ generally covers a whole interval [0, 2]. Therefore, instead of discrete set $\{\lambda_k\}$, the continuous (more

Table 6.7 The $\overline{N}_{\text{eff}}$ index and quasi-continuous NOON spectrum $\lambda(k)$ $(0 \leq k \leq 1)$ for infinite π-conjugated polymer chains

Polymer	$\bar{N}_{\text{eff}}$	Spectrum
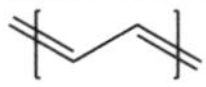 $(\eta = 0)$	0.092	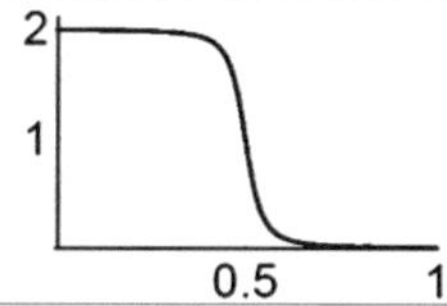
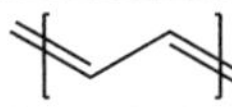 $(\eta = 0.07)$	0.076	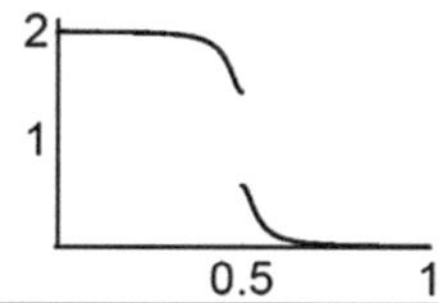
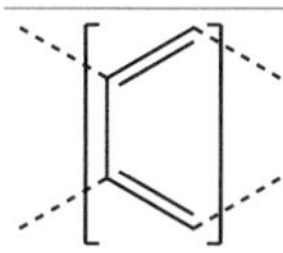	0.098	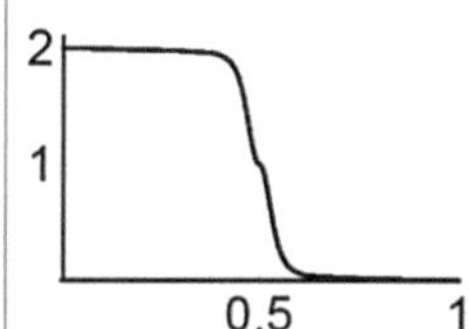
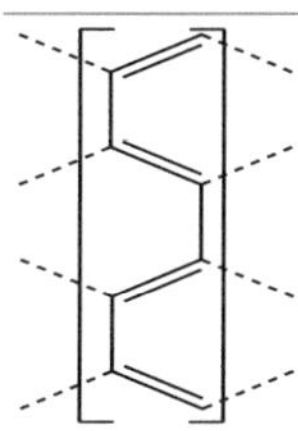	0.086	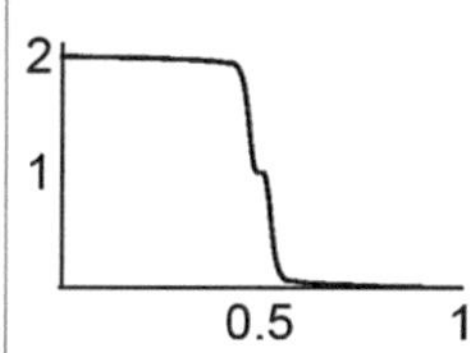
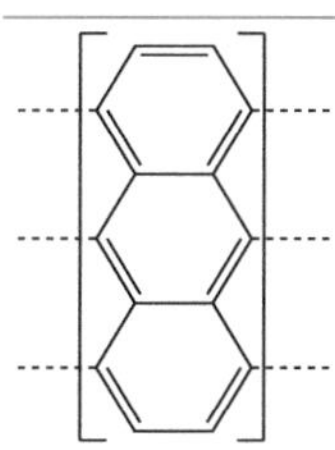	0.041	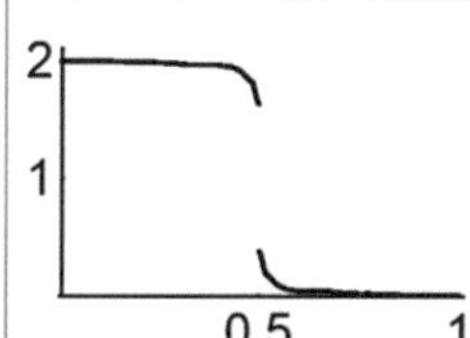

exactly, quasi-continuous) function $\lambda(k)$ of the continuous variable k makes its appearance. For convenience we make using the unity interval [0, 1] for continuous variable k.

A more general case is the polyene chain with alternating resonance integrals $\beta_{\mu,\mu+1} = [1 + (-1)^{\mu+1}\eta]\beta_0$, where η is usually small quantity (we put $\eta = 0.07$). The Hückel spectrum is of the form [110]: $\varepsilon(k) = \sqrt{2}\,[1 + \eta^2 + (1 - \eta^2)\cos \pi k]^{1/2}$, where $0 \leq k \leq 1$. This case is intractable analytically, but numerical computations are easily performed, and the results are given in Table 6.7 (the first two systems in the table). Another interesting example is the linear polyacene (the third system in Table 6.7), for

Table 6.8 TB energy, $\bar{E}^0$, QCTB energy, $\bar{E}$, both per electron, and the EUE characteristics $\bar{N}_{\mathrm{eff}}$, $\bar{N}_{\mathrm{eff}}^{\mathrm{bord}}$ for graphene nanoflakes. $\bar{N}_{\mathrm{eff}}^{\mathrm{bord}}$ is analogous to $\bar{N}_{\mathrm{eff}}$ for border atoms

Nanocluster		$\bar{E}^0$	$\bar{E}$	$\bar{N}_{\mathrm{eff}}$	$\bar{N}_{\mathrm{eff}}^{\mathrm{bord}}$
C_{1302}	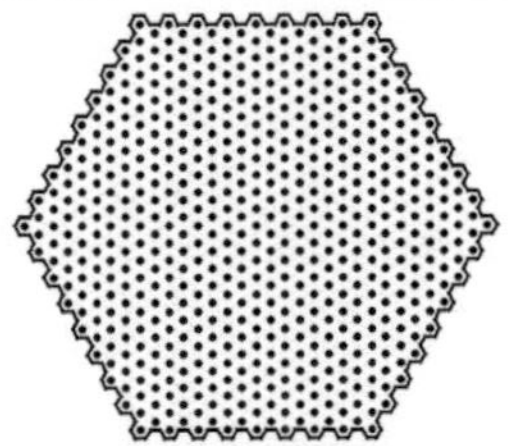	−1.553	−1.589	0.040	0.047
C_{1350}	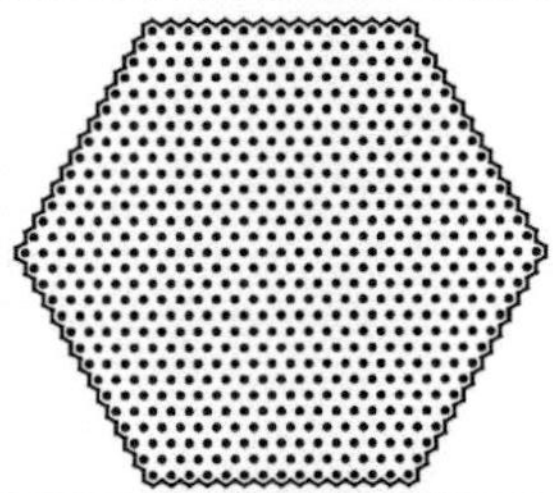	−1.551	−1.589	0.050	0.112
C_{1332}	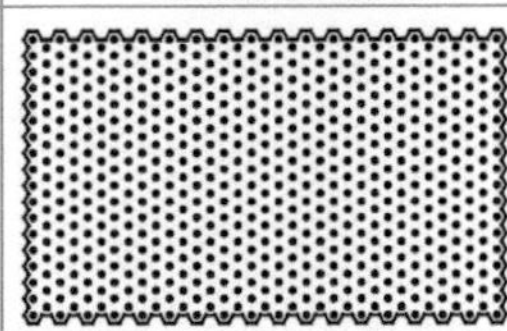	−1.550	−1.587	0.046	0.078

which in accordance with Coulson [111] we have $\varepsilon(k) = \left[1 \pm (9 + 8\cos\pi k)^{1/2}\right]/2$. In order to present a more complete comparison we add in the table the results for the graphene nanoribbon (4-ZGNR in the standard nomenclature) and for the poly(perianthracene) chain. The π-electron band structure of these two systems is computed by a code from [112].

As seen from Table 6.7, only the polyacetylene with alternating bonds and poly (perianthracene) molecules exhibit a gap in their NOON spectra. In contrast, the polyacene and 4-ZGNR demonstrate a quasi-continuous NOON spectrum covering the whole interval [0, 2]. Furthermore, crowding $\lambda(k)$ near a 'polyradical range', that is near $\lambda = 1$, is observed in these spectra. A significant difference, in the $\bar{N}_{\mathrm{eff}}$ index, between the 4-ZGNR and poly(perianthracene) can be simply understood in terms of Clar's aromatic sextet theory (for the latter see, e.g., [74, 113]).

Now we will discuss in brief the QCTB results for three graphene nanoclusters with $N \propto 10^3$, presented in Table 6.8. We only note that an unprecedented rise of interest in the graphene engineering researches generated the enormous literature in which recent books [74, 114, 115] only minimally reflect this graphene popularity. The first two systems in Table 6.8 are of a nanoflake family with the D_{6h} symmetry (hexagonal graphene nanoflakes). The cluster system, C_{1302}, is with the armchair-shaped edge, and the second, C_{1350}, with the zigzag-shaped edge. From the table we see that these two clusters have a small difference in energy stability (within QCTB), but a significant difference in the EUE characteristics. In zigzag-edge nanocluster C_{1350}, the third system in Table 6.8, more electrons are unpaired, and again these unpaired electrons are preferentially localized on edge atoms. It is revealed by localization index $\bar{N}_{\mathrm{eff}}^{\mathrm{bord}}$ (sum of atomic EUE occupancies divided by a number of the border atoms). On this account the zigzag edge atoms should be more unstable, or more reactive than the armchair edge atoms, and thereby the armchair nanoflakes be more stable in accordance with experiment (see [116], p. 382) and a model DFT study [117]. Chemical reactivity of graphene structures is a rather frequent issue discussed in current chemical literature [74, 118–122], and the principal inference we can make is that the major reactivity contribution comes from the edge states of nanoclusters. The very different models, from simplistic semiempirical to high-level nonempirical ones, predict the same qualitative trends. Notice that in the case of graphene nanoribbons with zigzag edges the strongly localized edge states were first reported almost 20 years ago [72] where the Hubbard π-UHF model was used. Apparently, in all models the characteristic effects of chemical topology are exhibited, and this fact demonstrates the practical usefulness of even naive models for studying large conjugated systems. We close this chapter by noting that the proposed π-electron QCTB scheme can be modified for an all valence-electron treatment by using the extended Hückel MO theory [123]. In this case the ionization potentials of 2p-electrons in the respective Wolfsberg-Helmholtz relation should be changed similar to Eq. (6.89).

6.13 Electron Unpairing in Strong Fields

The behavior of molecules under external perturbation shows the interesting, but not unexpected, fact that the electron unpairing greatly increases in strong fields. We consider here some representative examples carried over from [124, 125]. First, we discuss the effects of static electric fields for small molecules. A typical illustration is provided by an example of the rhombic cluster of Li_4 in an atomic-scale electric field ($\sim$0.1 atomic units). The results of the FCI/STO-3G calculations are shown in Fig. 6.5 where we plotted, in atomic units, the dipole moment d_x and N_{eff} as functions of the electric field strength **E**, and the static field is applied along the longest diagonal (x-direction) of the rhombus.

By inspecting the plots, we see a strong increase of the dipole moment in the field, but N_{eff} behaves more unpredictably, particularly in the region where the dipole moment curve undergoes a small inflection. A sharp maximum of N_{eff} in this region corresponds to a diradical state ($N_{\text{eff}} \cong 2.04$). Interestingly, in this extremal state the most unpaired atom (judging from D_A^{eff}) is the ending atom on the longest diagonal, whereas the opposite atom on the same diagonal has zero EUE density and net atomic charge +1 (that is, locally it is Li^+). This corresponds to the valence scheme of the form

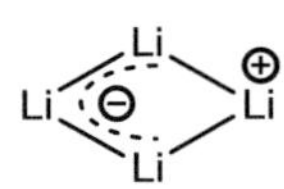

Practically the same field dependency is obtained within the CISD and EHF schemes. Furthermore, passing from the STO-3G basis set to the 6-31G basis set, we obtain similar plots. For other small systems, see [125]. As one can see from [124], the analogous treatment (at the FCI/PPP level) of small conjugated hydrocarbons, such as the naphthalene and biphenylene molecules, gives the results resembling those of the Li_4 case. For extended conjugated molecules the results are even more pronounced. Notice that in [124] we preferred using κ-index, Eq. (6.22), to using other EUE measures because the N_κ index requires no computations of D, and this gives a certain advantage when it is necessary to calculate, at the FCI/PPP level, the κ-index for hundreds or more electric-field points.

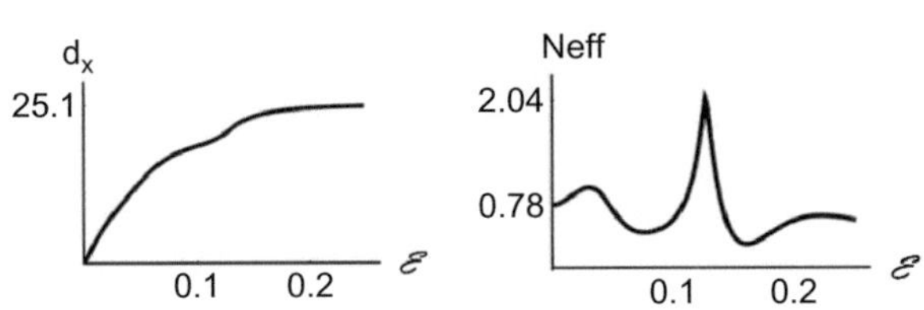

Fig. 6.5 Changes of dipole moment d_x and N_{eff} in a uniform electrostatic field of strength **E** for the rhombic Li_4 cluster at the FCI/STO-3G level

Table 6.9 Dipole moment d_x and EUE index N_{eff} for decacene $C_{42}H_{24}$ and Clar's structure $C_{42}H_{20}$ in a static electric field as functions of the strength field **E**, applied along the long molecular axis (x-axis) within π-EHF scheme

π -system	d_x	N_{eff}
$C_{42}H_{24}$ 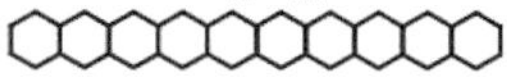	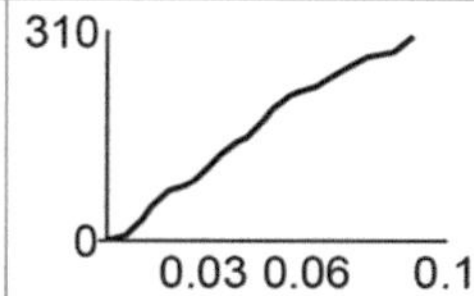	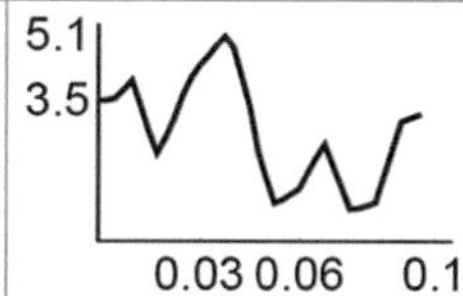
$C_{42}H_{20}$ 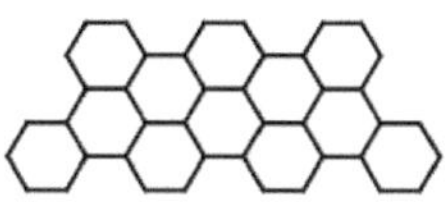	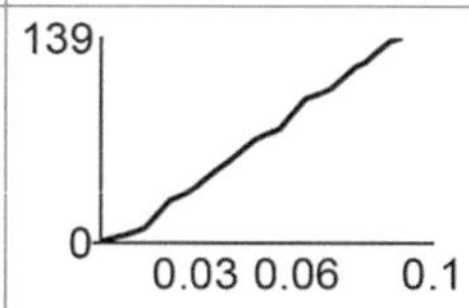	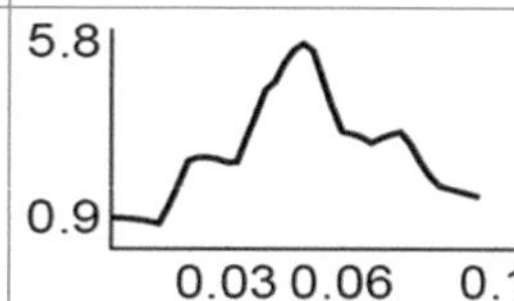

It is interesting to see a typical difference in behavior between strongly and slightly correlated systems in a strong static field. For definiteness, we examine two molecules with the same numbers of carbon atoms. These are the linear decacene molecule, $C_{42}H_{24}$, and Clar's type aromatic hydrocarbon $C_{42}H_{20}$ (see Table 6.9). Recall that Clar's hydrocarbon is described by a structural formula which consists exclusively of benzenoid rings interlinked by quasi-single bonds [126]. It is quite natural that the PAHs of the Clar type have significantly less diradical (polyradical) character. The data from Table 6.9 for the decacene and Clar's system $C_{42}H_{20}$ in zero electric field agree with this. However, the situation is somewhat different in the case of strong field. Namely, the maximal N_{eff} index of the Clar system is more than six times larger than that in the field's absence, and this value becomes still greater than the maximal N_{eff} index for the decacene molecule. In the latter, we observe only a modest increase of the electron unpairing in the field (the maximal N_{eff} is about 1.5 times larger than that in the field's absence). Note that the dipole moment shows the qualitatively similar behavior for both systems.

We would like to stress that the observed enhancing of EUE effects in the stable PAHs under strong fields is not confined to the static electric field. Somewhat similar effects were reported in [127] where the action of strong magnetic field on small acenes and antiaromatic systems was examined at the FCI/PPP level (within London's model of magnetic field effects). It turned out that in the strong magnetic field, aromatic molecules usually become diradicaloid and non-aromatic. Accordingly, the antiaromatic systems dramatically reduce their initial diradicaloid character, and thus lose their initial antiaromaticity. The interconnection between EUE and singlet-triplet splitting (STS) was also discussed for these π-problems [127]. At the ab initio level, a more extended study of STS and the polyradical character was presented in [11].

Let us return to the external field effects. Other cases are given in [27] where we studied small molecules in laser fields which generate transient diradical or more complex structures. Below we extend this study to the conjugated systems by using

Table 6.10 TD-UHF temporal behavior of dipole moment $d_x(t)$ and EUE index $N_{\text{eff}}(t)$ for the decacene molecule in the pulse laser field with enveloping function $\mathbf{E}(t)$ (all quantities in atomic units)

$\mathbf{E}(t)\cos(\omega t)$	$d_x(t)$	$N_{\text{eff}}(t)$
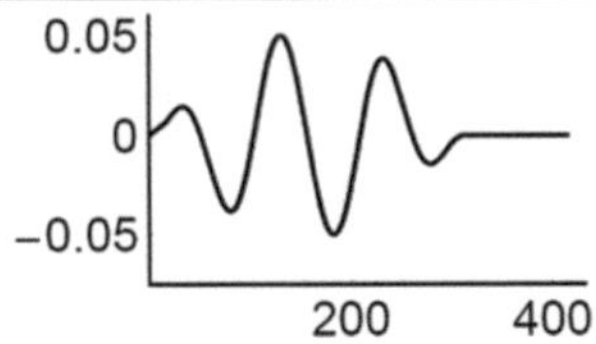	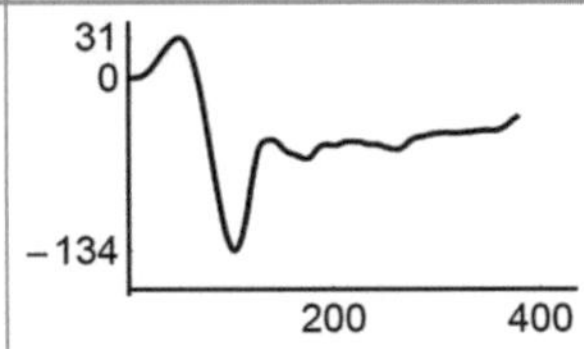	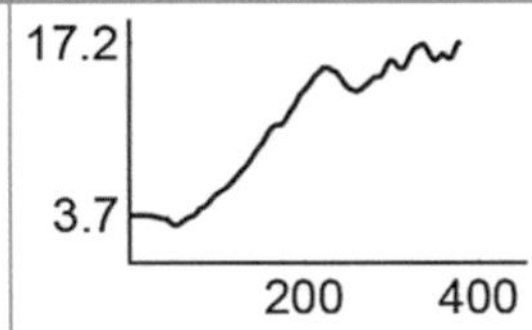

Total observation time is about 400

the π-electron time-dependent UHF (TD-UHF) method. We take again the decacene molecule, and the same pulsed laser field, as in [27], the first entry in Table 6.1. The corresponding pulse field is of the form $\mathbf{E}(t)\cos(\omega t)$, where $\mathbf{E}(t)$ is an enveloping function, and $\omega = 0.06$ atomic units. The pulse field and temporal dependencies of dipole moment, d_x, and N_{eff} index are given in Table 6.10. These data show that, unlike the static field (see Table 6.9), the pulse laser field produces the enormous electron unpairing in the linear decacene molecule. The similar results are obtained for other PAHs.

The above considered effects are implicitly connected with other strong perturbation effects reported in the literature. For instance, distorted graphene-like structures are really observed after irradiation generating lattice irregularities, e.g. vacancies. It leads to appearance the dangling carbon bonds, and thereby to the electron unpairing. The recent semiempirical study of surface states at a many-electron (UHF-type) level was given in [128]. In our context, paper [129] is presented even more relevant as giving the ab initio model of single defects and treating the EUE effects simultaneously. Summing this section we would like stress a nontrivial role of electron unpairing in various physical phenomena taking place in molecular materials.

6.14 In Search of Better EUE Measures

Now we address a difficult issue what is good and what is not for the EUE theory. The first discussions were started in papers [28, 29]. In the present work the Yamaguchi index [1] is treated mainly as a historically valuable first quantity for the EUE characterization, but not recommended for wide practical use (works [130] are a rare example of using $N_{\text{eff}}^{\text{Yam}}$ now). Instead, we advocate the H-G index [5] for which a clear hole-particle physical nature, Eq. (6.40), is established [12]. Nevertheless, one can encounter intrinsic interpretational problems when judging the adequacy of any quantitative EUE measure, and the N_{eff} index is not the exception. For instance, we obtained in the anthracene molecule a nonempirical estimate

$N_{\mathrm{eff}} \cong 1.9$. Given the well known stability of anthracene, is it a reasonable magnitude? Before going on, we must ones again consider a formal side of the issue.

For this, we will include into consideration the second index proposed by Head-Gordon in [5]. This index denoted by $N_{\mathrm{odd}}[2]$ (as in Appendix A) is defined as follows:

$$N_{\mathrm{odd}}[2] = \sum_k (2\lambda_k - \lambda_k^2)^2. \tag{6.94}$$

Equation (6.94) was intensively applied in papers [9, 11, 129, 131] as even more preferable tool for quantifying EUE. By using $N_{\mathrm{odd}}[2]$, a not so important part of the NOON spectrum (mainly due to dynamical electron correlation) is effectively cut off. Another way to approach the EUE problems is to use matrix invariants of the deviation matrix

$$\Delta D = D - 2\rho. \tag{6.95}$$

This matrix has a meaning of a hole-particle correction to 2ρ (see Appendix C for detail). Following the same procedure presented earlier for CIS states [76, 82, 84], we define the matrix modulus

$$|\Delta D| = [(\Delta D)^2]^{1/2} \tag{6.96}$$

and introduce the hole-particle invariants in form of traces of the $|\Delta D|$ powers:

$$N_{\mathrm{h-p}}[q] = \mathrm{Tr}|\Delta D|^q. \tag{6.97}$$

The formal analysis of $N_{\mathrm{h-p}}[q]$ is given in Appendix C. In particular, for ground state we have

$$N_{\mathrm{h-p}}[1] = N_{\mathrm{h-p}} = N_{\mathrm{eff}}. \tag{6.98}$$

The second-order index $N_{\mathrm{h-p}}[2]$ has a meaning of the squared norm of the density matrix deviation ΔD:

$$N_{\mathrm{h-p}}[2] = ||\Delta D||^2. \tag{6.99}$$

Our experience shows that the special case q = 3/4, which produced the hole-particle invariant $N_{\mathrm{h-p}}[4/3]$, can also serve as a good EUE measure. In this case, as in the case of the above N_{odd} [2] index, nonsignificant contributions from dynamical correlation are effectively suppressed. Apparently, the dynamical correlation has no direct relation to diradicality and polyradicality. However, the considered procedure may be not well suitable for dissociation states, as one example in [5] had shown this for the N_{odd} [2] measure. Nevertheless, for typical equilibrium molecular states, such a suppression seems to be quite sensible if one is

Table 6.11 Generalized EUE indices $N_{h-p}[q]$ and respective localization indices $\aleph^{loc}$ for the (5a, 4z), (5a, 5z), and (5a, 6z) periacene molecules within π-UHF scheme

Periacene	N_{h-p}	N_{odd} [2]	$N_{h-p}[4/3]$	$\aleph^{loc}_{h-p}$	$\aleph^{loc}_{odd}$ [2]	$\aleph^{loc}_{h-p}[4/3]$
(5a, 4z)	0.054	0.041	0.041	6.3	5.3	5.5
(5a, 5z)	0.057	0.042	0.042	8.4	7.1	7.0
(5a, 6z)	0.060	0.048	0.045	11.4	10.0	9.7

Here N_{h-p} coincides with $N_{eff}(t)$

interesting in an estimation of polyradicality. In Table 6.11 we present the results of using N_{h-p} (for the ground state $N_{h-p}=N_{eff}$), N_{odd} [2] and $N_{h-p}[4/3]$ for periacenes discussed in Sect. 6.11. We supplement the table by the values of average locality index $\aleph_{loc}$ (6.88) for the each EUE measure. It is seen that the N_{odd} [2] and $N_{h-p}[4/3]$ data differ nonsignificantly. With this, $N_{h-p}[4/3]$ is of the hole-particle origin, as the initial N_{h-p} index. Moreover, judging from the $\aleph_{loc}$ magnitudes in Table 6.11, both measures, N_{odd} [2] and $N_{h-p}[4/3]$ are associated with more localized EUE distributions than in the N_{eff} case. This fact gives additional support for using the special measures N_{odd} [2] and $N_{h-p}[4/3]$ for describing a polyradical character in conjugated hydrocarbons and related carbon-based networks.

Seemingly, it would be more preferable to perform EUE studies by invoking several quantification schemes. One needs also to take into account the specific features of the problem before deciding what the scheme should be taken as a more relevant in the problem context. For instance, if we are interesting in the EUE spatial localization, then the locality index $\aleph_{loc}$ (6.88) can be quite appropriate. If the localization measure of the NOON spectrum $\{\lambda_k\}$ near $\lambda = 1$ is in question then, then indices $N_{h-p}[q]$ and N_{odd} $[q]$ with large q, e.g., $q = 4$, are more preferable.

6.15 Concluding Remarks

The initial intent of this chapter was to provide a broad overview and a critical assessment of various trends in the theory of effectively unpaired electrons. In the process of preparing the manuscripts some accents were shifted, and we would unavoidably restrict ourselves to a narrow set of issues and examples for discussion. For instance, we only slightly touched on the electron unpairing analysis in structures with a spatial separation of molecular subunits. These are bichromophore systems, molecular dimers and complexes, radical and ion–radical pairs, etc. The recent papers [77, 78, 125] are dedicated just to these problems. Besides, many interesting systems, e.g., semiconductor quantum dots, fell beyond the scope of this review. Indeed, many-electron aspects of the multiple exciton generation (MEG) in quantum dots are closely related to the EUE theory, but only circumstantial evidence about EUE effects in MEG can be found in the current literature [127, 128].

Before closing, we briefly reiterate the basic points. The conventional formalism of density matrices was used here in such a way that it helped us to understand some essential features and interrelations between different EUE measures. In particular, in Sect. 6.5 we see that the hole-particle index, which was introduced in [15, 16], is identical, for ground states, to the previously proposed H-G index [5]. We also examined excited states where the situation is more complicated, so that the H-G and hole-particle indices should be considered concurrently.

We must admit that the EUE theory is a little trickier than it seems. Indeed, rigorously speaking, the EUE problem is not well-defined, because it does not have a unique solution [5]. Nevertheless, we tried to understand what are the most appropriate definitions of the EUE measures. Now, we could report only preparatory results until a subsequent, more substantial analysis. Nevertheless, possible directions to modifying the existing measures are discussed here. In particular, we could retain the same hole-particle description for the modified ('q-extended') EUE measures in which unwanted small contributions to EUE are suppressed, as it was previously done in the case of the second Head-Gordon index, Eq. (6.94). Another way to produce better EUE indices is to somehow estimate an average localization of the unpaired electron. This leads to a possible measure $\aleph_{loc}$, Eq. (6.88), giving the results which are closer to those expected from a 'common chemistry sense'.

In our specific applications the stress is put on large-scale carbon-containing molecules. For those we propose a simple semiempirical scheme [13] which has the same complexity as the ordinary Hückel method. It allows us to make quick and easy estimates of EUE effects even in gigantic conjugated networks which cannot be rigorously treated by high-level ab initio methods. It is important that the proposed elementary model yields the results which are qualitatively in agreement with the existing ab initio data [9, 11] for relatively small systems. The influence of external perturbations on EUE is another attractive field of application. It turns out that the systems which have a small polyradical character (hydrocarbons of the Clar type) become extremely polyradicaloid in a strong static electric field (Sect. 6.14). The laser field effects on the polyaromatic systems are even more pronounced, as preliminary studies show.

In sum, we would like to stress that during the last decade or so, the unpaired electron theory came up with new fruitful insights and techniques. In future, the semiempirical models and concrete results for large systems should be carefully examined by more rigorous approaches. We believe that a physical side of the existing simplified EUE theories remains, and plenty of new intriguing results for electrons in the conjugated molecules will be revealed.

Acknowledgment Many people influenced the author's work directly or indirectly, and most valuable were nice contacts with colleagues and coauthors Art Bochevarov, David Casanova, Anna Krylov, Hans Lischka, Felix Plasser, and Oleg Zhikol. The receipt of the ab initio π-NOON spectra for the (5a, 4z), (5a, 5z), and (5a, 6z) periacenes from H. Lischka and F. Plasser is also greatly acknowledged.

Appendix A: Duality Symmetry and Generalized EUE Indices

In this Appendix we clarify the cause for postulating symmetry relation (6.18). For this aim we introduce a formal operation which can be named the duality transformation and which is well known in multilinear algebra as the Hodge star operation, or Hodge dual [132]. In the RDM theory an equivalent transformation was applied in [19, 133], without recognizing it as a Hodge dual. The following simple example helps to explain this notion in the more familiar terms of many-electron state vectors.

We consider a two-electron problem in the basis of five spin-orbitals

$$\{|\chi_1\rangle, |\chi_2\rangle, |\chi_3\rangle|\chi_4\rangle, |\chi_5\rangle\}. \tag{A1}$$

Let the ket

$$|\Psi_{[2]}\rangle = |\chi_1\chi_2\rangle \tag{A2}$$

be the two-electron Slater determinant built from $|\chi_1\rangle$ and $|\chi_2\rangle$. By definition, the dual ket, $\left|\Psi^*_{[3]}\right\rangle$, is built up from the rest spin-orbitals, giving the three-electron determinant:

$$\left|\Psi^*_{[3]}\right\rangle = |\chi_3\chi_4\chi_5\rangle. \tag{A3}$$

In the same basis (A1), the maximal Slater determinant $|\Psi_{\max}\rangle$ is

$$|\Psi_{\max}\rangle = |\chi_1\chi_2\chi_3\chi_4\chi_5\rangle. \tag{A4}$$

It is not difficult understand that we can produce $\left|\Psi^*_{[3]}\right\rangle$ from $|\Psi_{\max}\rangle$ by annihilating in Eq. (A4) the state vector (A2). More exactly, apart from a prefactor we have

$$\left|\Psi^*_{[3]}\right\rangle = \langle\Psi_{[2]} \mid \Psi_{\max}\rangle. \tag{A5}$$

They say that the obtained three electron state $\left|\Psi^*_{[3]}\right\rangle$ is the Hodge dual of the two-electron state $|\Psi_{[2]}\rangle$.

Now consider the respective 1-RDMs. In notation of Sect. 6.6, we have from determinants (A2) and (A3) the usual D_1^{so} matrices in the form of projectors on occupied spin-orbitals of the respective determinants:

$$D_1^{\text{so}}(|\Psi_{[2]}\rangle) = |\chi_1\rangle\langle\chi_1| + |\chi_2\rangle\langle\chi_2|,$$

$$D_1^{\text{so}}(\Psi_{[3]}^*) = |\chi_3\rangle\langle\chi_3| + |\chi_4\rangle\langle\chi_4| + |\chi_5\rangle\langle\chi_5|.$$

We see that

$$D_1^{\text{so}}(\Psi_{[3]}^*) = I - D_1^{\text{so}}(\Psi^{[2]}) \tag{A6}$$

where unity operator I is a projector on all five spin-orbitals from Eq. (A1).

This line of reasoning can be directly extended to a general case including exact state vectors as well. It the general case we start with a r-dimensional spin-orbital basis $\{|\chi_k\rangle\}_{1\leq k\leq r}$ and build the respective maximal determinant $|\Psi_{\max}\rangle = |\chi_1 \ldots \chi_r\rangle$ (clearly, the only r-electron state vector is $|\Psi_{\max}\rangle \equiv |\Psi_{[r]}\rangle$). The given exact (or approximate) state-vector $\Psi_{[N]}$ produces the Hodge dual, as previously:

$$\left|\Psi_{[r-N]}^*\right\rangle = \langle\Psi_{[N]} \mid \Psi_{\max}\rangle. \tag{A7}$$

Accordingly, relation (A6) is generalized to be

$$D_1^{\text{so}}(\Psi_{[r-N]}^*) = I - D_1^{\text{so}}(\Psi_{[N]}). \tag{A8}$$

This is the duality transformation in terms of 1-RDM. The analogous relation for $D_2^{\text{so}}(\Psi_{[r-N]}^*)$ is somewhat more involved [133, 134]. The remarkable property of the Hodge duality transformation is its ability to preserve correlation operator Δ_2^{so} in Eq. (6.45), as it is first shown in [19]. The related expression is given in [135]. Thus, the other correlation matrices, e.g., D^{eff}, must be the same as well. It is worth mentioning in passing that in [128] and many subsequent papers, a somewhat inconvenient terminology is used for RDMs $D^{\text{so}}(\Psi_{[r-N]}^*)$—the latter are loosely identified with hole RDMs. Certainly, it leads to confusion and even misinterpretation, since generally such RDMs have no relation to the genuine, ‘physical’, hole RDMs discussed in Sect. 6.4 and in [16]. We prefer to refer to them as the dual RDMs [16].

We now have to sum over spin indices, making spin trace in Eq. (A8). As a result, the dual charge density matrix is yielded, viz.

$$D(\Psi_{[r-N]}^*) = 2 - D(\Psi_{[N]}). \tag{A9}$$

Then the NOON spectrum of the dual charge density matrix is simply a set $\{2 - \lambda_k\}$ where we imply that the initial NOON spectrum is the set $\{\lambda_k\}$. Recalling that EUE characteristics of the dual state (A7) should be the same as in the initial state $\Psi_{[N]}$, the identity

$$D^{\text{eff}}(\Psi_{[N]}) = D^{\text{eff}}(\Psi^{*}_{[r-N]}) \tag{A10}$$

is necessitated. Taking into account Eqs. (6.1), (6.5) and (6.5′) we arrive at the relation

$$\sum_k f(\lambda_k)|\varphi_k\rangle\langle\varphi_k| = \sum_k f(2-\lambda_k)|\varphi_k\rangle\langle\varphi_k|, \tag{A11}$$

from whence Eq. (6.18) immediately follows, that is

$$f(\lambda) = f(2-\lambda). \tag{A12}$$

The requirement (A12) allows us to specify a general dependence $\lambda^{\text{eff}} = f(\lambda)$, namely, λ^{eff} is a nonnegative definite function of argument $|\lambda - 1|$, with boundary values $f(0) = f(2) = 0$. Eqs. (6.7′) and (6.14) are evidently of this type. Rather general types of the functions can be proposed as 'q-extensions' of Eqs. (6.9) and (6.15). These are

$$N_{\text{odd}}[q] = \sum_k (1 - |\lambda_k - 1|^2)^q, \tag{A13}$$

$$N_{\text{eff}}[q] = \sum_k (1 - |\lambda_k - 1|)^q, \tag{A14}$$

where $q \geq 1$. We see that $N_{\text{odd}}[1]$ and $N_{\text{eff}}[1]$ produce the usual N_{odd} and N_{eff} measures, respectively. The choice $q = 2$ in Eq. (A13) leads to

$$N_{\text{odd}}[2] = \sum_k [1 - (\lambda_k - 1)^2)], \tag{A15}$$

which is the modified Head-Gordon index from [5]. This expression is trivially equivalent to Eq. (6.94).

Appendix B: Density Matrix and NOON for QCTB

We consider here in more detail the QCTB model described in Sect. 6.13. Having at hand the effective Hamiltonian matrices (6.91), we straightforwardly derive projector matrices ρ_α and ρ_β by using the well known expressions connecting Hamiltonians and respective projectors [19, 136, 137]). Let h be the Hermitian operator, such that exactly n eigenvalues of h lie below zero, and P be the projector on the corresponding eigenvectors. Then

$$P = (I - h/|h|)/2, \quad \text{(B1)}$$

where $|h| = [(h)^2]^{1/2}$ is the modulus of operator h. Further, let one-electron Hamiltonian matrix h^{δ} be defined as follows:

$$h^{[\delta]} = -\begin{pmatrix} \delta I & B \\ B^{+} & -\delta I \end{pmatrix}.$$

In particular, $h^{\alpha} = h^{[\delta]}$, $h^{\beta} = h^{[-\delta]}$. Then, by applying Eq. (B1) to $h = h^{[\delta]}$, we obtain the corresponding projector

$$P^{[\delta]} = \frac{1}{2}\begin{pmatrix} I + \delta(\delta^2 I + BB^{+})^{-1/2} & B(\delta^2 I + B^{+}B)^{-1/2} \\ (\delta^2 I + B^{+}B)^{-1/2}B^{+} & I - \delta(\delta^2 I + B^{+}B)^{-1/2} \end{pmatrix}. \quad \text{(B2)}$$

In derivation, the block-diagonal structure of $(h^{[\delta]})^2$ is used, that is

$$(h^{[\delta]})^2 = \begin{pmatrix} \delta^2 I + BB^{+} & 0 \\ 0 & \delta^2 I + B^{+}B \end{pmatrix}.$$

Equation (B2) was earlier derived by another technique for the special closed π-shells with alternating electronegativity [138]. Obviously, setting $\delta = 0$, we return to the Hall formula (6.90). By recalling Eq. (6.91) we have

$$\rho_{\alpha} = P^{[\delta]}, \quad \rho_{\beta} = P^{[-\delta]}. \quad \text{(B3)}$$

Putting together Eqs. (B2) and (B3), we get from Eq. (6.10) the main result:

$$D = \begin{pmatrix} I & B(\delta^2 I + B^{+}B)^{-1/2} \\ (\delta^2 I + B^{+}B)^{-1/2}B^{+} & I \end{pmatrix}. \quad \text{(B4)}$$

The problem of diagonalizing this D is a quite elementary, and the full NOON spectrum takes the form

$$\lambda_i = 1 + \varepsilon_i/\sqrt{\delta^2 + \varepsilon_i^2}, \quad \lambda_a = 1 - \varepsilon_a/\sqrt{\delta^2 + \varepsilon_a^2}, \quad \text{(B5)}$$

where $1 \le i, a \le n$, and nonnegative quantities $\varepsilon_i \equiv |\varepsilon_i|$, as well as $\varepsilon_a \equiv |\varepsilon_a|$, are eigenvalues of $(B^{+}B)^{1/2}$, that is $\{\varepsilon_i\}$ is the bipartite graph spectrum. From Eq. (B5) the main EUE indices within QCTB are easily deduced. For instance,

$$N_{\text{odd}} = 2\delta^2 \sum_{i=1}^{n} (\delta^2 + \varepsilon_i^2)^{-1}. \quad \text{(B6)}$$

Remark also an evident symmetry of the corresponding hole and particle occupancies, defined by Eq. (6.41′):

$$\{1-\varepsilon_i/\sqrt{\delta^2+\varepsilon_i^2}\}=\{1-\varepsilon_a/\sqrt{\delta^2+\varepsilon_a^2}\}. \tag{B7}$$

that follows from Eq. (B5). In other words, the hole and particle occupancy spectra are identical for this π-model.

As a matter of fact, the hole and particle occupancies are identical for any bipartite networks treated within π-approximation, up to FCI/PPP. This is a simple corollary of the generalized pairing theorem of McLachlan [94] stating that the π-electron charge density matrix of the alternant hydrocarbons is of the form

$$D=\begin{pmatrix} I & \partial \\ \partial^+ & I \end{pmatrix}, \tag{B8}$$

where the $2p_z$ AO basis set is ordered as in Eq. (6.89), and ∂ defines the inter-sublattice bond order matrix. Clearly, the corresponding NOON spectrum $\{\lambda_k\}$ is

$$\{1+\sqrt{\mu_i}\},\ \{1-\sqrt{\mu_a}\} \tag{B9}$$

where μ_i (or μ_a) are eigenvalues of $\partial^+\partial$, and $1\le i,\ a\le n$. As a result, the initial π-NOON spectrum is symmetrical in respect to the point $\lambda=1$. From Eq. (B9) we deduce that indeed the respective hole and particle π-occupancies, defined as in Eq. (6.41′), are identically the same:

$$\{1-\sqrt{\mu_i}\}=\{1-\sqrt{\mu_a}\}. \tag{B10}$$

Interestingly, an initio data [9, 11] approximately follow Eqs. (B9) and (B10). Incidentally, it follows, from this discussion, that the hole occupancy distribution $\{1-\sqrt{\mu_i}\}$ (generally $\{2-\lambda_i\}_{1\le i\le n}$) is sufficient for considering EUE problems. For instance, instead of plotting NOON spectrum $\{\lambda_k\}$, one can plot only hole occupancy spectrum $\{2-\lambda_i\}$ as even more suitable in the EUE context. This occupancy spectrum is in fact the second half of the typical π-NOON spectra which were presented in Tables 6.6, 6.7, and Fig. 6.4.

Appendix C: Generalized Hole-Particle Indices

Here we analyze the main EUE indices in terms of hole-particle quantities. We begin with the representation

$$D = 2\rho + \Delta D, \tag{C1}$$

where ρ is of the form (6.36), and $|\varphi_i\rangle$ are the natural orbitals of the state in question, so ΔD commutes with ρ. Then, using the same notation, as in Eq. (6.41), we obtain the spectral resolution

$$\Delta D = -\sum_{i \le n} \Delta_i |\varphi_i\rangle\langle\varphi_i| + \sum_{a > n} \lambda_a |\varphi_a\rangle\langle\varphi_a|, \tag{C2}$$

where

$$\Delta_i \equiv 2 - \lambda_i \tag{C3}$$

are new nonnegative quantities ($0 \le \Delta_i < 1$, and $i \le n$), and λ_a are related to 'virtual' natural orbitals. We see that correlation correction matrix ΔD has a clear hole-particle structure: Δ_i are the occupancy numbers for the holes, and λ_a are the same for the particles. In manipulations the identity

$$\sum_i \Delta_i = \sum_a \lambda_a \tag{C4}$$

will be useful as well. It follows from Eqs. (C1), (C2), and normalization (6.2).

Due to the diagonal form (C2) we trivially have the diagonal form of the matrix $|\Delta D|$ defined by Eq. (6.96):

$$|\Delta D| = \sum_i \Delta_i \; |\varphi_i\rangle\langle\varphi_i| + \sum_a \lambda_a |\varphi_a\rangle\langle\varphi_a|. \tag{C5}$$

But this is the same as the hole-particle density in Eq. (6.41), that is

$$D^{\mathrm{h-p}} = |\Delta D|. \tag{C6}$$

It is essential that under duality transformation (A9) the holes and particles in Eq. (C2) change place, so identity (A10) satisfies automatically for $D^{\mathrm{eff}} = D^{\mathrm{h-p}}$.

The appropriate q-extended ($q \ge 1$) hole-particle indices can be cast explicitly into the form

$$N_{\mathrm{h-p}}[q] = \mathrm{Tr}(|\Delta D|^q) = \sum_i \Delta_i^q + \sum_a \lambda_a^q. \tag{C7}$$

Particularly,

$$N_{\mathrm{h-p}}[2] = \sum_i \Delta_i^2 + \sum_a \lambda_a^2 = ||\Delta D||^2. \tag{C8}$$

The previously defined EUE indices can be rewritten in terms of the correlation-dependent quantities $\{\Delta_i, \lambda_a\}$:

$$N_{\text{odd}} = \sum_{1\le i\le n} (4\Delta_i + \Delta_i^2) + \sum_{a>n} \lambda_a^2, \tag{C9}$$

$$N_{\text{eff}} = \sum_i \Delta_i + \sum_a \lambda_a = 2 \sum_{1\le i\le n} \Delta_i, \tag{C10}$$

$$N_{\text{odd}}[2] = \sum_{1\le i\le n} [\Delta_i(2-\Delta_i)]^2 + \sum_{a>n} [\lambda_a(2-\lambda_a)]^2. \tag{C11}$$

where we used identity (C4).

For slightly correlated systems, the most important are the first order terms in Δ_i and λ_a. It gives $N_{\text{odd}} \cong 4\sum \Delta_i$, so

$$N_{\text{odd}} \cong 2N_{\text{eff}}, \tag{C12}$$

and this goes back to the rude estimation, Eq. (6.80). It is interesting that the exact interrelation $2N_{\text{eff}} - N_{\text{odd}} = ||\Delta D||^2$ is true. Likewise, the first-order estimation of the modified Head-Gordon index (6.94), that is Eq. (C11), is null:

$$N_{\text{odd}}[2] \cong 0. \tag{C13}$$

Indeed, Eq. (C11) contains only the second-order and higher-order terms:

$$N_{\text{odd}}[2] \cong 4\left(\sum_i \Delta_i^2 + \sum_a \lambda_a^2\right) = 4||\Delta D||^2. \tag{C14}$$

The above simple analysis now elucidates how small contributions from Δ_i and λ_a are essentially suppressed in the $N_{\text{odd}}[2]$ and $N_{\text{h-p}}[2]$ indices. As a rule, these small contributions appear mainly from dynamical correlations. For instance, MP2 (the Moller-Plesset second-order perturbation theory) normally produce the contributions of this kind. Evidently, they have no direct relation to diradicality and polyradicality, and the $N_{\text{odd}}[2]$ and $N_{\text{h-p}}[2]$ indices should be rather small without a significant contribution from non-dynamical correlation. This is a good property of the generalized indices such as (6.94) and (C8), and apparently, this is the basic reason why $N_{\text{odd}}[2]$ is systematically employed in papers [9, 11, 122, 124] for analyzing the unpaired electrons in large PAHs. At the same time, the dynamical correlation cannot fully ignored, and the problem of an optimal quantification remains.

References

1. Takatsuka K, Fueno T, Yamaguchi K (1978) Theor Chim Acta 48:175
2. McWeeny R, Kutzelnigg W (1968) Int J Quant Chem 2:187
3. Einstein A, Podolsky B, Rosen N (1935) Phys Rev 47:777

4. Afriat A, Selleri F (1999) The Einstein, Podolsky, and Rosen paradox in atomic, nuclear, and particle physics. Plenum Press, New York; McWeeny R (2000) Adv Quant Chem 36:365
5. Head-Gordon M (2003) Chem Phys Lett 372:508
6. Hachmann J, Dorando JJ, Aviles M, Chan GK-L (2007) J Chem Phys 127:134309
7. Yoneda K, Nakano M, Fukuda K, Champagne B (2012) J Phys Chem Lett 3:3338
8. Mizukami W, Kurashige Y, Yanai T (2012) J Chem Theor Comp 9:401
9. Plasser F, Pašalić H, Gerzabek MH, Libisch F, Reiter R, Burgdörfer J, Müller T, Shepard R, Lischka H (2013) Angew Chem Int Ed 52:2581
10. Casanova D (2014) J Comput Chem 35:944
11. Horn S, Plasser F, Müller T, Libisch F, Burgdörfer J, Lischka H (2014) Theor Chem Acc 133:1511
12. Luzanov AV (2014) J Struct Chem 55:799
13. Luzanov AV (2014) Funct Mater 21:437 [http://www.isc.kharkov.com/journal/contents/21-4/fm214-12.pdf]
14. Barnard AS, Snook IK (2011) Modelling Simul Mater Sci Eng 19:054001; Shi H, Barnard AS, Snook IK (2012) Nanotechnology 23:065707
15. Luzanov AV, Zhikol OA (2005) Int J Quant Chem 104:167
16. Luzanov AV, Prezhdo OV (2006) J Chem Phys 124:224109
17. Staroverov VN, Davidson ER (2000) Int J Quant Chem 77:316
18. Staroverov VN, Davidson ER (2000) Chem Phys Lett 330:161
19. Mestechkin MM (1977) Metod Matritsy Plotnosti v Teorii Molekul. Naukova Dumka, Kiev
20. Staroverov VN, Davidson ER (2000) J Am Chem Soc 122:7377
21. Lain L, Torre A, Bochicchio RC, Ponec R (2002) Chem Phys Lett 346:283; Alcoba DR, Bochicchio RC, Lain L, Torre A (2006) Chem Phys Lett. 429:286; Lain L, Torre A, Alcoba DR, Bochicchio RC (2009) Chem Phys Lett 476:101
22. Yamaguchi K, Kawakami T, Takano Y, Kitagawa Y, Yamashita Y, Fujita H (2002) Int J Quant Chem 90:370
23. Cheng M-J, Hu C-H (2003) Mol Phys 101(9):1319
24. Proynov EI (2006) J Mol Struct (Theochem) 762:159; Proynov E, Liu F, Kong J (2013) Phys Rev A 88:032510
25. Amos AT, Hall GG (1961) Proc Roy Soc A 263:483
26. Stück D, Baker TA, Zimmerman P, Kurlancheek W, Head-Gordon M (2011) J Chem Phys 135:194306
27. Luzanov AV (2013) Int J Quant Chem 113:2489
28. Bochicchio RC, Torre A, Lain L (2003) Chem Phys Lett 380:486
29. Head-Gordon M (2003) Chem Phys Lett 380:488
30. Döhnert D, Koutecký J (1980) J Am Chem Soc 102:1789; Gordon MS, Schmidt MW, Chaban G M, Glaesemann KR, Stevens WJ, Gonzalez C (1999) J Chem Phys 110:4199
31. Luzanov AV (2004) Kharkov University Bulletin. Chemical Series Issue 11(34):195 [http://chembull.univer.kharkov.ua/archiv/2004/12.pdf]
32. Luzanov AV, Umanski VE (1977) Theor Experim Chem 13:162
33. Luzanov AV, Pedash YF, Mohamad S (1990) Theor Experim Chem 26:513
34. Grobe R, Rzazewski K, Eberly JH (1994) J Phys B 27:L503
35. Luzanov AV, Prezhdo OV (2007) Mol Phys 105:2879
36. Bell RJ, Dean P, Hibbins-Butler DC (1970) J Phys C 32:111
37. Zyczkowski K (1999) Phys Rev A 60:3496
38. Luzanov AV (1989) Theor Experim Chem 25:19; Luzanov AV, Wulfov AL, Krouglov VO (1992) Chem Phys Lett 197:614
39. Helgaker T, Jørgensen P, Olsen P (2000) Molecular Electronic-Structure Theory. Wiley, New York
40. Purvis GD, Sheppard R, Brown BR, Bartlett RJ (1983) Int J Quant Chem 23:835
41. Kutzelnigg W, Smith VH (1968) Int J Quant Chem 2(531):553
42. Shavitt I, Bartlett RJ (2009) Many-body methods in chemistry and physics. Cambridge University Press, Cambridge

43. Nesbet RK (1958) Phys Rev 109:1632
44. Kumar K (1962) Perturbation theory and the nuclear many body problem. North-Holland, Amsterdam
45. Luzanov AV (1977) Theor Math Phys 30:232
46. Luzanov AV (2008) Int J Quant Chem 108:671
47. Luzanov AV, Prezhdo OV (2006) J Chem Phys 125:154106
48. Dirac PAM (1931) Cambr Phil Soc 27:240
49. Davidson ER (1976) Reduced density matrices in quantum chemistry. Academic, New York
50. Juhasz T, Mazziotti DA (2006) J. Chem. Phys. 125:174105
51. Mazziotti DA (ed) (2007) Advances in Chemical Physics, vol 134. Wiley, New York
52. Luzanov AV, Prezhdo OV (2005) Int J Quantum Chem 102:582
53. Roby KR (1974) Mol Phys 27:81
54. Clark AE, Davidson ER (2003) Int J Quantum Chem 93:384
55. Penney WG (1937) Proc Roy Soc A 158:306
56. Raos G, Gerratt J, Cooper DL, Raimondi M (1994) Chem Phys 186:233; Clark AE, Davidson ER (2001) J Chem Phys 115:7382; (2002) Mol Phys 100:373; (2002) J Phys Chem A 106:7456
57. Herrmann C, Reiher M (2006) J Comput Chem 27:1223; Podewitz M, Reiher M (2010) Adv Inorg Chem 62:177
58. Alcoba DR, Torre A, Lain L, Bochicchio RC (2011) J Chem Theory Comput 7:3560
59. Luzanov AV (2012) Int J Quantum Chem 112:2915
60. Ramos-Cordoba E, Matito E, Mayer I, Salvador P (2012) Phys Chem Chem Phys 14:15291
61. Matsika S, Feng X, Luzanov AV, Krylov AI (2014) J Phys Chem A 118:11943
62. Luzanov AV, Prezhdo OV (2011) J Chem Phys 135:094107
63. Cížek J, Paldus J (1967) J Chem Phys 47:3976
64. Kutzelnigg W, Smith VH (1964) J Chem Phys 41:896
65. Paldus J, Cížek J, Keating BA (1973) Phys Rev A 8:640
66. Mayer I (1980) Adv Quantum Chem 12:189
67. Klimo V, Tino J (1980) Mol Phys 43:477; Henderson TM, Tsuchimochi T, Scuseria GE (2012) J Chem Phys 136:164109; Luzanov
68. Hartmann H (1947) Z Naturforsch A 2:259
69. Ovchinnikov AA, Ukrainskii II, Kvenzel' GF (1973) Sov Phys Uspekhi 15:575
70. Misurkin IA, Ovchinnikov AA (1974) Mol Phys 27:237; Misurkin IA, Ovchinnikov AA (1977) Russ Chem Rev 46:96
71. Macêdo AMS, dos Santos MC, Coutinho-Filho MD, Macedo CA (1995) Phys Rev Lett 74:1851
72. Fujita M, Wakabayashi K, Nakada K, Kushakabe K (1996) J Phys Soc Jap 65:1920
73. Jiang D-E, Chen X-Q, Luo W, Shelton WA (2009) Chem Phys Lett 483:120; Jiang DE, Sumpter BG, Dai SJ (2007) Chem Phys 126:134701; Das M (2014) J Chem Phys 140:124317
74. Jiang DE, Chen Z (eds) (2013) Graphene chemistry: theoretical perspectives. Wiley, Puerto Rico
75. Jiménez-Hoyos CA, Rodríguez-Guzmán R, Scuseria GE (2014) J Phys Chem A 118:9925
76. Luzanov AV, Zhikol OA (2012) In: Leszczynski J, Shukla MK (eds) Practical aspects of computational chemistry I: an overview of the last two decades and current trends. Springer, Heidelberg, p 415
77. Minami T, Iton S, Nakano M (2013) J Phys Chem Lett 4:2133
78. Feng X, Luzanov AV, Krylov AI (2013) J Phys Chem Lett 4:3845
79. McWeeny R (1968) In: Reports on the Density Matrix Seminar. Queen's University. Ontario; p 25; Dacre PD, Watts CJ, Williams GRJ, McWeeny R (1975) Mol Phys 30:1203
80. Luzanov AV (1975) Theor Experim Chem 9:567
81. Ipatov A, Cordova F, Doriol LJ, Casida ME (2009) J Mol Struct (Theochem) 914:60
82. Luzanov AV, Zhikol OA (2010) Int J Quant Chem 110:902
83. Murrell JN, McEwen KL (1956) J Chem Phys 25:1143

84. Luzanov AV, Sukhorukov AA, Umanski VE (1976) Theor Experim Chem 10:354
85. Krylov AI, Sherrill CD, Head-Gordon M (2000) J Chem Phys 113:6509
86. Bendikov M, Duong H M, Starkey K, Houk KN, Carter EA, Wudl F (2004) J Am Chem Soc 126:7416; Son Y-W, Cohen ML, Louie SG (2006) Nature 444:347; Pisani L, Chan JA, Montanari B, Harrison NM (2007) Phys Rev B 75:064418; Hod O, Barone V, Scuseria GE (2008) Phys Rev B 77:035411
87. San-Fabian E, Moscardy F (2013) Int J Quantum Chem 113:815
88. Torres A, Guadarrama P, Fomine S (2014) J Mol Modeling 20:1
89. Frisch MJ, Trucks GW, Schlegel HB, Scuseria GE, Robb MA, Cheeseman JR, Montgomery JA Jr., Vreven T, Kudin KN, Burant JC, MillamJM, Iyengar SS, Tomasi J, Barone V, Mennucci B, Cossi M, Scalmani G, Rega N, Petersson GA, Nakatsuji H, Hada M, Ehara M, Toyota K, Fukuda R, Hasegawa J, Ishida M, Nakajima T, Honda Y, Kitao O, Nakai H, Klene M, Li X, Knox JE, Hratchian HP, Cross JB, Bakken V, Adamo C, Jaramillo J, Gomperts R, Stratmann RE, Yazyev O, Austin AJ, Cammi R, Pomelli C, Ochterski JW, Ayala PY, Morokuma K, Voth GA, Salvador P, Dannenberg JJ, Zakrzewski VG, Dapprich S, Daniels AD, Strain MC, Farkas O, Malick DK, Rabuck AD, Raghavachari K, Foresman JB, Ortiz JV, Cui Q, Baboul AG, Clifford S, Cioslowski J, Stefanov BB, Liu G, Liashenko A, Piskorz P, Komaromi I, Martin RL, Fox DJ, Keith T, Al-Laham MA, Peng CY, Nanayakkara A, Challacombe M, Gill PMW, Johnson B, Chen W, Wong MW, Gonzalez C, Pople JA (2004) Gaussian 03, Revision A.01. Gaussian, Inc., Wallingford
90. Gebhard F (1997) The Mott metal insulator transition: models and methods. Springer, Berlin; Yazyev OV (2013) Acc Chem Res 46:2319
91. SmeyersYG, Doreste-Suarez L (1973) Int. J. Quantum Chem 7:687
92. Murphy NC, Wortis R, Atkinson WA (2011) Phys Rev B 83:184206
93. Tyutyulkov N, Dietz F, Müllen K, Baumgarten M (1992) Chem Phys 163:55
94. McLachlan AD (1961) Mol Phys 4:49
95. Vonsovskii SV, Svirskii MS (1970) JETP 30:140; Vonsovskii SV (1974) Magnetism. vol 2. Wiley, New York
96. Majlis N (2000) The quantum theory of magnetism. World Scientific, Singapore
97. Luo Z, Kim S, Kawamoto N, Rappe AM, Johnson AT (2011) ACS Nano 5:9154
98. Girão EC, Cruz-Silva E, Liang L, Filho AGS, Meunier V (2012) Phys Rev B 85:235431
99. Fujihara M, Miyata Y, Kitaura R, Nishimura Y, Camacho C, Irle S, Iizumi Y, Okazaki T, Shinohara H (2012) J Phys Chem C 116:15141
100. Zhang X, Xin J, Ding F (2013) Nanoscale 5:2556
101. Lieb EH, Mattis DC (1962) J Math Phys 3:749; Langer WD, Mattis DC (1971) Phys Lett 36A:139; Lieb EH (1989) Phys Rev Lett 62:1201
102. Coulson CA, Rushbrooke GS (1940) Proc Cambridge Phil Soc 36:139; Brickstock A, Pople JA (1954) Trans Farad Soc 59:901
103. Kasteleyn P (1967) In: Harary F (ed) Graph theory and theoretical physics. Academic Press, London, p 43
104. Harary F (1972) Graph theory. Addison-Wesley, London; Cvetkovic´ DM, Doob M, Sachs H (1980) Spectra of graphs, theory and application. Academic Press, New York
105. Hall GG (1955) Proc Roy Soc A 229:251
106. Boykin TB (2009) J Comput Electron 8:142; Goringe CM, Bowler DR (1997) Hernández E Rep Prog Phys 60:1447
107. Mariscal M.M, Oviedo OA, Leiva EPM (eds) (2011) Metal clusters and nanoalloys: from modeling to applications. Springer, Berlin
108. Ivanov VV, Kisil IP, Luzanov AV (1996) J Struct Chem 37:537
109. Chien JCW (1984) Polyacetylene: chemistry, physics and materials science. Academic Press, New York
110. Lennard-Jones JE (1937) Proc Roy Soc A 158:280
111. Coulson CA (1948) Proc Phys Soc A 60:257
112. Luzanov AV (2002) J Struct Chem 43:711

113. Petersen R, Pedersen TG, Jauho A-P (2011) ACS Nano 5:523
114. Foa Torres LEF, Roche S, Charlie J-C (2014) Introduction to graphene-based nanomaterials: from electronic structure to quantum transport. Cambridge Uiversity Press, Cambridge
115. Wolf EL (2014) Graphene: a new paradigm in condensed matter and device physics. Oxford University Press, Oxford
116. Cresti A, Nemec N, Biel B, Nieble G, Triozon F, Cuniberti G, Roche S (2008) Nano Research 1:361
117. Wei D, Wang F (2012) Surf Sci 606:485
118. Baldoni M, Sgamellotti A, Mercuri F (2008) Chem Phys Lett 464:202; Selli D, Mercuri F (2014) Carbon 75:190; Baldoni M, Mercuri F (2015) Phys Chem Chem Phys 17:2088
119. Sharma R, Nair N, Strano MS (2009) J Phys Chem C 113:14771
120. Dinadayalane TC, Leszczynski J (2010) Struct Chem 21:1155
121. Wassmann T, Seitsonen AP, Saitta AM, Lazzeri M, Mauri F (2010) J Am Chem Soc 132:3440
122. Radovic R, Silva-Villalobos AF, Silva-Tapia AB, Vallejos-Burgos F (2011) Carbon 49:3471
123. Hoffmann R (1963) J Chem Phys 39:1397; Lowe JP (1978) Quantum chemistry. Academic, New York
124. Luzanov AV (2013) J Struct Chem 54:835
125. Luzanov AV (2013) Kharkov University Bulletin. Chemical Series Issue 22(45):9
126. Clar E, Zander M. (1958) J Chem Soc 1861; Clar E (1972) Aromatic sextet. Wiley, London
127. Luzanov AV (2013) J Struct Chem 54:277
128. Tyutyulkov N, Drebov N, Mu1llen K, Staykov A, Dietz F (2008) J Phys Chem C 112: 6232
129. Machado FBC, Aquino AJA, Lischka H (2014) ChemPhysChem 15:3334
130. Sheka EF (2012) Int J Quantum Chem 112:3076; Sheka EF (2015) Adv Quantum Chem 70:111
131. Cui Z-h, Lischka H, Mueller T, Plasser F, Kertesz M (2014) ChemPhysChem 15:165; Cui ZH, Lischka H, Beneberu, HZ, Kertesz M (2014) J Am Chem Soc 136:12958
132. Mardsen JE, Ratiu T, Abraham R (2001) Manifolds, tensor analysis, and applications. Springer, New York
133. Mestechkin MM (1969) Theor Math Phys 1:221
134. Ruskai MB (1970) J Math Phys 11:3218
135. Valdemoro C (1992) Phys Rev A 45:4462
136. Dirac PAM (1934) Proc Cambr Phil Soc 30:150
137. Davydov AS (1965) Quantum mechanics. Pergamon Press, Oxford [see Eqs. (61.12) and (61.29)]
138. Rebane TK (1963) Vestn Leningr Gos Univ 22:30; Mestechkin MM (1965) Theor Experim Chem 1:388
139. Li W, Chen F (2014) J Nanopart Res 16:2498
140. Löwdin P-O (1955) Phys Rev 97:1505
141. Fischer SA, Prezhdo OV (2011) J Phys Chem C 115:10006; Jaeger HM, Hyeon-Deuk K, Prezhdo OV (2012) Acc Chem Res 46:1280

Chapter 7
In Silico Assembly of Carbon-Based Nanodevices

Alfredo D. Bobadilla and Jorge M. Seminario

Abstract Carbon nanostructures are 0D, 1D and 2D nanomaterials with potential to enable new markets in the electronic industry due to their novel properties which have been recognized recently with the awarding of Nobel Prizes in Physics and Chemistry. However their very small size constitutes a great challenge in the manufacturing industry, demanding extraordinary and expensive efforts in experimentation. Thus, the best way to avoid unneeded trial-and-error experimentation is by using theoretical-computational tools for the molecular analysis and simulation of prospective devices and systems, allowing us to observe properties at the nanoscale that are practically difficult and sometimes impossible to observe experimentally. We decided to review in this Chapter the use of these tools in order to analyze several scenarios on the assembly and characterization of carbon-based nanodevices. In an in silico experiment, by using molecular dynamics, we analyzed the outcome of bombarding carbon nanotubes with argon ions and we found that for very high energies the type of defects created were almost exclusively single vacancy, which is important in the development of spin-based electronics. On the other hand, combining carbon nanostructures with DNA molecules offers the possibility of exploiting the chemical sensitivity of DNA and the transduction of electrical signals. Therefore, by using molecular dynamics, we predicted a stable structure for a non-covalent DNA junction with a carbon nanotube (CNT) and graphene as interface electrodes. The electronic structure calculations predicted that the DNA electronic structure is coupled to the carbon electron nanodevices, which allow the sensing of a chemical environment. Finally, in the field of drug-delivery,

A.D. Bobadilla · J.M. Seminario (✉)
Department of Chemical Engineering, Texas A&M University,
College Station, TX 77843, USA
e-mail: seminario@tamu.edu

A.D. Bobadilla · J.M. Seminario
Department of Electrical and Computer Engineering, Texas A&M University,
College Station, TX 77843, USA

J.M. Seminario
Department of Materials Science and Engineering, Texas A&M University,
College Station, TX 77843, USA

J. Leszczynski and M.K. Shukla (eds.), *Practical Aspects of Computational Chemistry IV*, DOI 10.1007/978-1-4899-7699-4_7

biological barriers and the immune system constitute challenges for the effective delivery of drugs to targeted areas of the human organism. Therefore, by using molecular dynamics, we predicted the structure and stability of maximum PEGylated carbon nanotubes. We found the size of the PEG-CNT complex to be smaller at conditions of maximum PEGylation and in the nanosized regime, which is an important requirement for the effective delivery of drugs.

7.1 Introduction

Carbon nanostructures can exhibit unique properties that include extremely high mechanical strength, high thermal conductivity, and excellent chemical and thermal stability. Furthermore, carbon nanostructures can be complexed with other molecules showing different functional capabilities. Because nanoscale devices are comparable in size to molecules and are much smaller than biological cells, potential applications of these devices are on single-molecule analytical sensors, biosensors, single cell diagnostics, implantable devices and drug-delivery.

7.1.1 Engineering Carbon Nanostructures

Irradiation of carbon nanostructured materials with electron or ion beams constitutes a novel technique to engineer the structure and properties of these nanomaterials with high precision [1–4]. Guo et al. [5] got cutting a carbon nanotube (CNT) by opening a window in a PMMA thin film covering CNT, and then exposing this zone to reactive oxygen plasma. Under optimized conditions, ~25 % of the tubes are completely cut among ~2500 devices tested. Other technique used to engineer nanostructured materials is electrical breakdown, which occur in SWCNT depending on the contact resistances, nanotube length, applied gate voltage, heat transfer to the contacts and its structural perfection [6–8]. Hadeed and Durkan [9] controlled the size and position of a gap in gold-palladium nanowires by changing the electromigration and joule heating regimes which determine the electrical breakdown of the nanowire.

7.1.2 A Peptide Covalent Bond Between Carbon Nanotube and DNA

Guo et al. [10] developed a method to connect a gapped carbon nanotube (CNT) with single DNA molecules. In their method, a gap is created in the carbon nanotube by oxidative cutting with reactive ion etching; all the CNT is covered with a PMMA film but a small CNT zone is exposed to oxygen plasma by selectively opening a small window on the film with e-beam lithography. During the etching process, carbon nanotube is functionalized with carboxyl groups. Amine functionalization of DNA molecules allowed a covalent amide linkage to CNT functionalized with carboxyl groups. Using this method they obtained 10 working devices out of 370 that were tested. This low yield on successful interconnection of carbon nanotube with DNA is due to an inherent difficulty on creating a covalent bond. Interestingly they found a dsDNA molecule does not conduct electricity when a base mismatch is present.

7.1.3 A Carbon Nanotube-DNA Origami Junction

A nucleic acid-labelled single-walled carbon nanotube (NL-SWCNT) can be aligned along patterns of the corresponding complementary single-stranded DNAs (ssDNA) 'hooks' on DNA origami. Nucleic acid labels are partially protected by complementary strands before they bind to the hooks on DNA origami. Each linker also has a 40-base poly-thymine nucleobase domain that adsorbs onto the SWCNT sidewall via vdW forces. The protection strand prevents adsorption of the labelling domain onto SWCNT. By using this technique, two carbon nanotubes can be aligned on a DNA origami structure, with a carbon nanotube on each side of the DNA origami template, and an orthogonal orientation between carbon nanotubes [11].

7.1.4 Observation of Electrical Gating by ssDNA Upon Binding to Carbon Nanotube

It has been reported that a DNA wrapped carbon nanotube device can change from metallic behavior in dry conditions to semiconductor behavior in wet conditions [12]. Ouellette [13] analyzed the time evolution of electrical current in a suspended carbon nanotube positioned in a microfluidic channel through which DNA molecules were allowed to flow. Electrical spikes were observed when DNA molecules were present in the microfluidic channel; DNA molecules constantly flowed through the microfluidic channel and electrostatic screening of van der Waals interaction due to ions was minimized. Spikes current levels are below or above the original CNT current level depending on the type of DNA molecule tested.

Those spikes are produced every time a DNA molecule reaches the carbon nanotube surface, and a different change in current level, positive or negative spikes, is observed for different DNA molecules. A DNA sequence dependence is further suggested by the known correlation between nucleobases polarizability and

CNT-DNA interaction strength reported by Gowtham et al. [14, 15], and by studies on the electrical property of graphene upon interaction with nucleobases [16–18].

7.1.5 *Carbon Nanotubes for Drug-Delivery*

Exposure of biological cells to carbon nanomaterials can lead to loss of cell viability [19]. However carbon nanotubes can be linked to a wide variety of molecules, including biological molecules and polymers, and functionalized carbon nanotubes have shown low toxicity and are not immunogenic [20]. Carbon nanotubes also show a high propensity to cross cell membranes and this process is endocytosis-independent [21]. Another important property they show is excretion, they can be excreted via biliary pathway without causing obvious toxic effects to normal organs [22]. And the ultrahigh surface area of these one-dimensional polyaromatic macromolecules allows for efficient loading of chemotherapy drugs [23].

7.2 Irradiation-Induced Defects in a Silica-Supported Carbon Nanotube

Carbon nanotubes are typically synthesized with poor control of length and chirality [24, 25]. After carbon nanotubes have been selected and deposited on a substrate, an important step in the fabrication of carbon nanodevices is the patterning to reach an optimal size or dimensions. A common tool to pattern carbon nanostructures is the selective exposure to ion beams, with this technique it has been possible reaching the 10 nm range [26, 27].

By using molecular dynamics with a reactive force field [28–30] as implemented in LAMMPS [31], we were able to analyze the side effects of ion bombardment on a silica-supported single walled carbon nanotube. A reactive force field enables simulating the breaking and formation of covalent bonds. Apart from observing the effective removal of carbon atoms, we found the possibility of undesired effects on the carbon nanotube sidewall, on the substrate as well as at the interface between the carbon nanotube and the substrate (Fig. 7.1). We highlight the main types of atomic defect found on carbon nanotube sidewall, vacancy defects and chemisorption.

Atomic defects on carbon nanostructures produced during the fabrication process are typically not reported. Atomic defects are not visible by characterization techniques typically employed such as AFM, SEM, light microscopy and Raman spectroscopy. The atomic structure can be visible by STM [32–35] and TEM [36, 37] but they are tedious and heavily time consuming, and they are typically not employed in the characterization of carbon-based electron nanodevices reported in journal publications. Molecular simulation tools offer an alternative to visualizing and predicting the nanostructure at atomic detail.

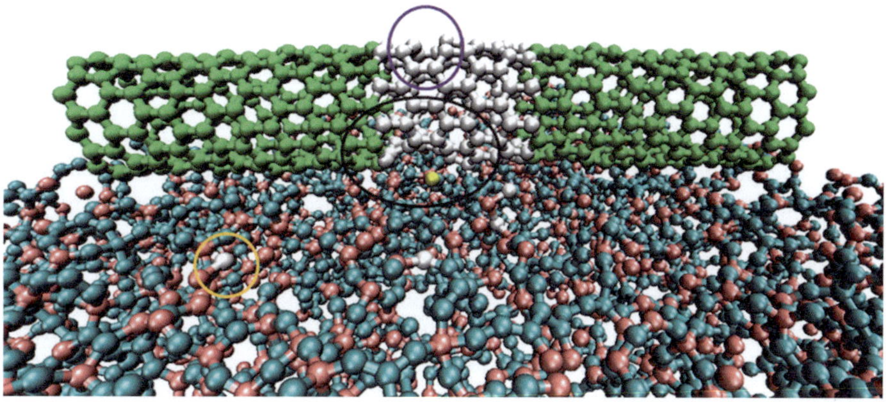

Fig. 7.1 Several types of defects generated on carbon nanotube wall and silicon dioxide substrate after argon atom collision. Single vacancy (*purple circle*), kink (complex) defect (*black circle*), carbon chemisorption and doping on SiO_2 substrate (*orange circle*) [38]

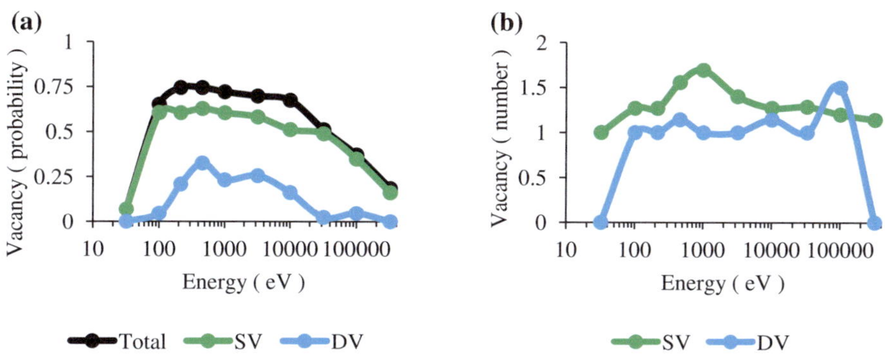

Fig. 7.2 **a** Probability of single and double vacancy at different argon beam energy levels. **b** Average number of single and double vacancies [38]

Vacancy defects represent the effective removal of carbon atoms from the nanotube sidewall. Vacancy defects were always present and with maximum probability of occurrence around 320 eV; however the probability for this type of defect was close to null at beam energy higher than 32 keV (Fig. 7.2).

In graphitic surfaces, like in the carbon nanotube sidewall, carbon atoms are covalently bonded in the sp^2 configuration forming benzene-like rings, with electrons delocalized in the whole surface, in what is called a graphene plasmon. Vacancy defects disrupt this electronic structure configuration giving the possibility of unpaired electrons or dangling bonds and the emergence of magnetic properties that could enable the development of spintronic devices. The magnetic properties will depend on the chirality of the carbon nanotube and the configuration of the vacancy defect [39–42].

We also predicted the possibility of chemisorption of carbon and oxygen atoms to the carbon nanotube sidewall. Oxygen atoms were ejected from the substrate after the collision of the argon atom beam with the silicon dioxide substrate. And when a carbon atom was effectively removed from the nanotube sidewall, it ended up doping the substrate or chemisorbed on other sector of the nanotube sidewall. We denoted chemisorption as the adsorption by covalent bonding of an atom to the nanotube sidewall. It could be considered the opposite of a vacancy defect.

We found the probability of chemisorption defects to be maximum around beam energy of 100 eV and a close to null probability of occurrence for beam energy higher than 10 keV. The probability for chemisorption on the inner side of carbon nanotube was always small, less than 0.15 (Fig. 7.3).

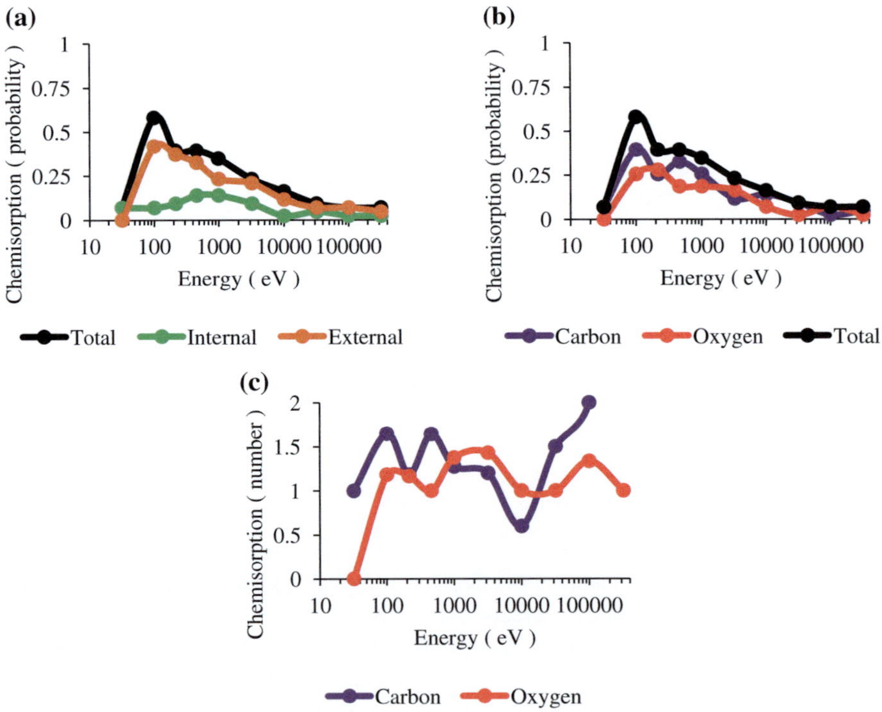

Fig. 7.3 **a** Probability of chemisorption on CNT wall, and probability of chemisorption on internal side of CNT wall or external side of CNT wall. **b** Probability of carbon and oxygen chemisorption. **c** Average number of carbon and oxygen atoms chemisorbed on CNT wall at different argon beam energy levels [38]

The effect of oxygen chemisorptions on semiconducting single wall carbon nanotubes is mainly a reduction in the energy bandgap and the degree of change depends again on chirality and defect configuration [43–45].

7.3 Gating Mechanism of DNA Wrapping on Carbon Nanotube

DNA add to the set of tools of nanotechnology due to the recent development of DNA self-assembly techniques [46–48] as well as to the fact that DNA is a versatile biological molecule, it can be used for diverse tasks, for example as a catalyzer, chemical sensor and computational device.

DNA has been found to adopt a helical structure on carbon nanotube sidewall [49, 50], affecting the electrical property of carbon nanotube devices [12, 13]. We analyzed the electronic structure of each molecule as well as the hybrid of carbon nanotube and DNA. We used the CHARMM force field [51] as implemented in LAMMPS [31] to obtain the atomic structure; and then obtained the electronic structure by DFT calculations in Gaussian 09 [52].

We found DNA effectively wraps carbon nanotube due to the pi-pi interaction between the nanotube sidewall and the nucleobases (Fig. 7.4). We observed discrete

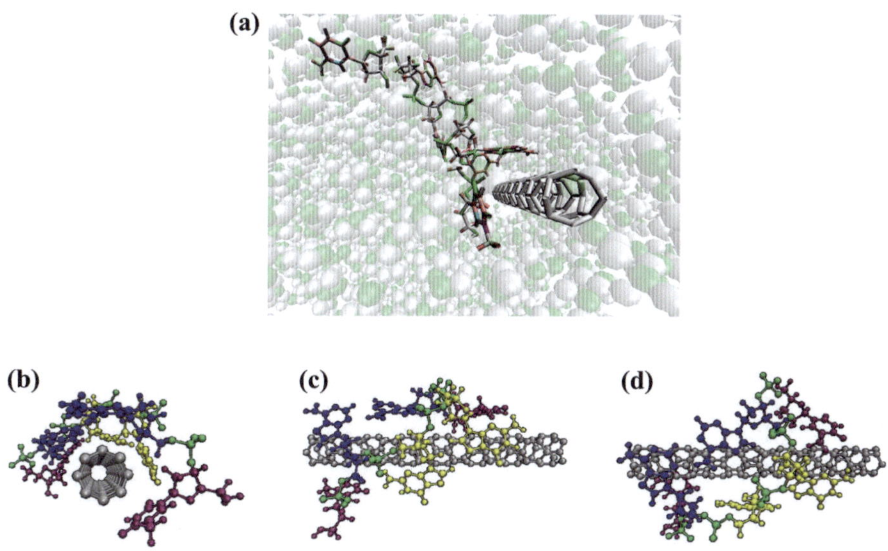

Fig. 7.4 Molecular dynamics of CNT-DNA interaction. **a** Snapshot at t = 2.6 ns: DNA molecule approaching CNT at the initial stage of the process at 340 K. The high temperature accelerates the van der Waals attraction between DNA bases and the carbon nanotube surface. **b** Front view of DNA-CNT helical wrapping after 12.2 ns, at the final stage of equilibration at 300 K. **c** and **d** show two side views. DNA molecule is color coded: thymine (*violet*), adenine (*blue*), guanine (*yellow*), phosphate groups (*green*) [53]

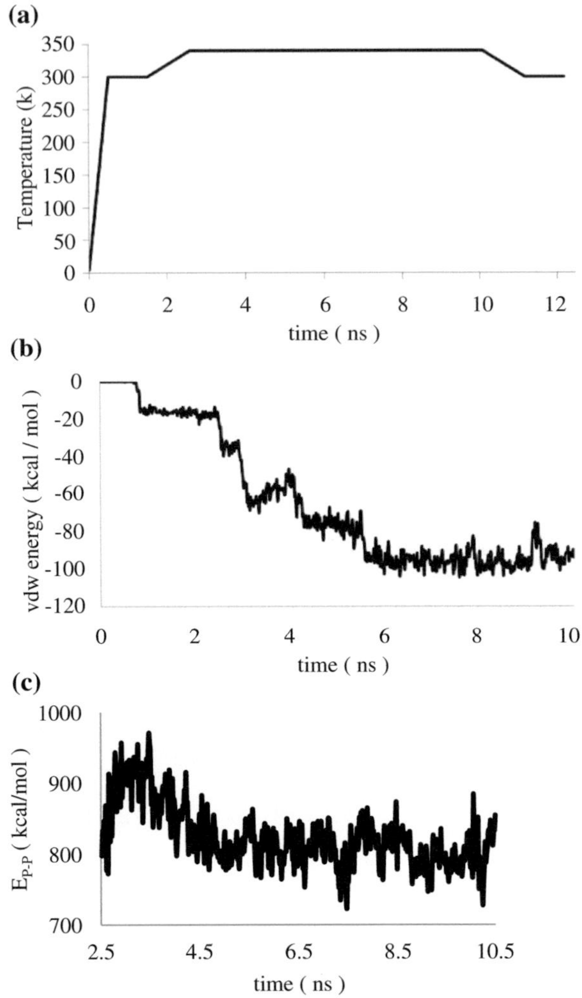

Fig. 7.5 **a** Input temperature: the equilibration process at 340 K begins at t = 2.6 ns and finishes at t = 10.1 ns. **b** Energy jumps are observed in van der Waals (vdW) energy as each nucleotide base binds to the carbon nanotube surface. **c** Electrostatic energy of phosphate atoms in the ssDNA backbone during equilibration process at 340 K [53]

jumps in the vdW energy of interaction every time a nucleobase adsorbed on the nanotube sidewall, and a decrease in the electrostatic energy due to phosphate-phosphate atom charge interaction in the backbone of DNA (Fig. 7.5).

We obtained the atomic structure from molecular dynamics for the CNT molecule (not shown), DNA molecule (not shown) as well as for the hybrid (Fig. 7.4). We performed electronic structure calculations in vacuum as well as in solvent conditions. For the solvent conditions, we also tested an alternative method, the PCM method (Polarizable Continuum Model) [54], which models the solvent as a

Table 7.1 HOMO, LUMO and gap energies (eV) of DNA and CNT as isolated molecules and when they are part of the complex DNA-CNT

System		Vacuum			PCM solvent			Water molecules		
		HOMO	LUMO	Gap	HOMO	LUMO	Gap	HOMO	LUMO	Gap
Isolated	DNA	−5.14	−1.17	3.97	−5.37	−1.09	4.27	−5.93	−4.22	1.71
	CNT	−5.00	−4.59	0.41	−5.09	−4.66	0.43	−4.48	−4.04	0.44
Complex	DNA	−4.89	−1.48	3.41	−5.50	−1.55	3.95	−5.98	−3.38	2.60
	CNT	−5.00	−4.58	0.43	−5.09	−4.66	0.43	−5.54	−5.15	0.39

Calculations are done in vacuum, under PCM solvent, and under actual water molecules in the Hamiltonian. All systems are run as uncharged and singlets [53]

polarizable continuum. The PCM method is originally intended to avoid the heavy computational cost of including water molecules in the electronic structure calculations. We found the PCM method is not accurate on the prediction of the electronic structure of the hybrid CNT-DNA in water solvent (Table 7.1).

When the carbon nanotube is alone, the carbon nanotube HOMO-LUMO energy levels increased by about 0.5 eV when changing from vacuum to water solvent conditions.

When the carbon nanotube is wrapped by the DNA molecule, the opposite took place, the carbon nanotube HOMO-LUMO energy levels decreased by about 0.5 eV when changing from vacuum to water solvent conditions. A detail of molecular orbital energies for the CNT-DNA hybrid is given in Fig. 7.6 and a schematic representation of the device in Fig. 7.7.

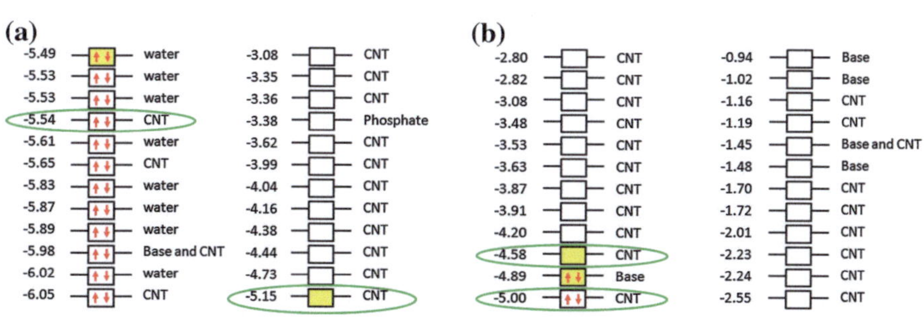

Fig. 7.6 Molecular orbital energies (eV) for the CNT-DNA nanostructure in **a** water and **b** vacuum conditions. *Highlighted squares* indicate HOMO and LUMO of the complex system. *Green ellipses* indicate carbon nanotube HOMO and LUMO. Orbitals are localized in the carbon nanotube (CNT), water molecules (water), bases (Base), and Phosphate groups (Phosphate) [53]

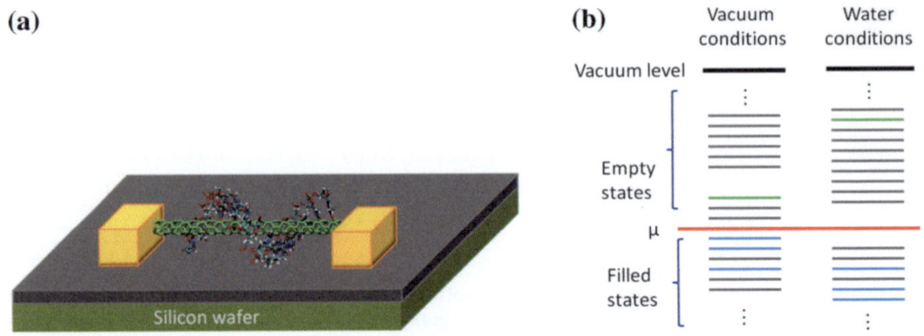

Fig. 7.7 **a** Carbon nanotube-DNA based transistor and **b** its corresponding energy level diagram for vacuum conditions and water solvent conditions. *Gray lines* correspond to CNT energy states, *green lines* correspond to DNA molecule energy states, and *cyan lines* correspond to molecular orbitals shared by CNT and DNA. μ represents the gold electrode work function. Energy levels corresponding to water molecules are omitted for clarity [53]

The gating mechanism in the HOMO-LUMO gap caused by DNA wrapping on carbon nanotube explains the observed change in the electrical property of carbon nanotube upon interaction with DNA molecules in water solvent.

7.4 Assembly and Electron Transport Characteristic of a DNA-Graphene Junction

We analyzed a device with potential applications in DNA sequencing and chemical sensing. Nanoscale ribbon patterns can be created on graphene by selective exposure to oxygen plasma or ion beam. A very small gap (~1 nm) between graphene electrodes is feasible by an electrical breakdown technique [55, 56]. We used the CHARMM force field [51] as implemented in LAMMPS [31] to analyze the directed assembly process of a small strand of DNA (GAG) in a gapped graphene. The GENIP program [57–59] was used to perform electron transport calculations of the hybrid structure in water solvent.

We observed at room temperature all nucleobases had a preference for binding on a single graphene electrode (Fig. 7.8). We therefore needed to assist the assembly process to position the DNA molecule in the gap between graphene electrodes. The DNA backbone is negatively charged due to the phosphate groups, we therefore opted to apply an electric field. An electric field of 0.8 V/nm allowed overcoming the binding strength of DNA to graphene and 0.3 ns was enough time to succeed on the positioning of DNA in the gap (Fig. 7.9). If the electric field was applied for longer time, the DNA molecule abandoned the gap.

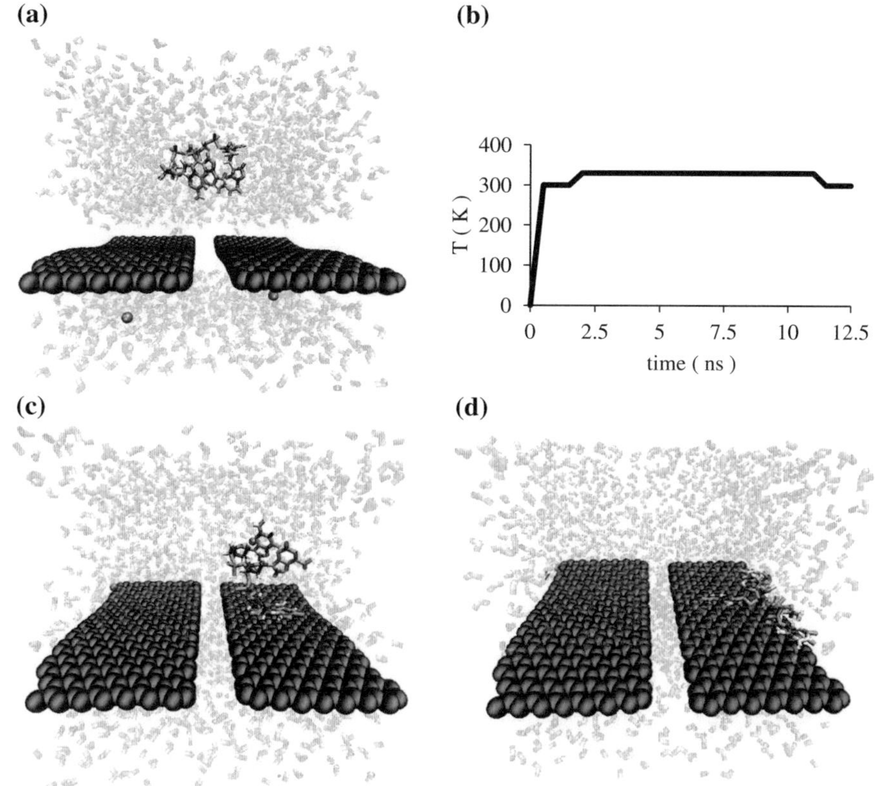

Fig. 7.8 ssDNA (GAG) absorption to graphene nanoribbon. **a** Initial structure, **b** Time evolution of temperature, **c** ssDNA absorption on graphene at room temperature, **d** Final ssDNA conformation during equilibration at 330 K with three nucleobases absorbed on graphene. Molecular dynamics (MD) simulation performed with periodic boundary conditions [60]

To analyze the stability of the DNA junction we performed molecular dynamics above room temperature, at 330 K, for 10 ns. During all the process we observed two nucleobases are absorbed to the left graphene electrode and one nucleobase was absorbed on the right electrode. Therefore we considered the DNA junction was stable.

The system size is relatively big and the associated computational cost prohibitively high for quantum chemistry and electron transport calculations. We therefore reduced the system by inducing the evaporation of water molecules. The evaporation process (not shown) was performed gradually and in a time span of 45 ns, during that process the graphene ribbon size was also reduced (not shown).

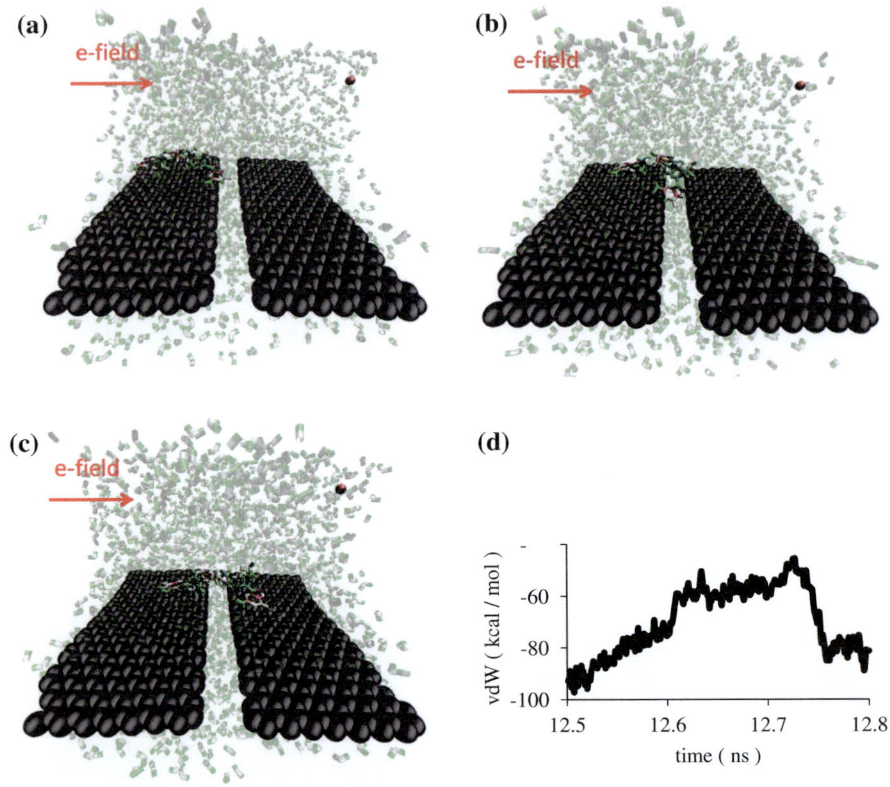

Fig. 7.9 Electric field assisted positioning of ssDNA on graphene gap at room temperature. ssDNA conformation at **a** 12.6 ns, **b** 12.7 ns and **c** 12.8 ns under the application of an electric field with magnitude 0.08 V/Å to phosphate atoms in the DNA backbone. **d** van der Waals (vdW) energy of interaction between graphene nanoribbons and ssDNA molecule. Electric field is applied at t = 12.5 ns and stopped at t = 12.8 ns. Simulation performed with periodic boundary conditions [60]

To analyze the stability of the final structure we performed molecular dynamics at room temperature for 7 ns. During that process the junction was stable and we only observed a change in conformation, the final structure had the central nucleobase positioned in the nanogap (Fig. 7.10).

We then obtained the electronic structure by DFT calculations in Gaussian 09 [52]. We observed the effect of ssDNA (GAG) on the gapped graphene is a gating mechanism, we observed a slight change in the energy levels of the HOMO-LUMO gap when ssDNA fills the nanogap as well as an increase in magnitude of the electron transmission probability function (Fig. 7.11).

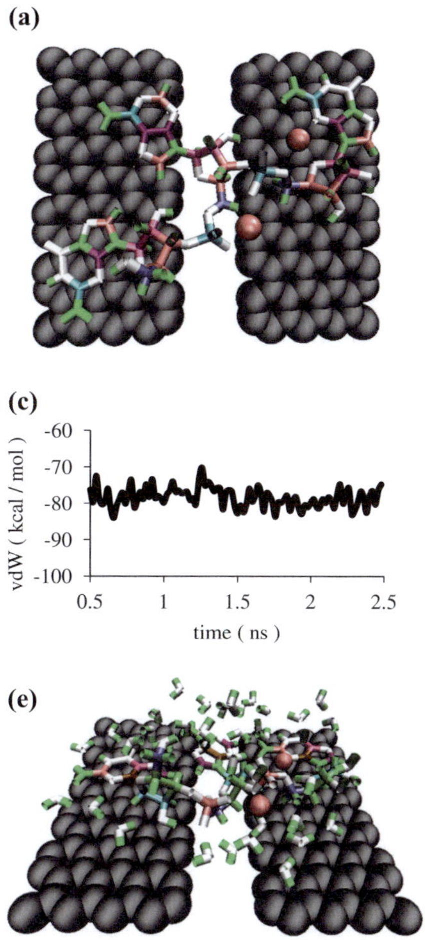

Fig. 7.10 Reduced graphene length and nonperiodic boundary conditions. **a** Structure at t = 1.5 ns with two nucleobases absorbed on left graphene electrode, **b** Structure at t = 2.5 ns with one nucleobase positioned in the nanogap zone, **c** vdW energy for graphene-ssDNA interaction during the initial 2 ns of equilibration at room temperature, **d** vdW energy for graphene-ssDNA interaction during the last nanosecond of equilibration at room temperature, and (i) Final conformation at 1 K. Water molecules are omitted in **a** and **b** for visualization purposes [60]

For the electron transport calculations we included gold atoms covalently bonded to carbon atoms at the opposite edges of graphene electrodes, these gold atoms acted as interface to the electronic structure of bulk gold electrodes (Fig. 7.12).

We analyzed the effect of the electric field (associated with a bias voltage) in the DOS (density of states) and in the TF (electron transmission probability function). The electric field caused the splitting of peaks or energy bands in the DOS and a

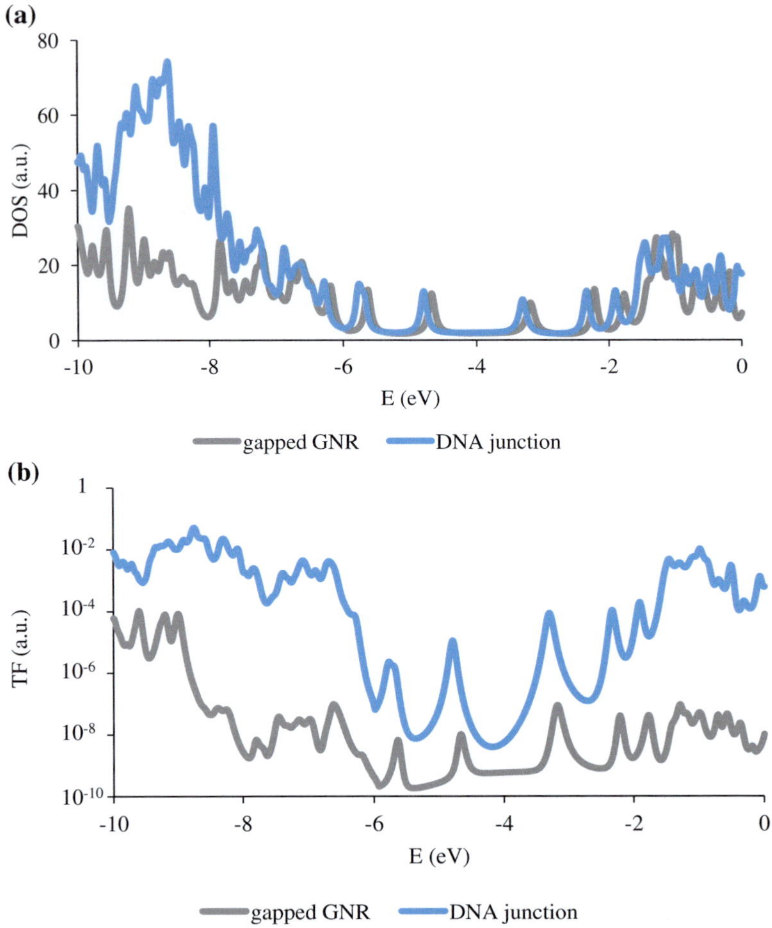

Fig. 7.11 Effect of DNA molecule on DOS and TF of gapped graphene at zero bias voltage. **a** Density of states (DOS in arbitrary units, a.u.) and **b** electron transmission probability function (TF in arbitrary units, a.u.) for gapped GNR (*gray colored*) and GNR-DNA junction (*cyan colored*) [60]

significant increase in magnitude of the TF (Fig. 7.13). The gapped graphene and the DNA junction were significantly different in the TF function at 1.4 V.

We analyzed as well the effect of the electric field on the HOMO-LUMO energy levels of gapped graphene and DNA junction. We observed the gating mechanism induced by the DNA molecule is not affected by the electric field (Fig. 7.14), and the electric field was effective on reducing the HOMO-LUMO energy gap which is expected to produce a non-zero electrical current through the junction.

The tunneling current of the gapped graphene was non-zero only after a bias voltage larger than 1.5 V is applied. While the DNA junction had a non-zero electrical current for bias voltage large than 0.5 V. But the effect of DNA on the

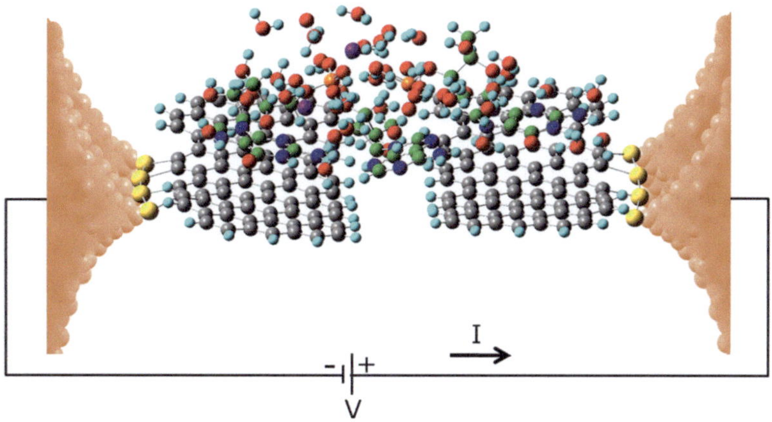

Fig. 7.12 Graphene-DNA junction. Interfacial gold atoms (*yellow*), bulk electrodes (*orange*), graphene carbon atoms (*grey*), and DNA carbon atoms (*green*) [60]

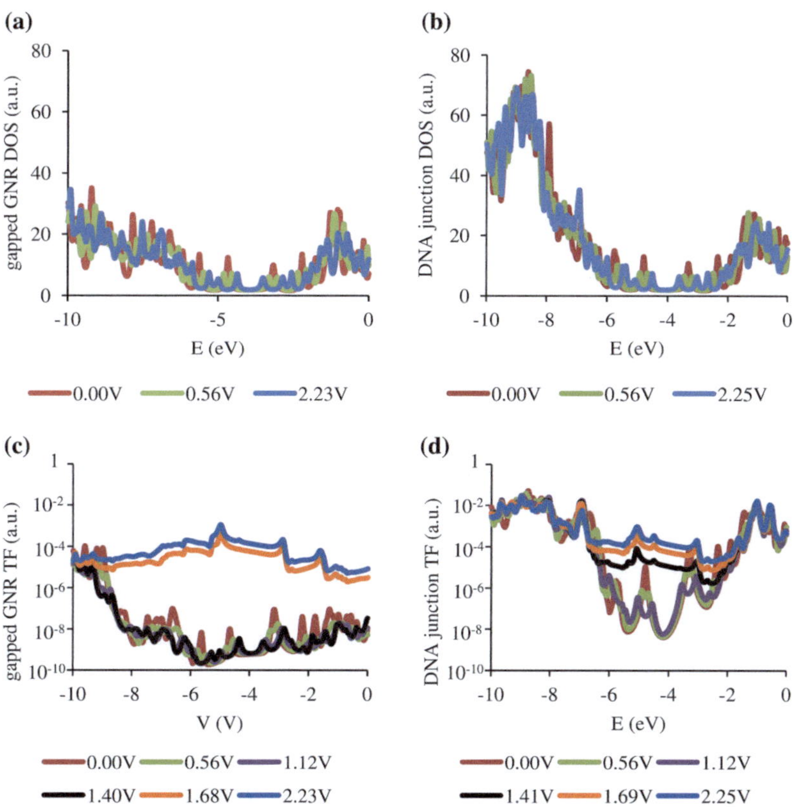

Fig. 7.13 Density of states (DOS) for **a** gapped GNR, **b** DNA junction, **c** electron transmission function (TF) for gapped GNR, and **d** DNA junction [60]

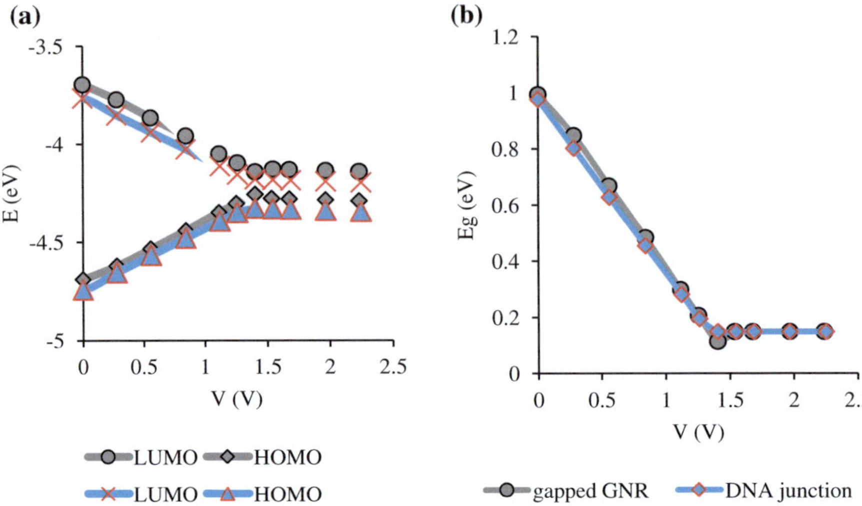

Fig. 7.14 Bias voltage dependence of **a** HOMO, LUMO and **b** HOMO-LUMO gap. Gapped GNR (*gray*), DNA junction (*cyan*) [60]

gapped graphene only reflected in a detectable electrical signal at a bias voltage of 1.4 V when the magnitude of electrical current through the DNA junction was 1.6 nA (Fig. 7.15) and the relative difference of current levels $(I_{DNA} - I_{GNR})/I_{GNR}$ was maximum (Fig. 7.16).

7.5 The Hydrodynamic Volume of Maximum PEGylated Carbon Nanotube

Drug molecules are typically of hydrophobic character [61, 62] and naturally absorb on the hydrophobic (nonpolar) surface of carbon nanotube [63–65]. To protect the drug from the immune response in the human body, a polymer protection is typically needed [63, 65, 66].

We analyzed the structure of a carbon nanotube functionalized with polyethylene glycol (PEG). A maximum amount of polymer ensure a better protection of the drug, we therefore analyzed the scenario of maximum PEGylation. The main requirement is a small volume of the drug carrier to ensure an effective penetration of the natural biological barriers and an optimal circulation through the human body.

To obtain the polymer structure we performed molecular dynamics (MD) simulations [31, 51, 67, 68] with 1, 40 and 208 PEG molecules. The initial PEG structure is obtained from a crystallographic database [69] and modified to perform a MD simulation in water at room temperature (now shown). The final structure

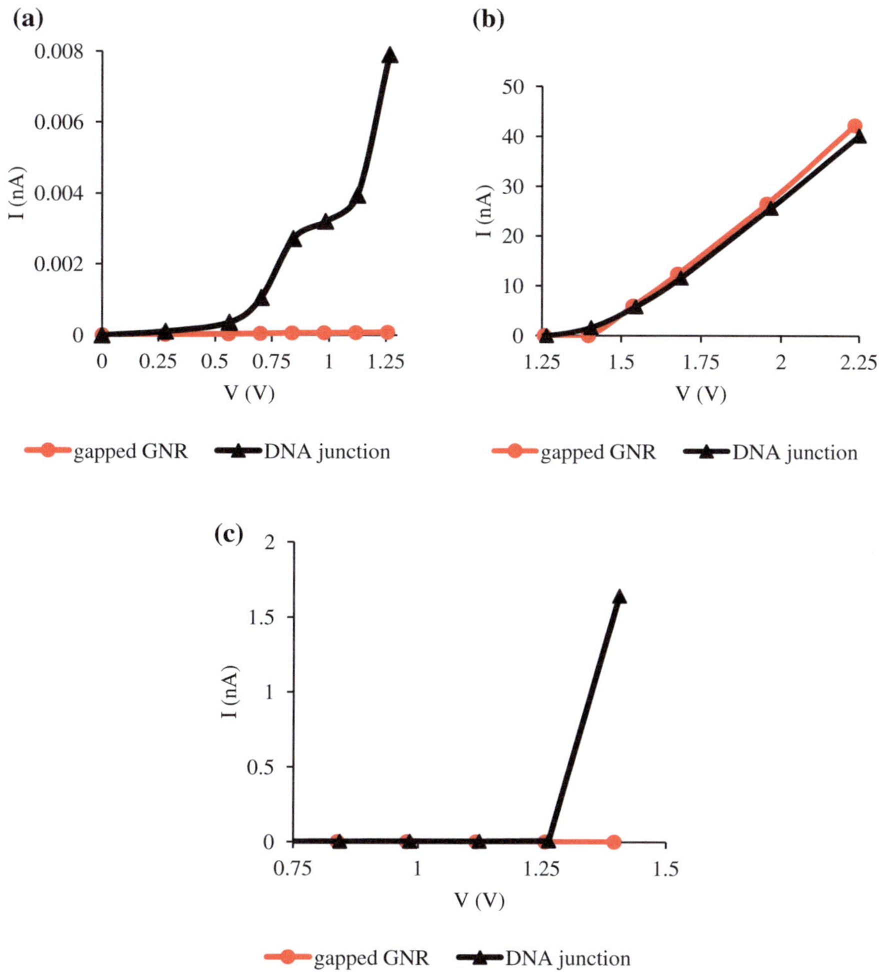

Fig. 7.15 Current-voltage characteristic at several bias voltage ranges for the GNR-DNA junction (*black*) and gapped GNR (*red*). **a** Low bias voltage range (0–1.25 V), **b** high bias voltage range (1.25–2.25 V), and **c** transition between electrically nonconductive (<0.01 nA) to conductive state (>1 nA) for DNA junction [60]

served as input to build a system of carbon nanotube (CNT) functionalized with 40 PEGs (Fig. 7.17), and 40 was about the maximum amount possible to attach.

To validate the result of this simulation we compared the radius of the complex to that reported in experiments on AFM measurements of a similar complex, equal diameter of carbon nanotube (1 nm) and equal molecular weight (5 kDa) of polyethylene glycol. The radius of the complex we obtained (7.6 nm, Fig. 7.18) is very close to the experimental value of 7 nm [71].

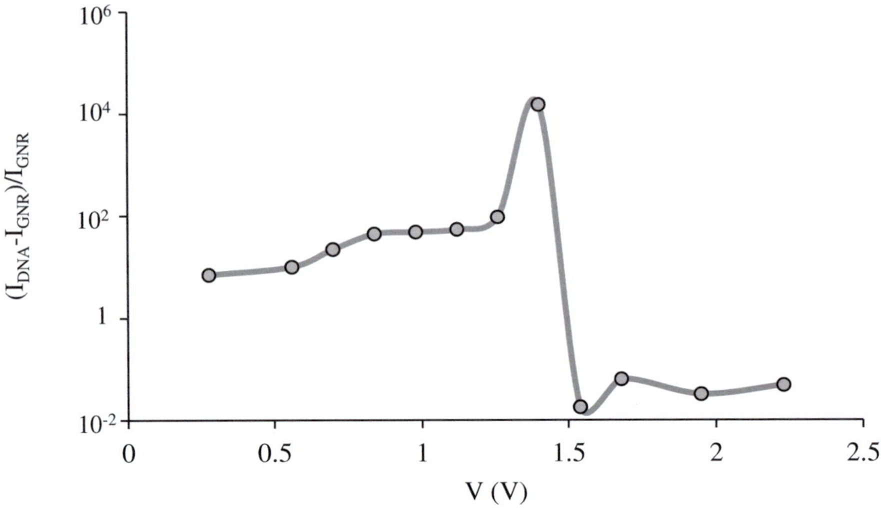

Fig. 7.16 Relative difference of the currents for the gapped GNR (I_{GNR}) and DNA junction (I_{DNA}) [60]

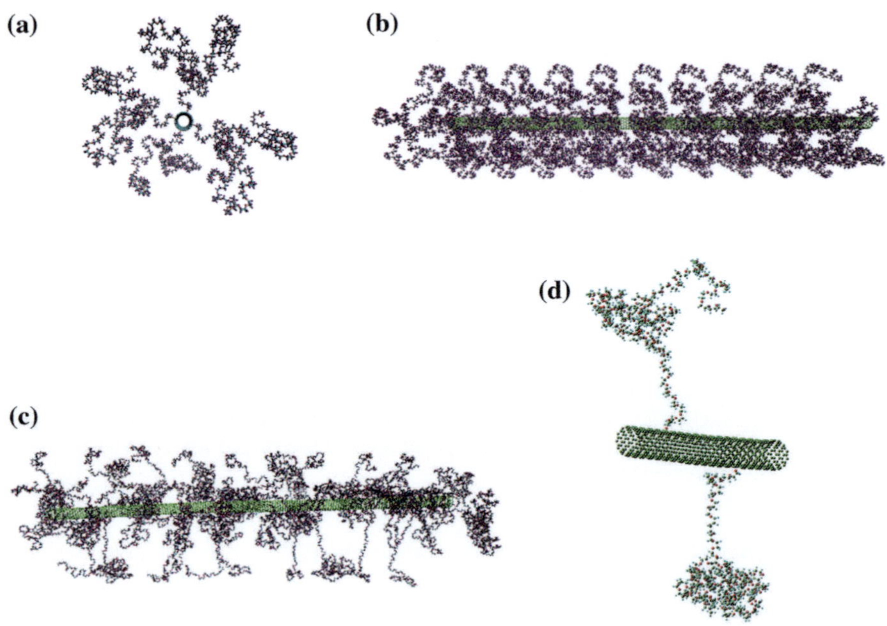

Fig. 7.17 **a** Side view of initial structure. **b** PEG-CNT initial structure, including 40 PEG molecules, before energy minimization. **c** PEG-CNT final structure at 300 K for MD simulation. Water molecules are omitted for visualization purposes **d** PEG molecules attached to the CNT sidewall show globular and linear segments. Structure was extracted from PEG-CNT final conformation at 300 K [70]

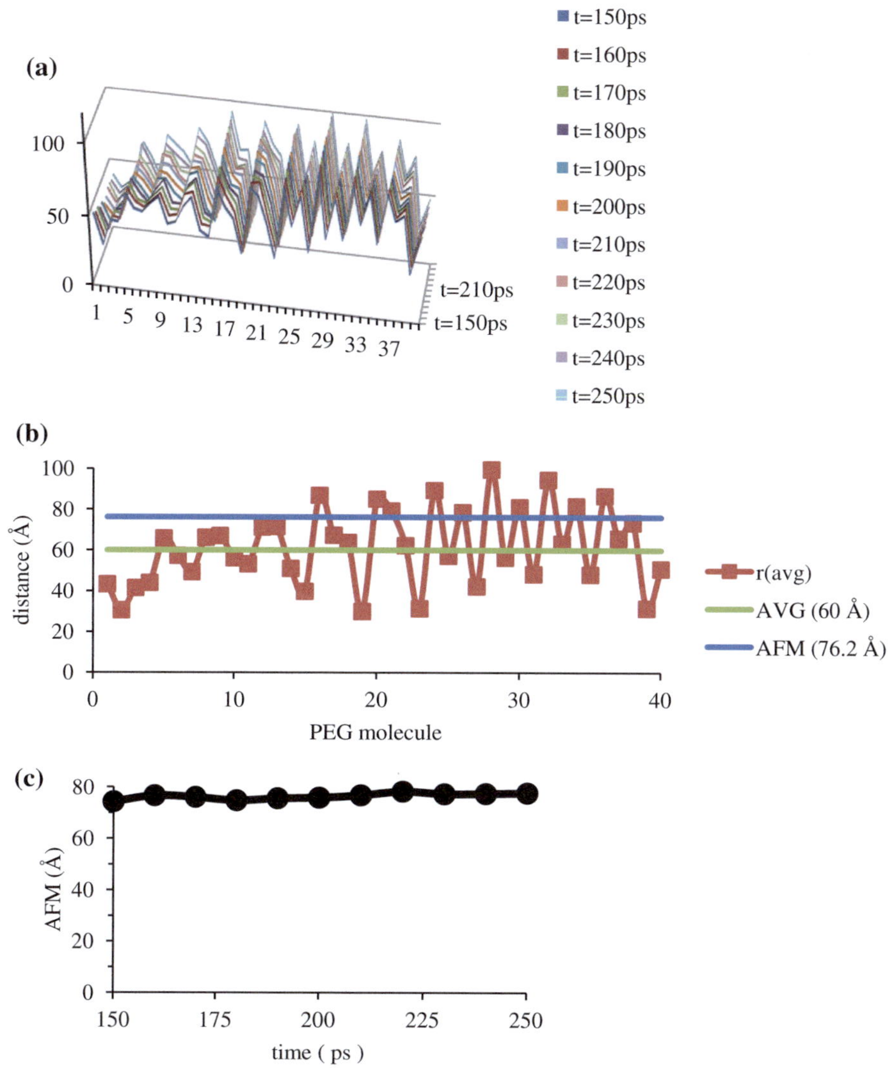

Fig. 7.18 Estimation of polymer radial thicknes (*AFM* value) wrapping CNT. **a** Time evolution of *rmax*(*t*), maximum distance between a PEG atom and CNT sidewall, for every PEG molecule. *Vertical axis* is in distance units (Å), *horizontal axis* is an identifier for PEG molecules with values 1, 2, ..., 40, and curves are color coded for different time (picoseconds), **b** *AFM* value estimation from *r*(*avg*) values. We calculate *r*(*avg*) for every PEG molecule as the average over time of *rmax*(*t*) value in a 100 ps MD run. *AVG* is the average value of *r*(*avg*) and *AFM* is the average value of *r*(*avg*) when *r*(*avg*) is larger than AVG (60 Å), **c** AFM. value estimation from instantaneous *rmax*(*t*) values in a 100 ps MD run [70]

To build a PEG-CNT complex with maximum PEGylation we take a polymer molecule from the CNT-40PEG system as an initial reference for the CNT-208PEG system. The 'linear fragment' of the polymer (Fig. 7.17d) is elongated to enable an increase in the amount of PEG molecules attached to the CNT. After the elongation, 208 PEG molecules is about the maximum amount possible to attach to the CNT sidewall (Fig. 7.19).

To characterize the complex we estimated the radius of the PEG-CNT complex ($\sim r_{uv}$) as well as the end-to-end distance of PEG molecules (h_i) (Fig. 7.20). We found the maximum radius of complex decreases from 10 nm (for CNT-40PEG) to 7 nm (for CNT-208PEG). And the maximum end-to-end PEG distance decreased from 7.5 to 5 nm (Figs. 7.21a, b and 7.22a, b). They reflected an overall reduction in the volume of the PEG-CNT complex; therefore, maximum PEGylation is not only beneficial for optimal protection of the drug molecule but to achieve a small volume of the carrier and an optimal circulation through the human body. As a reference for the volume of the complex with maximum PEGylation (CNT-208PEG), the maximum diameter was 14 nm and the length 40 nm (Fig. 7.22c, d).

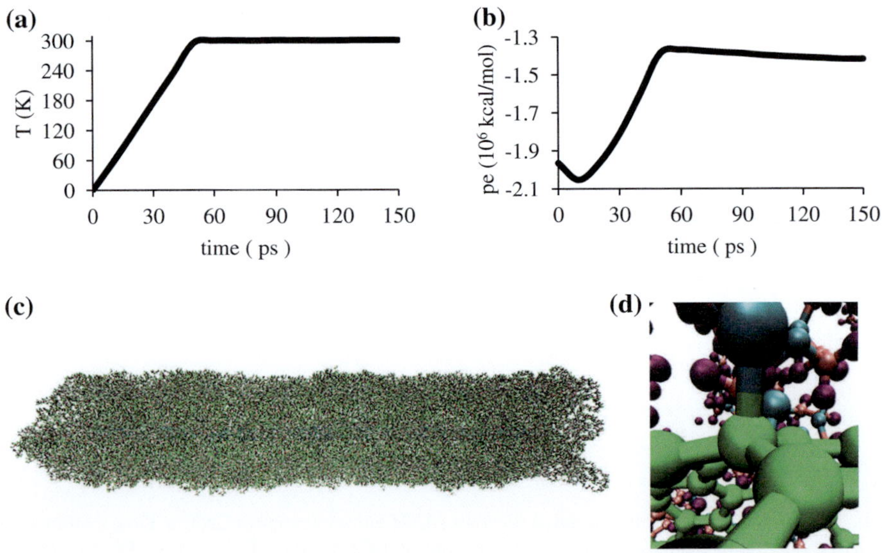

Fig. 7.19 Time evolution of **a** temperature and **b** total potential energy for PEG-CNT construct with 208 PEG molecules. **c** PEG-CNT structure at t = 150 ps and room temperature, and **d** tetrahedral structure at a PEG-CNT link [70]

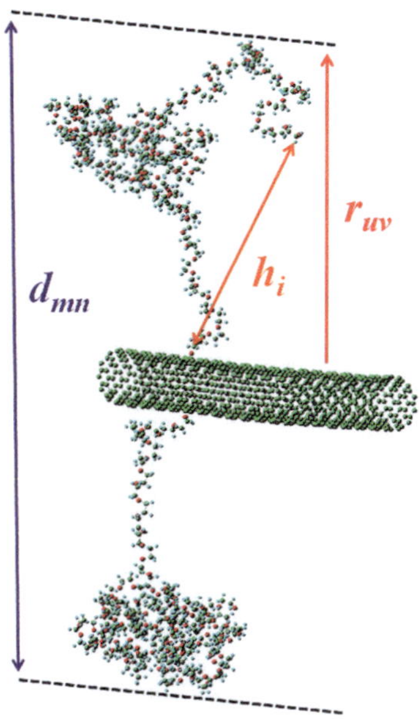

Fig. 7.20 Distances calculated in the PEG-CNT complex. d_{mn} is the maximum distance between atoms of opposite PEG molecules attached to a same CNT ring and along a direction orthogonal to the CNT axis, m is an index for every CNT ring and n is an index for every pair of opposite PEG molecules attached to the ring. h_i is the end-to-end distance for every PEG molecule. r_{uv} is the maximum distance between a PEG atom and the CNT sidewall, u is an index for every CNT ring and v is an index for every PEG molecule attached to the ring [70]

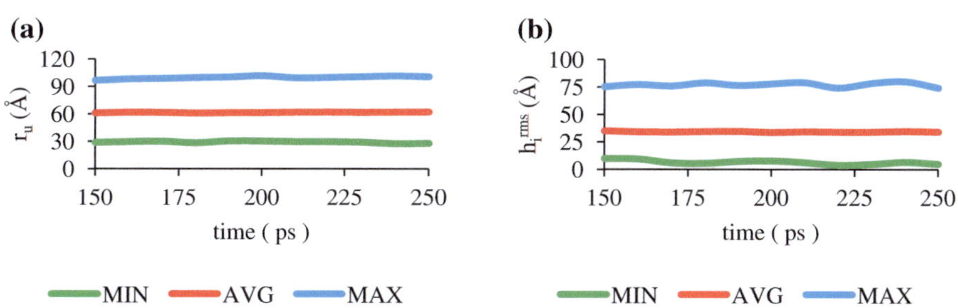

Fig. 7.21 Estimates for a PEG-CNT construct with 40 PEG molecules. **a** Maximum distance between an atom of PEG and CNT sidewall, and **b** root mean squared end-to-end distance for a PEG molecule [70]

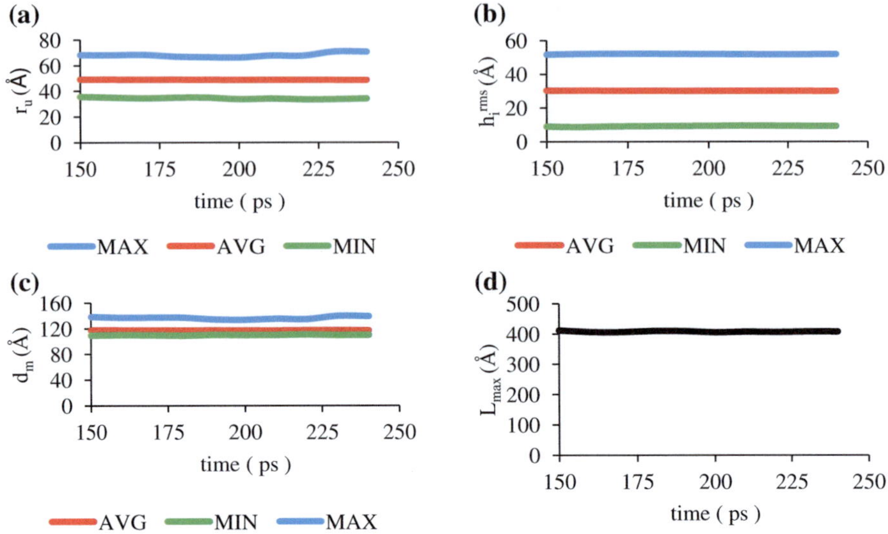

Fig. 7.22 Estimates for a PEG-CNT complex including 208 PEG molecules. **a** Maximum distance between a PEG atom and CNT sidewall, **b** root mean squared end-to-end distance, **c** width of the PEG-CNT complex has a maximum value of 136.5 Å (averaged during the last 50 ps), and **d** the maximum length of the PEG-CNT complex is 407 Å (averaged during the last 50 ps) [70]

7.6 Summary and Conclusions

We analyzed the effects of irradiation on carbon nanotube and we found a predominance of single vacancy defects upon exposure to very high energy argon ion. These defects resembled magnetic impurities. Then, by using molecular dynamics, we were able to predict the structure of irradiated carbon nanotubes, DNA-carbon hybrid nanostructures and PEGylated carbon nanotube. This allowed a detailed analysis of the atomic structure, otherwise difficult and sometimes impossible to attain by conventional imaging tools. Furthermore, the structure found by molecular dynamics can serve as input for electronic structure calculations and we suggested the possibility of novel electron devices. Finally, by exploiting the strong van der Waals interaction in electron-rich pi-systems, we suggested the assembly of novel nanostructures based on DNA and carbon nanostructures, we found that DNA always induces a gating mechanism on the electronic structure of carbon; therefore, the chemical sensitivity of DNA can be effectively coupled to the electrical property of the carbon nanodevice.

References

1. Krasheninnikov AV, Banhart F (2007) Engineering of nanostructured carbon materials with electron or ion beams. Nat Mater 6:723–733
2. Park J-Y, Yaish Y, Brink M, Rosenblatt S, McEuen PL (2002) Electrical cutting and nicking of carbon nanotubes using an atomic force microscope. Appl Phys Lett 80:4446–4448
3. Weng L, Zhang L, Chen YP, Rokhinson LP (2008) Atomic force microscope local oxidation nanolithography of graphene. Appl Phys Lett 93:093107
4. Yuzvinsky TD, Mickelson W, Aloni S, Begtrup GE, Kis A et al (2006) Shrinking a carbon nanotube. Nano Lett 6:2718–2722
5. Guo X, Small JP, Klare JE, Wang Y, Purewal MS et al (2006) Covalently bridging gaps in single-walled carbon nanotubes with conducting molecules. Science 311:356–359
6. Seidel RV, Graham AP, Rajasekharan B, Unger E, Liebau M et al (2004) Bias dependence and electrical breakdown of small diameter single-walled carbon nanotubes. J Appl Phys 96:6694–6699
7. Collins PG, Hersam M, Arnold M, Martel R, Avouris P (2001) Current saturation and electrical breakdown in multiwalled carbon nanotubes. Phys Rev Lett 86:3128–3131
8. Collins PG, Arnold MS, Avouris P (2001) Engineering carbon nanotubes and nanotube circuits using electrical breakdown. Science 292:706–709
9. Hadeed FO, Durkan C (2007) Controlled fabrication of 1–2 nm nanogaps by electromigration in gold and gold-palladium nanowires. Appl Phys Lett 91:123120
10. Guo X, Gorodetsky AA, Hone J, Barton JK, Nuckolls C (2008) Conductivity of a single DNA duplex bridging a carbon nanotube gap. Nat Nanotechnol 3:163–167
11. Maune HT, Han S-P, Barish RD, Bockrath M, Goddard IIA et al (2010) Self-assembly of carbon nanotubes into two-dimensional geometries using DNA origami templates. Nat Nanotechnol 5:61–66
12. Cha M, Jung S, Cha M-H, Kim G, Ihm J et al (2009) Reversible metal—semiconductor transition of ssDNA-decorated single-walled carbon nanotubes. Nano Lett 9:1345–1349
13. Ouellette L (2008) Chemical and biological sensing with carbon nanotube in solution. Ph.D., Physics, Cornell University, Ithaca, New York
14. Gowtham S, Scheicher RH, Pandey R, Karna SP, Ahuja R (2008) First-principles study of physisorption of nucleic acid bases on small-diameter carbon nanotubes. Nanotechnology 19:125701
15. Gowtham S, Scheicher RH, Ahuja R, Pandey R, Karna SP (2007) Physisorption of nucleobases on graphene: density-functional calculations. Phys Rev B 76:033401
16. Luis AA, Jacob G, Christian W, Nicholas K (2012) Aviram–Ratner rectifying mechanism for DNA base-pair sequencing through graphene nanogaps. Nanotechnology 23:135202
17. Min SK, Kim WY, Cho Y, Kim KS (2011) Fast DNA sequencing with a graphene-based nanochannel device. Nat Nanotechnol 6:162–165
18. Dontschuk N, Stacey A, Tadich A, Rietwyk KJ, Schenk A et al (2015) A graphene field-effect transistor as a molecule-specific probe of DNA nucleobases. Nat Commun 6
19. Foldvari M, Bagonluri M (2008) Carbon nanotubes as functional excipients for nanomedicines: II. Drug delivery and biocompatibility issues. Nanomed-Nanotechnol 4:183–200
20. Bianco A, Kostarelos K, Prato M (2005) Applications of carbon nanotubes in drug delivery. Curr Opin Chem Biol 9:674–679
21. Liu Z, Chen K, Davis C, Sherlock S, Cao Q et al (2008) Drug delivery with carbon nanotubes for in vivo cancer treatment. Cancer Res 68:6652–6660
22. Pastorin G, Wu W, Wieckowski S, Briand J-P, Kostarelos K et al (2006) Double functionalisation of carbon nanotubes for multimodal drug delivery. Chem Commun 1182–1184

23. Chen J, Chen S, Zhao X, Kuznetsova LV, Wong SS et al (2008) Functionalized single-walled carbon nanotubes as rationally designed vehicles for tumor-targeted drug delivery. J Am Chem Soc 130:16778–16785
24. Ren ZF, Huang ZP, Xu JW, Wang JH, Bush P et al (1998) Synthesis of large arrays of well-aligned carbon nanotubes on glass. Science 282:1105–1107
25. Zhu HW, Xu CL, Wu DH, Wei BQ, Vajtai R et al (2002) Direct synthesis of long single-walled carbon nanotube strands. Science 296:884–886
26. Abbas AN, Liu G, Liu B, Zhang L, Liu H et al (2014) Patterning, characterization, and chemical sensing applications of graphene nanoribbon arrays down to 5 nm using helium ion beam lithography. ACS Nano 8:1538–1546
27. Bell DC, Lemme MC, Stern LA, Williams JR, Marcus CM (2009) Precision cutting and patterning of graphene with helium ions. Nanotechnology 20:455301
28. Tersoff J (1989) Modeling solid-state chemistry: interatomic potentials for multicomponent systems. Phys Rev B 39:5566
29. Tersoff J (1988) New empirical approach for the structure and energy of covalent systems. Phys Rev B 37:6991
30. Ziegler JF, Biersack JP, Littmark U (1985) The stopping and range of ions in matter. Pergamon, New York
31. Plimpton S (1995) Fast parallel algorithms for short-range molecular dynamics. J Comput Phys 117:1–19
32. Krasheninnikov AV, Nordlund K, Sirviö M, Salonen E, Keinonen J (2001) Formation of ion-irradiation-induced atomic-scale defects on walls of carbon nanotubes. Phys Rev B 63:245405
33. Osváth Z, Vértesy G, Tapasztó L, Wéber F, Horváth ZE et al (2005) Atomically resolved STM images of carbon nanotube defects produced by irradiation. Phys Rev B 72:045429
34. Osváth Z, Tapasztó L, Vértesy G, Koós AA, Horváth ZE et al (2007) STM imaging of carbon nanotube point defects. Phys Status Solidi A 204:1825–1829
35. Tolvanen A, Buchs G, Ruffieux P, Gröning P, Gröning O et al (2009) Modifying the electronic structure of semiconducting single-walled carbon nanotubes by ion irradiation. Phys Rev B 79:125430
36. Huang PY, Ruiz-Vargas CS, van der Zande AM, Whitney WS, Levendorf MP et al (2011) Grains and grain boundaries in single-layer graphene atomic patchwork quilts. Nature 469:389–392
37. Gómez-Navarro C, Meyer JC, Sundaram RS, Chuvilin A, Kurasch S et al (2010) Atomic structure of reduced graphene oxide. Nano Lett 10:1144–1148
38. Bobadilla AD, Seminario JM (2014) Argon-beam-induced defects in a silica-supported single-walled carbon nanotube. J Phys Chem C 118:28299–28307
39. Berber S, Oshiyama A (2006) Reconstruction of mono-vacancies in carbon nanotubes: Atomic relaxation vs. spin polarization. Phys B 376–377:272–275
40. Orellana W, Fuentealba P (2006) Structural, electronic and magnetic properties of vacancies in single-walled carbon nanotubes. Surf Sci 600:4305–4309
41. Zanolli Z, Charlier JC (2010) Spin transport in carbon nanotubes with magnetic vacancy-defects. Phys Rev B 81:165406
42. Farghadan R, Saffarzadeh A (2010) The effect of vacancy-induced magnetism on electronic transport in armchair carbon nanotubes. J Phys-Condens Matter 22:255301
43. Barone V, Heyd J, Scuseria GE (2004) Effect of oxygen chemisorption on the energy band gap of a chiral semiconducting single-walled carbon nanotube. Chem Phys Lett 389:289–292
44. Liu HJ, Chan CT, Liu ZY, Shi J (2005) Density functional study of oxygen adsorption on 4Å carbon nanotubes. Phys Rev B 72:075437
45. Khorrampour R, Esrafili MD, Hadipour NL (2009) Density functional theory study of atomic oxygen, O_2 and O_3 adsorptions on the H-capped (5,0) single-walled carbon nanotube. Physica E 41:1373–1378
46. Rothemund PWK (2006) Folding DNA to create nanoscale shapes and patterns. Nature 440:297–302

47. Adleman L (1994) Molecular computation of solutions to combinatorial problems. Science 266:1021–1024
48. Winfree E, Liu F, Wenzler LA, Seeman NC (1998) Design and self-assembly of two-dimensional DNA crystals. Nature 394:539–544
49. Johnson RR, Johnson ATC, Klein ML (2008) Probing the structure of DNA—carbon nanotube hybrids with molecular dynamics. Nano Lett 8:69–75
50. Cathcart H, Nicolosi V, Hughes JM, Blau WJ, Kelly JM et al (2008) Ordered DNA wrapping switches on luminescence in single-walled nanotube dispersions. J Am Chem Soc 130:12734–12744
51. Brooks BR, Bruccoleri RE, Olafson BD, States DJ, Swaminathan S et al (1983) CHARMM: a program for macromolecular energy, minimization, and dynamics calculations. J Comput Chem 4:187–217
52. Frisch MJ, Trucks GW, Schlegel HB, Scuseria GE, Robb MA et al (2009) Gaussian 09, Revision B.01, Wallingford CT
53. Bobadilla AD, Seminario JM (2011) DNA—CNT interactions and gating mechanism using MD and DFT. J Phys Chem C 115:3466–3474
54. Miertus S, Scrocco E, Tomasi J (1981) Electrostatic interaction of a solute with a continuum. A direct utilizaion of AB initio molecular potentials for the prevision of solvent effects. Chem Phys 55:117–129
55. Prins F, Barreiro A, Ruitenberg JW, Seldenthuis JS, Aliaga-Alcalde N et al (2011) Room-temperature gating of molecular junctions using few-layer graphene nanogap electrodes. Nano Lett 11:4607–4611
56. Wei D, Liu Y, Cao L, Wang Y, Zhang H et al (2008) Real time and in situ control of the gap size of nanoelectrodes for molecular devices. Nano Lett 8:1625–1630
57. Derosa PA, Seminario JM (2000) Electron transport through single molecules: scattering treatment using density functional and green function theories. J Phys Chem B 105:471–481
58. Agapito LA, Bautista EJ, Seminario JM (2007) Conductance model of gold-molecule-silicon and carbon nanotube-molecule-silicon junctions. Phys Rev B 76:115316
59. Jauregui LA, Salazar-Salinas K, Seminario JM (2009) Transverse electronic transport in double-stranded DNA nucleotides. J Phys Chem B 113:6230–6239
60. Bobadilla AD, Seminario JM (2013) Assembly of a noncovalent DNA junction on graphene sheets and electron transport characteristics. J Phys Chem C 117:26441–26453
61. Di L, Kerns EH, Li SQ, Petusky SL (2006) High throughput microsomal stability assay for insoluble compounds. Int J Pharm 317:54–60
62. Lipinski CA (2000) Drug-like properties and the causes of poor solubility and poor permeability. J Pharmacol Toxicol 44:235–249
63. Berlin JM, Leonard AD, Pham TT, Sano D, Marcano DC et al (2010) Effective drug delivery, in vitro and in vivo, by carbon-based nanovectors noncovalently loaded with unmodified paclitaxel. ACS Nano 4:4621–4636
64. Hummer G, Rasaiah JC, Noworyta JP (2001) Water conduction through the hydrophobic channel of a carbon nanotube. Nature 414:188–190
65. Leng LC, Qi LH, Ru TH, Ye L (2010) Delivery of paclitaxel by physically loading onto poly (ethylene glycol) (PEG)-graftcarbon nanotubes for potent cancer therapeutics. Nanotechnology 21:065101
66. Stephenson JJ, Hudson JL, Leonard AD, Price BK, Tour JM (2007) Repetitive functionalization of water-soluble single-walled carbon nanotubes. Addition of acid-sensitive addends. Chem Mater 19:3491–3498
67. Vorobyov I, Anisimov VM, Greene S, Venable RM, Moser A et al (2007) Additive and classical drude polarizable force fields for linear and cyclic ethers. J Chem Theory Comput 3:1120–1133
68. Lee H, Venable RM, MacKerell AD, Pastor RW (2008) Molecular dynamics studies of polyethylene oxide and polyethylene glycol: hydrodynamic radius and shape anisotropy. Biophys J 95:1590–1599

69. French AC, Thompson AL, Davis BG (2009) High-purity discrete PEG-oligomer crystals allow structural insight. Angew Chem Int Edit 48:1248–1252
70. Bobadilla AD, Samuel ELG, Tour JM, Seminario JM (2013) Calculating the hydrodynamic volume of poly(ethylene oxylated) single-walled carbon nanotubes and hydrophilic carbon clusters. J Phys Chem B 117:343–354
71. Hong G, Wu JZ, Robinson JT, Wang H, Zhang B et al (2012) Three-dimensional imaging of single nanotube molecule endocytosis on plasmonic substrates. Nat Commun 3:700

Chapter 8
Computational Mechanochemistry

Przemyslaw Dopieralski and Zdzislaw Latajka

Abstract Mechanochemistry is a field with a long history, but only about a decade ago it was transferred from macroscopic milling-type experiments down to the molecular level. The Bochum group headed by Professor Dominik Marx established a general theoretical framework for describing such mechanical manipulation of covalent bonds in terms of force transformed potential energy surfaces. The original static approach has been there recently extended to dynamical simulations of mechanochemical reactions in solution in order to explore both solvation and thermal activation effects what is going to be discussed in this chapter. First of all the isotensional ab initio metadynamics yields force transformed free energy landscapes, which were used successfully to unravel the complex mechanochemistry of force-induced ring-opening of cyclopropanes in the gas phase. After that the step forward from isolated systems into solvated was made and by using a minimal molecular model, which is diethyl disulfide in water, we have most recently published an explanation of the biphasic, Janus-faced behavior of the reaction rate, the so-called "reactivity switch", as a function of force that has been observed experimentally in stretched proteins. At the end the simulations on protein are shown, where the drastic topological changes of the free energy profiles along the S-S-C-C dihedral angle as the external forces increases are qualitatively similar to those observed for simple molecular model.

8.1 Introduction to Mechanochemistry. What Is It?

Mechanochemistry history has started a year 1892 with Carey Lea [1], when he proved empirically that by use of mechanical force onto AgCl, AgBr and AgI those materials can be dissociated. It is also noted that Lea was the first who published

P. Dopieralski (✉) · Z. Latajka
Faculty of Chemistry, University of Wroclaw, Joliot–Curie 14,
50-383 Wroclaw, Poland
e-mail: przemyslaw.dopieralski@chem.uni.wroc.pl

Z. Latajka
e-mail: zdzislaw.latajka@chem.uni.wroc.pl

J. Leszczynski and M.K. Shukla (eds.), *Practical Aspects of Computational Chemistry IV*, DOI 10.1007/978-1-4899-7699-4_8

his original findings in this branch of chemistry, but mankind used mechanical brute force long before Lea to induce fire and thus produce heat for thermochemistry. In recent years the term mechanochemistry has been videly used to describe mechanically induced chemistry and when these processes are connected with breaking of covalent bonds it is usually called as a covalent mechanochemistry (CMC), in opposition to "non-covalent mechanochemistry" [2–4]. Shear stress used onto some chemical substances as a solids were developed into grinding, ball-miling, compresion or mastication techniques [5, 6], but only a decade ago the macroscopic milling-type experiments were transferred down to the molecular level, where it became possible to exert the required nano-Newton forces onto molecules in rather controlled way, thus triggering and influencing site-specific chemical reactions. This new chapter has started with the seminal paper "How Strong is a Covalent Bond?" published in 1999 by Gaub et al. [7]. Nowadays more sophisticated and advanced techniques are used for manipulation in micro scale: atomic force microscope (AFM) [8, 9] in a force-clamp mode [10–13] optical and magnetic tweezers [14] or sonochemistry. [15] As those techniques have been already reviewed at least few times [16–18], we are not going to spent any line for discussion and doubling the same information already accessible from cited reviews.

Why there is an growing interest in a covalent mechanochemistry (CMC)? [16–18]. Mostly because of the potential use of such "smart" materials in applications that range from drug delivery to camouflage systems and self-healing materials [19, 20]. At this moment it is hard to overestimate the potential and importance of mechanochemistry as it has a great impact in different fields such as synthetic chemistry and materials science [21].

We refer the reader to available reviews—two excellent experimentally oriented reviews on mechanochemistry [16, 17] and recent one where theoretical/computational mechanochemistry is discussed in great details [18].

Our goal here is to guide the reader through the leatest theoretical framework (based on two computational examples), which was implemented in Car–Parrinello Molecular Dynamics package CPMD [22] by Marx laboratory [23, 24]. Our implemented new theoretical method is able to investigate for the first time force-transformed Free Energy Surface (FES) and explore possible reaction pathways as a function of applied external constant force at a finite temperature, thus including thermal fluctuations. All limitations which apply when static zero temperature perspective is used like lack of proper solvation effects are now gone. Presented novel approach has been succesfully used to study mechano-reactivity of isolated cyclopropanes and fully solvated disulfides, providing deep understaing of the chemistry in the heart of the conducted experiments, what is going to be disccused in Sect. 8.3.

8.2 Computations

Phenomenological model to describe the effect of external force onto the rate of bond dissociation was firstly presented by Eyring et al. [25] and Zhurkov [26] but is usually atrributed to Bell [27]. Unfortunatelly Bell model assumes that potential energy landscape of the reaction does not change with applied force, thus Bell model as simple one dimensional model is valid only for one dimensional processes—where the force acts exactly in line with reaction coordinte. The tilted potential energy profile model is an improvment of the Bell's model and now the structures of the reactants and trasition states of the reaction can be affected by mechanical external force. Concept was already presented by Eyring [25] but only Evans and Ritchie [28] combine the model with reaction kinetics theory to describe bond dissociation reaction.

After short historical recall let know introduce electronic structure based approaches. Very first computer simulations in this field imposed distance constraint and carried out constraint minimization for different values of this distance showing this way that external force can be simulated by simple geometry constraints—COGEF (constrained geometries simulate external force) that was firstly introduced by Beyer [29]. Those are so-called isometric calculations in contrast to isotensional approach where not distance but desired force is fixed and this formalism EFEI (external force is explicitly included) was firstly introduced by Marx et al. [30] and Martinez et al. [31] and followed by Wolinski and Baker [32]. It has been also shown and proven that EFEI potential is the Legendre transform of the COGEF potential. For more details see [18]. In here we will excusively elaborate about EFEI fromalism as the one which mimics the isotensional experiments.

8.2.1 Explicit External Force Approach

Previous work [30, 33, 34] on CMC was mostly focused on static descriptions of the molecular response for applied external constant force in vacuo providing "isotensional" conditions. The central concept of new formalism introduced by Marx and co-workers [30] is the force transformed Potential Energy Surface (PES), which, given an external constant force $\mathbf{F}_0$, is rigorously defined as

$$V_{\mathrm{EFEI}}(\mathbf{x}, \mathbf{F}_0) = V_{\mathrm{BO}}(\mathbf{x}) - \mathbf{F}_0\mathbf{q}(\mathbf{x}), \tag{8.1}$$

where $V_{\mathrm{BO}}(\mathbf{x})$ is the usual Born–Oppenheimer PES as a function of all nuclear cartesian coordinates $\mathbf{x}$, and $\mathbf{q}$ is the mechanical coordinate, i.e. a structural parameter (a generalized coordinate in terms of $\mathbf{x}$) on which the force acts. By locating the stationary points of this function, in which the "External Force is Explicitly Included" (EFEI), one can evaluate, without invoking any approximation, properties such as reactant and transition state structures as functions of $\mathbf{F}_0$.

8.2.2 Explicit External Force at Finite Temperatures

As thermal activation effects are not negligible at room temperature, and as they must eventually override mechanical activation effects at some point, in Marx laboratory

[23] recently the EFEI approach has been implemented in the CPMD code [22]. Importantly, when combined with such a techniques like "blue moon" constrained molecular dynamics [35, 36] (thermodynamic integration [37])—which is proper generalization of COGEF method—or ab initio metadynamics technique [38, 39], this methodology allow for the first time to properly study force-transformed free energy curves and respectively force-transformed free energy surfaces (FT–FES).

This methodology is the basis for exploring possible reaction pathways as a function of applied tensile external constant force at a finite temperature as disussed on some selected examples in the following chapter.

8.3 Isotensionl Ab Initio Molecular Dynamics

Static zero temperature perspective is able to provide great insight into the mechanochemistry but unfortunately neglecting finite-temperatures effects and influence of proper solvation effects, which are simplyfied to microsolavation or continuum approximations at zero temperature static view, is in many cases (in some complex reactions) simply not possible to obtain right mechanisms and not only the energies. Thus in here we will exclusively elaborate about three recent ab initio molecular dynamics computations performed in domain of mechanochemistry.

8.3.1 Mechano-Stereochemistry of Cyclopropane Ring-Opening Reactions

The model system chosen to explore the mechanochemistry of *g*DCCs is 1,1–dichloro–2,3–dimethylcyclopropane: its cis and trans isomers and the four possible distinct reaction products of the corresponding ring-opening processes are depicted in Fig. 8.1. Our simulations [23] (both at 0 K and at 300 K) have revealed that the ring-opening of these molecules to yield the corresponding 2,3–dichloroalkenes proceeds via a concerted disrotatory mechanism, whereby the breaking of the C–C bond takes place in concert with the C–Cl bond cleavage and the subsequent Cl migration. For cis *g*DCC there are two possible pathways as indicated in Fig. 8.1: the "outward pathway" which passes through TS–I and the "inward pathway" via TS–III. On the basis of the difference between the activation energies of both pathways (about 5 kcal/mol, see Fig. 8.1), it has been concluded that the thermal ring-opening of cis *g*DCC occurs via a "disrotatory outward mechanism", whose BO–PES features a TS of C_s symmetry (TS–I) and a bifurcation point along the IRC after the TS is left behind. By virtue of this topological feature, the migrating Cl atom can move either to the C atom on the right side (thus yielding the Z,R–alkene) or to the left (leading to Z,S–alkene), see Fig. 8.1. Given the topology of the underlying PES, the ring-opening of cis *g*DCC is expected to yield the two enantiomeric alkenes

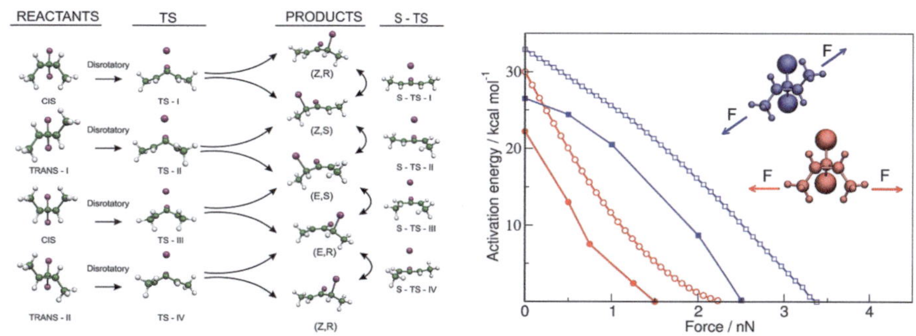

Fig. 8.1 *Left* Scheme showing the involved chemical species, i.e. all reactants (cis; trans–I and trans–II being enantiomers), transition states (TS–I to TS–IV; S–TS–I to S–TS–IV), and products (Z,R; Z,S; E,S; E,R). The *arrows* connecting the reactants with the distinct products via the corresponding TSs represent the reaction paths obtained from IRC mapping and ab initio trajectory shooting starting from the TSs (see text). The second set of TSs (S–TS) belongs to interconversion reactions between selected products as indicated. For simplicity all structures correspond to the stationary points at zero force, the Z,R product is reproduced twice for clarity, and the Cl atoms are colored violet. *Right* Force-dependence of activation energies $\Delta E^{\ddagger}(F_0)$ (*open symbols*) and free energies $\Delta A^{\ddagger}(F_0)$ at 300 K (*filled symbols*) of the disrotatory ring-opening of cis (*red circles* for the "outward" pathway) and trans (*blue squares*) 1,1–dichloro–2,3–dimethylocyclopropanes. The stretching force is applied to the C atoms of the two terminal methyl groups as indicated in the inset. Reprinted with permission from Ref. [40]

with equal probability. The disrotatory ring-opening of trans *g*DCC at zero force, in its turn, implies either the TS–II or the TS–IV according to Fig. 8.1. The computed activation free energies were found to be lower (by about 4 kcal/mol) for the ring-opening of cis *g*DCC than trans *g*DCC, as shown in Fig. 8.1. As justified in Ref. [23], of all the reaction pathways plotted in Fig. 8.1, only two of them are relevant for fully describing the mechanochemistry of *g*DCCs: the disrotatory outward mechanism of cis *g*DCC and the disrotatory ring-opening of trans *g*DCC passing through TS–II. As shown in Fig. 8.1, the force-dependence featured by the activation free energy, $\Delta A^{\ddagger}$, and the activation energy, $\Delta E^{\ddagger}$, for the two considered pathways follows a similar general trend. The values of $\Delta A^{\ddagger}(F_0)$ have been obtained from the FT-FESs shown in original paper. [23] Despite the similar trends, the values of $\Delta A^{\ddagger}(F_0)$ are much lower by about 6–7 kcal/mol over the whole range of forces, which results in a dramatic rate acceleration due to finite-temperature and entropy effects not accounted for by $\Delta E^{\ddagger}(F_0)$. With reference to the selectivity with which the tensile load might promote preferentially the ring-opening of one isomer, both the qualitatively different shapes of the energy curves associated with the cis and trans reactants and the corresponding rupture forces (1.5 and 2.5 nN, respectively, at 300 K) reflect the fact that the external forces enhance the ring-opening of cis *g*DCCs more efficiently. This is somehow contradictory with the experimentally found lack of selectivity [41]. As argued in Ref. [23], the most plausible suggestion for resolving such apparent contradiction would be to speculate that sufficiently large forces (forces on the order of 2 nN would suffice, according to the data in Fig. 8.1) were generated in the sonochemical experiments as to reach the barrierless regime for the ring-opening reactions of both isomers.

8.3.2 S_N2 at Sulfur: Disulfide Bond Reduction

Fernandez [12, 13], by applying AFM (Atomic Force Microscopy) in a force-clamp mode, demonstrated that the resulting stretching force leads to protein unfolding and accelerates disulfide cleavage in the presence of nucleophiles, such as hydroxide ions, thiol or phosphine reagents. It has been observed that the reaction rate changes with the magnitude of the applied force. As we have recently communicated to *Nature Chemistry*, the nucleophilic substitution at sulfur, "S_N2 at S", was found to feature the experimentally observed two distinct regimes of disulfide reactivity below 500 pN and above, which are attributed by us to one and the same reaction, which is disulfide-bond reduction—based on simple molecular model. This has been traced back to be a manifestation of the subtle interplay between two antagonistic effects: on the one hand side the pulling force increases exponentially the rate of the reaction, but on the other hand side stretching out the molecule causes a conformational change due to distorting the molecule. In particular, the dihedral angle S-S-C-C increases from about 70° ("open conformation") to 170° ("closed conformation") upon pulling, which is partially blocking the reaction cone for nucleophilic attack at the sulfur sites (this has been discussed in much detail in our article [24] and the extensive Supporting Information attached to this paper so that we only explain here the key conclusions).

The results from our simulations are presented in Fig. 8.2. Based on two different free energy sampling techniques we have arrived at quite similar results which show the overall acceleration effect on the reaction when the system is exposed to tensile stress. What is more important, however, is the nonlinear nature of this effect as a result of a conformational distortion from "open" to "closed" conformers. One can easily see from Fig. 8.2 inset in panel **a** that at zero force the reaction cone is really symmetrical and fully open (see red sphere in Fig. 8.2 inset in panel **a**), whereas at a force of 1.5 nN (cf. violet sphere) the reaction cone is in fact no more cone-shaped and no longer favoring the ideal 180° attacking angle S–S$\cdots$$OH^-$, This different approaching angle from the one required for S_N2–type reactions imposes an energetic penalty and leads to the nonlinear dependence of the activation free energies with applied force. This finding not only explains the aforementioned AFM experiment, but will be of general importance to mechanochemistry as such: force may not only accelerate by making free energy landscapes flatter, but might as well grow new barriers due to molecular distortions that must be overcome.

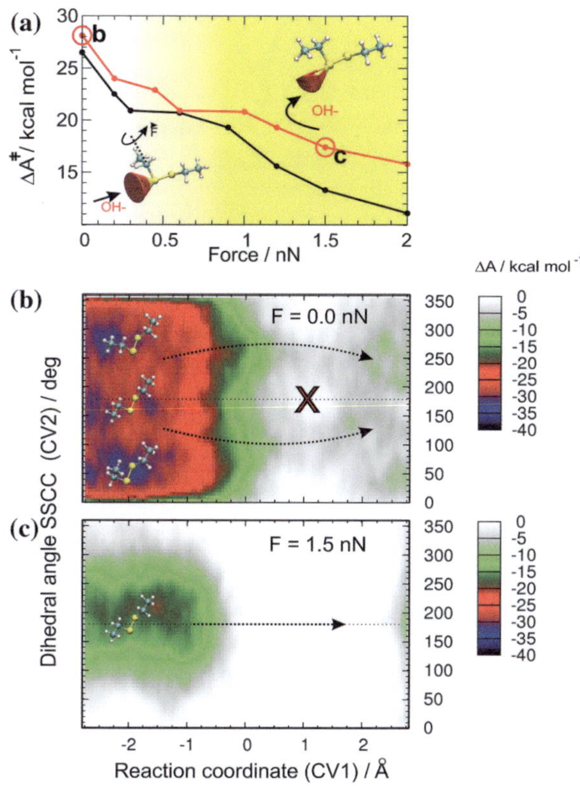

Fig. 8.2 Activation free energies as a function of external force (panel a) for nucleophilic cleavage of diethyl disulfide in bulk water by OH^- from isotensional ab initio metadynamics sampling (*red*) and thermodynamic integration (*black*) at $T = 300$ K. Free energy surfaces in the absence of force (panel **b**) and at a force of 1.5 nN (panel **c**). CV1 is the difference between the S–S distance and the distance of the attacked sulfur, $S^\star$, with respect to the attacking hydroxyl oxygen (thus negative (positive) values correspond to reactant (product) states respectively). CV2 is the S-$S^\star$-C-C dihedral angle on the side of the reaction center (where the *dotted horizontal lines* mark 180° and thus the so-called *closed* conformation of diethyl disulfide, see text). The *arrows* and *crosses* highlight schematically the energetically preferred and disfavored pathways, respectively, according to the topology of the underlying free energy landscape. *Red circles* in panel a mark points for which free energy surfaces are shown in panels **b** and **c**. Reprinted from ref. 24. Copyright 2013 NPG

8.3.3 From Simple Diethyl Disulfide Model into Protein

Given that our model system was smaller than the protein employed in the force-clamp AFM experiments, the very first thing that needs to be proven is that the observed force-induced conformational rearrangement in the diethyl disulfide molecule can also take place in the protein and, thus, that the mechanism extracted from this model is responsible for the experimental findings reported for the protein. Force-field molecular dynamics simulations carried out on the same system used in

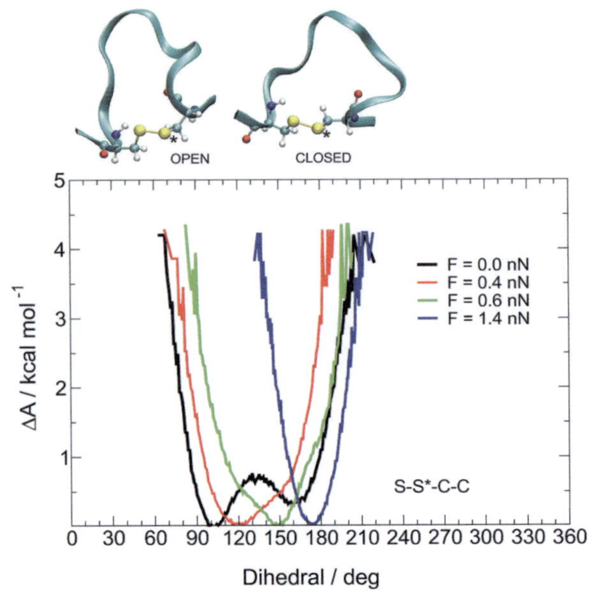

Fig. 8.3 Free energy profiles along the S-S⋆-C-C dihedral angle of the disulfide bridge of the I27 protein as a function of external force. The profile has been obtained from force–field molecular dynamics simulations carried out with GROMOS at $T = 300\,\mathrm{K}$. The model system in these simulations comprises the I27 protein and SPC water molecules solvating the protein in a tetragonal box. The relevant S-S-C-C conformers of the disulfide bridge embedded in the protein are shown on *top* using a partial and schematic representation. Courtesy of Padmesh Anjukandi

the experiments, that is a computational model of the mutated 27th immunoglobulin-like domain of cardiac titin (I27), show that this is indeed the case. As displayed in Fig. 8.3, the drastic topological changes of the free energy profiles along the S-S⋆-C-C dihedral angle as the external forces increases are *qualitatively similar* to those observed for simple molecular model—diethyl disulfide. In particular, at zero force, the *open* conformer of the disulfide bridge embedded in the protein is slightly more stable than the *closed* one, in agreement with the conformational properties of the diethyldisulfide molecule [42]. The χ_2 and χ_2' angles have been found to change from conformations that are open to nucleophilic attack to sterically hindered, so-called closed states upon exerting tensile stress. In view of the growing evidence of the importance of C-C-S-S dihedrals in tuning the reactivity of disulfides. [24] In our latest work we present a systematic study of the conformational diversity of disulfides as a function of tensile stress [42]. With the help of force-clamp metadynamics simulations, we show that tensile stress brings about a large stabilization of the closed conformers, thereby giving rise to drastic changes in the conformational free energy landscape of disulfides. Statistical analysis shows that native thiol–disulfide interchange proteins (TDi), disulfide oxidoreductases (DO) and interchain immunoglobulin (Ig) protein disulfides prefer open conformations, whereas the intrachain disulfide bridges in Ig proteins favor closed conformations. Correlating mechanical stress with the distance between the two α–carbons of the disulfide moiety reveals that the strain of intrachain Ig protein disulfides corresponds to a mechanical activation of about 100 pN (see in Fig. 8.4) Such mechanical activation leads to a severalfold increase of the rate of the elementary redox S_N2 reaction step. All these findings constitute a step forward towards achieving a full understanding of functional disulfides.

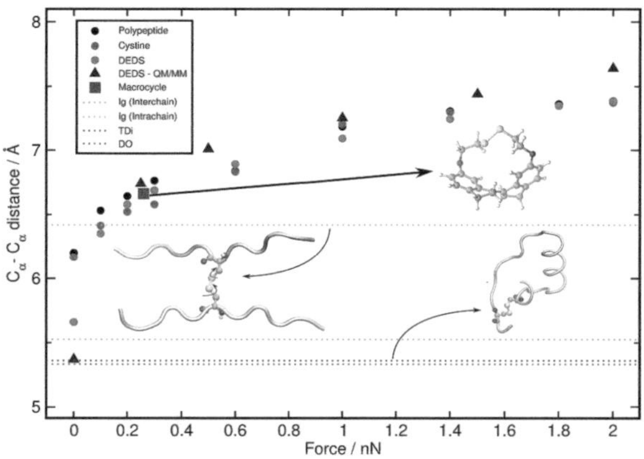

Fig. 8.4 Dependence of the computed average distance between the C_α–atoms as a function of F_0 for the polypeptide model (*black circles*), cystine (*red circles*), and DEDS (*green circles*) and obtained from force field equilibrium (at zero force) and force clamp MD (for $F_0 > 0$ nN) simulations. Computational reference data for DEDS obtained from QM/MM simulations are shown by GEOMETRY FILE/created by CPMD *brown triangles* and the experimental reference based on the strained macrocycle [43] is marked by a *violet square*. The *horizontal blue, pink* and *orange dotted lines* are the average C_α–C_α distances of disulfide bonds in TDi, DO and interchain Ig proteins, respectively, whereas the *cyan dotted line* corresponds to intrachain Ig proteins. Reprinted from Ref. [42]

Conclusion

Prompted by recent controversial issues connected with disulfides reactivity we decided to investigate the disulfide bond breaking reaction mechanism with and without applied external force in bulk aqueous environments. Before doing so, our thermodynamic approach has been fully validated based on *cis*– and *trans*-1,1-dichloro-2,3-dimethylcyclopropane (*g*DCC) molecular systems [23] and was supplemented by ab initio trajectory shooting simulations operating on the FT–PESs in order to dissect genuinely dynamical effects on branching ratios as a function of force (see Sect. 8.3.1). Based on these methods we have unveiled the mechanisms of force-induced ring-openings of cis versus trans *g*DCCs, which rationalizes puzzling experimental findings [41].

This work is the basis of InSiDE article on HPC applications carried out on JUGENE platform.

In our next project we have investigated with much effort and in great detail the mechanochemical hydrolysis mechanism of the simplest disulfide model system in aqueous solution—substitution at sulfur. For the first time in a calculation, the enigmatic biphasic behavior of the reaction rate as a function of applied force obtained from force-clamp AFM experiments, see [13], has been reproduced in silico! [24]

Only this allowed us to extract the molecular reason behind this nonlinear effect, which is different from *all* the possibilities suggested in the earlier literature based on plausibility arguments. Our results are shortly reviewed in Sect. 8.3.2 and the corresponding article is already published in Nature Chemistry.

Thus, next our methodology was used to carry out force-field molecular dynamics simulations on the same system as used in the experiments, that is a computational model of the mutated 27th immunoglobulin-like domain of cardiac titin (I27) and we have succesfully proved that the mechanism extracted from the minimal model is responsible for the experimental findings reported for the protein.

Acknowledgments All work covered in this chapter as well as a all original publications on which it is based have been carried out in a tight and longstanding collaboration with Professor Dominik Marx and his research group (Lehrstuhl fuer Theoretische Chemie, Ruhr-Universitat Bochum, 44780 Bochum, Germany). It gives us great pleasure to thank Professor Dominik Marx for giving us opportunity to work with him on covalent mechanochemistry and Jordi Ribas–Arino and Padmesh Anjukandi for fruitful collaboration.We are grateful to National Science Center Poland under Grant No. 2014/13/B/ST4/05009 and Ministry of Science and Higher Education Poland under Grant No. 627/STYP/9/20l4 for financial support.

References

1. Lea Carey M (1892) Disruption of the silver haloid molecule by mechanical force. Philos Mag 34:46–50
2. Janshoff A, Neitzert M, Oberdörfer Y, Fuchs H (2000) Force spectroscopy of molecular systems-single molecule spectroscopy of polymers and biomolecules. Angew Chem Int Ed 39:3212–3237
3. Samorì B (2000) Stretching single molecules along unbinding and unfolding pathways with the scanning force microscope. Chem Eur J 6:4249–4255
4. Bustamante C, Chemla YR, Forde NR, Izhaky D (2004) Mechanical processes in biochemistry. Annu Rev Biochem 73:705–748
5. Kaupp G (2009) Mechanochemistry: the varied applications of mechanical bond-breaking. CrystEngComm 11:388–403
6. Friscic T (2010) New opportunities for materials synthesis using mechanochemistry. J Mater Chem 20:7599–7605
7. Grandbois M, Beyer M, Rief M, Clausen-Schaumann H, Gaub HE (1999) How strong is a covalent bond? Science 283:1727–1730
8. Binnig G, Quate CF, Gerber C (1986) Atomic force microscope. Phys Rev Lett 56:930–933
9. Giessibl FJ (2003) Advances in atomic force microscopy. Rev Mod Phys 75:949–983
10. Wiita AP, Ainavarapu SRK, Huang HH, Fernandez JM (2006) Force-dependent chemical kinetics of disulfide bond reduction observed with single-molecule techniques. Proc Natl Acad Sci USA 103:7222–7227
11. Wiita AP et al (2007) Probing the chemistry of thioredoxin catalysis with force. Nature 450:124–127
12. Ainavarapu SRK, Wiita AP, Dougan L, Uggerud E, Fernandez JM (2008) Single-molecule force spectroscopy measurements of bond elongation during a bimolecular reaction. J Am Chem Soc 130:6479–6487
13. Garcia-Manyes S, Liang J, Szoszkiewicz R, Kuo TL, Fernandez JM (2009) Force-activated reactivity switch in a bimolecular chemical reaction. Nat Chem 1:236–242
14. Moffitt JR, Chemla YR, Smith SB, Bustamante C (2008) Recent advances in optical tweezers. Annu Rev Biochem 77:205–228
15. Cravotto G, Cintas P (2007) Forcing and controlling chemical reactions with ultrasound. Angew Chem Int Ed 46:5476–5478
16. Beyer MK, Clausen-Schaumann H (2005) Mechanochemistry: the mechanical activation of covalent bonds. Chem Rev 105:2921–2948

17. Caruso MM et al (2009) Mechanically-induced chemical changes in polymeric materials. Chem Rev 109:5755–5798
18. Ribas-Arino J, Marx D (2012) Covalent mechanochemistry: theoretical concepts and computational tools with applications to molecular nanomechanics. Chem Rev 112:5412–5487
19. Davis DA et al (2009) Force-induced activation of covalent bonds in mechanoresponsive polymeric materials. Nature 459:68–72
20. Wietor J-L, Sijbesma R (2008) A self-healing elastomer 47:8161–8163
21. Lenhardt JM, Craig SL (2009) Force probes in a bottle. Nat Nanotech 4:284–285
22. Hutter J et al CPMD *Program Package*
23. Dopieralski P, Ribas-Arino J, Marx D (2011) Force-transformed free energy surfaces and trajectory shooting simulations reveal the mechano-stereochemistry of cyclopropane ring-opening reactions. Angew Chem Int Ed 50:7105–7108
24. Dopieralski P et al (2013) The Janus-faced role of external forces in mechanochemical disulfide bond cleavage. Nat Chem 5:685–691
25. Kauzmann W, Eyring H (1940) The viscous flow of large molecules. J Am Chem Soc 62:3113–3125
26. Zhurkov SN (1965) Kinetic concept of the strength of solids. Int J Fract Mech 1:311–323
27. Bell GI (1978) Models for the specific adhesion of cells to cells. Science 200:618–627
28. Evans E, Ritchie K (1997) Dynamic strength of molecular adhesion bonds. Biophys J 72:1541–1555
29. Beyer MK (2000) The mechanical strength of a covalent bond calculated by density functional theory. J Chem Phys 112:7307–7311
30. Ribas-Arino J, Shiga M, Marx D (2009) Understanding covalent mechanochemistry. Angew Chem Int Ed 48:4190–4193
31. Ong MT, Leiding J, Tao H, Virshup AM, Martinez TJ (2009) First principles dynamics and minimum energy pathways for mechanochemical ring opening of cyclobutene. J Am Chem Soc 131:6377–6379
32. Wolinski K, Baker J (2009) Theoretical predictions of enforced structural changes in molecules. Molec Phys 107:2403–2417
33. Ribas-Arino J, Shiga M, Marx D (2009) Unravelling the mechanism of force-induced ring-opening of benzocyclobutenes. Chem Eur J 15:13331–13335
34. Ribas-Arino J, Shiga M, Marx D (2010) Mechanochemical transduction of externally applied forces to mechanophores. J Am Chem Soc 132:10609–10614
35. Carter E, Ciccotti G, Hynes JT, Kapral R (1989) Constrained reaction coordinate dynamics for the simulation of rare events. Chem Phys Lett 156:472–477
36. Sprik M, Ciccotti G (1998) Free energy from constrained molecular dynamics. J Chem Phys 109:7737–7744
37. Kirkwood JG (1935) Statistical mechanics of fluid mixtures. J Chem Phys 3:300–313 . http://dx.doi.org/doi/10.1063/1.1749657
38. Laio A, Parrinello M (2002) Escaping free-energy minima. Proc Natl Acad Sci USA 99:12562–12566
39. Iannuzzi M, Laio A, Parrinello M (2003) Efficient exploration of reactive potential energy surfaces using Car-Parrinello molecular dynamics. Phys Rev Lett 90:238302–1–4
40. Dopieralski P, Ribas-Arino J, Marx D (2012) Mechanochemistry of covalent bond breaking from first principles simulations. In: Binder K, Münster G, Kremer M (eds) NIC Symposium 2012 Proceedings, vol 45. Forschungszentrum Jülich GmbH, pp 115–122
41. Lenhardt JM, Black AL, Craig SL (2009) gem-dichlorocyclopropanes as abundant and efficient mechanophores in polybutadiene copolymers under mechanical stress. J Am Chem Soc 131:10818–10819
42. Anjukandi P, Dopieralski P, Ribas-Arino J, Marx D (2014) The effect of tensile stress on the conformational free energy landscape of disulfide bonds. PLOS One 9, e108812–1–7
43. Kucharski TJ et al (2009) Kinetics of thiol/disulfide exchange correlate weakly with the restoring force in the disulfide moiety. Angew Chem Int Ed 48:7040–7043

Chapter 9
Hydrogen Bond and Other Lewis Acid–Lewis Base Interactions—Mechanisms of Formation

Sławomir J. Grabowski

Abstract The hydrogen bond formation leads to numerous structural changes in interacting sub-systems. These changes are a consequence of a redistribution of electron charge density being a result of complexation. This is important that similar transformations are observed for other Lewis acid–Lewis base interactions. In general, for such interactions, including the hydrogen bond, an electron charge shift is observed from the Lewis base unit to the Lewis acid. This leads to further processes such as a change of polarizations of bonds, rehybridization of atoms, and numerous others. The transformations being the result of complexation are reflected in changes of geometrical, energetic and topological parameters. The results of ab initio calculations as well as of the Quantum Theory of 'Atoms in Molecules' (QTAIM) and Natural Bond Orbitals (NBO) approaches are presented here for selected types of Lewis acid–Lewis base interactions. Experimental X-ray and neutron diffraction measurements' results on organic crystal structures are analyzed to support the ideas presented.

9.1 Which Interactions Are Classified as the Lewis Acid–Lewis Base Ones?

The term "noncovalent interactions" is often used in literature for such interactions as hydrogen bond, dihydrogen bond, halogen bond and others to indicate that corresponding atom-atom contacts differ from typical covalent bonds. However this term does not seem to be relevant and there are numerous reasons of its unsuitability. First of all it covers the variety of interactions [1], the mentioned above

S.J. Grabowski (✉)
Kimika Fakultatea, Euskal Herriko Unibertsitatea and Donostia International Physics Center (DIPC), P.K. 1072, 20080 Donostia, Spain
e-mail: s.grabowski@ikerbasque.org

S.J. Grabowski
IKERBASQUE, Basque Foundation for Science, 48011 Bilbao, Spain

J. Leszczynski and M.K. Shukla (eds.), *Practical Aspects of Computational Chemistry IV*, DOI 10.1007/978-1-4899-7699-4_9

hydrogen and halogen bonds, as well as weak interactions where systems are mainly stabilized by dispersive forces; such as, for example, the methane dimer, interactions between noble gas atoms or interactions between hydrogen molecules; there are other numerous examples. On the other hand interactions between simple ions, known in the classical literature as ionic bonds may be de facto also classified as noncovalent interactions; one can mention a classical example of NaCl moiety.

The next problem concerns the meaning of the term "covalent". Pauling has pointed out early on that the covalent bond is "the sharing of a pair of electrons by the two bonded atoms" [2]. The more recent definition recommended by IUPAC is not as restrictive as the Pauling one since it states that the covalent bond is "a region of relatively high electron density between nuclei which arises at least partly from sharing of electrons and gives rise to an attractive force and characteristic internuclear distance" [3]. The IUPAC definition shows that some of very strong intermolecular interactions could be classified as covalent bonds. One can mention here the low barrier hydrogen bonds (LBHBs) analyzed in earlier and recent studies [4–6]. Besides the covalent character (or covalency) of inter- and intramolecular hydrogen bonds as well as of other interactions was discussed in numerous studies [7, 8]. Hence according to those studies the classification of the hydrogen bond as a noncovalent interaction seems to be confusing. The term "covalent character" is often related to the hydrogen bonded system stabilization connected with the electron charge shifts being the result of complexation. This stabilization is reflected in meaningful attractive interactions which are defined in different ways in various decomposition schemes of the energy of interaction [8]. The delocalization, charge transfer, polarization or induction terms are often used in literature for the interaction energy terms related to the electron charge shifts [9–11]. Certainly, those attractive interactions are accompanied by other attractive contributions such as electrostatic or dispersive ones.

It is worth mentioning that the covalent character of hydrogen bond was analyzed early on by Pauling [2], Coulson [12] as well as by Pimentel and McClellan [13]. Since the stabilization of the hydrogen bond and of other interactions is partly ruled by covalency thus the use of the term "noncovalent interactions" seems to be misleading. Besides it was pointed out earlier here that the latter term covers broad class of interactions, in principle all which are not classified as covalent bonds in the classical Lewis [14] or Pauling [2] definitions' meaning.

This is why the use of the term Lewis acid–Lewis base interactions seems to be useful since it covers hydrogen bond, halogen bond and the other ones such as pnicogen [15–21], hydride [22, 23] or chalcogen [24–26] bonds recently investigated. The latter term requires the strict determination of Lewis acid and Lewis base centers which if there are in contact may lead to the stabilization of the system. Besides weak, mainly dispersive interactions for such systems as noble gases or particularly the dimer of methane are excluded here. What is common for the Lewis acid–Lewis base interactions? There is the electron charge shift from the Lewis base to the Lewis acid subunit [27], often the meaningful shift is observed for very strong interactions—this is discussed in the next sections of this chapter. This shift is related to attractive interactions such as the charge transfer and polarization (if one refers to the Kitaura-Morokuma decomposition scheme of the energy of interaction [9]). The Lewis acid and Lewis base centers are usually characterized by the positive and negative electrostatic potentials at the molecular surfaces, respectively [28]. Hence the electrostatic interaction is also important for the Lewis acid–base interactions.

Which interactions may be classified as the Lewis acid–Lewis base (La–Lb) ones? First of all, the hydrogen bond which is the most often analyzed interaction since its key role in numerous chemical, physical and biochemical processes is very well known [29–33]. It is often identified as A–H…B where the A–H proton donating bond is the part of the Lewis acid subunit while B marks the Lewis base center possessing at least one free electron pair. The H…B contact is exactly that one between the Lewis acid and base centers; the H-atom hemisphere is characterized by the positive electrostatic potential while the B-center by the negative electrostatic potential. Scheme 9.1 shows an example of the water dimer linked through the hydrogen bond.

It is worth to mention that different kinds of the hydrogen bond may be classified as La–Lb interactions; also those where π-electrons play the role of the electron donors, i.e. A–H…π hydrogen bonds [13, 31, 34]. Such molecules as ethylene, acetylene, benzene, their derivatives and numerous π-electron systems may act as the Lewis base units; they do not possess the free electron pairs from the classical point of view but the π-electron systems may be treated as multicenter proton acceptors here. Similarly the A–H…σ hydrogen bonds with σ-electrons acting as the Lewis base may be classified as La–Lb interactions [35–38]. Such systems with the molecular hydrogen acting as the Lewis base sub-unit were analyzed both theoretically [35–38] and experimentally [39–43]. The dihydrogen bond being the sub-class of the hydrogen bonds is another example where the H…H contact between a protic hydrogen being the Lewis acid center and a hydric hydrogen playing the role of the Lewis base center is observed [44].

σ-hole bonds are also classified as the Lewis acid–Lewis base interactions [28, 45–47]. A σ-hole bond is defined as an interaction between a covalently-bonded atom of Groups IV–VII and an electron donating center. “It involves a region of positive electrostatic potential, labeled a σ-hole, on the extension of one of

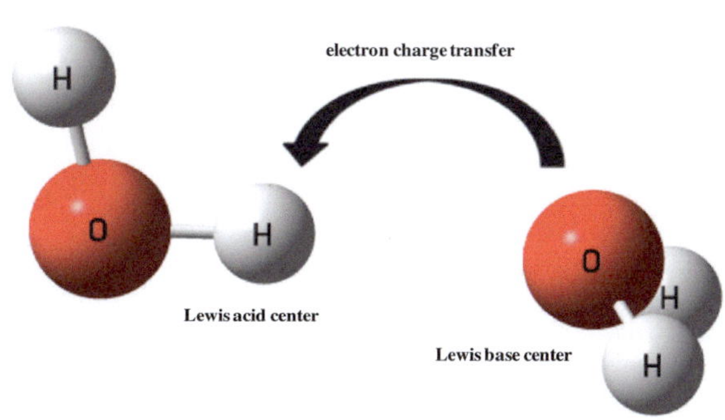

Scheme 9.1 The water dimer linked through the O–H…O hydrogen bond

the covalent bonds to the atom" [47]. Considering the Groups IV–VII atoms as the Lewis acid centers interacting with the Lewis bases the following σ-hole bonds may be indicated; tetrel (Group IV) [48–53], pnicogen (V) [15–21], chalcogen (VI) [24–26] and halogen (VII) [54–58] bonds. The mentioned here Lewis acid centers are usually classified as the electronegative atoms possessing the negative atomic charges. However the parts of molecular surfaces related to these atomic centers are characterized by the positive electrostatic potential regions attributed to the σ-holes [28, 46, 47].

This is worth to mention that there are also regions of the molecular surfaces at these atoms characterized by the negative electrostatic potential [28, 46, 47]. This is why the Groups V–VII atoms possess a dual character and the same center may act simultaneously as the Lewis acid and as the Lewis base. This dual character is not related to the sp^3 hybridized tetrel atoms not possessing lone electron pairs thus they act only as the Lewis acid centers (of course there are exceptions where the dual character is revealed, like, for example, for carbenes). Figure 9.1 presents the fragment of the crystal structure of 2-amino-5-bromopyridinium propynoate where the bromine center acts simultaneously as the Lewis acid and the Lewis base since it is involved in C–Br...O halogen bond and the C–H...Br hydrogen bond, respectively. The C–Br...O arrangement close to linearity is a consequence of the σ-hole location (the positive electrostatic potential) at the molecular surface, in the elongation of the C–Br bond. Similarly there is the belt of the negative electrostatic potential around the Br-atom (approximately in the direction perpendicular to C–Br bond) owing the lone electron pairs of bromine atom. Hence the characteristic arrangement of atoms for the C–H...Br–C hydrogen bond bridge is observed (Fig. 9.1).

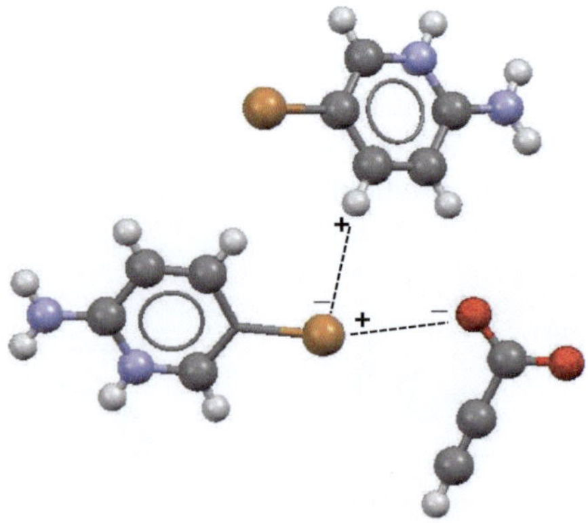

Fig. 9.1 The fragment of the crystal structure of 2-amino-5-bromopyridinium propynoate with intermolecular interactions (*broken lines*); C–H...Br hydrogen bond and C–Br...O halogen bond; *black circles* correspond to carbon atoms, *grey* to hydrogens, *red* to oxygens, *blue* to nitrogens and *brown* to bromines, the positive and negative regions of the electrostatic potential for atoms being in contact are designated by + and –, respectively; this structure and next two structures (Figs. 9.2 and 9.3) were taken from the Cambridge Structural Database [59]

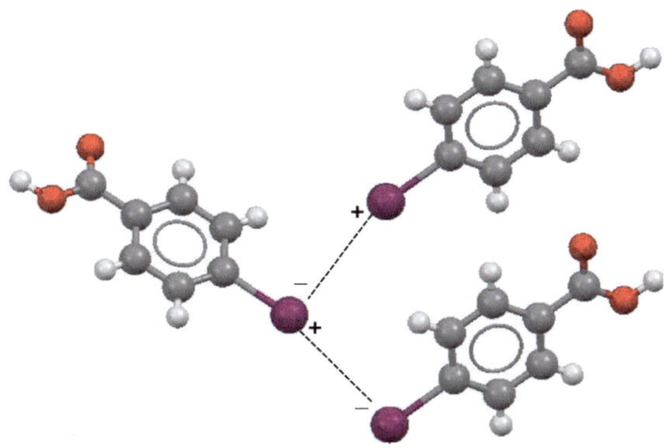

Fig. 9.2 The fragment of the crystal structure of 4-iodobenzoic acid with C–I…I intermolecular interactions (dihalogen bonds -broken lines); black circles correspond to carbon atoms, grey to hydrogens, red to oxygens and violet to iodines, the positive and negative regions of the electrostatic potential for atoms being in contact are designated by + and –

The mentioned here dual character of atoms of Groups V–VII, i.e. that σ-holes are found in conjunction with regions of the negative electrostatic potential, results in the existence of interactions between symmetrically equivalent atoms. Thus the halogen-halogen (dihalogen) bonds [60] are observed as well as chalcogen-chalcogen and pnicogen-pnicogen interactions. This is often observed in crystal structures where the equivalent atoms (related by symmetry) interact between themselves. Figure 9.2 presents the crystal structure of 4-iodobenzoic acid where dihalogen bonds are observed; note that the same I-center acts as the Lewis base and the Lewis acid here and for this center two C–I…I contacts are observed (perpendicular and linear arrangements). Such two simultaneous interactions occur for I-centers of all equivalent 4-iodobenzoic acid molecules in the crystal structure.

Figure 9.3 presents the fragment of the crystal structure of bis (2,3,5,6–tetrafluoro–4-trifluoromethylphenyl)–di-tellurium. The Te-center is the Lewis acid since the σ-hole exists in the elongation of the C–Te bond and it is the Lewis base owing to the lone electron pairs. Note that also dihydrogen bond as a sub-set of hydrogen bond interactions [38] is classified as the La–Lb interaction. However in this case H-atoms do not possess the dual character; the protic and hydric H-atoms being in contact are characterized by the positive and negative electrostatic potential hemispheres, respectively. Hence the dihydrogen bond, where different kinds of atoms are in the contact, can not be formed between the symmetrically equivalent centers.

The hydrogen bond was suggested to belong to the σ-hole bond class of interactions since for the σ-hole bonds as well as for the hydrogen bond there is the electron charge shift within the Lewis acid unit from the Group IV–VII atom (Z-atom) and from the H-atom, respectively, to the center bonded with this atom (Y-center in Scheme 9.2) [61]. This leads to the reduction of electron charge at the Z-atom, in the elongation of the Y–Z bond, i.e. to the σ-hole region at Z-atom often characterized by the positive electrostatic potential.

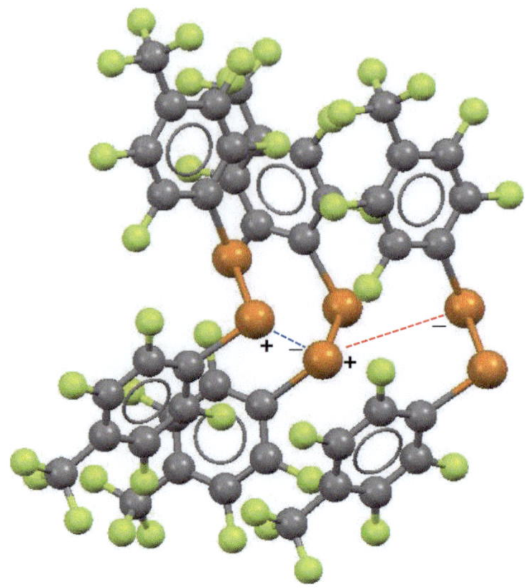

Fig. 9.3 The fragment of the crystal structure of bis(2,3,5,6–tetrafluoro–4-trifluoromethylphenyl)–di-tellurium; with Te…Te intermolecular interactions (chalcogen–chalcogen interactions—*broken lines*: *red* one—central atom acts as the Lewis acid, *blue* one—central atom is the Lewis base); *black circles* correspond to carbon atoms, *green* to fluorines and *orange* to seleniums, the positive and negative regions of the electrostatic potential for atoms being in contact are designated by + and –

Scheme 9.2 The electron charge shift for Y–Z bonds; the case of CF_3Cl and H_2O molecules is presented; *arrows* show the electron charge shifts

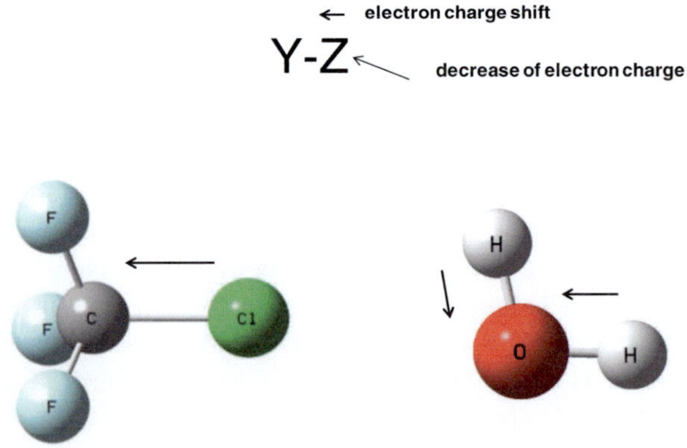

Figure 9.4 presents the electrostatic potential (EP) maps for F_3CCl and H_2O molecules. One can see that in a case of chlorine center there is the region of the positive electrostatic potential in the elongation of the C–Cl bond (σ-hole) and "the belt" of the negative EP around this atom being the consequence of the existence of the lone electron pairs. In a case of water molecule the whole hemispheres of H-atoms are characterized by the positive EP.

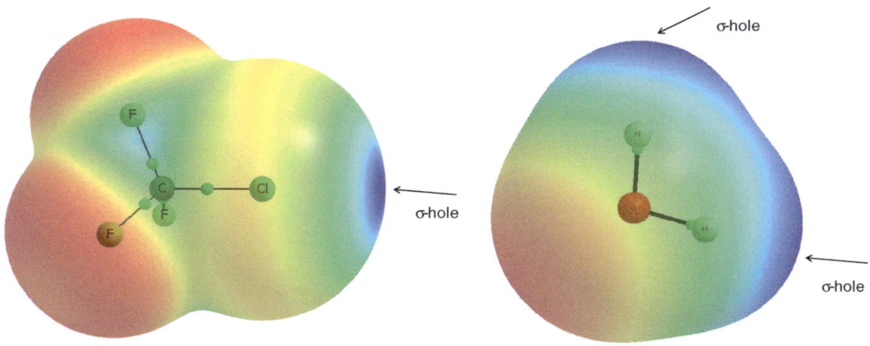

Fig. 9.4 Computed electrostatic potential (EP) on the 0.001 au molecular surface (MP2/aug-cc-pVTZ calculations) of F_3CCl (*left*) and H_2O (*right*). *Blue color* corresponds to the maximum and *red* one to the minimum EP for each case considered separately. *Arrows* show the regions of positive EP (σ-holes)

It was pointed out that the hydrogen bond is rationalized differently than the other σ-hole bonds [61]. This difference was also analyzed recently [62]. For example, in CF_3X molecules (X designates the Cl, Br or I halogen atom) the contribution of halogen atom to the σ_{CX} bond is mainly a p-orbital which is colinear with the C–X bond; this leads to the mentioned above depletion of the electron charge (σ-hole) at the X halogen center. Such depletion may be reduced if X = F since fluorine contributes more s-character into the σ_{CX} bond—the more spherical F-atomic hybrid orbital reduces the σ-hole [61]. This is why the C–F…B (B designates the Lewis base center) halogen bonds are rather rare. In a case of the A–H…B hydrogen bond the p-character of the H-atom orbital contributing to the σ_{AH} bond is meaningless if any. Thus the electron charge shift for A-H bonds potentially forming further A–H…B hydrogen bond bridges is ruled by the other mechanisms. It was explained that the hydrogen nucleus attracts less the electron cloud than any other heavier atom nucleus, thus for the A–H bond there is the natural electron charge shift from the H-center to the more electronegative A-atom. This effect is observed experimentally, for example, bond lengths to hydrogen measured by microwave spectroscopy depending on the positions of the nuclei are longer than those measured for the same bond by techniques that depend on diffraction by the electrons [63].

One can also compare the neutron diffraction and X-ray diffraction experiments where there is the diffraction of neutrons on nuclei and the diffraction of X-rays on electrons, respectively [64, 65]. As a consequence the positions of nuclei and the positions of maxima of the electron density are determined in neutron diffraction and X-ray diffraction measurements, respectively. Thus in a case of neutron diffraction the results correspond to the common understanding of molecular structure where the bond length is the distance between atoms' nuclei. In a case of X-ray results the positions of electron density local maxima are practically the same as the positions of nuclei but only for non-hydrogen atoms. For the H-atom, as it was described earlier

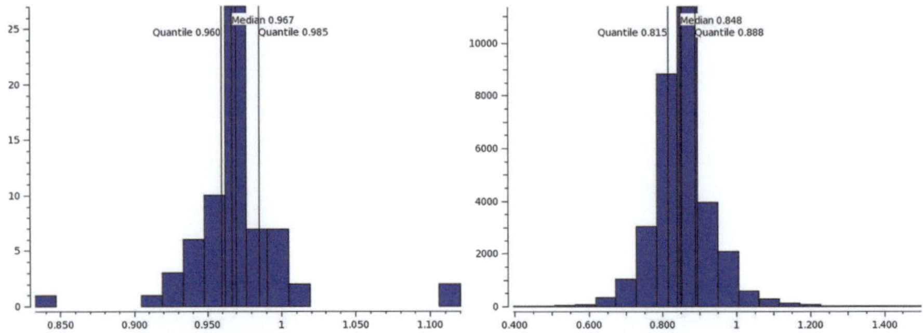

Fig. 9.5 Histograms of the O–H bond length for water molecules detected in crystal structures, neutron diffraction (*left*) and X-ray diffraction (*right*) results are presented. The horizontal axes correspond to O–H bonds (length in Å), the vertical axes to number of observations, based on the results from the Cambridge Structural Database [59]

here, the position of H-atom electron charge density local maximum is shifted to the heavy atom bonded with the hydrogen. Figure 9.5 presents histograms of O–H bond lengths for water molecules detected in crystal structures. The neutron diffraction and X-ray diffraction results are presented. The low precision of positions of H-atom's electron density maxima and other traits of X-ray technique (like spherical approximation of H-atom's thermal motions) results these positions are not directly comparable with the neutron diffraction H-nuclei positions but Fig. 9.5 roughly shows that the X-ray O–H bonds are much shorter than the neutron diffraction counterparts. The number of X-ray observations is large in comparison with the neutron diffraction ones since the total number of crystal structures stored into Cambridge Structural Database is equal to 686,944 while only 1616 of them are the neutron diffraction measurements (CSD summary statistics—6 January 2014 [66]).

The regions of low electron density situated at a planar part of the molecular system containing the atomic center considered were also analyzed [47]. It was pointed out that such regions exist for the boron center in the BX_3 planar molecules (X = F, Cl, Br, I) as well as for the sulfur in SO_2 and the nitrogen in FNO_2, the other examples of Cl_2CO, F_2CO and F_2SiO molecules were discussed where the carbon and silicon centers are characterized by the electron charge depletion. Such regions of the low electron density were named as the π-holes and the corresponding interactions with the Lewis base centers as the π-hole bonds.

The Lewis acid properties of the triel Z atoms (Z = B, Al, Ga, In, Tl) were analyzed recently in triel trifluorides, ZF_3 [67]. The strong acidic properties of the triel atoms in those planar molecules are connected with their electron structures since the triel atom possesses six electrons in the outer shell thus it is electron deficient. This electron structure is connected with the existence of the vacant p-orbital situated perpendicularly to the plane of the molecule. Consequently it results in the strong electrophilic properties of the Z-center. The octet rule is not fulfilled here what is often named as the hypovalency in the literature [68]. The similar electronic structure and consequently the electron deficiency is observed for

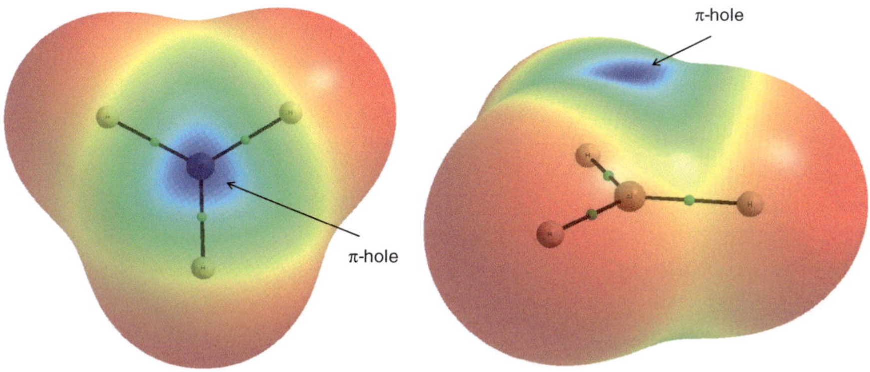

Fig. 9.6 Computed electrostatic potential (EP) on the 0.001 au molecular surface (MP2/aug-cc-pVTZ calculations) for the AlH_3 molecule (two projections). *Blue color* corresponds to the maximum (positive) and *red* one to the minimum (negative) EP. *Arrows* show the region of positive EP (π-hole)

the triel trihydrides and other triel trihalides what results in the positive electrostatic potential at the Z-atom in the direction perpendicular to the molecule. Figure 9.6 presents the electrostatic potential (EP) map for the AlH_3 molecule (two projections) where one can see the regions of the positive (π-hole) and negative EP attributed to the Al-center and to the H-atoms, respectively.

The results of the MP2/aug-cc-pVTZ calculations performed on complexes of the ZF_3 molecules with one and with two ligands playing the role of the Lewis base centers led to the conclusion that the Lewis acid–Lewis base interactions result here in the tetrahedral or the trigonal bipyramidal structures. The tetrahedral structure is the result of the interaction of the ZF_3 molecule with one ligand and the complex formed is linked through a very strong triel…Lewis base interaction which may be named as the triel bond [67]. In this case in the initial step of the reaction the lone electron pair of the Lewis base is directed to one of lobes of the vacant p-orbital of the Z-center. The octet rule is obeyed in the tetrahedral complex formed. Also two Lewis base ligands being directed to two lobes of the vacant p-orbital of the Z-center may interact with the ZF_3 molecule. The latter leads to the pentacoordinate Z-atom structure, in a case of strong interactions both Z…Lewis base contacts possess characteristics of covalent bonds what leads to ten electrons in the outer shell of the Z-atom (hypervalency [68]). Figure 9.7 presents two complexes; BF_3–NCH and GaF_3–$(NCH)_2$ representing the tetravalency of the boron atom (tetrahedral structure) and pentavalency of the gallium atom (trigonal bipyramide structure), respectively.

The results of calculations presented here are confirmed by experiment, especially the analogues interactions exist in the crystal structures containing triel trihydrides and triel trihalides [67]. Figure 9.8 presents species of two crystal structures of (1,1,3,3-tetramethylguanidine)-gallane and of (acetonitrile-N)-trichlorido-gallium (III). For the first structure one can see the result of the interaction between the

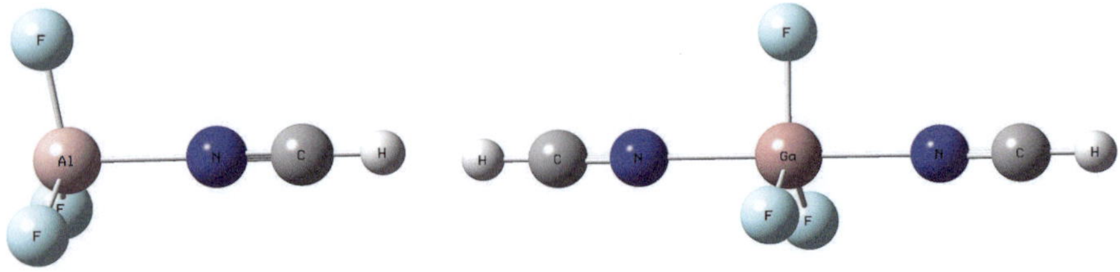

Fig. 9.7 The structures of the AlF_3–NCH (*left*) and of GaF_3-$(NCH)_2$ (*right*) complexes

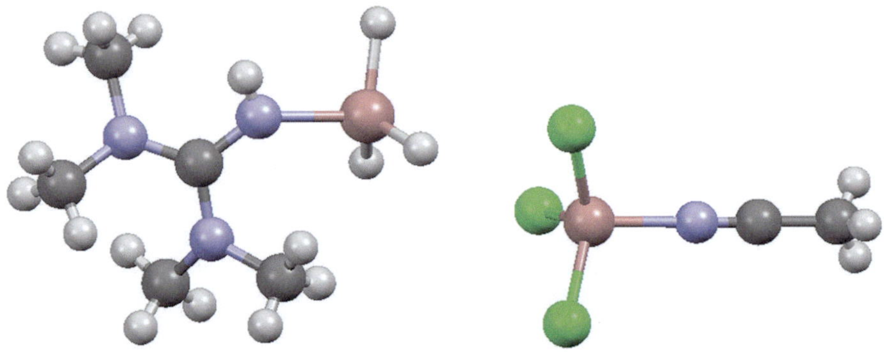

Fig. 9.8 The fragments of the crystal structures of (1,1,3,3-tetramethylguanidine)-gallane (*left*) and of (acetonitrile-N)-trichlorido-gallium(III) (*right*); *black circles* correspond to carbon atoms, *green* to chlorine atoms, *grey* to hydrogens, *pink* to gallium and *blue* to nitrogens, the structures taken from the Cambridge Structural Database [59]

gallium trihydride and the nitrogen center of the 1,1,3,3-tetramethylguanidine, the link through the Ga…N triel bond is the result of complexation; for the second structure the result of the interaction between the gallium trichloride and the acetinitrile is observed. Thus the same type of interaction, i.e. Ga…N triel bond exists for both crystal structures. This is interesting that the gallium center for both structures corresponds to the arrangement close to the tetrahedral one thus both Ga…N contacts may be treated as covalent bonds.

The pentacoordinated triel atoms are also observed in numerous crystal structures. Figure 9.9 presents two examples, two species of the tris(hydrido)-bis(trimethylamine)-aluminium and of the trichloro-bis(trimethylamine)-aluminium(III) crystal structures. In both cases the trigonal bipyramid arrangement for the aluminium center is observed. In one case the species observed in the crystal structure may be treated as the interaction between AlH_3 and nitrogen Lewis base centers, in the second case that is the interaction of $AlCl_3$ with the same ligands.

One can see that the diversity of La–Lb interactions is observed not only in theoretical results of calculations but also in the real systems analyzed experimentally; it is possible to enumerate their different kinds, often differencing in

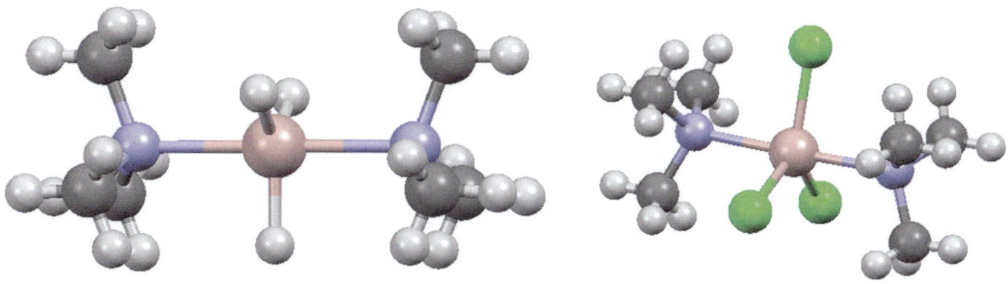

Fig. 9.9 The fragments of the crystal structures of tris(hydrido)-bis(trimethylamine)-aluminium (*left*) and of trichloro-bis(trimethylamine)-aluminium(III) (*right*); *black circles* correspond to carbon atoms, green to chlorine atoms, *grey* to hydrogens, *pink* to aluminiums and *blue* to nitrogens, the structures taken from the Cambridge Structural Database [59]

numerous characteristics. Not all of them were described here; for example, the special attention is needed for the beryllium bonds [69–72] where the Be-atom center acts as the Lewis acid. Also the lithium bond [62, 73–75] was not described here. However few additional types of La–Lb interactions are discussed in the next sections of this chapter. It seems that at least few properties are common for all La–Lb interactions, and that these interactions are ruled by the same or similar mechanisms. One can mention significant or at least noticeable electron density shift from the Lewis base to the Lewis acid unit as a common feature of interactions labeled here as La–Lb pairs (instead of the commonly used term "noncovalent interactions"). It seems that this name is also problematic since other interactions as for example those between ions or van der Waals interactions are also characterized by the charge transfer component. However in a case of hydrogen bond, halogen bond or other interactions mentioned earlier here the charge transfer is the main term or at least comparable with the other attractive interactions while for ionic or van der Waals contacts this term is much less important than the electrostatic or dispersive contributions. Nevertheless there is no doubt that the terminology of interactions should be ordered and specified in the near future.

9.2 Mechanisms of Hydrogen Bond Formation

The recent IUPAC definition covers various interactions often classified as hydrogen bonds. It is as follows: "the hydrogen bond is an attractive interaction between a hydrogen atom from a molecule or a molecular fragment X–H in which X is more electronegative than H, and an atom or a group of atoms in the same or a different molecule, in which there is evidence of bond formation" [76]. One can see that the X–H…Y designation was applied in this definition to describe the hydrogen bonded system. The A–H…B designation is used in this chapter to emphasize that the A–H bond is a fragment of the Lewis acid unit and that B is the Lewis base center; besides X in this chapter and in numerous studies is usually reserved for the halogen atom.

In other words one may state that the hydrogen bond is the Lewis acid–Lewis base interaction which leads to the formation of the A–H…B link where A–H and B play the role of the Lewis acid (electron accepting) and Lewis base (electron donating) centers, respectively. Further the numerous properties of this "link" may be specified, as for example that B may be mono-center (atom of any species or ion) or multicenter (π-electron or σ-electron system).

In early studies on hydrogen bonds it was assumed that the H…B distance should be shorter than the sum of the corresponding van der Waals radii (of H and B atoms), sometimes it was stated that the H…B distance should be shorter at least by 0.3 Å than the mentioned above sum [77]. This is often observed for the A and B centers characterized by the high electronegativity; i.e. for the O–H…O, N–H…O, N–H…N, F–H…O links, and others. However the C–H…O, C–H…N and in general C–H…B (B designates here, as before, the Lb center) contacts were found in crystal structures and classified as hydrogen bonds [77, 78]. It was found that for numerous such systems the H…B distance is close to the corresponding sum of van der Waals radii or even exceeds it; it was explained that the hydrogen bond is partly long range electrostatic interaction being attractive far beyond this sum [31]. Desiraju has pointed out that the hydrogen bond is an interaction without borders [79] which may be treated in terms of its electrostatic, van der Waals, and covalent components. The term "without borders" may be understood in a following way; there is the smooth crossing between van der Waals interactions and very weak hydrogen bonds as well as there is the smooth crossing between very strong hydrogen bonds and covalent bonds.

Since the hydrogen bond is a complex phenomenon thus the question arises if for such complex interaction the common mechanisms of its formation may be determined? This question seems to be justified since almost all characteristics indicated in the literature before as attributed to the hydrogen bond may be contested. The characteristics often pointed out in the early literature that for the A–H…B hydrogen bond the A and B centers should be highly electronegative and that the H…B distance should be shorter than the corresponding sum of van der Waals radii are not justified as it was described earlier here. Another characteristic often treated before as a signature of the hydrogen bond formation is that the complexation leads to the elongation of the A–H bond with the concomitant red-shift of the corresponding stretch frequency. However the later studies showed that this is not a rule since sometimes the shortening of the A–H proton donating bond with the corresponding blue-shift is observed [80–83].

It seems that the starting point to analyze the mechanism of the hydrogen bond formation may be related to the electron charge transfer from the Lewis base to the Lewis acid since this is common for La–Lb interactions [27]. The mechanism of the hydrogen bond formation based on the latter property was proposed in terms of the Natural Bond Orbitals (NBO) approach [68]. It was explained that the A–H…B hydrogen bond is a combination of two effects: the hyperconjugative A–H bond weakening and the rehybridization-promoted A–H bond strengthening [84].

The first effect is very well known and it was discussed in early and recent studies [68, 85, 86]. It is connected with the electron charge transfer from the lone pair of

B Lewis base center to the antibonding σ* orbital of the A–H bond. The interaction energy of the corresponding $n_B \rightarrow \sigma^*_{AH}$ orbital-orbital overlap is often presented in studies on the hydrogen bonded systems and it often correlates with the other parameters related to the hydrogen bond strength [87]. The mentioned above overlap refers to the 3c–4e (3 center and 4 electron) system, however different other types of hydrogen bond may be also considered [38]. For the A–H…π, A–H… σ and A–H… H–B hydrogen bonds following orbital-orbital interactions were found as the most important ones; $\pi_B \rightarrow \sigma^*_{AH}$, $\sigma_B \rightarrow \sigma^*_{AH}$ and $\sigma_{BH} \rightarrow \sigma^*_{AH}$, respectively. The latter A–H… H–B interaction, dihydrogen bond, is classified as a special kind of the hydrogen bond where the H…H interatomic contact is observed between the protic positively charged H(A)-atom and the hydric negatively charged hydrogen—H(B) [88–90].

The second effect, the rehybridization-promoted A–H bond strengthening, was described in literature recently [84] and it is connected with the further redistribution of the electron charge being the result of complexation. It leads to the increase in the s-character of the A-atom hybrid orbital in the A–H bond. This redistribution of the electron charge is connected with its shift from the hydrogen atom of the A–H bond to the further parts of the Lewis acid unit, especially to the A-center. This is why the A–H…B hydrogen bond formation is connected with the increase of the positive charge of H-atom and increase of the negative charge of the A-atom [68, 84]. However there are exceptions, the F–H…SH_2 complex is an example of such situation where the hydrogen bond formation leads to the decrease of the positive charge of H-atom [91]. The definition proposed recently by Weinhold and Klein is in line with the above descriptions of the hydrogen bond formation since it is stated that the hydrogen bond is "commonly originating in the $n_B \rightarrow \sigma^*_{AH}$ donor–acceptor interaction between the lone pair n_B of the Lewis base and the hydride antibond σ^*_{AH} of the Lewis acid" [86]. This definition refers to the 3 center–4 electron A–H…B systems and may be slightly modified to be proper for the other types of hydrogen bond mentioned earlier here [38].

The processes accompanying the hydrogen bond formation and explained in terms of the NBO method correspond to descriptions based on the Quantum Theory of Atoms in Molecules (QTAIM) [92, 93] results. For example, eight QTAIM criteria of the existence of the hydrogen bond were proposed by Koch and Popelier [94]; the criterion of the existence of the H…B bond path with the corresponding bond critical point is the most often checked one in studies on various interactions. It was pointed out that for the hydrogen bond the electron density at the H…B BCP should be situated within the range of (0.002, 0.04) au; the range for the laplacian of the electron density at BCP was also proposed [94]. The decrease of the electron charge (the increase of the positive charge) of the hydrogen atom as an effect of the A–H…B hydrogen bond formation and the simultaneous decrease of the hydrogen atom volume were also listed among these criteria. The latter dependence between the atomic charge and its volume seems to be obvious since this volume as well as the charge are related to the electron atomic basin within the QTAIM approach [93].

Different Lewis acid–Lewis base interactions were analyzed recently (MP2/6-311++G(d,p) calculations) [95], among them hydrogen bonds in the

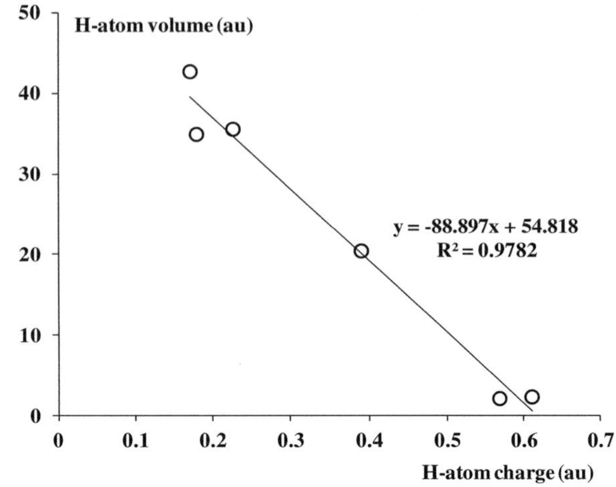

Fig. 9.10 The correlation between the atom volume and its integrated charge for the H-atom in the A–H…B hydrogen bonded systems

following systems; $F_3CH\ldots NH_3$, $F_3CH\ldots OH^-$, $(HCCH)_2$ (T-shaped dimer), $HCCH\ldots NH_3$, $(H_2O)_2$ and $NH_4^+\ldots NH_3$ and numerous correlations between NBO and QTAIM parameters were discussed. For example, Fig. 9.10 presents the correlation between the QTAIM integrated H-atom charge and its QTAIM volume for those hydrogen bonds. This correlation seems to be important since the protic H-atom in complexes considered belongs to various Lewis acid species (fluoroform, water, acetylene, and ammonia cation) and neither its charge nor its volume are normalized here to be related to the H-atom in the corresponding La species not involved in any interaction.

Figure 9.11 presents for the same sample of hydrogen bonds the correlations between the electron charge shift from the Lewis base to the Lewis acid and different

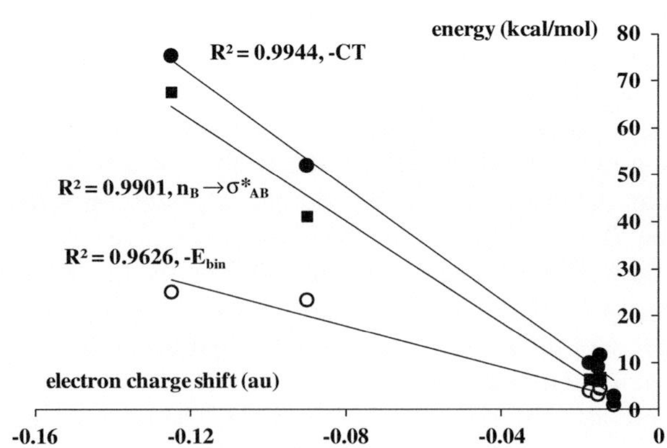

Fig. 9.11 The correlation between the electron charge shift (au) and energy (kcal/mol); *open circles* are related to $-E_{bin}$ energy (E_{bin} is the binding energy), *full circles* to –CT (charge transfer DFT/NEDA energy) while *full squares* correspond to the part of CT, i.e. the $n_B \rightarrow \sigma^*_{AH}$ orbital-orbital interaction energy

energy terms. Note, that the electron charge shift is the sum of QTAIM atomic charges of the Lewis acid unit within the complex. The Lewis base–Lewis acid electron charge shift is observed here like for most of the La–Lb interactions. Hence the charge of Lewis acid unit in the complex is negative. The following energies are presented in figure, the binding energy being the difference between the energy of the complex and the sum of energies of monomers in their energetic minima (the BSSE correction included), the charge transfer energy calculated within the DFT/NEDA decomposition scheme [96] and the $n_B \rightarrow \sigma^*_{AH}$ orbital-orbital interaction energy. For all energy types good correlations are observed. However these are only rough dependencies from the statistical point of view since only six complexes are considered; besides two points corresponding to the greatest electron density shifts are outside the remaining part of the sample. The tendencies observed in Fig. 9.11 may indicate the importance of the electron charge density shift for the hydrogen bonded complexes since this shift coincides not only with the orbital-orbital interaction or with the total charge transfer but even with the binding energy.

It was mentioned earlier here that the hydrogen bond formation leads to the rehybridization process. This is connected with the increase in s-character of A-atom hybrid orbital of the proton donating A–H bond [84]. The latter is in agreement with the Bent rule [97] which states that atoms maximize their s-character in hybrid orbitals directed towards electropositive substituents and they maximize their p-character towards electronegative substituents. The Bent rule may be applied to the A–H...B hydrogen bonded systems since for them the complexation is connected with the increase of the positive charge of the hydrogen atom; thus one can state that the H-atom is more electropositive after the complexation than it was before. It seems that the increase of the s-character is the common feature of hydrogen bonds, except of rare cases when the positive charge of the hydrogen decreases in the process of complexation (FH...SH_2 complex mentioned above here [91]).

The MP2/6-311++G(3df,3pd) calculations were performed recently for the hydrogen bonded complexes of acetylene and fluoroform [98]. For all of them the increase in the s-character of the C-hybrid orbital participating in the C–H proton donating bond was observed as a result of complexation. This is important that such an increase does not depend on the blue- or red-shifted hydrogen bond is considered but it is related to the strength of interaction. Figure 9.12 presents relationships between the binding energy and the s-character for these two series of complexes. The greater increase of the s value is observed for stronger interactions. For example, the s-character for the carbon hybrid orbital in the isolated acetylene molecule is equal to 48.1 % while this value for the HCCH...F^- complex amounts 54.3 %. For the isolated CF_3H molecule the s-character for the C-atom hybrid orbital is equal to 30.4 % and in the F_3C–H...F^- complex it is equal to 36.6 %. This is worth to mention that the s-character described here correlates not only with the binding energy but also with other measures of the hydrogen bond strength; and different correlations between the s-character and the other parameters were found for numerous samples of the hydrogen bonded systems [95, 98].

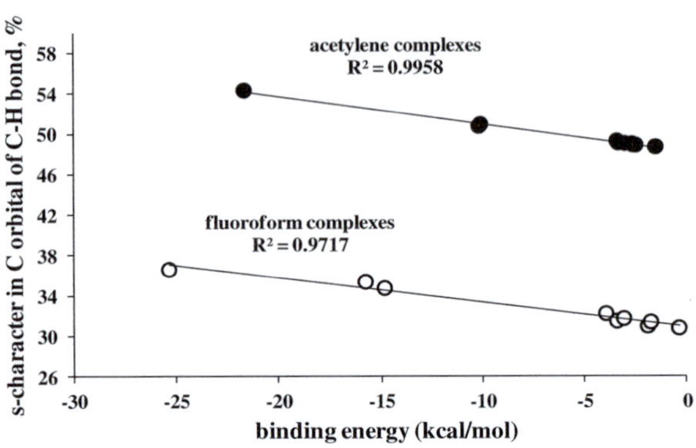

Fig. 9.12 The relationship between the binding energy (kcal/mol) and the s-character in the C orbital of the C–H bond (%) for acetylene (*full circles*) and fluoroform (*open circles*) complexes

The QTAIM approach mentioned here before is a very useful tool to describe La–Lb interactions since it is possible to have a deeper insight into changes in the electron charge distribution being the result of complexation; especially the changes of hydrogen bonded systems. For example, the position of the C–H proton donating bond critical point (BCP) was analyzed recently [98] for the mentioned earlier here complexes of acetylene and fluoroform. It is possible to decompose the C–H bond into two radii defined by the position of BCP (see Fig. 9.13). Hence there is the radius of the carbon atom which is the distance between BCP and the C-atom

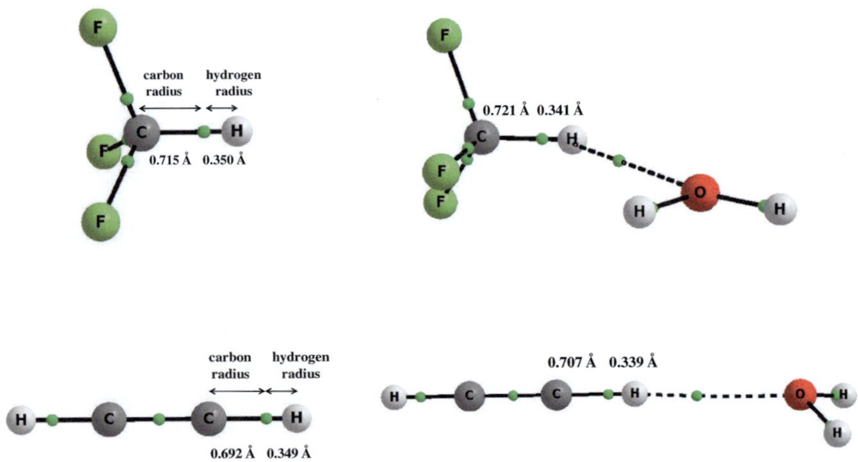

Fig. 9.13 The molecular graphs of the $CF_3H...OH_2$ and $HCCH...OH_2$ complexes (*right*) as well as of the isolated CF_3H and HCCH molecules (*left*). *Big circles* correspond to attractors while small ones to the bond critical points, *solid* and *broken lines* are the bond paths connecting pairs of atoms. The atomic radii of the C–H proton donating bond are presented

attractor and the radius of the hydrogen atom being the distance between BCP and the H-atom attractor.

It was found for the complexes of acetylene and fluoroform that the A–H…B hydrogen bond formation results in the increase of the radius of proton donator (A) and the decrease of the radius of H-atom of the A–H proton donating bond. If the H-radius decrease outweighs the A-radius increase thus one can observe the blue shifted hydrogen bond, for the reverse situation there is the red shifted hydrogen bond. This finding is in line with the mechanisms of the formation of the hydrogen bond proposed earlier in literature and described here [84, 86, 94]. It was explained before that the hydrogen bond formation results in the increase of the positive charge of H-atom what is connected with the decrease of its volume; the H-atom radius is related to its volume thus it also decreases. Similarly the negative charge of A-atom increases thus the concomitant increase of its volume is observed; the latter results in the increase of A-atom radius. This is important that there is no difference between the blue and red-shifted hydrogen bonds; in both cases the same changes of the mentioned above radii are observed. This is interesting that these radii correlate with the binding energy and other interaction strength parameters [98]. Thus for stronger H-bond interactions the greater elongation of the A-radius and the greater shortening of the H-radius are observed.

Figure 9.13 presents discussed above parameters for the $CF_3H…OH_2$ and $HCCH…OH_2$ complexes. For the first one the slight shortening by 0.001 Å of the C–H proton donating bond as a result of complexation is observed (from 1.085 to 1.084 Å) with the concomitant stretch frequency corresponding to the blue shift. For the second complex there is the elongation of the C–H proton donating bond from 1.062 to 1.067 Å with the concomitant stretch frequency corresponding to the red shift. One can see that for both complexes there is the increase of the C-atom radius and the decrease of the H-atom radius of the C–H bond after the formation of the hydrogen bond. Note that the sums of the atoms' radii are smaller than the bond lengths. It was discussed in the previous section than the attractor of the H-atom (the maximum of the electron density) is shifted towards the heavier atom; hence the C–H nucleus-nucleus distance is greater than the corresponding C–H attractor-attractor distance.

9.3 Mechanisms Accompanying Formation of Lewis Acid–Lewis Base Links

The question arises if the mechanism of formation of the hydrogen bond described earlier here is similar or the same one for other Lewis acid–Lewis base interactions; particularly if two effects existing in the A–H…B hydrogen bonds: the hyperconjugative A–H bond weakening and the rehybridization-promoted A–H bond strengthening [84] have their analogies in the La-Lb interactions. The reply to this question seems to be very interesting since the protic hydrogen of the A–H bond

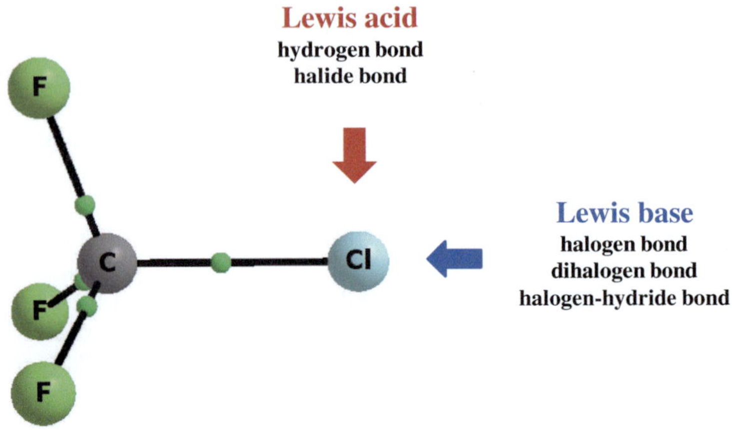

Scheme 9.3 The molecular graph of CF_3Cl molecule, the directions of possible interactions with Lewis acids and Lewis bases are shown

acts only as the Lewis acid center because its whole hemisphere is characterized by the positive electrostatic potential while the pnicogen, chalcogen and halogen centers possess the dual character and may act as La and Lb centers simultaneously. For example, on one hand the halogen center (and particularly chlorine atom presented in Scheme 9.3) may form hydrogen bonds or halide bonds with Lewis acid centers and on the other hand it forms halogen, dihalogen, halogen-hydride bonds with Lewis bases.

The MP2/6-311++G(d,p) calculations have been performed recently on the complexes of chlorotrifluoromethane, CF_3Cl, with different species possessing properties of Lewis acids and Lewis bases [99, 100]. The CF_3Cl species was chosen since the chlorine is characterized by the region of positive EP in the extension of the C–Cl bond and by the region of negative EP in the direction perpendicular, or nearly so, to this bond due to the lone electron pairs of chlorine (see Fig. 9.4). This is why the CF_3Cl moiety may interact through the chlorine center with Lewis acids and with Lewis bases. The complexes of CF_3Cl with following moieties acting as the Lewis bases; CH_2O, Cl^-, HMgH, FCl, OH^- and CH_3Cl were analyzed. Figure 9.14 presents selected examples of those complexes. One can see the classical linear C-Cl…O halogen bond for the CF_3Cl…OCH_2 complex. This linearity is a result of the restricted region of the positive EP at the Cl-atom in the elongation of the C–Cl bond. The CF_3Cl…OH^-, CF_3Cl…Cl^-, and CF_3Cl…FCl complexes (not presented in Fig. 9.14) are other examples of the typical halogen bonds; two first interactions, C–Cl…O^- and C–Cl…Cl^-, are assisted by the negative charge thus they may be named as the negatively charge assisted halogen bonds. For the complexes of chlorotrifluoromethane with Cl^- and FCl one can expect the term dihalogen bond since the Cl…Cl^- and Cl…F connections are formed here. However it seems that the term halogen bond is proper for both interactions since the chlorine anion as well as the fluorine center act here as the Lewis bases and they are not characterized by the dual character. The chlorine anion

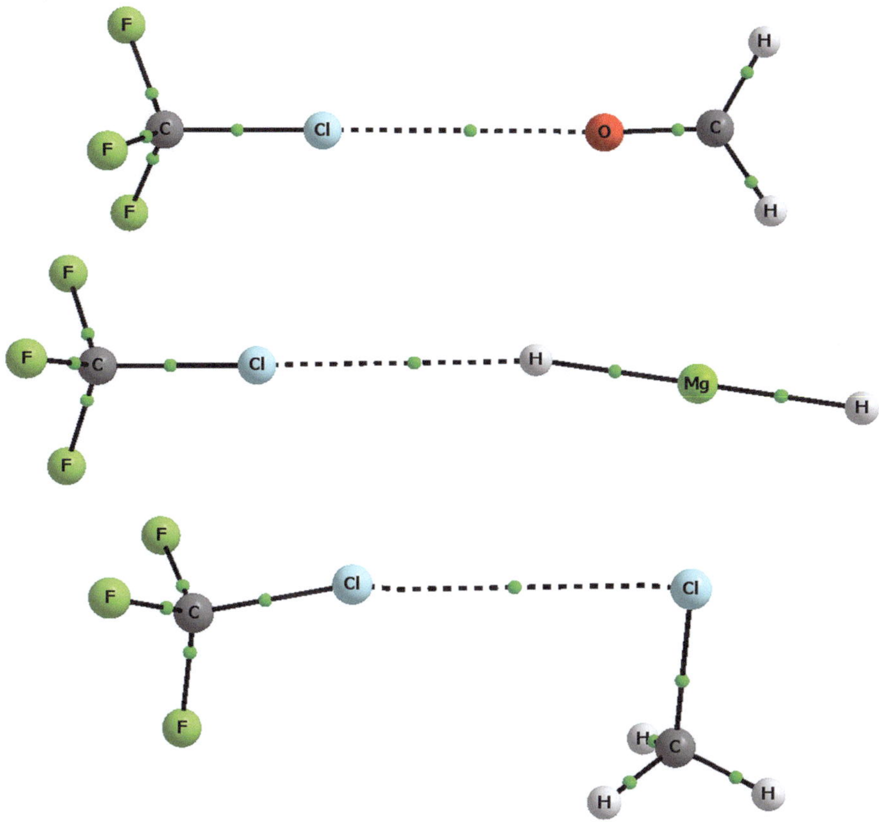

Fig. 9.14 The molecular graphs of the $CF_3Cl...OCH_2$, $CF_3Cl...HMgH$ and $CF_3Cl...ClCH_3$ complexes; *big circles* correspond to attractors, small ones to bond critical points; *solid* and *broken lines* to bond paths

does not possess the regions of the positive EP, similarly as the F-center in the FCl molecule where the whole hemisphere of fluorine is characterized by the negative EP. It seems that the term dihalogen bond is proper for the halogen-halogen stabilizing contacts if both halogen atoms are characterized by the dual character; i.e. if both centers may act as the Lewis base and as the Lewis acid. This is why the dihalogen bond may be formed between the same kind molecules like in a case of 4-iodobenzoic acid (see Fig. 9.2). The $CF_3Cl...ClCH_3$ complex (see Fig. 9.14) represents such a situation of the dihalogen bond. The CF_3Cl species acts here as the Lewis acid through the region of the positive EP situated in the elongation of the C–Cl bond while Cl–CH_3 acts as the Lewis base through the chlorine negative region of EP being perpendicular to the C–Cl bond.

Figure 9.14 presents the $CF_3Cl...HMgH$ complex where the hydric H-atom of MgH_2 acts as the Lewis base center, the intermolecular Cl…H interaction (where electronegative Cl atom acts as the Lewis acid and the H-atom as the Lewis base!)

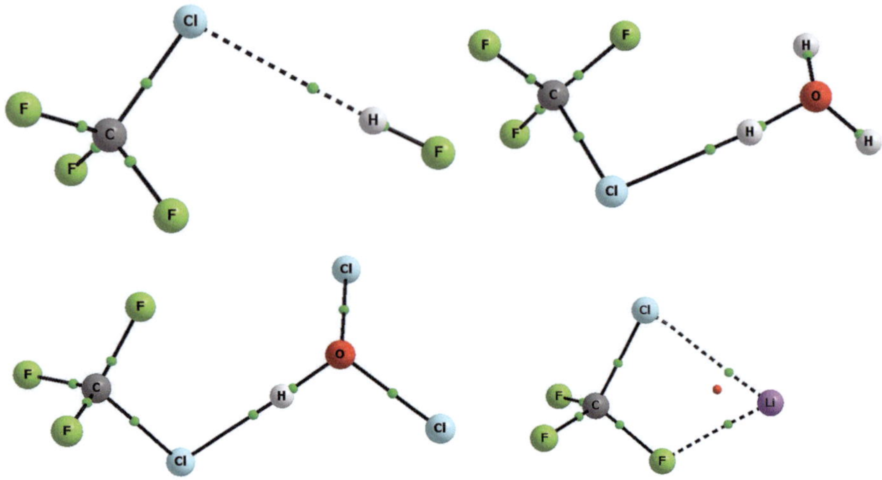

Fig. 9.15 The molecular graphs of the $CF_3Cl...HF$, $CF_3Cl...H_3O^+$, $CF_3Cl...HCl_2O^+$ and $CF_3Cl...Li^+$ complexes; *big circles* correspond to attractors, small ones to bond critical points; *solid* and *broken lines* to bond paths

was named as the halogen-hydride bond [27]. It is important that this interaction is very rare since in the Cambridge Structural Database [59] where over 600,000 crystal structures are collected [66] only in few cases interactions which may be classified as halogen-hydride ones were found [27].

It was mentioned earlier here that the chlorine center of CF_3Cl may act also as the Lewis base; its complexes with HF, H_3O^+, HCl_2O^+ and Li^+ were analyzed (Fig. 9.15). There is the F–H...Cl hydrogen bond for the $CF_3Cl...HF$ complex; for the H_3O^+ and HCl_2O^+ species the O–H...Cl positively charge assisted hydrogen bonds exist while for the complex of CF_3Cl with the Li + cation there are halide bonds. Figure 9.15 shows two intermolecular $Cl...Li^+$ and $F...Li^+$ bond paths for the latter complex corresponding to two $C–Cl...Li^+$ and $C–F...Li^+$ halide bonds. One can see that in all cases where chlorine acts as the Lewis base, the Lewis acid unit is directed perpendicularly, or nearly so, to the C–Cl bond, i.e. it is directed to the region of the negative EP of the chlorine center.

Table 9.1 presents selected geometrical and energetic parameters of the complexes described above. One can see that longer intermolecular contacts are observed for complexes where CF_3Cl acts as the Lewis acid than in a case where the chlorine is the Lewis base center. Only, for the first group of complexes, relatively small Cl...O distance of 2.46 Å is observed for the $CF_3Cl...OH^-$ complex where the halogen bond is assisted by the negative charge and where the strongest interaction is detected (E_{bin} = −13.8 kcal/mol). The shorter H...Cl distances for the hydrogen bonded systems are justified since the H-atom being in the intermolecular contact is characterized by a small van der Waals radius. Also for the $CF_3Cl...Li^+$ complex characterized by the strong interaction (E_{bin} = −12.1 kcal/mol) the small Cl...Li distance amounting 2.48 Å is observed.

Table 9.1 MP2/6-311++G(d,p) results; the chlorine—Lewis acid/Lewis base center distance (Cl…A(B), in Å), ΔC–Cl the change of bond length (Å) as a result of complexation (+ denotes increase and − decrease), C–Cl…B(A) angle (degrees) as well as binding energy, E_{bin} (corrected for BSSE, in kcal/mol), E_{NBO} (kcal/mol) is the orbital-orbital interaction energy described in the text; C–Cl bond length for F_3CCl species not involved in any interaction amounts 1.746 Å

Species	Cl…B(A)	ΔC–Cl	C–Cl…B(A)	E_{bin}	E_{NBO}
CF_3Cl as Lewis acid					
OCH_2	3.059	−0.006	179.8	−1.4	0.6
Cl^-	3.098	−0.025	180.0	−7.3	5.5
OH^-	2.460	−0.012	175.5	−13.8	15.7
FCl	2.990	−0.002	175.6	−0.4	0.7
$ClCH_3$	3.467	−0.001	168.8	−0.7	0.9
MgH_2	2.964	−0.004	169.4	−1.1	0.7
CF_3Cl as Lewis base					
HF	2.592	+0.012	96.6	−0.1	1.5
H_3O^+	1.979	+0.049	94.1	−7.5	36.3
HCl_2O^+	1.787	+0.077	95.6	−9.0	82.4
Li+	2.482	+0.027	78.7	−12.1	22.3

If the CF_3Cl species acts as the Lewis base the elongation of the C–Cl bond is observed as a result of complexation (Table 9.1). However if the CF_3Cl acts as the Lewis acid, in the C–Cl…B halogen bonds and related interactions; dihalogen bond and halogen-hydride bond, in all cases considered, the shortening of the C–Cl bond is detected. This is slightly surprising since for the A–H…B hydrogen bonds the red shift corresponding to the elongation of the proton donating bond is usually observed and the shortening of the A–H bond with the simultaneous blue shift is not so often. However it was described in the literature that the blue shift for halogen bonds occurs more often than in a case of hydrogen bonds [99, 100]. One can see that for the halogen bond and related interactions the linear, or nearly so, C–Cl…B arrangements are observed since this angle is situated within the 168–180° range (Table 9.1). In a case of hydrogen bonded systems and the halide interaction the range of 78–97° is observed, close to perpendicular C–Cl…A (A designates Lewis acid center) arrangement. This is in line with the description presented before here that the negative EP of chlorine concerns the region being perpendicular to the C–Cl bond while the region of positive EP is situated in the extension of the C–Cl bond.

The last column of Table 9.1 presents E_{NBO} values which correspond to the orbital-orbital interaction energies calculated within the NBO scheme. The complexes of CF_3Cl with HF, H_3O^+ and HCl_2O^+ are linked through the hydrogen bonds and the lone pair (n) → σ^* interaction is detected for them (see previous section of this chapter). This is the n(Cl) → σ^*_{FH} interaction for the complex with HF molecule and the n(Cl) → σ^*_{OH} interaction for the complexes with H_3O^+ and HCl_2O^+. In two latter cases the meaningful orbital-orbital interaction energies of 36 and 82 kcal/mol are observed. In a case of halide bond (C–Cl...Li^+) the lone pair → lone pair interaction, n(Cl) → $n(Li)^*$, occurs.

The n(B) → σ^*_{CCl} interaction (B corresponds to the Lewis base center) is observed if the CF_3Cl species acts as the Lewis acid. The O, Cl, O, F and Cl atoms act as the Lewis base centers for the CF_3Cl complexes with OCH_2, Cl^-, OH^-, FCl and $ClCH_3$, respectively. This is a very similar type of orbital-orbital interaction as that one usually observed for the A–H...B hydrogen bonds (n(B) → σ^*_{AH}); in both cases, of the hydrogen bond and of the halogen bond, we have the antibonding orbital corresponding to the Lewis acid moiety, hydride σ^*_{AH} orbital and halide σ^*_{AX} orbital, respectively (σ^*_{CCl} for the CF_3Cl molecule considered here). In a case of the CF_3Cl...HMgH complex there is the σ_{MgH} → σ^*_{CCl} orbital-orbital interaction, however even here one can see the same antibonding σ^*_{CCl} orbital as for other interactions where the chlorine plays a role of the Lewis acid center.

Table 9.2 presents other characteristics of the CF_3Cl complexes. It is obvious that there is the electron charge shift to the CF_3Cl species if the latter one acts as the Lewis acid (positive ET values in Table 9.2) and a withdrawing of electrons from the CF_3Cl if it acts as the Lewis base (negative ET values). One can see that the greatest absolute ET values are observed for the charge assisted interactions. It is interesting to analyze the change of atomic charges being a result of complexation. For the A–H...B hydrogen bond the increase of the positive charge of the H-atom and the increase of the negative charge of the A-atom are observed [68] (there are exceptions described in the previous parts of this chapter). The analogues to the hydrogen bond systems are those linked through the halogen bond and related dihalogen and halogen-hydride bonds where for the C–Cl...B interaction the increase of the positive charge of Cl-atom and the increase of the negative charge of

Table 9.2 NBO parameters for the F_3CCl complexes; ET (in milielectrons) is the amount of electron charge transferred from the Lewis base to the Lewis acid (positive values if F_3CCl is the Lewis acid, negative values if it acts as the Lewis base

Species	ET	C-charge	Cl-charge	Pol(%C)	s(%C)
OCH_2	1.5	0.9844	0.0318	47.76	27.27
Cl^-	54.2	0.9426	0.0967	50.18	29.96
OH^-	123.4	0.9162	0.1060	51.88	31.54
FCl	2.2	0.9904	0.0154	47.29	26.90
$ClCH_3$	6.0	0.9906	0.0116	47.28	26.89
MgH_2	2.6	0.9882	0.0234	47.53	27.16
HF	−5.6	1.0021	−0.0129	46.38	25.95
H_3O^+	−122.7	1.0318	0.0241	44.28	23.84
HCl_2O^+	−221.7	1.0507	0.0652	42.23	22.25
Li^+	−69.2	1.0122	0.0110	45.99	25.4
F_3CCl*	–	0.9942	0.0073	47.0	26.62

C-charge and Cl-charge NBO charges (in au). Pol(%C) is the C–Cl bond polarization (% at C-atom), s(%C) is the percentage s-character in C orbital of the C–Cl bond. The last line of the table (*) presents characteristics of the CF_3Cl monomer not involved in any interaction

C-atom are observed (Table 9.2). The Cl-atom charge in the CF_3Cl free molecule is equal to +0.007 au and it increases for all halogen, dihalogen and halogen-hydride bonded complexes. The greatest increase is observed for the strongest interaction with the OH^- ion where the Cl-atom charge is equal to +0.106 au. The situation for the CF_3Cl species interacting with the Lewis acids is not so clear. However in such a case hydrogen bondsHydrogen bond or halide bond are formed and one should analyze the changes of characteristics of the H- or Li- atoms.

For the free CF_3Cl molecule the charge of the carbon atom is equal to +0.994 au and it decreases for all complexes linked through halogen bond and related interactions, similarly as in the A–H…B hydrogen bonds where the decrease of the positive charge (increase of negative charge) of the A-atom is usually observed. The positive C-atom charge increases if the CF_3Cl species acts as the Lewis base, in other words there is the outflow of the electron charge in this case to the Lewis acid unit (HF, H_3O^+, HCl_2O^+ or Li^+), partly from the carbon center.

There are other consequences of the complexation if chlorine in the CF_3Cl species acts as the Lewis acid center, the outflow of the electron charge from chlorine and its accumulation at the carbon center leads to the increase of the polarization of the C–Cl bond (see Table 9.2); the greatest polarization of 51.9 % (% of the electron charge at the carbon center) is observed for the $CF_3Cl…OH^-$ complex where the strongest interaction occurs. And the increase of the s-character is also detected, the same effect was observed for A-center in A–H…B as a result of the hydrogen bond formation. One can see that the hyperconjugative and rehybridization processes known for A–H…B hydrogen bonds are also observed for A–X…B halogen bonds (and related interactions). This is interesting that the polarization of the C–Cl bond decreases if the chlorine in CF_3Cl species acts as the Lewis base, similarly s-character of the hybrid orbital of carbon atom in C–Cl decreases.

The changes being the result of complexation and concerning polarization, s-character, atomic charges of the C–Cl bond, etc. correlate with the binding energy

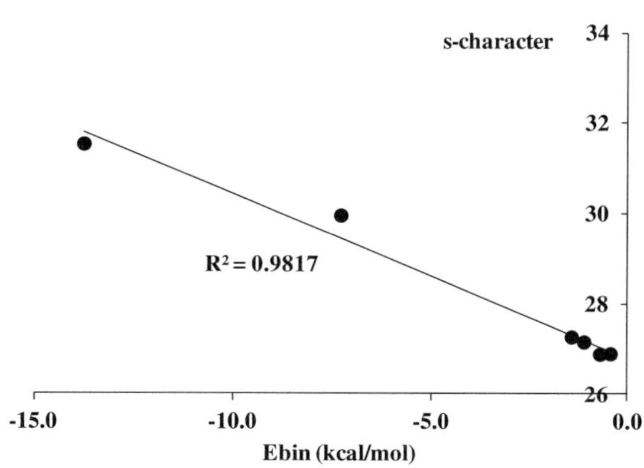

Fig. 9.16 The correlation between the binding energy (kcal/mol) and the s-character of the C-hybrid orbital in C–Cl bond; the sample of the CF_3Cl complexes where chlorine acts as the Lewis acid center

and with other descriptors of the strength of interaction [99, 100]. Figure 9.16 presents an example of such correlation between the binding energy and the s-character for the CF_3Cl complexes described above. Only complexes where chlorine plays the role of the Lewis acid center are taken into account here. One can see an excellent linear dependence; however only six complexes are considered here thus from the statistical point of view it may be treated only as a tendency; besides two points corresponding to the strongest interactions are far from the remaining part of the sample (it is no use that those remaining four points also well correlate—the sample is too small from the statistical point of view). However it seems that the s-character may be treated as a measure of the strength of interaction since such correlations were detected for other larger samples of complexes, particularly those linked through the hydrogen bond [98].

The complexes linked by more distinct interactions were analyzed recently; various types of hydrogen bonds were taken into account as well as halogen, hydride and dihydrogen bonds (MP2(full)/6-311++(3df,3pd) calculations were performed) [101]. It was found that mechanisms of formation of various complexes are very similar in spite of the diversity of interactions considered. Figure 9.17 presents examples of NH_4^+…HBeH and NH_4^+…HMgH complexes linked through the dihydrogen bond (DHB). This interaction was analyzed before [88–90] thus one may state, being in agreement with those earlier descriptions that the dihydrogen bond is the Lewis acid–Lewis base interaction characterized by the H…H contact where one of H-atoms plays the role of the Lewis acid center (protic H-atom) and

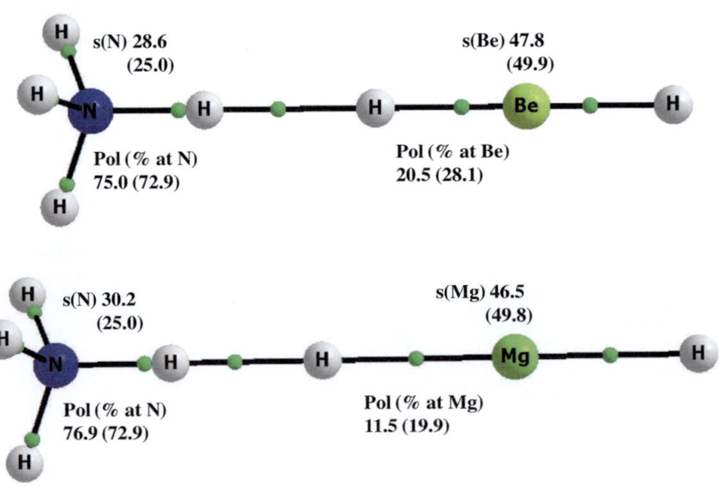

Fig. 9.17 The molecular graphs of NH_4^+…HBeH (*up*) and NH_4^+…HMgH (*down*) complexes; *small circles* correspond to bond critical points, big ones to attractors, *solid lines* represent bond paths, s designates s-character of the hybrid orbital (for example N-hybrid orbital of the N–H bond); Pol is the polarization of the bond considered (% of the electron charge at the atom indicated); the values in parentheses concern monomers not involved in dihydrogen bonds

the second hydrogen (hydric) is the Lewis base center. The hydrogen atoms being in contact possess the opposite charges. It is often designated as A–H…H–B where the protic hydrogen belongs to the Lewis acid unit (A) and the hydric hydrogen to the Lewis base (B). One can see that this interaction may be treated as a sub-class of hydrogen bonds since it is characterized by the A–H proton donating bond, as for the classical hydrogen bonds, and the negatively charged hydric hydrogen center, similarly as the B-center for A–H…B hydrogen bonds.

In general the same mechanisms of the electron charge shift are observed for DHB as for the hydrogen bond. The QTAIM analysis of the atomic charges of NH_4^+…HBeH and NH_4^+…HMgH complexes show that the complexation leads to the increase of the positive charge of the protic H-atom of the NH_4^+ cation and the increase of the negative charge of the N-atom. In a case of Be–H or Mg–H bond of the Lewis base unit being in contact with the protic hydrogen of ammonia ion there is the outflow of the electron charge since for both atoms of the bond the decrease of the negative charge is observed. The electron charge shifts for these complexes lead to the other changes (Fig. 9.17), the increase of the s-character of the N-hybrid orbital of the N–H proton donating bond as well as the increase of the polarization of this bond (% of the electron charge at the N-atom)—the same changes as those usually observed for the A–H…B hydrogen bonds as well as for the A–X…B halogen bonds. One can also observe the changes of the s-character and of the polarization for the electron donating Be–H and Mg–H bonds which are reverse to those detected for the proton donating N–H bond. The s-character of Mg or Be hybrid orbital decreases as well as the polarization of the Mg–H or Be–H (% of the electron density at the Mg or Be atom) decreases.

This is worth mentioning that the $\sigma \rightarrow \sigma^*$ interaction is the main orbital-orbital overlap for DHBs [38]; particularly in a case of the BeH_2 and MgH_2 complexes discussed here these are the following interactions; $\sigma_{BeH} \rightarrow \sigma^*_{NH}$ and $\sigma_{MgH} \rightarrow \sigma^*_{NH}$. The similar $\sigma \rightarrow \sigma$ orbital-orbital interactions occur for the A–H…σ hydrogen bonds where the σ-electrons of molecular hydrogen play the role of the Lewis base [38]. For the typical hydrogen bond the $n(B) \rightarrow \sigma^*$ interaction is often treated as its signature [68].

The numerous results presented here for different La–Lb interactions show that at least few common characteristics may be indicated for them; especially for the Lewis acid bond being in contact with the Lewis base center. This is the A–H bond for the A–H…B hydrogen bonds and for the A–H…H–B dihydrogen bonds as well as the A–X bond for the A–X…B halogen bonds. Scheme 9.4 shows selected characteristics of the A–Y…B La–Lb interaction, where B is the Lewis base center while A–Y is the bond of the Lewis acid unit.

The changes presented in Scheme 9.4 were analyzed in several studies; for example the hydrogen and halogen bonds were analyzed in several complexes and these changes were detected as common for both kinds of interactions [95]. Particularly the $(H_2O)_2$ and CF_3Cl…NH_3 complexes linked through the O–H…O hydrogen bond and the C–Cl…N halogen bond, respectively, were compared [95]. It was found that the shortening of the H…O and Cl…N intermolecular distances leads to the same electronic and structural changes what comes from the same

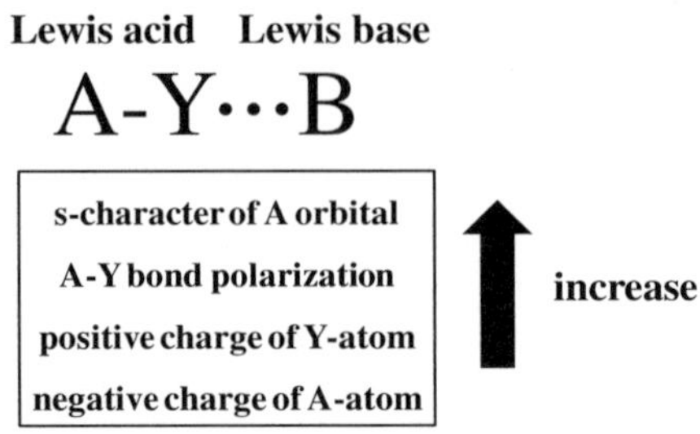

Scheme 9.4 The change (increase) of few parameters of the A–Y bond involved in the A–Y…B La–Lb interaction as a result of complexation

mechanisms of the complex formation. The positive charge of H or Cl-atom increases, its volume decreases; the negative charge of O and C atoms connected with H and Cl atoms, respectively, increases what leads to the increase of their volumes. The corresponding s-character of the hybrid orbital of O (and C) of the O–H (and C–Cl) bond increases what is in agreement with the Bent rule [97]. This is important that these changes follow changes of the intermolecular distance (H…O and Cl…N) from the distances greater than those corresponding to energetic minima up to exactly those of the equilibrium states. Hence one may say that these changes correspond to the process of complex formation (the formation through the hydrogen and halogen bonds).

9.4 The Electron Charge Shifts in a Case of Cooperativity Effects

The cooperativity effect is not strictly defined; however, very often it is understood as the enhancement (or diminution) of interaction in the complex if the additional species interacts with one of components of this complex forming the next interaction [102, 103]. This effect is analyzed most often for hydrogen bond interactions; for example Jeffrey analyzed the cooperativity in crystal structures claiming that it is connected with σ-bonds which can interrelate in a chain or in a cycle; the case of O–H bonds is an example since such (R)O–H…(R)O–H…(R)O–H.. chain or cycle may be formed here [30]. This effect occurs very often in crystal structures where the translational symmetry requires the reduplication of the system through the intermolecular interactions, often through the hydrogen bonds. However there are other experimental evidences of the existence of cooperativity effects; as for the H_3N…HF and H_3N…HF…HF systems which were analyzed by the microwave and theoretical techniques [104]. It was found that for the H_3N…HF…HF complex the H…N interaction is stronger than in the complex containing one HF molecule, besides the H…N distance is by 0.21(6) Å shorter for the former system than for the latter one.

There are different studies where the enhancement of the hydrogen bond due to additional interactions is observed; however it was also found that sometimes the weakening of the hydrogen bond occurs as a result of the cooperativity [103]. Additionally the number of studies where the other La–Lb interactions, not only hydrogen bonds, are analyzed in terms of the cooperativity effects increases in recent years [102]. For example, the electron density shifts were analyzed recently for dyads and triads linked through hydrogen and/or halogen bonds and the calculations were performed at the MP2(full)/6-311++G(d,p) level [105]. Figure 9.18 presents systems selected from those analyzed before [105]; HF and C_2H_2 monomers, C_2H_2...HF and C_2H_2...Cl^- dyads, and the HF...C_2H_2...Cl^- triad. In a case of the C_2H_2...HF system there is the FH...π hydrogen bond since this is the T-shaped complex where the HF molecule is perpendicular to the acetylene species. For the C_2H_2...Cl^- linear complex the C–H...Cl^- hydrogen bond is formed, and for the mentioned above triad we have both kinds of the hydrogen bond (see Fig. 9.18). Different parameters are presented in Fig. 9.18 thus one can see the changes being the result of complexations and the cooperativity effects occurring in a case of triad.

From one side it is possible to analyze the influence of the C–H...Cl^- hydrogen bond on the FH...π interaction and from the other side, vice versa, the influence of the latter hydrogen bond on the former one. Let us consider the first case. For the

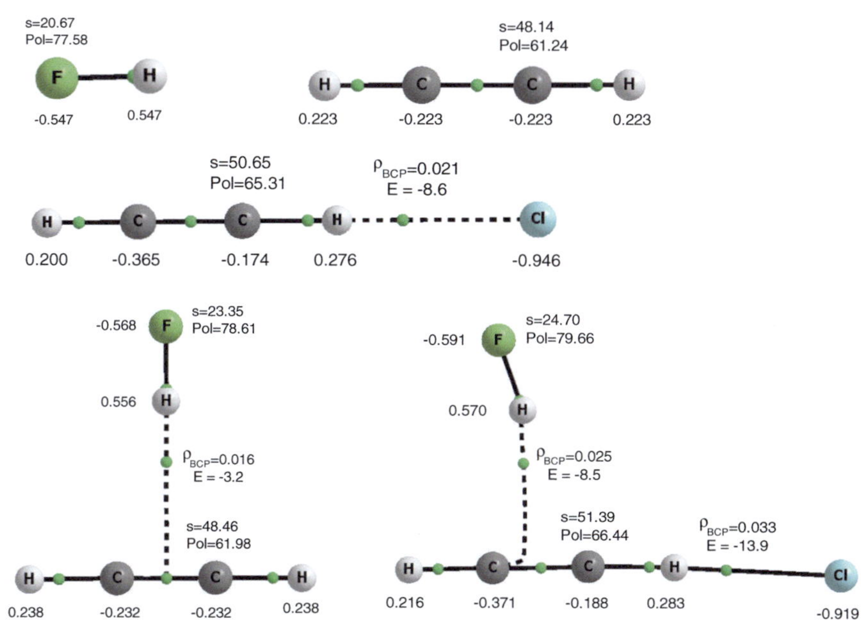

Fig. 9.18 The molecular graphs of HF, C_2H_2, C_2H_2...Cl^-, C_2H_2...HF and HF...C_2H_2...Cl^- species, s-character, polarization of the bond participating in an interaction, and NBO charges (in au, presented below the *pictures*) are given. The electron density at BCP corresponding to interaction, ρ_{BCP} (au), is included and the binding energy, E (kcal/mol)

C_2H_2...HF T-shaped complex the complexation leads to the increase of the positive charge of hydrogen atom of the HF molecule from 0.547 to 0.556 au and the increase of the negative charge of the F-atom from −0.547 to −0.568 au. Additionally the polarization of the H–F bond and s-character of the F-hybrid orbital increase since for the isolated HF molecule they are equal to 77.58 and 20.67 %, respectively, while these values in the C_2H_2...HF complex amount 78.61 and 23.35 %, respectively. The binding energy for this complex is equal to −3.2 kcal/mol (corrected for BSSE); there is the bond path corresponding to the intermolecular H...π contact; π denotes the bond critical point of the C≡C bond of acetylene. The electron density at this BCP, ρ_{BCP}, is equal to 0.016 au. It is worth mentioning that the ρ_{BCP} value is often treated as a measure of the hydrogen bond strength since it correlates for numerous samples of complexes with the binding energy; especially, such correlations were observed for the homogeneous (structurally related) samples of complexes [8].

For the HF...C_2H_2...Cl^- triad the additional C–H...Cl^- interaction enhances the F–H...π hydrogen bond, thus the positive charge of the H-atom of HF molecule amounts 0.570 au while the negative charge of fluorine is equal to −0.591 au. The polarization and s-character are greater than for the isolated HF as well as than for the C_2H_2...HF complex, 79.66 and 24.70 %, respectively. One can see that the hyperconjugative and rehybridization processes known for the hydrogen bonded systems exist also for the FH...π hydrogen bond considered here and that these processes are enhanced in a case of cooperativity. The binding energy related to the F–H...π interaction in the triad is equal to −8.5 kcal/mol, the corresponding ρ_{BCP} value is equal to 0.025 au. One can see that evidently, the additional C–H...Cl^- interaction enhances the F–H...π one. The binding energy corresponding to the F–H...π hydrogen bond in this triad is calculated as the difference between the energy of the triad in its energetic minimum and the sum of the energies of HF monomer and HCCH...Cl^- dyad; the monomer and the dyad correspond to their energetic minima, BSSE correction was included here.

Let us consider the influence of FH...π hydrogen bond on the C–H...Cl^- interaction. One can see that the H and C atomic charges for the isolated acetylene molecule are equal to 0.233 au and −0.233 au, respectively. For the C_2H_2...Cl^- linear complex the positive charge of the hydrogen being in contact with the chlorine ion increases to 0.276 au but the carbon of the C–H proton donating bond is less negative (−0.174 au) than in the isolated acetylene. However the next carbon is much more negative (−0.365 au); this means that there is the electron charge transfer from the chlorine ion to acetylene as a result of complexation and next the triple C≡C bond is an excellent electron charge transmitter. The polarization of the C–H bond and the s-character of the carbon hybrid orbital for the isolated acetylene are equal to 61.24 and 48.14 %, respectively; the corresponding values for the C_2H_2...Cl^- complex amount 65.31 and 50.65 %, respectively. The binding energy for this complex is equal to −8.6 kcal/mol (corrected for BSSE); the electron density at the BCP corresponding to the H...Cl bond path is equal to 0.021 au.

Let us look at the HF...C_2H_2...Cl^- triad to analyze the influence of the F–H...π interaction on the C–H...Cl^- hydrogen bond. For the acetylene molecule in the triad there are the following atomic charges, 0.283 au for hydrogen, more than for isolated acetylene and more than in the C_2H_2...Cl^- diad, and the C-atoms charges of −0.188 au and −0.371 au, more negative than for the dyad. The electron charge transmission effect through the C≡C bond described earlier for the dyad is also detected here. If one considers the C–H proton donating bond of the C–H...Cl^- hydrogen bond in the triad thus the polarization of the bond and s-character of carbon are equal to 66.44 and 51.39 %, respectively, more than for the isolated acetylene and more than for the C_2H_2...Cl^- complex. The binding energy corresponding to the C–H...Cl^- hydrogen bond in this triad is calculated as the difference between the energy of the triad in its energetic minimum and the sum of the energies of Cl^- ion and the C_2H_2...HF dyad; the monomer and the dyad correspond to their energetic minima, BSSE correction was also included here. It is equal to −13.9 kcal/mol and the electron density at the corresponding BCP amounts 0.033 au; hence there is the enhancement of the H...Cl^- interaction in the triad in comparison with the dyad. One can see that also in this case the cooperativity may be treated as an effect which enhances the hyperconjugative and rehybridization processes.

This is worth to mention that the same dependences are detected for the HF... HCCCl...Cl^- triad where the F–H...π hydrogen bond is enhanced by the C–Cl... Cl^- halogen bond and vice versa, the C–Cl...Cl^- interaction is enhanced by the F–H...π hydrogen bond [105]. Figure 9.19 presents the molecular graphs of the HF

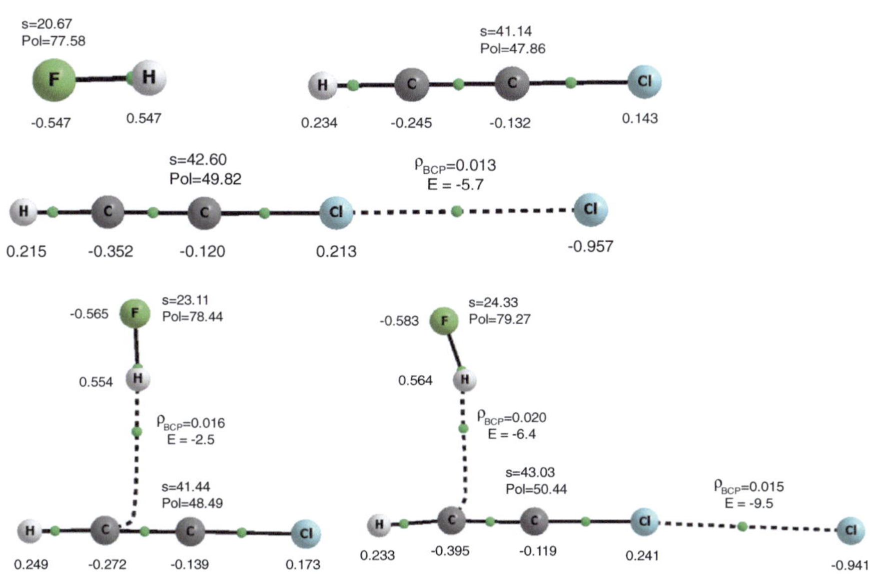

Fig. 9.19 The molecular graphs of HF, HCCCl, HCCCl...Cl^-, HCCCl...HF and HF...HCCCl... Cl^- species, s-character, polarization of the bond participating in an interaction, and NBO charges (in au, presented below the *pictures*) are given. The electron density at BCP corresponding to interaction, ρ_{BCP} (au), is included and the binding energy, E (kcal/mol)

and HCCCl monomers, the HCCCl...HF and HCCCl...Cl$^-$ dyads, and the HF... HCCCl...Cl$^-$ triad. The similar changes as those found for the HF...C_2H_2...Cl$^-$ triad are observed here for the s-character and polarization (see Fig. 9.19). Let us mention only binding energies and the values of the electron densities at BCPs. For the HCCCl...HF dyad the binding energy is equal to −2.5 kcal/mol and the value of the corresponding ρ_{BCP} is equal to 0.016 au; the increase of the latter values are observed in a case of the additional interaction with Cl$^-$ anion in the HF... HCCCl...Cl$^-$ triad; to −6.4 kcal/mol (the absolute value increases) and 0.020 au. In a case of the HCCCl...Cl- dyad linked through the halogen bond the binding energy and ρ_{BCP} are equal to −5.7 kcal/mol and 0.013 au, respectively; they are equal to −9.5 kcal/mol and 0.015 au in the triad.

One can see that if there is any system possessing the Lewis acid and Lewis base properties, like C_2H_2 or HCCCl considered here, and if this system acts as the Lewis base within the complex (the C_2H_2...HF and HCCCl...HF complexes considered here) or as the Lewis acid (the C_2H_2...Cl$^-$ and HCCCl...Cl$^-$ complexes considered here) thus the additional interaction where the C_2H_2 or HCCCl species acts as the Lewis acid or Lewis base, respectively, enhances the former interaction. The other triads linked through the hydrogen and halogen bonds were also analyzed in terms of the cooperativity effects and the same clear conclusions were stated for them [105].

9.5 Summary

There are different types of the Lewis acid–Lewis base (La–Lb) interactions; one may mention the hydrogen, lithium, pnicogen, chalcogen, triel bonds and numerous others. The part of these interactions is classified as the σ-hole bonds, or the π-hole bonds. However for all of them there is the characteristic electron charge transfer from the Lewis base unit to the Lewis acid. One may state that such an electron charge transfer, if it is not negligible, is a signature of the La–Lb interactions for which contacts between the atomic centers characterized by regions of the opposite electrostatic potentials are observed. In such a way the systems stabilized mainly by the dispersive forces such as the dimer of methane or noble gas species are not classified as La–Lb complexes.

It is interesting to know if there are other characteristics common for all Lewis acid–Lewis base interactions. It is difficult to point out general statements. Even for one type of La–Lb interaction, like for example for the hydrogen bond, it is difficult to indicate common characteristics. However few trends or mechanisms which most often are observed for the large part of La–Lb interactions could be listed. Few common characteristics are observed for the A–Y bond of the Lewis acid unit if this bond is in contact with the Lewis base center (in other words for the A–Y...B interactions, see Scheme 9.4). The increase of the polarization of the A–Y bond is the result of complexation (% of the electron charge density calculated at A-center) what comes from the increase of the positive charge of Y-atom and the increase of the negative charge of A-center; additionally the increase of the s-character of the A-hybrid orbital of the A–Y bond is observed. All those changes may be described as coming from two general mechanisms firstly described for the A–H…B hydrogen bonds [84, 86]; the hyperconjugative A–Y(A–H) bond weakening and the rehybridization-promoted A–Y (A–H) bond strengthening.

Acknowledgments Financial support comes from Eusko Jaurlaritza (IT588-13) and the Spanish Office for Scientific Research (CTQ2012-38496-C05-04). Technical and human support provided by Informatikako Zerbitzu Orokora—Servicio General de Informática de la Universidad del País Vasco (SGI/IZO-SGIker UPV/EHU), Ministerio de Ciencia e Innovación (MICINN), Gobierno Vasco Eusko Jaurlanitza (GV/EJ), European Social Fund (ESF) is gratefully acknowledged.

References

1. Hobza P, Müller-Dethlefs K (2010) Non-covalent interactions, theory and experiment. Royal Society of Chemistry, Thomas Graham House, Science Park, Milton Road, Cambridge
2. Pauling L (1960) The nature of the chemical bond, 3rd edn. Cornell University Press, Ithaca, New York, p 8
3. International Union of Pure and Applied Chemistry, Compendium of Chemical Terminology, Gold Book, Version 2.3.3 2014-02-24, p 344, http://goldbook.iupac.org/C01384.html
4. Warshel A (1991) Computer modeling of chemical reactions in enzymes and solution. Wiley, New York
5. Perrin CL, Nielson JB (1997) Annu Rev Phys Chem 48:511
6. Nadal-Ferret M, Gelabert R, Moreno M, Lluch JM (2014) J Am Chem Soc 136:3542
7. Sobczyk L, Grabowski SJ, Krygowski TM (2005) Chem Rev 105:3513
8. Grabowski SJ (2011) Chem Rev 11:2597
9. Morokuma K, Kitaura K (1980) Molecular interactions. In: Ratajczak H, Orville-Thomas WJ (eds) vol 1. Wiley, New York, pp 21–66
10. Sokalski WA, Roszak S, Pecul K (1988) Chem Phys Lett 153:153
11. Sokalski WA, Roszak S (1991) J Mol Struct (THEOCHEM) 234:387
12. Coulson CA (1952) Valence. Oxford University Press, Oxford
13. Pimentel GC, McClellan AL (1960) The hydrogen bond. W.H.Freeman and Company, San Francisco and London
14. Lewis GN (1923) Valence and the structure of atoms and molecules. Chemical Catalog, New York
15. Sundberg MR, Uggla R, Viñas C, Teixidor F, Paavola S, Kivekäs R (2007) Inorg Chem Commun 10:713
16. Tschirschwitz S, Lönnecke P, Hey-Hawkins E (2007) Dalton Trans 1377
17. Murray JS, Lane P, Politzer P (2007) Int J Quant Chem 107:2286
18. Bauer S, Tschirschwitz S, Lönnecke P, Franck R, Kirchner B, Clark ML, Hey-Hawkins E (2009) Eur J Inorg Chem:2776
19. Del Bene JE, Alkorta I, Sánchez-Sanz G, Elguero J (2011) J Phys Chem A 115:13724
20. Scheiner S (2011) Chem Phys Lett 514:32
21. Guan L, Mo Y (2014) J Phys Chem A 118:8911
22. Rozas I, Alkorta I, Elguero J (1997) J Phys Chem A 101:4236
23. Grabowski SJ, Sokalski WA, Leszczynski J (2006) Chem Phys Lett 422:334
24. Sanz P, Mó P, Yañez M (2002) J Phys Chem A 106:4661
25. Wang W, Ji B, Zhang Y (2009) J Phys Chem A 113:8132
26. Alikhani E, Fuster F, Madebene B, Grabowski SJ (2014) Phys Chem Chem Phys 16:2430

27. Lipkowski P, Grabowski SJ, Leszczynski J (2006) J Phys Chem A 110:10296
28. Politzer P, Murray JS (2013) ChemPhysChem 14:278
29. Jeffrey GA, Saenger W (1991) Hydrogen bonding in biological structures. Springer, Berlin
30. Jeffrey GA (1997) An introduction to hydrogen bonding. Oxford University Press, New York
31. Desiraju GR, Steiner T (1999) The weak hydrogen bond in structural chemistry and biology. Oxford University Press Inc., New York
32. Scheiner S (1997) Hydrogen bonding: a theoretical perspective. Oxford University Press, New York
33. Grabowski SJ (ed) (2006) Hydrogen bonding—New insights, vol 3: Challenges and advances in computational chemistry and physics, Leszczynski J (ed). Springer
34. Nishio M, Hirota M, Umezawa Y (1998) The CH/π interaction, evidence, nature, and consequences. Wiley-VCH, New York
35. Szymczak JJ, Grabowski SJ, Roszak S, Leszczynski J (2004) Chem Phys Lett 393:81
36. Grabowski SJ, Sokalski WA, Leszczynski J (2006) Chem Phys Lett 432:33
37. Grabowski SJ (2007) J Phys Chem A 111:3387
38. Grabowski SJ (2013) J Phys Org Chem 26:452
39. Jucks KW, Miller RE (1987) J Chem Phys 87:5629
40. Moore DT, Miller RE (2003) J Chem Phys 118:9629
41. Moore DT, Miller RE (2003) J Phys Chem A 107:10805
42. Moore DT, Miller RE (2004) J Phys Chem A 108:1930
43. Bieske EJ, Nizkorodov SA, Bennett FR, Maier JP (1995) J Chem Phys 102:5152
44. Custelcean R, Jackson JE (2001) Chem Rev 101:1963
45. Murray JS, Lane P, Politzer P (2009) J Mol Model 15:723
46. Politzer P, Murray JS, Clark T (2010) Phys Chem Chem Phys 12:7748
47. Politzer P, Murray JS, Clark T (2013) Phys Chem Chem Phys 15:11178
48. Bundhun A, Ramasami P, Murray JS, Politzer P (2013) J Mol Model 19:2739
49. Mani D, Arunan E (2013) Phys Chem Chem Phys 15:14377
50. Grabowski SJ (2014) Phys Chem Chem Phys 16:1824
51. Bauzá A, Mooibroek TJ, Frontera A (2013) Angew Chem Int Ed 52:12317
52. McDowell SAC (2014) Chem Phys Lett 598:1
53. Li Q, Guo X, Yang X, Li W, Cheng J, Li H-B (2014) Phys Chem Chem Phys 16:11617
54. Metrangolo P, Resnati G (2001) Chem Eur J 7:2511
55. Cavallo G, Metrangolo P, Pilati T, Resnati G, Sansotera M, Terraneo G (2010) Chem Soc Rev 39:3772
56. Wang L, Gao J, Bi F, Song B, Liu C (2014) J Phys Chem A 118:9140
57. Metrangolo P, Resnati G (eds) (2008) Halogen bonding, fundamentals and applications. Springer, Berlin
58. Grabowski SJ (2014) Chem Phys Lett 605–606:131
59. Wong R, Allen FH, Willett P (2010) J Appl Crystallogr 43:811
60. Formigué M, Batail P (2004) Chem Rev 104:5379
61. Murray JS, Riley KE, Politzer P, Clark T (2010) Aust J Chem 63:1598
62. Lipkowski P, Grabowski SJ (2014) Chem Phys Lett 591:113
63. Clark T (2013) WIREs Comput Mol Sci 3:13
64. Tsirelson VG, Ozerov RP (1996) Electron density and bonding in crystals. Institute of Physics, Bristol
65. Coppens P (1997) X-Ray charge densities and chemical bonding. Oxford University Press, IUCr
66. http://www.ccdc.cam.ac.uk/Lists/ResourceFileList/2014_stats_entries.pdf
67. Grabowski SJ (2014) ChemPhysChem 15:2985
68. Weinhold F, Landis C (2005) Valency and bonding, a natural bond orbital Donor–Acceptor perspective. Cambridge University Press, Cambridge
69. Yañez M, Sanz P, Mó O, Alkorta I, Elguero J (2009) J Chem Theor Comput 5:2763
70. Martín-Sómer A, Lamsabhi AM, Mó O, Yañez M (2012) Comput Theor Chem 998:74
71. Mó O, Yañez M, Alkorta I, Elguero J (2012) J Chem Theor Comput 8:2293

72. Albrecht L, Boyd RJ, Mó O, Yañez M (2012) Phys Chem Chem Phys 14:14540
73. Kollman PA, Liebman JF, Allen LC (1970) J Am Chem Soc 92:1142
74. Ammal SSC, Venuvanalingam P (1998) J Chem Soc, Faraday Trans 94:2669
75. Ammal SSC, Venuvanalingam P (2000) J Phys Chem A 104:10859
76. Arunan E, Desiraju GR, Klein RA, Sadlej J, Scheiner S, Alkorta I, Clary DC, Crabtree RH, Dannenberg JJ, Hobza P, Kjaergaard HG, Legon AC, Mennucci B, Nesbitt DJ (2011) Pure Appl Chem 83:1637
77. Taylor R, Kennard O (1982) J Am Chem Soc 104:5063
78. Suttor DJ (1963) J Chem Soc 1105
79. Desiraju GR (2002) Acc Chem Res 35:565
80. Pinchas S (1955) Anal Chem 27:2
81. Schneider WG, Bernstein HJ (1956) Trans Faraday Soc 52:13
82. Trudeau G, Dumas JM, Dupuis P, Guerin M, Sandorfy C (1980) Topics of Curr Chem 93:91
83. Hobza P, Havlas Z (2000) Chem Rev 100:4253
84. Alabugin IV, Manoharan M, Peabody S, Weinhold F (2003) J Am Chem Soc 125:5973
85. Reed AE, Curtiss LA, Weinhold F (1988) Chem Rev 88:899
86. Weinhold F, Klein R (2012) Mol Phys 110:565
87. Alkorta I, Elguero J, Grabowski SJ (2008) J Phys Chem A 112:2721
88. Richardson TB, de Gala S, Crabtree RH (1995) J Am Chem Soc 117:12875
89. Wessel J, Lee JC Jr, Peris E, Yap GPA, Fortin JB, Ricci JS, Sini G, Albinati A, Koetzle TF, Eisenstein O, Rheingold AL, Crabtree RH (1995) Angew Chem Int Ed Engl 34:2507
90. Crabtree RH, Siegbahn PEM, Eisenstein O, Rheingold AL, Koetzle TFA (1996) Acc Chem Res 29:348
91. Platts JA, Howard ST, Bracke BRF (1996) J Am Chem Soc 118:2726
92. Bader RFW (1985) Acc Chem Res 18:9
93. Bader RFW (1990) Atoms in molecules, a quantum theory. Oxford University Press, Oxford
94. Koch U, Popelier PLA (1995) J Phys Chem 99:9747
95. Grabowski SJ (2013) Phys Chem Chem Phys 15:7249
96. Glendening ED (2005) J Phys Chem A 109:11936
97. Bent HA (1961) Chem Rev 61:275
98. Grabowski SJ (2011) J Phys Chem A 115:12789
99. Grabowski SJ (2011) J Phys Chem A 115:12340
100. Grabowski SJ (2012) J Phys Chem A 116:1838
101. Grabowski SJ (2013) J Mol Model 19:4713
102. Alkorta I, Blanco F, Deyà PM, Elguero J, Estarellas C, Frontera A, Quiñonero D (2010) Theor Chem Acc 126:1
103. Kar T, Scheiner S (2004) J Phys Chem A 108:9161
104. Hunt SW, Higgins KJ, Craddock MB, Brauer CS, Leopold KR (2003) J Am Chem Soc 125:13850
105. Grabowski SJ (2013) Theor Chem Acc 132:1347

Chapter 10
Iodine Containing Drugs: Complexes of Molecular Iodine and Tri-iodide with Bioorganic Ligands and Lithium Halogenides in Aqueous Solutions

Gulnara A. Yuldasheva, Georgii M. Zhidomirov, Jerzy Leszczynski and Aleksandr I. Ilin

Abstract This chapter reviews the results of molecular modeling of iodine-containing drugs. They are active ingredients of mixtures that in aqueous solutions consist of molecular iodine, bio-organic ligands, and potassium and lithium halogenides. In these drugs molecular iodine is in such an active form that after oral administration it minimizes toxic effects in humans. Previously it was shown that the active complex (AC) of the drugs contains molecular iodine that is located inside α-helix of dextrin and is coordinated by lithium halides and polypeptides (LiI_5-α-dextrin polypeptide). In these types of complexes the electronic structure of the I_2 molecule is different from the electronic structure of I_2 in complexes with organic ligands, or in its free state. Interestingly, in the AC the molecular iodine exhibits acceptor properties with respect to polypeptides, and donor properties with respect to lithium halide. Our group was the first to propose the molecular model of active complexes of the iodine-containing drugs. This was based on the results of calculations performed using the DFT-B3PW91/midi approach. Model system of the water-glycine KI_3-LiCl-ethanol was considered in this study. The calculations of the spectral parameters of the proposed structures are in good agreement with the experimental data of UV and IR spectral investigations. We have shown that α-dextrins ensure the presence in the studied mixtures of the three active centers located within the α-dextrin helix: molecular iodine coordinated lithium halogenides and polypeptides, triiodide, and lithium halogenides. Using UV spectroscopy, the interaction of α-dextrin-LiCl(I)-I_2-polypeptid with the AGA nucleotide triplet was investigated. Comparison of the quantum chemical calcula-

G.A. Yuldasheva (✉) · A.I. Ilin
Scientific Centre for Anti-infectious Drugs, Almaty, Kazakhstan
e-mail: yuldasheva57@rambler.ru

G.M. Zhidomirov
Boreskov Institute of Catalysis SB RAS, Novosibirsk, Russia

J. Leszczynski
Interdisciplinary Nanotoxicity Center, Jackson State University, Jackson, MS, USA

J. Leszczynski and M.K. Shukla (eds.), *Practical Aspects of Computational Chemistry IV*, DOI 10.1007/978-1-4899-7699-4_10

tions carried out for electronic transitions obtained for the structure that models the interaction of α-dextrin-LiCl(I)-I_2-polypeptid with the nucleotide triplet indicates that the DNA nucleotides can displace polypeptide and form stable complexes with molecular iodine and lithium halogenides. In such structures, molecular iodine binds both the nucleotide triplet and lithium halogenides. We have shown that the presence of molecular iodine is vital for activity of compounds that inhibit the active site of HIV-1 integrase. Iodine prevents the active site of integrase from the formation of a complex with HIV DNA and inhibits the active complex of integrase and viral DNA, becoming the center of another nucleoprotein complex, and binding together the active site of integrase and viral DNA.

10.1 Introduction

Iodine is known for its broad spectrum of antimicrobial and antiviral activity; it kills all known microflora. The need to unveil the details of the interaction of iodine complexes with biomolecules has grown considerably in recent years. On the one hand, this is due to the search for the interrelationship between the characteristics of donor-acceptor complexes and physiological and biological activity of their constituent molecules. On the other hand, it is because of successful developments of a number of biologically active compounds and drugs containing iodine complexes.

There are some general outlines related to composition of complex compounds of iodine. The following example demonstrates components of a species that represents such a complex. They include: DI_2 outer-sphere complexes, formed due to the donor-acceptor interaction of iodine with organic compounds (D); iodine halogenides of organic cations such as $CtXI_2$ (Ct is an organic cation, XI_2 (X = Cl, Br, I)), iodine-containing complex anions, and organic cations polyiodide CtI_{2n+1} (n = 2–4), in which the complexant is iodide or triiodide, while iodine molecules serve as ligands.

A great structural variety of solid and liquid polyiodides has been demonstrated in [1]. This review covers research activities related to the chemistry of polyiodide for the period from 1970 to 2002. The formation of anions containing more than one molecule of iodine is considered as resulting from the interaction between the components of basic building blocks (I^-, I_2, I_3^-).

Higher polyiodide ions have a more complicated structure than I_3^-. For example, I_5^- anion is an angular species, which can be regarded as formed by the coordination of two iodine molecules by iodide ion. The heptaiodide ion can be described as $[(I_3^-)\,2I_2]$, $[(I_3^-)\,3I_2]$, $[(I_5^-)\,I_2]$, with a more or less distorted pyramidal geometry or Z-shape.

In recent works [2, 3] the methods of molecular dynamics and X-ray photoelectron spectroscopy were used to study the effect of the type of solvent on the I_3^- geometric structure. One finds the solvent effects on the I_3^- to be unusually strong as it is a highly polarizable species. Protic solvent such as water, ethanol, and methanol that can form hydrogen bonds provide interactions that lead to the I_3^- geometry with two unequal bonds and asymmetric distribution of charges. However, for the solvents such as xenon, tetrahydrofuran, methyltetrahydrofuran and acetonitrile the I_3^- adopts only geometry with two equal bonds.

Of a particular interest is the study of the structure/function/properties in the poly halogenide molecules of cations of nitrogen-containing heterocycles that exhibit high biological activity. Among such species are diodhlorid (I_2Cl^-), diodbromid (I_2Br^-), and triiodide (I_3^-) N-cetyl (decyl) pyridinium that were synthetized and isolated with the aim to synthetically produce drugs with a predetermined ability to release molecular iodine. X-ray analysis of triiodide and iodine halogenides of N-cetylpyridinium shows that the crystal structure of organic polyhalides has a layered packaging with uniformly alternating layers of nitrogen-containing cations and iodine halogenide anions [4].

By now a significant amount of research related to the interaction of halogen molecules as electron acceptors with organic ligands and polymers has been carried out [5–11]. Halogen molecules form charge transfer complexes with organic ligands. In such complexes a transfer of electron density from the lone pair electron of the heteroatom of the organic ligand (σ-donors), or a transfer of π-electron density of the ligand (π-donor) to the molecule of halogen occurs. Alkylamines, nitrogen heterocycles, nitriles, alcohols, ethers, carbonyl compounds, sulfides, selenides, and organophosphorus compounds are classified as σ-donors, while arenes and some of heterocycles fall under the category of π-donors.

Interestingly, a three-stage mechanism of interaction of iodine with organic bases has been proposed based on the results of quantum chemical and spectrophotometric studies [12–14]. The first, fast stage involves the formation of the D-I-I outer-sphere complex. At the second, slower stage the [D-I^+] I^- inner-sphere complex is formed, and at the third stage the inner-sphere complex interacts with molecular iodine to form [D-I^+(I_3^-)].

At the stage of formation of the D-I-I outer-sphere complex a charge transfer from the lone electron pair of the heteroatom of the organic ligand to the molecule of iodine occurs, and this leads to the polarization of the I-I bond, or even to its rupture. The I_2 molecule in the D-I-I complex already displays donor properties and can transfer its charge to another I_2 molecule.

The stability of the D-I-I complex and the cleavage of the I-I bond depend on the donor activity of the ligand and the solvent. The formation of a complex between I_2 and three pentadentate Schiff bases, 1,3-bis (salicylideneamino)-2-propanol, 1,3-bis (2-hydroxy-1-naphthylideneamino)-2-propanol, and 1,3-bis [1-(pyridine-2-yl) methylideneamino]-2-propanol results only in polarization of the I-I bond [13]. The formation of a complex between pyridine and molecular iodine (I_2) leads to the polarization of the I-I bond in some non-polar solvents such as liquid alkanes and is accompanied by cleavage of the I-I bond in low-density polyethylene [14].

Spectroscopic analysis has demonstrated that many nitrogen-containing heterocyclic drugs form charge transfer complexes with iodine [15, 16]. UV-spectrophotometric study provides an estimation method therapeutic drugs possessing antifungal, antihistamine, β-adregenic, antidepressant properties from study of their complex with iodine [15].

Interestingly, the value of formation constant K_c of a complex with iodine has been used as a measure for evaluation of anti-thyroid activity of such compounds. The complexes of morpholine with iodine were shown to be of the n-σ type with a 1:1 stoichiometry. A strong donor-acceptor interaction was found in this complex ($Kc = 1261 \pm 12\ mol^{-1}$ at 20 °C in CCl_4). This is considerably higher than analogous characteristics of complexes of aromatic compounds with iodine. The high value of the formation constant for this complex indicates that morpholine could serve as a starting point for the synthesis of novel anti-thyroid drugs [16].

Pharmaceutical compounds containing complexes of molecular iodine have a wide spectrum of antibacterial and antiviral activity, including HIV. However, all known iodine-containing drugs are characterized as highly toxic substances and therefore in fact are not used in medical practice for parenteral application.

Among the drugs that comprise iodine complexes the species containing not only molecular iodine complexes with organic ligands, but also potassium and lithium halogenides stand out considerably [17–19]. These drugs have a low toxicity and can be used internally. Armenicum is a name of a drug used to treat HIV infection [17, 18]. Recently a new anti-infective drug with anti-HIV action has been patented (AID) [19]. Armenicum is composed of the LiI_5-α-dextrin complex. The AID contains polypeptides along with the LiI_5-α-dextrin complex (LiCl(I)-I_2 DP).

Recently we have shown that the active complex (AC) of preparations contains molecular iodine located inside the α-dextrin helix and coordinated by lithium halogenides and polypeptides [20]. In the AC [17–20], molecular iodine forms a complex with the polypeptides and halogenides of alkali metals. This complexed molecular iodine shows acceptor properties with respect to polypeptides and donor properties towards lithium halogenides.

Moreover, for the first time we proposed a model of active complexes (AC) of the iodine-containing drugs [21]. This has been a result of investigation of the following systems: water-glycine (a), water-glycine-KI_3 (b), water-glycine-KI_3-LiCl-ethanol (c). The study was performed using a combination of experimental techniques (UV-IR spectroscopy), and quantum-chemical calculations (DFT-B3PW91/midi level).

As shown by the results of this combined investigation in an aqueous solution the zwitterionic form of amino acids solvated by water molecules is more stable than the neutral form [22–27]. The activation energy required for the neutral form of glycine to transfer into the zwitterionic in water amounts to 16.85 kcal/mol [22].

However, the localization of negative and positive charges at the ends of the zwitterion creates conditions for the formation of strong hydrogen bonds not only with water molecules, but also among zwitterions. The results of an X-ray diffraction test for the crystals of L-DL-alanine in their zwitterionic form have been recently published [28]. In the crystal structure of these alanine forms there is a common head-to-tail structural element, where the carboxyl group is bounded with the protonated amino group by hydrogen bonds.

Interestingly, also the possibility of formation of a zwitterion cluster in water has been presented [21]. In this cluster glycine molecules are bonded by strong hydrogen bonds and are located one under the other, so that the protonated amino group is always under or above the carboxyl group.

The first solvation shell of eight molecules of glycine was proposed for such a cluster. With involvement of all hydrogen bonds one derives a structure where two water molecules interact with a Gly-Gly pair, while Gly-Gly pairs are bonded with each other by one molecule of water.

In the water-glycine-KI_3-LiCl-ethanol system the cluster of glycine zwitterions splits iodide ion I_3^- into I^- and I_2 fragments (I^- interacts with the protonated amino group, while I_2 interacts with the carboxyl group). This creates conditions for the formation of iodine complex compound, in which molecular iodine exhibits the acceptor properties with respect to glycine and the donor properties with respect to LiCl-ethanol complex. The UV and IR spectra of the systems (a–c) are in good agreement with the results of quantum-chemical calculations (at the DFT (B3PW91/midi level). The results reveal spectral characteristics of structures that simulate the effect of the zwitterionic form of glycine on the complex compounds of iodine in the model systems (a–c).

The analysis of UV spectra indicates that in an aqueous solution of KI-I_2-amylose the number of rings of the dextrin helix may affect the iodine-triiodide equilibrium [29]. With increase in the number of rings ($N \geq 15$) UV spectra detect existence of complexes of iodine and triiodide. The α-dextrin which is a part of the drugs [17–20] has 15 or more rings, therefore complexes of iodine and triiodide could exist inside the helix.

As shown in [21], the cluster of glycine zwitterions, like the dextrin helix, creates conditions for the presence of I_2, I^-, I_3^- in solution. It implies that the discussed previously mixture (c) can be considered as a model for the drugs [17–20].

This chapter reviews the effect of alkali metal halogenides and zwitterionic form of glycine on the iodine-triiodide equilibrium in aqueous solution and the interaction of I_2, and I_3^- with lithium halogenides (LiCl). It reveals and analyzes the results obtained by experimental studies using UV and IR spectroscopy, as well as the outcomes of theoretical investigations. Computational studies have been carried out using quantum-chemical methods of DFT-B3PW91/midi, explicit inclusion of solvent molecules, as well as by application of COSMO continuum model [30].

10.2 Computational Details

An adequate description of the structure and electronic properties of iodine complexes with the transfer of charge and polyiodide ions requires taking into account of electron correlation. A detailed comparative analysis of various quantum chemical approaches including CCSD(T) and DFT (BLYP, BPW91, B3LYP, B3PW91) for calculations of dissociation energies, bond lengths, and harmonic frequencies for polyiodide anions and evaluation of the obtained data based on agreement with experimental IR and Raman Spectroscopy studies is presented in [31]. Comparison of the results obtained by DFT-methods with the results obtained by the CCSD(T) approach and experimental data shows that the most reliable results are produced by the DFT-B3PW91 level of theory. Based on these conclusions the DFT-B3PW91/midi approximation was selected in our investigations of molecular structures and characteristics of the complexes containing lithium halides and iodide anion.

For the systems (a–c) with full geometry optimization at the DFT-B3PW91/midi level the spatial structure of the most stable complexes of iodine and triiodide was calculated. Using the optimized structures as the references, in the next step vibrational frequencies required for the interpretation of IR spectra were calculated at the same level of theory.

COSMO [30] was used to model effect of water solution. Gibbs free energy (ΔG) of the most stable structures was defined in terms of the COSMO model. To study electronic properties of the considered system the wavelength of the specify electronic transitions were calculated by the TD-DFT/B3PW91 approach.

In solving the equation for eigenfunctions and eigenvalues in the framework of the TD-DFT method the convergence of eigenvalue matrix diagonalization depends on the size of the matrix. In the case of complexes containing six molecules of glycine the application of TD-DFT/B3PW91 has not resulted in convergence of calculations. Therefore, to overcome this problem and to calculate the wavelength of electron transitions for the most stable structures obtained with full geometry optimization a model system that approximates the studied species was considered. It consists of two pairs of Gly-Gly, interacting with I^- and I_2 that were modeled without geometry optimization. The calculations were performed using GAUSSIAN09 [32].

10.3 Results and Discussion

Interestingly, the solvent effects and the role of organic compounds on the equilibria in iodine triiodide solutions were studied as early as at the beginning of last century [33]. Though this subject has already a long history it is still considered to be important and draws attention of various research groups [34–40].

The mixtures containing of iodine-triiodide ion complexes represent complicated chemical systems. Their equilibrium reflects the entire spectrum of complex and varied interactions of iodine and its anions with solvent, dissolved metal ions, and organic or bioorganic compounds present in the solution.

The spectrophotometric properties of I^-, I_2 and I^-/I_2 mixture for system LiI-I_2 were studied in numerous solvents in order to reveal the influence of various solutions on their characteristics. Among the studied solvents were: 1,2-dichloroethane (DCE), acetone (AC), acetonitrile (ACN), ethanol (EtOH), methanol (MeOH), tertiary-butanol (t-BuOH), dimethylformamide (DMF), propylene carbonate (PC), 3-methoxypropionitrile (MePN), dimethylsulfoxide (DMSO), dioxane (DIO), and pyridine (PY) [40]. It was concluded that the basicity of the solvent is the main factor promoting the formation of solvent-I^+ complex, particularly in the presence of a high amount of I_2.

In most Lewis basic solvents tiriodid ion prevails. Triiodide ion is easily formed from I^- and I_2, and the higher the basicity of the solvent, the easier the triiodide ion is determined by UV-spectra. However, in DMF and PC triiodide ion is unstable.

Iodine, although is poorly soluble in water, it is readily soluble in solution in the presence of alkali metal iodides. This is due to the formation of polyiodide complexes.

The thermal effect of the I_3^- forming reaction in an aqueous solution in the presence of alkali metal iodides was evaluated early 60th of the last Century by the colorimetric method [41]. The entropy and enthalpy of the I_3^- formation were calculated. By means of the Raman scattering the related equilibrium rates of forward and reverse reactions of I_3^- formation at the temperature of 298 K and ionic strength of 0.02 were evaluated. They amount to $6.2 \times 10^9\ M^{-1}c^{-1}$, and $8.5 \times 10^6\ M^{-1}c^{-1}$, respectively [42]. These data indicate that in the presence of alkali metal ions the equilibrium is shifted towards the formation of the triiodide ion, so that in comparison I_3^-, molecular iodine and I^- in aqueous solution could hardly be detected. This was confirmed by the results of quantum-chemical calculations. They revealed that in such a solution the equilibrium is shifted toward the formation of poliiodide anions [38].

The polar glycine zwitterion is able to shift the iodine-triiodide equilibrium. We have considered two possible ways of triiodide ion interactions with a cluster that consists of six glycine molecules (Ia, IIa). In the structure Ia, a cluster of glycine zwitterions splits the triiodide ion into I^- ion and molecular iodine. The negatively charged iodine ion interacts with the amino groups of two glycine molecules from different Gly-Gly pairs, and molecular iodine is complexed by the carboxyl group. In the structure represented by IIa, the triiodide ion interacts with the amino groups of two glycine molecules (see Fig. 10.1, structure Ia, IIa).

The geometry of model structures was optimized at the DFT-B3PW91/midi level. The influence of solvent on the stability of model structures was taken into account not only by way of constructing the first solvation shell, but also applying the continuum model. The COSMO method was used to calculate the total energies and Gibbs free energies for the optimized structures. When developing molecular geometry for structures Ia and IIa, it was assumed that I_3^-, I_2, and I^- displaced two water molecules from the coordination sphere. The charge distribution in the cluster zwitterions of glycine was obtained using the COSMO continuum model. Calculations showed that the positive charges on the NH_3^+ hydrogen groups and negative charges on the COO^- groups were distributed unevenly.

Initial assumptions have been made for the location of the mixtures components and starting geometries of the complexes and were set up accordingly. In the IIa mixture the triiodide ion was attached to the hydrogen atom of the amino group with the highest positive charge. As the result of geometry optimization triiodide position was not changed, but in the optimized structure the triiodide interacts with

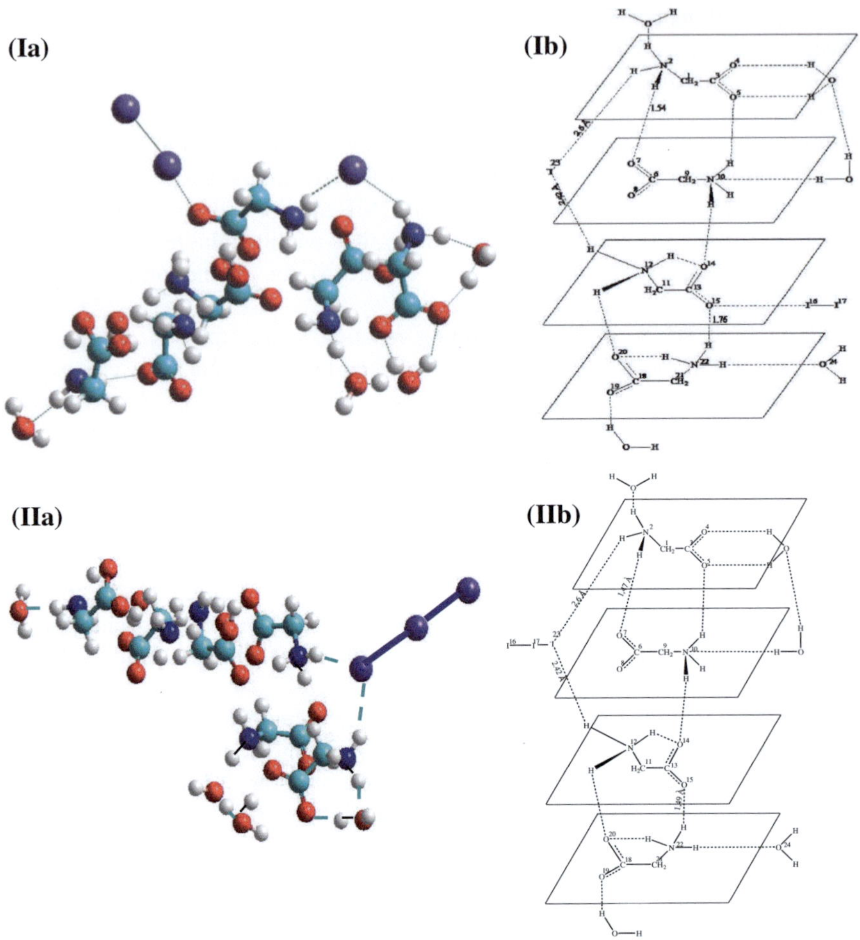

Fig. 10.1 The structure of molecular iodine and triiodide complexes in an aqueous glycine-KI-I_2 solution. Color code: *blue balls*—carbon atoms, *dark blue balls*—nitrogen atoms, *red balls*—oxygen atoms, *violet balls*—iodine atoms

the hydrogen atoms of the two amino groups of Gly-Gly pairs (Fig. 10.1, structure IIa).

When the initial molecular geometry of the complex Ia was set up the data obtained from the previous calculation were taking into account. It includes the distribution of negative charges on the COO-groups in the complex IIa, as calculated by the COSMO model. The molecular iodine was attached to the oxygen atom of the COO^- group with the highest negative charge. As the result of the geometry optimization the position of the molecular iodine did not change.

In contrast with the structure Ia, the enhancement of hydrogen bonds in pairs Gly-Gly was observed in the optimized IIa structure. This is evidenced by a decrease in the interatomic distances r (O_7-H) and r (O_{15}-H) (Fig. 10.1). In the optimized structures Ia and IIa in the Gly-Gly pair, that did not interact with I_3^- and I^-, a transition from the glycine zwitterionic form to the neutral one was observed and a new O-H bond was formed. A new O-H bond was formed in the Gly-Gly pair that did not interact with I_3^- and I^-. We observe a transition the zwitterionic form of glycine to the neutral one.

The small difference in the Gibbs free energy ($\Delta G = G(Ia) - G(IIa) = -4.77$ within B3PW91/midi approximation, and −3.55 kcal/mol value obtained within COSMO approach) for the complexes Ia and IIa makes it possible to assume that these two complexes can be detected by UV and IR methods in an aqueous solution of glycine-KI_3 mixture.

The UV spectrum of the system (b) is presented and discussed in [21]. There are two fairly broad bands recorded at 212–257 nm and 257–319 nm in the UV spectrum of the aqueous glycine-KI_3 solution (Fig. 10.2).

Two distinct bands in the UV-spectrum of experimental system (b) 212–257 nm, 257–319 nm are interpreted in the literature. 212–257 nm correspond to charge transfer transitions in iodine complexes with organic ligands [7]. 257–319 nm correspond to transitions in triiodide complexes with the dextrins [29].

The experimental data could be used to evaluate, validate, and select computational techniques able to accurately predict the investigated species. The wavelength corresponding to the electronic transitions were calculated by the TD-DFT/B3PW91 approach. This approach has some challenges. When solving the equation for the eigenfunctions and eigenvalues within the TD-DFT method the convergence of diagonalization of the eigenvalue matrix depends on the size of the matrix. In our calculations in the case of complexes Ia-IIa no convergence could be achieved using the TD-DFT/B3PW91 method. Therefore some additional approximations were necessary in order to predict the wavelength of the electronic

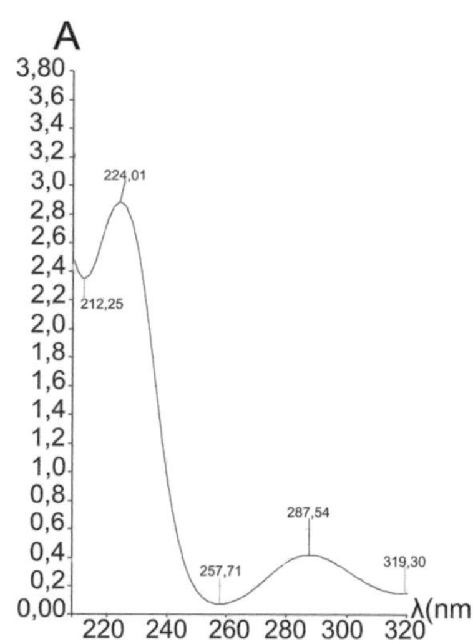

Fig. 10.2 Experimental UV spectrum of system water-glycine-KI_3 (b)

transitions in the structures Ia, IIa; two Gly-Gly pairs interacting with I^- and I_2 (structure Ib, Fig. 10.1); or I_3^- (structure IIb Fig. 10.1). The electronic transitions were calculated for structures Ib and IIb based on optimized geometry of structures Ia and IIa.

The absence of spatial symmetry in our cluster structures greatly interferes with high quality interpretation of calculated optical transitions. Generally speaking, all the electron terms involved in transitions are characterized by a mixed orbital composition of atoms in the structure. However, among the many found transitions first of all we are interested in transitions associated with a change in the electronic states of iodine atoms. It is such transitions that are critical in the diagnosis of the type of binding of the iodine component in the complex. In this case, being guided by the local symmetry of some iodine fragments, we felt comfortable to introduce a symbolic representation of the π- and σ-orbitals of iodine structures. This approximation is justified by the fact that the main contribution to the MO involved in our transitions is made by the orbitals of iodine atoms. Another locally selected group of orbitals is orbitals corresponding to lone pairs electron of heteroatoms of nitrogen and oxygen. Hereinafter we will use these terms to characterize the transitions in question in terms of quality, keeping in mind the conditional nature of assigned notations. Another important fact characterizing our transitions involving iodine atoms is their relatively high oscillator strength associated with the fact that they to a greater or lesser extent are charge-transfer transitions.

The predicted electronic transitions in the Ib complex, described in [21], indicate the formation of a coordination bond between molecular iodine and the glycine carboxyl group. This results in their contribution towards short wave band at 212–257 nm (Table 10.1).

The transitions at 226 and 266 nm can be interpreted as the charge transfer transition from n-pair of oxygen atom to the coordination bond O^{15}-I^{16}-I^{17}. Under the coordination the partial negative charge is transferred to the molecular iodine, as evidenced by the appearance 228 nm corresponds a charge transfer transition from the molecular iodine to the protonated glycine group.

The electronic transitions in the IIb complex make a contribution to a longer-wave region at 257–319 nm (Table 10.2). They correspond to the transitions between the orbitals in the triiodide ion.

Table 10.1 Calculated (TD-DFT/B3PW91 level) wavelengths of electronic transitions (λ, nm) and oscillator strength (I^λ) for the Ib complex [21]

$\lambda^{theor.}$	Assignment	I^λ
266	$^1A \rightarrow {}^1A$ (n(O^{14}, O^{15}) → *(O^{15}-I^{16}-I^{17}))	0.1105
228	$^1A \rightarrow {}^1A$ ((O^{15}-I^{16}-I^{17}) → *($^+N^{22}H_3$, $O^{24}H_2$))	0.1208
227	$^1A \rightarrow {}^1A$ (n(O^7, O^8) → *($N^{22}H_3$ $O^{24}H_2$))	0.1343
226	$^1A \rightarrow {}^1A$ (n (O^{14}, O^{15}) → *(O^{15}-I^{16}-I^{17})) $^1A \rightarrow {}^1A$ ((O^{14}-C^{13}-O^{15}-I^{16}-I^{17}) → *(O^{15}-I^{16}-I^{17})) $^1A \rightarrow {}^1A$ ((O^{14}-C^{13}-O^{15}-I^{16}-I^{17}) → *($^+N^{22}H_3$, $O^{24}H_2$))	0.9881

Table 10.2 Calculated (TD-DFT/B3PW91 level) wavelengths of electronic transitions (λ, nm) and oscillator strength (I^{λ}) for the IIb complex

$\lambda^{theoret.}$	Assignment	I^{λ}
270	$^1A \rightarrow {}^1A$ $((I^{17}\text{-}I^{16}\text{-}I^{23}) \rightarrow *(I^{17}\text{-}I^{16}\text{-}I^{23}))$	0.6430
259	$^1A \rightarrow {}^1A$ $((O^7\text{-}C^6\text{-}O^8)\ (O^{19}\text{-}C^{18}\text{-}O^{20}) \rightarrow *(I^{17}\text{-}I^{16}\text{-}I^{23}))$ $^1A \rightarrow {}^1A$ $((I^{17}\text{-}I^{16}\text{-}I^{23}) \rightarrow *(I^{17}\text{-}I^{16}\text{-}I^{23}))$	1.0892

The bands at 259 and 270 nm can be attributed to the transition between the occupied and the excited orbitals of triiodide. The electronic transition between the orbital with main contribution bond O^7-C^6-O^8, O^{19}-C^{18}-O^{20} of the carboxyl groups and excited orbital of triiodide makes a contribution at 259 nm. Theoretical frequency 270 nm corresponds to the peak (287 nm) at the region 257–319 nm, within the error of applied computational method.

In the recent work [21] only the band at 212–257 nm was interpreted. Interestingly, the calculated data of electronic transitions for the Ib-IIb complexes indicate that the iodine-triiodide equilibrium in the solution glycine-KI-I_2 is reflected in the UV spectrum of the system (b). It is demonstrated by the presence of two bands at 212–257 nm and 257–319 nm, corresponding to the spectrum of iodine-triiodide complexes.

The experimental IR spectrum for the mixture: glycine-KI-I_2 is provided in [21]. For the Ia and IIa structures of their vibrational spectra are predicted at the DFT-B3PW91/midi level. The possibility of the presence of small amounts of water molecules in the structure of solid amino acids is established by IR spectroscopy [43]. Following this suggestion when calculating the vibrational frequencies the water molecules were not eliminated from the computational models of Ia and IIa structures. The details of vibrational frequency analysis for the Ia complex are provided in [21].

In Table 10.3 the most intense IR absorption bands in the system glycine-KI-I_2 are compared with the theoretically calculated frequencies for the Ia, and IIa complexes. One notices that the vibrational frequencies of the two complexes are in good agreement with the experimental frequencies.

In the IR spectra of the aqueous solution of glycine-KI-I_2 in the range of 1600–1300 cm^{-1} the intensive absorption bands at 1580, 1497 and 1407 cm^{-1}, 1322 cm^{-1} are observed. The symmetric and antisymmetric deformation vibrations of NH_3 and symmetric and antisymmetric vibrations of the carboxyl group make a contribution at 1580, and 1497 cm^{-1}. The bands at 1407, and 1322 cm^{-1} may be attributed to the symmetric valence vibration of the carboxyl group.

When comparing the experimental vibrational frequencies of the aqueous glycine solution and the aqueous solution of the system glycine-KI-I_2 as discussed in [21], it was noted that the coordination of the carboxyl group with molecular iodine results in the reduction of the valence vibration frequency of the carboxyl group (from 1332 to 1322 cm^{-1}). As can be seen from Table 10.3, no shift to the short-wave region in the IIb complex is observed. The band at 1322 cm^{-1} is wider than the bands at 1580 and 1497 cm^{-1}, therefore the vibrational frequencies of the

Table 10.3 Experimental (for the system: water-glycine-KI-I_2) and calculated (DFT-B3PW91/midi level) vibrational IR frequencies (cm^{-1}) for complex Ia, IIa and their assignment

Experimental	Ia	Ia Assignment	IIa	IIa Assignment
1580	1592	$\delta(^{+}N^{12}H_3)$ asym $\nu(C^{13}O^{14}O^{15})$ asym	1600	$\delta(^{+}N^{12}H_3)$ asym $\delta(^{+}N^{2}H_3)$ asym
1497	1507	$\delta(^{+}N^{12}H_3)$ sym $\delta(C^{11}H_2)$ sym $\nu(C^{13}O^{14}O^{15})$ asym	1480	$\delta(^{+}N^{12}H_3)$ sym $\nu(C^{6}O^{7}O^{8})$ asym
1407	1410	$\nu(C^{3}O^{4}O^{5})$ sym $\delta(C^{1}H_2)$ sym	1405	$\delta(^{+}N^{2}H_3)$ sym $\nu(C^{3}O^{4}O^{5})$ sym $\nu(C^{1}$-$C^{3})$
1322	1307	$\nu(C^{13}O^{14}O^{15})$ sym	1346	$\nu(C^{18}O^{19}O^{20})$ sym $\delta(^{+}N^{22}H_3)$ sym
1111	1083	$\delta(C^{1}H_2)$ sym $\nu(C^{3}O^{4}O^{5})$ sym	1091	$\delta(^{+}N^{2}H_3)$ $\delta(C^{1}H_2)$
2123	–	–	2123	ν(H-N^{22}-H) asym
2650–2740	2632	$\nu(N^{10}$-H) $\nu(O^{25}$-H) $\nu(O^{26}$-H)	2733	$\nu(N^{22}$-H) $\nu(O^{24}$-H)
2876	2920	$\nu(N^{10}$-H) $\nu(O^{25}$-H) $\nu(O^{26}$-H)	–	
2962	2961	$\nu(N^{2}$-H)	2939	$\nu(N^{2}$-H) $\nu(O^{27}$-H)
3003	3007	$\nu(N^{2}$-H)	3015	$\nu(N^{12}$-H) $\nu(N^{10}$-H)
3090	–	–	3190	$\nu(N^{22}$-H)
3157	3159	$\delta(C^{2}H_2)$ asym	–	
3228	3234	$\nu(N^{22}$-H)	–	

carboxyl group in the complexes Ia (1307) and IIa (1346) fall within the range of its interval.

In the 2500–3500 cm^{-1} region, the intensive vibrations can be distinguished at 2606, 2692, 2822, 2962, 3003, 3156, and 3228 cm^{-1}. They may be attributed to the valence vibrations of N-H bonds of the protonated amino group. The valence vibrations of the O-H groups of the two water molecules bound by hydrogen bonds with NH_3 and COO^{-}-groups also make a contribution within the broad band at 2650–2740 cm^{-1}. It was concluded that the vibrations in this area are observed for all amino acids comprising interactions between NH_3-H_2O and COO-H_2O [43].

The enhancement of hydrogen bond r(O_7-H) in IIa results in the reduced frequency of the valence symmetric vibration of the carboxyl groups (C^6-O^7-O^8) (the frequency at 1480 cm^{-1} corresponds to the experimental frequency at 1496 cm^{-1} in IIa), and the valence vibration frequency of N^2-H (the frequency at 2939 cm^{-1} corresponds to the experimental frequency at 2961 cm^{-1} in IIa).

Based on the obtained data we conclude that the zwitterionic form of glycine significantly influences the iodine-triiodide equilibrium. In the presence of glycine in a solution containing potassium halogenides and molecular iodine, molecular iodine and triiodide complexes can be detected both in the UV and IR spectra. The calculated spectral characteristics for the complexes of molecular iodine and triiodide provide the most complete assignment and description of the experimental UV-IR characteristics for the system (b).

For the water-glycine-KI_3-LiCl-ethanol system (system c), which is of a vital importance since it models the composition of the drugs [17–20], various possible ways of coordination of the molecular iodine and the triiodide ion with lithium halogenides were examined. Based on the developed model geometries the total energies for the complexes IIIa-Va were calculated. In the complex III(a), studied in [21], the molecular iodine interacts with glycine and LiCl-ethanol, revealing its acceptor properties with respect to glycine and the donor properties to LiCl-ethanol. The complexes IV (a) and V (a) correspond to the two possible interaction types of LiCl-ethanol with the triiodide ion-cluster of the glycine zwitterions complex. In one case, the triiodide ion and LiCl-ethanol are spatially separated; I_3^- interacts with the amino groups of two glycine molecules, and LiCl-ethanol with an oxygen atom of the carboxyl group (complex IVa). However, in the second developed model the LiCl-ethanol complex is coordinated with an oxygen atom of the glycine carbonyl group and triiodide ion complex Va (Fig. 10.3).

The initial molecular geometry for the complex IVa was based on the previously determined geometry of the complex IIa. Given the charge distribution in the complex IIa, evaluated using the COSMO model, LiCl was attached to the oxygen atom with the highest negative charge of the COO^- group. A Gly molecule, belonging to one of the two Gly-Gly pairs of Gly-Gly assembly, was assumed to interact with triiodide.

After optimization of the molecular geometry the resulted geometry of IVa displayed the distinctive arrangement of water molecules (Fig. 10.3). In this optimized structure the triiodide interacts with the amino groups of the two Gly-Gly pairs, without participation of water molecules. The location of coordination of LiCl-ethanol has been also altered. A LiCl-ethanol fragment coordinates the carboxy group of the third Gly-Gly pair, which does not interact with I_3^-. A new OH-bond was formed in the glycine molecule, clear of coordination in the Gly-Gly pair that interacted with LiCl-ethanol.

The results of calculations have shown that the most stable complex (IVa) is formed when the ion triiodid and LiCl-ethanol are spatially separated. The Gibbs free energies of complexes IIIa and IVa differ only by 1.47 kcal/mol within B3PW91/midi level predictions and by 5.66 kcal/mol within COSMO approximation. Therefore these two complexes can be detected at the same time in the system (c) using UV and IR methods.

The UV spectrum of the system (c) is presented and discussed in [21]. In the UV spectrum of the aqueous glycine-KI_3-LiCl-ethanol solution it is possible to distinguish three fairly broad bands at 212–257 nm, 257–319 nm, 320–380 nm.

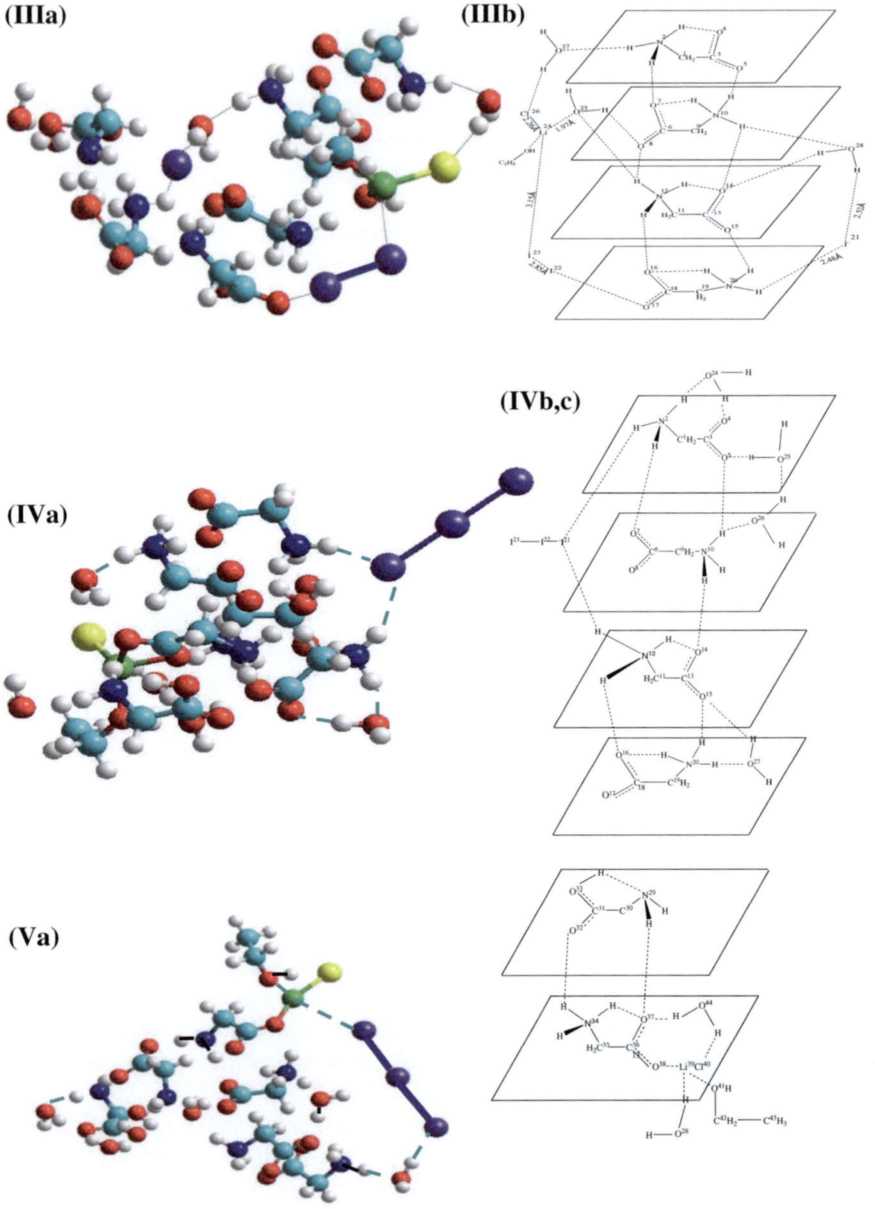

Fig. 10.3 The molecular structures of the possible molecular iodine-triiodide complexes in the aqueous glycine-KI_3-LiCl-ethanol solution. Color codes: *blue balls*—carbon atoms, *dark blue balls*—nitrogen atoms, *red balls*—oxygen atoms, *violet balls*—iodine atoms, *yellow balls*—chlorine atoms, *green balls*—lithium atoms

Table 10.4 Calculated (TD-DFT/B3PW91 level) wavelengths of electronic transitions (λ, nm) and oscillator strength (I^{λ}) for the IIIa complex [21]

$\lambda^{theor.}$	Assignment	I^{λ}
272	$^1A \rightarrow {}^1A\ ((O^{14}\text{-}C^{13}\text{-}O^{15}\text{-}I^{16}\text{-}I^{17}) \rightarrow *(O^{15}\text{-}I^{16}\text{-}I^{17}))$	0.0867
259	$^1A \rightarrow {}^1A(I^- \rightarrow {}^*({}^+N^{20}H_3))$	0.02081
235	$^1A \rightarrow {}^1A\ ((O^{14}\text{-}C^{13}\text{-}O^{15}\text{-}I^{16}\text{-}I^{17}) \rightarrow *(O^{15}\text{-}I^{16}\text{-}I^{17}))$	0.7280
226	$^1A \rightarrow {}^1A\ (\pi\ (O^{14}\text{-}C^{13}\text{-}O^{15}\text{-}I^{16}\text{-}I^{17}) \rightarrow *({}^+N^{20}H_3))$	0.0294

Since the attempts to optimize molecular geometry of IIIa failed in order to calculate UV spectrum we developed molecular models that provide all significant characteristics of the studied species. At the first step the IIIb complex, comprising of two Gly-Gly pairs, which interact with both I^- and the I_2LiCl-ethanol complex was separated out. The IVa complex was divided into the IVb complex, comprising two Gly-Gly pairs, interacting with I_3^-' and IVc complex, consisting of one Gly-Gly pair, where one of the carboxyl groups is coordinated with LiCl-ethanol. Using the TD-DFT/B3PW91 method, the wavelengths of electronic transitions for such constructed IIIb-IVb,c complexes were calculated.

As shown in [21], the electronic transitions in the IIIb complex make contributions to two bands at 212–257 nm and 257–319 nm (Table 10.4). Coordination of molecular iodine with glycine also takes place in the solution (c). However, in this system the molecular iodine interacts with the glycine carboxyl group and LiCl, revealing the acceptor properties with respect to glycine and donor properties to LiCl. This resulted weakening iodine bond. Predicted the transitions between occupied and excite MO of the coordination bond $O^{15}\text{-}I^{16}\text{-}I^{17}$ (235 and 272 nm) is accompanied by a shift to longer wave region, as compared to the IIb complex. In the IIIb complex, I^- interacts only with one protonated amino group, therefore the charge transfer to $^+N^{20}H^3$ group is increased, thus giving rise to the charge-transfer transition at 259 nm.

The electronic transitions in the IVb complex (Table 10.5) make contributions to the longer-wavelenght bands at 257–319 nm and 320–380 nm. This is an indication of the presence of a triiodide complex in the solution [29].

The electronic transitions in triiodide between occupied and excited orbitals make contributions to bands at 376, 352 and 258 nm. The transfer of partial electron density from the carboxyl group to the excite orbital of I_3^- causes the appearance of

Table 10.5 Calculated (TD-DFT/B3PW91 level) wavelengths of electronic transitions (λ, nm) and oscillator strength (I^{λ}) for the IVb complex

$\lambda^{theor.}$	Assignment	I^{λ}
376	$^1A \rightarrow {}^1A\ ((I^{21}\text{-}I^{22}\text{-}I^{23}) \rightarrow *(I^{21}\text{-}I^{22}\text{-}I^{23}))$	0.0671
352	$^1A \rightarrow {}^1A\ ((I^{21}\text{-}I^{22}\text{-}I^{23}) \rightarrow *(I^{21}\text{-}I^{22}\text{-}I^{23}))$	0.1252
310	$^1A \rightarrow {}^1A\ ((I^{21}\text{-}I^{22}\text{-}I^{23}) \rightarrow *(I^{21}\text{-}I^{22}\text{-}I^{23}))$ $^1A \rightarrow {}^1A\ ((O^{17}\text{-}C^{18}\text{-}O^{16}) \rightarrow *(I^{21}\text{-}I^{22}\text{-}I^{23}))$	0.0555
266	$^1A \rightarrow {}^1A\ (n\ (O^{17},O^{16}) \rightarrow *(I^{21}\text{-}I^{22}\text{-}I^{23}))$	0.0607
258	$^1A \rightarrow {}^1A\ ((I^{21}\text{-}I^{22}\text{-}I^{23}) \rightarrow *(I^{21}\text{-}I^{22}\text{-}I^{23}))$ $^1A \rightarrow {}^1A\ ((I^{21}\text{-}I^{22}\text{-}I^{23}) \rightarrow *(I^{21}\text{-}I^{22}\text{-}I^{23}))$	2.1374

the band at 310 nm. The transition at 266 nm is due to the charge transfer from n-pair of oxygen atoms of the carboxyl group to the excite orbital of triiodide.

In the IVc complex, it is possible to distinguish only one weak band at 211 nm. The corresponding electronic transition is attributed to the charge-transfer transition from the water molecule $O^{28}H_2$ to the Li^+ ion.

Our calculations are not always able to reproduce the oscillator strengths of UV spectra, because in our calculations of frequencies by TD-DFT we select a very small cluster of glycine molecules (only four molecules). Probably for that reason, the theoretical intensity transitions of 258 or 259 nm do not correspond to the experimental ones.

Interestingly, based on the results of calculations one uncovers novel details concerning considered systems. It is revealed that the most complete and accurate description of the UV spectrum of the system (c) corresponds to the assumption that in this system two complexes may be formed at the same time: IIIa, in which molecular iodine interacts with the glycine zwitterion and LiCl-ethanol, and complex IVa, comprising spatially separated triiodide and LiCl-ethanol.

Also, a comparison of the experimental frequencies in the IR spectrum of the system (c) with the calculated frequencies for the IIIa, and IVa model complexes resulted in similar conclusions. An analysis of experimental and theoretically predicted vibrational frequencies indicates that in the aqueous glycine-KI_3-LiCl-ethanol solution the formation of these two complexes is feasible.

Experimental IR spectrum for the system (c) is discussed in details in the recent work [21]. For the IIIa and IVa structures the theoretical data are analyzed. The vibrational frequencies and their interpretation for the IIIa structure are also presented in this paper [21].

Table 10.6 displays experimental, the most intense bands in the IR spectrum of the system (c). They are compared with the frequencies calculated for the IIIa, and IVa complexes. The calculated vibrational frequencies are in good agreement with the experimental data.

In comparison with the system (b), two additional bands at 1605 and 1523 cm^{-1} appear in IR spectrum of the system (c). These bands are associated with changes in the structure of the complex upon the LiCl-ethanol addition.

The frequencies at 1603 cm^{-1} in the IIIa complex, and 1618 cm^{-1} in the complex IVa, may be attributed to the experimental band at 1605 cm^{-1}. The band at 1603 cm^{-1} can be interpreted as an asymmetric deformation vibration of $^{+}N^{20}H_3$ groups in the glycine molecule which interacts with I^- ion. The frequency at 1618 cm^{-1} corresponds to the asymmetric valence vibration of the carboxyl group O^{32}-C^{31}-O^{33} (complex IVc).

In the IIIa complex the band at 1523 cm^{-1} corresponds to the frequency at 1527 cm^{-1}, which can be attributed to the asymmetric deformation vibrations of the $^{+}N^{10}H_3$ group, which interacts with I^- through the water molecule. In the IVa complex there is a vibration at 1515 cm^{-1}, corresponding to symmetric deformation vibrations of the $^{+}N^{10}H_3$ group coordinated with triiodide.

Thus a comparison of the theoretical spectral characteristics of the most stable model complexes of iodine and triiodide for the systems (b, and c) with the

Table 10.6 Experimental (for the system: water-glycine-KI-I_2-LiCl-ethanol) and calculated (DFT/B3PW91/midi level) vibrational IR frequencies for complex IIIa, and and IVa (cm^{-1}) and their assignment

Experiment	IIIa	IIIa Assignment	IVa	IVa Assignment
1605	1603	$\delta(^+N^{20}H_3)$ asym	1618	$\nu(O^{33}$-H) $\nu(C^{31}O^{32}O^{33})$ asym
1581	1577	$\delta(^+N^{10}H_3)$ asym $N^{10}H_3$-$O^{28}H_2$ $\nu(C^{16}O^{18}O^{17})$ asym	1577	$\delta(^+N^2H_3)$ sym N^2H_3-$O^{24}H_2$ $\delta(^+N^{12}H_3)$ sym $\delta(^+N^{10}H_3)$ sym $N^{10}H_3$-$O^{26}H_2$ $\nu(C^6O^8O^7)$ asym
1523	1527	$\delta(^+N^{10}H_3)$ sym $N^{10}H_3$-$O^{28}H_2$	1515	$\delta(C^{11}H_2)$ sym $\delta\delta(^+N^{12}H_3)$ sym
1498	1512	$\delta(^+N^2H_3)$ sym $\delta(C^1H_2)$ sym	1489	$\delta(^+N^{10}H_3)$ sym $N^{10}H_3$_$O^{26}H_2$ $\delta(C^9H_2)$ sym
1408	1400	$\delta(^+N^{20}H_3)$ sym $\nu(C^{16}O^{18}O^{17})$ sym	1404	$\nu(C^6O^8O^7)$ asym δ(H-C^9-N^{10}-H)
1322–1350	1333	$\nu(C^{16}O^{18}O^{17})$ sym	1357	$\nu(C^3O^4O^5)$ sym $\delta(C^1H_2)$ sym $\nu(C^1$-$C^3)$
2630–2720	2656	$\nu(N^{12}$-H) $\nu(O^{25}$-H)	2725	$\nu(O^{24}$-H) $\nu(N^2$-H)
2800–2900	2905	νN^{20}-H	2904	$\nu(N^{20}$-H)
3156	3136	$\nu(C^{19}H_2)$ asym	3153	$\nu(N^{34}$-H) ν(H-N^{29}-H) sym
3228	3230	$\nu(O^{28}$-H) $\nu(N^{10}$-H)	3200	$\nu(N^2$-H) $\nu(N^{17}$-H)

experimental data indicates that the polar structure of the glycine zwitterions significantly affects the iodine-triiodide equilibrium. Where only alkali metal halogenides are present, the equilibrium is shifted towards the formation of triiodide. Therefore, in comparison with I_3^- the molecular iodine and I^- cannot be detected in such a system. However, this changes when amino acid is added to an aqueous solution and then the iodine and triiodide complexes can be detected by the IR-UV spectroscopy.

In the system (c), that closely models the composition of a drug [17–20], the glycine zwitterion cluster forms two complexes. In one of them the molecular iodine may be coordinated by both the lithium halogenides and zwitterions cluster, showing the acceptor properties with respect to glycine, and the donor characteristics with respect to lithium halogenides. In the other complex the triiodide and LiCl-ethanol are spatially separated, while triiodide interacts with protonated amino groups, and LiCl-ethanol assembly interacts with carboxy glycine zwitterions. The calculated spectral characteristics of the complexes in question provide an adequate description of the experimental UV and IR spectra of system (c), indicating appropriate selection of the computational model. Thus one assumes that involvement of α-dextrins in the mixture ensure the presence of the three active centers located within the α-dextrin helix: molecular iodine coordinated lithium halogenides and polypeptides, triiodide and lithium halogenides.

Interestingly, the authors of papers [44, 45] have shown that the presence in solution of the drug I_2 and I_3^- species affect its biological properties, such as toxicity, and irritation.

Drugs made using aqueous solutions of iodine complexes with organic compounds (povidone, polysaccharides) also may contain ions I^-, I_2, I_3, I_5^-, I_6^{2-}, HOI, OI^-, IO_2^- influence of initial concentrations of molecular iodine and iodide, and also pH (0-14) on the equilibrium concentration of molecular iodine and the above ions was studied using an aqueous solution of iodine and iodide ion [45]. Based on the studies, the authors have come to the conclusion, that the aqueous solutions of complexes of iodine with organic compounds actually contain only I^-, I_3^- and I_2.

There is general agreement that free molecular iodine represents the real active iodine species which is based on the frequently observed positive correlation between equilibrium concentration of I_2 and the rate of microbial extermination [46]. Furthermore, it is an important parameter for toxicity related features like skin irritation. Triiodide is non-toxic and does not have antimicrobial properties. Importantly, selection of equilibrium concentrations of molecular iodine and triiodide can reduce toxicity while retaining antimicrobial properties.

10.4 The Mechanism of Anti-HIV Action

Efficient combination of experimental and computational techniques results in shedding a light on complex phenomena. In the recent work we proposed a mechanism of anti-HIV effects of drugs [20]. In the active site of biomolecule the molecular iodine, located inside the dextrin helix and coordinated by peptides and lithium halides, is hardly inaccessible for interaction with bio-organic ligands. Only bioorganic ligands whose donor activity is higher than that of amides can compete for complexing with iodine. We have shown that the nucleotides of viral DNA are able to compete with amide [20].

UV spectroscopy is one of the most widely used methods for studying the structure of DNA. UV spectra are sensitive to the type of nucleotide base and to the formation of hydrogen or coordination bonds within the DNA. For this reason the UV experiments are often used to investigate the interaction of drugs with DNA nucleotides [47–51].

Authors of paper [48] studied the self-complementary duplexes A and B and compared them to their single strands, in order to compare helix or G-quadruplex. Clear differences in the UV spectra of the double helix and G-quadruplex are observed, as compared to the corresponding single strands. The red shift upon higher-order structure formation is attributed to the absorption of clusters of bases interacting via hydrogen bonds (Watson–Crick type for the duplex and Hoogsteen type for the quadruplex) and via base stacking. Theoretical calculations of the electronic spectra on large DNA structures (200 atoms and more) also reveal a red shift upon double-helix and G-quadruplex structure formation.

This approach was also used in our recent project. Using UV spectroscopy, we investigated the interaction of LiCl(I)-I_2-α-dextrin-peptide with the AGA nucleotide triplet. UV spectra of aqueous solutions of the AGA nucleotide triplet (system d) of the LiCl(I)-I_2-α-dextrin-peptide (system f) and the aqueous solution of the LiCl(I)-I_2-α-dextrin-peptide complex, and the AGA nucleotide triplet (system e) are presented and discussed in [52].

Comparison of the UV spectra of aqueous solutions of systems (d-e) indicates that in (e) the nucleotide triplet interacts with the LiCl(I)-I_2-α-dextrin-peptide complex. The band at 287 nm observed in system (f) disappears in the UV spectrum of system (e). In addition, the band recorded at 256 nm in system (d), becomes less intense and changes shape in (e).

The molecular structures of molecular iodine complexes with nucleotides adenosine (VI) and guanosine (VII) and LiCl were studied using computational methods. The DFT/B3PW91 method with the 6-31G ** basis set for the atoms of C, N, O, H, Li, Cl and the midi basis set for I was applied. The considered complexes were fully optimized (Fig. 10.4).

In addition, UV spectrum of the studied complexes was simulated by theoretical methods. The TD-DFT/B3PW91 approach was used to calculate the energies of electronic transitions in complexes VI and VII (Tables 10.7 and 10.8).

The splitting of the experimental band at ~200 nm can be described by electronic transitions of 207, 210, and 219 nm in complex VI. Transitions between π-orbitals of I_2 and π*-orbitals of the nucleotide base contribute to these frequencies. The electronic transition due to the transfer of electron density from lithium chlorine to I_2 also contributes to the 210 nm frequency, while the transition between the occupied orbitals of the coordination bond N-I and unoccupied orbitals of I_2

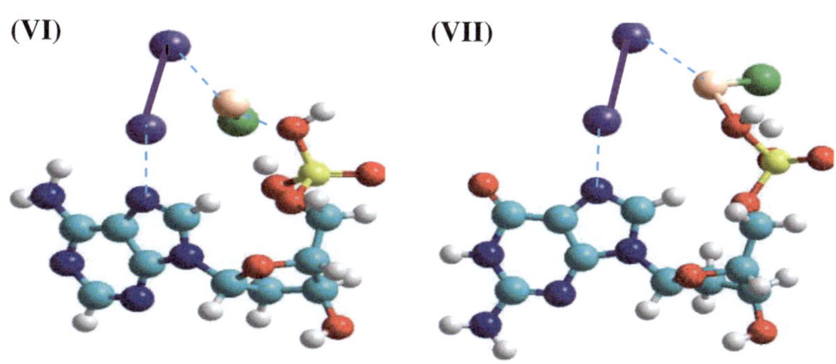

Fig. 10.4 Complexes of molecular iodine with adenosine (VI), guanosine (VII) and lithium chlorine. Color codes: *blue balls*—carbon atoms, *dark blue balls*—nitrogen atoms, *red balls*—oxygen atoms, *violet balls*—iodine atoms, *orange balls*—lithium atoms, *green balls*—chlorine atoms, *yellow balls*—phosphor atoms

Table 10.7 Calculated (TD-DFT/B3PW91 level) wavelengths of electronic transitions (λ, nm) and oscillator strength (I^{λ}) in complex VI

$\lambda^{theor.}$	Assignment	I^{λ}
244	$^1A \rightarrow {}^1A$ (π (nucleot.) $\rightarrow \pi^*$ (nucleot.))	0.1319
235	$^1A \rightarrow {}^1A$ ($\pi(I_2) \rightarrow \pi^*$ (I_2)) $^1A \rightarrow {}^1A$ (π (nucleot.) $\rightarrow \pi^*$ (I_2))	0.3325
227	$^1A \rightarrow {}^1A$ (π (nucleot.) $\rightarrow \pi^*$ (I_2))	0.1017
219	$^1A \rightarrow {}^1A$ (π (nucleot.) $\rightarrow \pi^*$ (I_2)) $^1A \rightarrow {}^1A$ ((N-I-I) $\rightarrow \pi^*$ (I_2))	0.1410
210	$^1A \rightarrow {}^1A$ ($\pi(I_2) \rightarrow \pi^*$ (nucleot.)) $^1A \rightarrow {}^1A$ ((Cl-Li-I) $\rightarrow \pi^*$ (I_2))	0.3493
207	$^1A \rightarrow {}^1A$ ($\pi(I_2) \rightarrow \pi^*$ (nucleot.)) $^1A \rightarrow {}^1A$ (π (nucleot.) $\rightarrow \pi^*$ (nucleot.))	0.2503

Table 10.8 Calculated (TD-DFT/B3PW91 level) wavelengths of electronic transitions (λ, nm) and oscillator strength (I^{λ}) in complex VII

$\lambda^{theor.}$	Assignment	I^{λ}
244	$^1A \rightarrow {}^1A$ ($\pi(I_2) \rightarrow \pi^*$ (I_2))	0.1515
234	$^1A \rightarrow {}^1A$ ($\pi(I_2) \rightarrow \pi^*$ (nucleot.))	0.1155
215	$^1A \rightarrow {}^1A$ ($\pi(I_2) \rightarrow \pi^*$ (I_2))	0.6343

contributes to the 219 nm transition. Complex VII also displays the 215 nm transition between occupied and unoccupied orbitals of I_2. This transition contributes to the ~200 nm band.

Contributing to the 223 nm band in complex VI are the following two transitions: 227 and 235 nm. They could be considered as transitions between occupied and unoccupied orbitals of the nucleotide base of adenosine, and transitions between occupied and unoccupied orbitals of I_2. In complex VII contributions to the said band are from the 234 nm transition that corresponds to the transfer of electron density from the molecular iodine onto π-orbitals of the nucleotide base of guanosine.

Comparison of the quantum chemical calculations for electronic transitions for the structure modeling the interaction of LiCl(I)-I_2-α-dextrin-peptide complex with the nucleotide triplet indicates that the DNA nucleotides can displace polypeptide and form stable complexes with molecular iodine and lithium halogenides. Interestingly, in such structures, molecular iodine binds both nucleotide triplet and lithium halogenides.

Thus, the conclusion initially derived only on the basis of quantum-chemical calculations about the possibility of interaction of viral DNA nucleotides with LiCl (I)-I_2-α-dextrin complexes, that are part of drugs [17–20], is further confirmed by the results of complementary experimental investigations. They include UV-spectroscopy study.

10.5 Conclusion

The unique composition of drugs [17–20] assures a presence in their composition of an iodine complex, whose structure is unique and differs from the structure of iodine complexes in other known drugs. In the active site species the molecular iodine interacts with the bio-organic ligands (polypeptides) and lithium halogenide and exhibits acceptor properties with respect to polypeptides and donor properties in relation to lithium halide. The of existence such iodine complex is the probable cause of the low toxicity of the drugs.

In the active site of biomolecules molecular iodine is being held inside the α-dextrin helix and coordinated by lithium halides and polypeptides. It is hardly accessible for interactions with other bio-organic ligands and only the nucleotides of viral DNA are able to compete with polypeptides for formation of chemical bonds with the iodine.

Previously we have shown that the molecular iodine may be assigned to compounds that inhibit the active site of HIV-1 integrase [52, 53]. It prevents the active site of integrase from the formation of a complex with HIV DNA and inhibits the active complex of integrase and viral DNA. Such complex becomes the center of another, larger nucleoprotein species, binding the active site of integrase and viral DNA.

Remarkably, the presence in solution of the drug of I_2 molecules and I_3^- ions affect drug's biological properties such as toxicity, irritation [44, 46]. Selection of equilibrium concentrations of molecular iodine and triiodide can reduce toxicity while retaining antimicrobial properties. The reviewed studies could shed a light on such phenomena. One assumes that involvement of α-dextrins in the mixture ensure the presence of the three active centers located within the α-dextrin helix: molecular iodine coordinated lithium halogenides and polypeptides, triiodide and lithium halogenides. We believe that the existence of two complexes of iodine and triiodide reduces toxicity of the drugs [17–20].

References

1. Svensson PH, Kloo L (2003) Chem Rev 103:1649
2. Zhang FS, Lynden-Bell RM (2005) Eur Phys J D 34:129
3. Jena NK, Josefsson I, Eriksson SK, Hagfeldt A, Siegbahn H, Bjorneholm O, Rensmo H, Odelius M (2015) Chem Eur J 21:4049
4. Shilov GV, Kasheva ON, Deshenko OA, Chernovenc MS, Simonen CC, Goleva VE, Pishev AI (2002) Russ J Phys Chem 76:1295
5. Grozema FC, Zijstra RJ, Swart M, Duijnen PT (1999) Int J Quantum Chem 75:709
6. Su JT, Zewail AH (1998) J Phys Chem A 102:4082
7. Engel PS, Duan S, Whitmire KH (1998) J Org Chem 63:5666
8. Esseffar M, Bouah W, Lamsabbi A, Abboud JM, Notario R, Yanez M (2000) J Am Chem Soc 122:2300

9. Daga V, Hadjikakou SK, Hadjliadis N, Kubieki M, Santos JHZ, Butler IS (2002) Eur J Inorg Chem 7:1718
10. Antoniadis CD, Hadjikakou SK, Hadjiliadies N, Kubieki M, Butler IS (2004) Eur J Inorg Chem 21:4324
11. Moulay S (2013) J Polym Eng 33(5):389
12. Nessreen A, Al-Hashimi, Hussein YHA (2010) Spectrochim Acta Part A 75:198
13. Khouba Z, Benabdallah T, Maschke U (2014) Spectrochim Acta Part A Mol Biomol Spectrosc 125:61
14. Karlsen EM, Spanget-Larsen J (2009) Chem Phys Lett 473:227
15. Karuma T, Neelima K, Venkateshwarlu G, Swamy PY (2006) J Sci Ind 65:808
16. Lagoce JF, Jambut-Absil AC, Buxeraud J, Moesch C, Raby C (1990) Chem Pharm Bull 38 (8):2172
17. International Application (2000) Ilyin Alexandr, Gab-rielyan Emil, Mkhitaryan, Levon Antiviral and antibacte-rial pharmaceutical preparation “Armenicum” And its use for treatment of infectious diseases. Patent No.: PCT/AM 2000/000002
18. Mkhitaryan LM, Davtyan TK, Gabrielyan ES, Gevorkyan LA (2007) Int J Biotechnol 9 (3,4):301
19. Ilin AI, Kulmanov ME (2004) Bactericide and viricidal pharmaceutical preparation for prophylaxis and treatment mono and mixtinfections, manner of prophy-laxis o/or treatment viral, bacterium and mixtinfection. Patent No.: 15116
20. Yuldasheva GA, Zhidomirov GM, Ilin AI (2012) Biotechnol Appl Biochem 59(1):29
21. Yuldasheva GA, Zhidomirov GM, Leszczynski J, Ilin AI (2013) J Mol Struct 1033:321
22. Nagaoka M, Okuyama-Yoshida N (1998) J Phys Chem A 102(42):8202
23. Grasjo J, Andersson E, Forsberg J, Duda L, Henke E, Pokapanich W, Bjrneholm O, Andersson J, Pietzsch A, Hennies F, Rubensson JE (2009) J Phys Chem B 113(49):16002
24. Hwang TK, Eom G-Y, Choi M-S, Jang S-W, Kim J-Y, Lee S (2011) J Phys Chem B 115 (33):10147
25. Aikens CH, Gordon MS (2006) J Am Chem Soc 128(39):12835
26. Rahaman O, van Duin ACT, Goddard WA, Doren DJ (2011) J Phys Chem B 115(2):249
27. Tajkhorshid E, Jalkanen KJ, Suhai S (1998) J Phys Chem B 102(30):5899
28. Minkov VS, Chesalov UA, Boldyreva EV (2010) J Struct Chem 51(6):1091
29. Thoma JA, French D (1960) J Am Chem Soc 82:4144
30. Klamt A, Schuurmann, GJ (1993) Chem Soc Perkin Trans 2:799
31. Shart SB, Gellene GI (1997) J Phys Chem A 101:2192
32. Frisch MJ, Trucks GW, Schlegel HB, Scuseria GE, Robb MA, Cheeseman JR, Scalmani G, Barone V, Mennucci B, Petersson GA, Nakatsuji H, Caricato M, Li X, Hratchian HP, Izmaylov AF, Bloino J, Zheng G, Sonnenberg JL, Hada M, Ehara M, Toyota K, Fukuda R, Hasegawa J, Ishida M, Nakajima T, Honda Y, Kitao O, Nakai H, Vreven T, Montgomery JA Jr, Peralta JE, Ogliaro F, Bearpark M, Heyd JJ, Brothers E, Kudin KN, Staroverov VN, Kobayashi R, Normand J, Raghavachari K, Rendell A, Burant JC, Iyengar SS, Tomasi J, Cossi M, Rega N, Millam JM, Klene M, Knox JE, Cross JB, Bakken V, Adamo C, Jaramillo J, Gomperts R, Stratmann RE, Yazyev O, Austin AJ, Cammi R, Pomelli C, Ochterski JW, Martin RL, Morokuma K, Zakrzewski VG, Voth GA, Salvador P, Dannenberg JJ, Dapprich S, Daniels AD, Farkas O, Foresman JB, Ortiz JV, Cioslowski J, Fox DJ (2009) Gaussian 09, Revision C.01. Gaussian Inc., Wallingford
33. Parsons CL, Corliss HP (1910) J Am Chem Soc. 32(11):1367
34. Immel S, Lichtenthaler FW (2000) Starch 52(1):1
35. Refat MS, Killa HMA, Hamada MA, El-Sayed MY (2011) Bull Chem Soc Ethiop 25(1):137
36. Refat MS, Killa HMA, El-Maghraby A, El-Sayed MY (2012) Bul Chem Commun 44(1):74
37. Burgess AE, Davidson JC (2012) J Chem Educ 89:814–816
38. Calabrese VT, Arshad Khan A (2000) J Phys Chem A 104:1287
39. Gerasimova GV, Variamova TM, Mushtakova SP (2008) Russ J Phys Chem Desember 82 (12):2235
40. Kebede Z, Lindquist SE (1999) Sol Energy Mater Sol Cells 57:259

41. Stern JH, Passchier AA (1962) J Phys Chem 66:752
42. Turner DH, Flynn GW, Sutin N, Beitz JV (1972) J Am Chem Soc 94(5):1554
43. Cherenkova UA, Kotova DL, Kricanova TA, Selemenev BF (2008) Sorpt Chromatogr process 8(2):314
44. Shart SB, Gellene GI (1997) J Phys Chem A 101:2192
45. Gottardi W (1999) Arch Pharm Pharm Med Chem 332:151
46. Gottardi W (1991) in Disinfection, Sterilization and Preservation (Ed.:S.S.Block), Lea & Febiger, Philadelphia, chapter 8
47. Eaman AS, AL-Rubaee (2014) Int J Sci Nat 5(1):22
48. Rosu F, Gabelica V, Pauw ED, Antoine R, Broyer M, Dugourd P (2012) J Phys Chem A 116:5383
49. Wani A, Mushtaq S, Ahsan H, Ahmad R (2012) Asian J Biomed Pharm Sci 2(13):23
50. Perveen F, Qureshi R, Ansari FL, Kalsoom S, Ahmed S (2011) J Mol Struct 1004:67
51. Sirajuddin M, Saqib Ali, Badshah A (2013) J Photochem Photobiol B Biol 124:1
52. Yuldasheva G, Argirova R, Ilin A (2015) AIDS Clin Res 6(6):2
53. Yuldasheva G, Zhidomirov G, Ilin A (2011) Nat Sci 3(7):573

Chapter 11
Detailed Atomistic Modeling of Si(110) Passivation by Atomic Layer Deposition of Al_2O_3

Andrey A. Rybakov, Alexander V. Larin, Daniel P. Vercauteren and Georgy M. Zhidomirov

Abstract Typical structural defects were studied theoretically in the course of O → Al → O atomic depositions on the basic Si(110) surface. The defects were determined by analyses of the band gap states and projected densities of the *s*- and *p*-states after the deposition aimed to form a Si(110)/SiO_X/AlO_Y/γ-Al_2O_3 slab. The extent of Si(110) passivation after every deposition step was studied by scanning the band structure calculated using Density Functional Theory with periodic boundary conditions. The atomic structure of the optimized Si(110) surface was compared to the one of Si(100) for which more information is available. Our modeling reproduces most features of the use of trimethylaluminium or any other organic ligand as Al precursor along O_2 plasma assisted atomic layer deposition (PA ALD) when the organic ligands are completely oxidized so that their participation can be neglected in the deposition as already shown experimentally. The final oxidation step corresponds to the junction of the slab deposited over Si(110) with a γ-Al_2O_3 fragment, whose super cell (SC) parameters have been selected to lead to the minimum mismatch. Different examples of either non-satisfactory or

A.A. Rybakov · A.V. Larin · G.M. Zhidomirov
Department of Chemistry, Moscow State University, Leninskie Gory, GSP-2, Moscow Russia119992
e-mail: rybakovy@gmail.com

G.M. Zhidomirov
e-mail: zhidomirov@mail.ru

A.V. Larin (✉)
OOO Plasmonika, Ural Building, 100 Novaya Str, Skolkovo, Moscow Region, Russia143025
e-mail: nasgo@yandex.ru

D.P. Vercauteren
University of Namur, Rue de Bruxelles 61, Namur 5000, Belgium
e-mail: daniel.vercauteren@unamur.be

G.M. Zhidomirov
Boreskov Institute of Catalysis, SO RAN, Novosibirsk, Russia 630090

J. Leszczynski and M.K. Shukla (eds.), *Practical Aspects of Computational Chemistry IV*, DOI 10.1007/978-1-4899-7699-4_11

accurate junction of the oxidized Si(110) slab and γ-Al_2O_3 fragment (under two different forms) are discussed aiming to develop a route for understanding the dominant defect types at the interface. Such theoretical work should be the first step for the elaboration of computational tools for the passivation of silicon with amorphous oxides. The latter are mainly formed at the conditions of the PA ALD depositions. The list of formed typical defects at the Si(110)/SiO_X/AlO_Y/γ-Al_2O_3 boundary is presented and characterized by the projected density of states and respective band structure around the band gap.

11.1 Introduction

Specific geometries of fully coordinated atoms and groups of atoms, whose states can appear in the band gap or at the edges of conduction and valence zones crucial for the electronic properties, have already been widely analyzed using different models of amorphous Si (*a*-Si) bulk or Si/SiO_2 boundaries [1–7]. Many outstanding results were, for example, obtained using the tight binding (TB) approximation and Bethe models. The later works continued the discussions on the localization problem of the electronic states in *a*-Si ([8–12] and the Refs. therein) or hydrogenated *a*-Si:H [8] started by Anderson for amorphous materials [13], involving more elaborated cluster [8] and periodic [9–12, 14, 15] models. A discussion of the evolution of methods *versus* the localization-delocalization problem is presented in [16]. The energy of the states relative to the mobility edge indeed determines the charge transport mechanism, being of prime importance for solar energetics. In [15], the authors considered the penetration of O atoms to the deeper Si (001) layers (involving a 4th layer for a total of 14) with up to 16 O atoms per 2×2 super cell (SC) by comparing various optimized geometries mainly using Local Density Approximation (LDA) with General Gradient Approximation (GGA)/PW91 corrections and combined ultra-soft/norm-conserving pseudopotentials. Charged states of nitrogen defects were involved in the super-lattice model of Si(100)/SiO_2 interface [14].

It is known that charge transport becomes more complex in multi-component systems obtained at the Si/X or Si/SiO_2/X boundaries passivated with an X oxide. Al_2O_3 remains the best passivation material for both *n*- and *p*-silicon as compared to other known candidates (SiO_x, *a*-SiN_x:H, *a*-Si:H) [17]. Al_2O_3 is usually obtained in amorphous state by various deposition methods. The states of both fully and under-coordinated atoms have however rarely been discussed for alternate Al/O depositions upon atomic layer deposition (ALD) conditions. The identification of fully coordinated atoms and consideration as reactive species is important to develop for example kinetic Monte-Carlo (KMC) schemes [18, 19] oriented towards the passivation problems with amorphous oxides. It is therefore important to recognize the arising defects from the geometrical (bonds and angles) and/or topological (atomic/hybridization types of the 1st and, may be, 2nd neighbors) points of view because no other information is available regarding KMC modeling of any large system.

After the KMC studies [18, 19], two experimental facts were discovered: (1) as-deposited amorphous Al_2O_3 transforms to the γ-Al_2O_3 state only after annealing [20]; (2) full oxidation of TMA ligands takes place in the course of O_2 plasma assisted (PA) ALD [21, 22]. In parallel, upon annealing amorphous Al_2O_3 achieves the desired parameters of the Al_2O_3/Si surface (low interface density of 10^{11} $eV^{-1}cm^{-2}$, low surface recombination velocity near 100 cm/s, and long effective time-of-life of 1 ms) irrespective of thermal ALD (T ALD) or PA ALD [20]. Hence, a separation of the low temperature deposition and high temperature annealing is necessary. The crucial importance of the annealing step for Al_2O_3 as compared, for example to *a*-SiN_x:H [23], should thus define the strategy of modeling. To our best of knowledge, this has not been addressed in the series of attempts of KMC modeling of Al_2O_3/Si deposition [18, 19] performed before the appearance of [20] and for which traditional (long and complex) kinetic schemes for T ALD from TMA/H_2O have been realized. The authors in [18, 19] considered a crystalline Al_2O_3 lattice to calculate the events and rate constants. But at deposition temperatures lower than 600 °C, Al_2O_3 can be obtained in the amorphous form only regardless the type of substrate [24]. The accuracy of the replacement of an amorphous oxide model by a crystalline one for rate constant calculations [18, 19] will depend on the chosen oxide. The absence of successful KMC results for the Al_2O_3 growth does not confirm such possibility for the Al_2O_3 case even if additional reasons can be behind this result.

The second tip, i.e., oxidation of TMA ligands in O_2 plasma [21, 22], has been confirmed also after the first attempt of KMC modeling of Al_2O_3/Si deposition upon O_2 PA ALD [18, 19]. It allows to select the simplest adequate computational model for PA ALD. Detailed results of mass spectrometry (CH_x^+, $C_2H_x^+$, $CO_2{}^+$, CO^+) and optical emission (H, O, $O_2{}^+$, CO) spectroscopy in O_2 plasma conditions [22] have shown that CH_3 ligands of adsorbed TMA are completely oxidized by O_2 plasma at room temperature (at the deposition temperatures of 70 and 200 °C) with the formation of CO, CO_2, H_2O, and possibly H and/or OH at the oxidized Si/SiO_2 surface. A way for the formation of observed small quantities of higher C_2H_x hydrocarbons has also already been proposed [25]. As a result, all the specifically adsorbed species that can be hardly deleted are present in negligible concentrations after each purging step of the ALD cycle. These experimental findings of Kessels group [21, 22] have thus opened a route to the simplest possible solid state kinetic scheme compared to the traditional ones [26–31] applied for the T ALD modeling, the latter including multi-step decomposition of TMA adsorbed at the surface as performed in [27]. The full oxidation of the TMA ligands in O_2 PA ALD [21, 22] minimizes the number of participating atomic species (Si, O, Al, H, from which Si participates at the earliest steps) in the growing Al_2O_3 slab only. Hence, our aim is to analyze the possible computational schemes of Al_2O_3/Si deposition with the participation of Si, O, and Al atoms (without H).

Formally, the presented work has repeated the approach developed in [7, 32] wherein distorted Si-Si and Si-O bonds were modeled at the Si(111)/SiO_2 [7] and Si

(100)/SiO_2 [7, 32] interfaces. The computational aspects and hierarchy of all obtained models are described in Sect. 11.2. The initial Si model for oxidation is described in Sect. 11.3.1, while Sects. 11.3.2 and 11.3.3 are devoted to the successive O and Al deposition/oxidation steps, respectively. The trend to keep—$(Al\ldots Al)_n$− chains is assigned to the formation of boehmite fragments on the deposited amorphous slab. In Sect. 11.3.4, we discuss the results of joining the Si (110) slabs with pre-deposited O/Al layers considering several fragments of Al_2O_3 obtained in the previous steps to accelerate the growth of the passivated Si(110)/SiO_2/AlO_X/γ-Al_2O_3 layer. The geometries of all observed defects at the interface are discussed together with the assignments of their states in the band gap.

11.2 Computational Details

The isolated Si(110) and γ-Al_2O_3 slabs were constructed as parts of the unified supercell (SC). The results of the interaction between the parts were studied at the Density Functional Theory (DFT) GGA/PBE level with periodic boundary conditions (PBC) using the VASP code [33, 34]. The junction of the SC parts was realized after a series of successive O → Al → O depositions on the Si(110) plane (Scheme 11.1) to form a final Si(110)/SiO_X/AlO_Y/γ-Al_2O_3 slab. We therefore will first discuss the Si(110) slab model.

Possible slab candidates for the Si(110) part of the reactive system were sought throughout the SCs with sizes ($n \times 3.8608$, $m \times 5.46$) proportional to the ones of the elementary cell, i.e., 3.8608 and 5.46 Å. The (n, m) values were selected as the minimal mismatch with the second part of the slab, γ-Al_2O_3. The final Si(110) system contains eight layers of Si atoms (Fig. 11.1). Free valences of the three-coordinated Si atoms in the deepest layer were capped by H atoms. Hydrogens were placed at a distance of 1.5 Å perpendicularly to the Si(110) surface to simulate the whole crystal. The deepest Si and H layers were kept fixed along the geometry optimization.

The γ-Al_2O_3 fragment was cut parallely to the (110) plane of the γ-Al_2O_3 bulk from the structure constructed on the basis of the (113) SC (Fig. 11.2). Two Al vacancies were admitted in the initial γ-Al_2O_3 at the octahedral sites as providing the most stable configuration *versus* the other vacancy types [35, 36]. The sizes of the γ-Al_2O_3 (113) slabs are 7.9931 and 16.8269 Å. The latter values correspond to the best (n, m) selection, (2, 3) relative to the Si(110) SC parameters ($2 \times 3.8608 = 7.7216$ and $3 \times 5.46 = 16.38$ Å). These Si(110) SC parameters are pretty close to the SC parameters of Al_2O_3 with a mismatch of 3.4 and 2.7 % only. The last SC sizes of Al_2O_3 were re-scaled to exactly 7.7216 and 16.38 Å, like for the SC of Si, including seven atomic layers (16 Al and 30 O atoms) of the Al_2O_3 slab. We did not conserve an exact Al/O ratio of 2/3 to achieve a better oxidation step with a small number of successive deposition steps. For coherence, a separate computation was also performed for the stoichiometric Al_2O_3 slab of six layers with 16 Al and 24 O atoms.

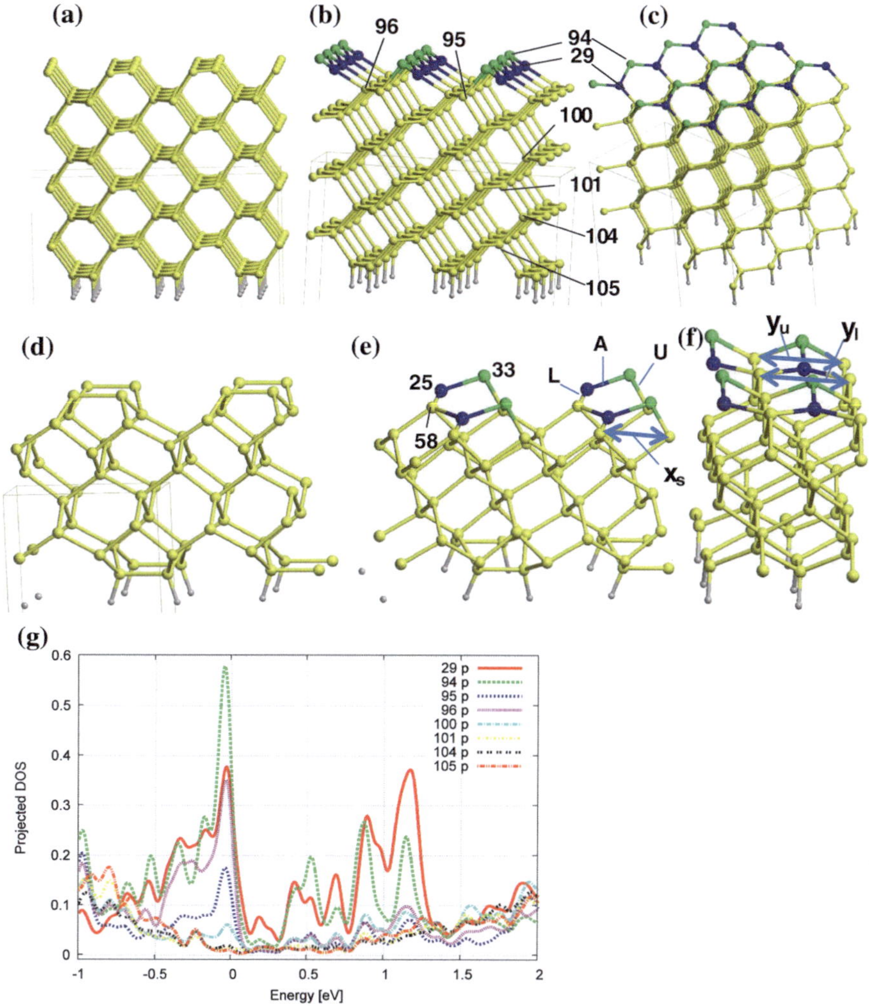

Fig. 11.1 Illustration of the Si(sp^2) formation when optimizing the geometry of the clean (110) (**a**) and (100) (**d**) Si planes. Views of the Si(sp^2) (*blue*) and Si(sp^3) (*green*) atoms in the two upper layers after the optimization of the (110) (**b, c**) and (100) (**e, f**) planes. Other Si atoms are in sp^3 states. Si and H atoms are in *yellow* and *grey*. *p*-type PDOS values (**g**) with atomic labels given in (**b, c**) for selected Si atoms for the (110) plane. Bond notations (A, L, U, y_u, y_l, x_s) in (**e, f**) are used in Table 11.2. Some of the atomic labels are used in Fig. 11.3 and Table 11.3

A model with PBC requires a vacuum interval for the slabs modeling. The vacuum interval was varied between 80 and 83 Å for the different models, so that the total cell height (z-period) was around 100 Å. The algorithm to model the deposition procedure was then simplified according to several experimental arguments. Considering the case of PA ALD with O_2, we admitted that the stage of complete H deletion happens at the pre-oxidation step. Hence, we added O atoms

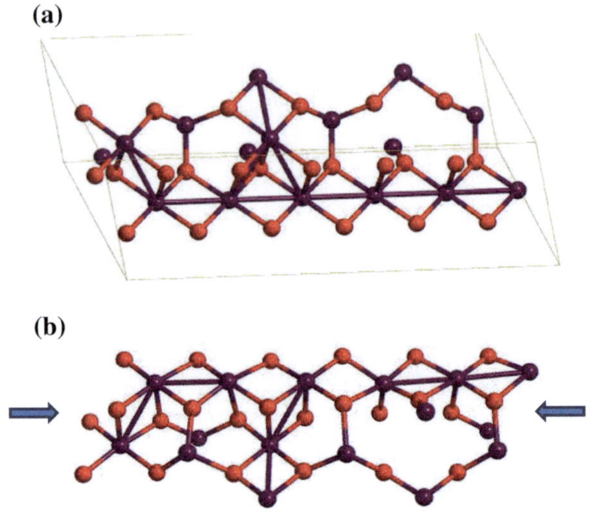

Fig. 11.2 Geometry of the (113) supercell (SC) of the **γ**-Al_2O_3 of *I4*$_1$*/amd* group with two octahedral Al vacations before (**a**) and after (**b**) geometry optimization. Al and O atoms are in *violet* and *red*

without the parallel presence of H atoms and thus did not consider the defects formed owing to the participation of hydrogens. The upper (110) surface, optimized without H atoms, was further used for the successive O → Al → O depositions. To check the optimization tools, i.e., VASP [33, 34] and PBE/PAW level, we optimized similarly small 2 × 2 SCs for the traditional Si(100) surface (Fig. 11.1d–f) and compared the structures to the known data (Sect. 11.3.1).

In all cases, we tried to locate the O atoms so that a complete passivation of all the undercoordinated Si atoms could be achieved. One should note that such deposition method still allows many different ways to improve the quality of the passivation by subsequent atomic transformations, which then would require the search of transition states (TS) and analyses of the relative barriers. But as the search of these successive steps and realization of such procedure are related to the domain of KMC methods, we will not discuss these issues herein.

Our series of constructed structural and compositional models involves eight (I to VIII) classes (Scheme 11.1): (I) the first model was obtained via the direct reaction between Si(110) and $Al_{16}O_{30}$; (II) a series of 8, 12, 18, and 20 O atoms per SC was after added over the Si(110) slab leading to seven different structural models; the way to add the oxygens is discussed below; (III) $Al_{16}O_{30}$ was then added to model II, with 8 oxygens only; (IV) a series of 4, 6, and 7 Al atoms (after the 12, 18, and 20 O atoms, respectively) were also added to model II leading to six different models; (V) an $Al_{16}O_{30}$ slab was added over model IV leading to six different models; (VI) 18 O atoms per SC were added to model IV (after the 18 O and 6 Al atoms) leading to one model; (VII) $Al_{16}O_{30}$ was added to model VI; (VIII) 9 O or 11 O atoms were withdrawn (under the forms of stable O_2 and/or O_3 species) from model VI leading to two different compositional models. The geometry of each of the 25 models obtained by either covering of different atomic layers or

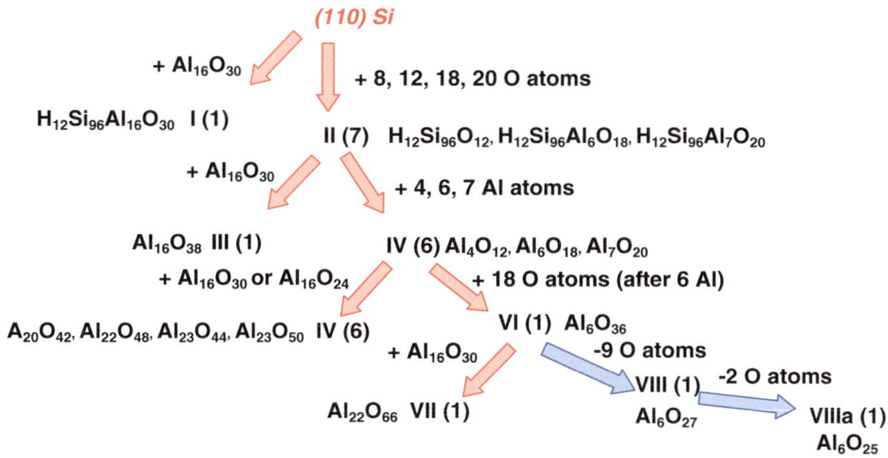

Scheme 11.1 Constructed slab models with successive O and Al deposition including (classes I, III, V, VII) or not (classes II, IV, VI, VIII) capping $Al_{16}O_{30}$ fragment. Number of the models for each class is given between parentheses. The "$H_{12}Si_{96}$" part is omitted in the titles of the slab classes from III to VIII for shortness

merging with the $Al_{16}O_{30}$ slab was optimized using VASP. As final step, we expected to obtain an Al/O ratio which exceeds the 1/3 experimental value related to the large number of ALD cycles as mentioned in [37] but that is smaller than the 2/3 ratio as in [17, 20, 38]. Each class thus contains different number of models whose formation was principally conditioned by the variable initial geometries of deposited atoms. The numbers in parentheses in Scheme 11.1 indicate the quantity of models in each class.

The assignment of defect states after each optimization step was realized using the gradient-corrected GGA/PBE functional and the projected-augmented wave (PAW) method on the basis of the comparison of the *s*- and *p*-projected density of states and band structure. The energy cut-off was set to 500 eV. The Brillouin zone *k*-sampling was restricted to the Γ–point for the geometry optimization, and was (8 4 1) for the density of states (DOS) calculation. Bands were calculated along the Γ-(0.5, 0) - (0.5, 0.5) - (0, 0.5) – Γ line on the (0, 0, 0) plane in reciprocal space. Figures were realized with MOLDRAW2.0 [39].

In order to compare the relative ΔU energies of the 25 different optimized $H_mSi_nAl_kO_p$ systems, we considered the following reaction scheme:

$$H_mSi_n + (p/2 - 3k/4)O_2 + (k/2)Al_2O_3 \rightarrow H_mSi_nAl_kO_p \quad (11.1)$$

(p/2–3 k/4) being positive for all the obtained models. The energies computed at the PBE/PAW level are –6.770, –9.858, –36.977, and –5.425 eV for H_2, O_2, Al_2O_3, and Si [40], respectively. The total energy value for Al_2O_3, –295.816 eV, was obtained from the full geometry optimization of bulk γ-Al_2O_3 with a SC containing the $Al_{16}O_{24}$ unit.

11.3 Results and Discussion

11.3.1 Si Slabs

The optimized geometry of the Si(110) slab is shown in Fig. 11.1. One observes chains of alternated three-coordinated Si(*sp*^{2}) and Si(sp^3) atoms (Fig. 11.1b, c). The deviation of the Si29(sp^2) atom (in blue in Fig. 11.1b, c) relative to the plane containing its three neighbors is around 0.02 Å, thus confirming the sp^2 hybridization *versus* 1.18 Å for the three-coordinated Si94(*sp*^{3}) defect. The largest peak in the *p*-type projected DOS (PDOS) belongs to Si94 while the three-coordinated Si29(*sp*^{2}) and fully coordinated Si96(*sp*^{3}), i.e., the nearest neighbor of Si29, possess similar *p*-densities (Fig. 11.1g). The *p*-PDOS value decreases steeply for the fully coordinated Si atoms from the top to the bottom layer (with the hydrogenated side) (Fig. 11.1g). Negligible variations in the PDOS were obtained with more accurate schemes of Si-H arrangements (the optimization of H positions in the direction of the deleted Si-Si bond) in the lowest layer and hence the capping scheme of H atoms was not modified to take into account more tiny relaxation effects.

Two types of Si(sp^2) and Si(sp^3) three-coordinated atoms were obtained for the non-oxidized totally dehydrogenated Si(110) and Si(100) slabs by geometry optimization (Table 11.1). A common feature of both Si hybridization types at the (100) and (110) surfaces is the similar Si-Si bond lengths for the sp^2 and sp^3 types (Table 11.1), in agreement with earlier computations [4–6, 27, 41, 42]. Our optimized models showed a satisfactory agreement with the experimental distances of 2.28 and 2.32 Å for the Si(100)2 × 1 surface [42] (Table 11.1). Our values obtained for the Si25-Si33, Si25-Si58, and Si53-Si bonds (noted as A, L, and U in Fig. 11.1e) are slightly overestimated (Table 11.2). Their relative order is nevertheless coherent with the experimental A, L, and U values [42–44]. Let us note a small difference of the Si-Si bond lengths between the sp^2 and sp^3 states with the different DFT methods applied at the PBC level (Table 11.1). The data for the (100) surface were included to demonstrate the similarity between the Si(sp^3) or Si (sp^2) types for both the (100) or (110) surfaces optimized using the same procedure[1] and thus to evaluate the different bulk restrictions imposed on the sp^2 geometry (Table 11.1). A plain Si(sp^2) type geometry for the dangling bond (DB) Si atom was obtained for a completely unconstrained cluster so that a non-plain sp^2 geometry could be due to the bulk restrictions [41], which vary between the (110) and (100) forms. That is why the Si(sp^2) geometry for our (110) model also deviates from a planar one. Regarding the similar deviations from planarity for (100) or (110) surfaces (the angle δ in Table 11.1, δ being the deviation of the Si atom from the plane with its three Si neighbors), we can consider that the deviation

[1]Only one type of Si(sp^3) three-coordinated atom was obtained for the Si(111) plane with the same procedure so that we did not considered it. A Si(111) 7 × 7 superstructure was observed experimentally in [40] but cannot be modeled with the smaller SC considered herein.

Table 11.1 Optimized structural parameters at different theory levels of the slab models containing Si(sp^n) atoms (n = 2, 3) with relative formation energies ΔU (eV)[a)] from crystalline Si and H_2: Si-Si bond lengths (Å), valence angles α = X-Si-Y (°), deviation angle δ (°) of central Si (sp^n) atom from the plane containing three neighbor atoms, and δ' = sin(δ)/sin(δ_T) [41], the angle δ_T of 35.26° corresponding to an ideal sp^3 hybridization (or δ' = 1)

Slab	Method	ΔU	sp^n	Si-Si	α	δ	δ'
$H_{12}Si_{96}(110)$	PBE	38.407	sp^3	2.310, 2.342, 2.342	84.5, 84.5, 123.8	38.11	1.069
			sp^2	2.342, 2.342, 2.345	116.7, 116.7, 123.8	10.63	0.320
$H_8Si_{64}(100)$	PBE	19.464	sp^3	2.304, 2.380, 2.380	94.2, 94.23, 122.5	36.93	1.041
			sp^2	2.342, 2.342, 2.304	115.6, 115.7, 127.7	6.30	0.190
	PW91	–	sp^2	2.321, 2.321, 2.302	115.7, 115.7, 127.9	5.72	0.173
	PBEsol	–	sp^2	2.311, 2.311, 2.304	114.5, 114.5, 129.1	8.50	0.256
	PBEsol[b]	–	sp^2	2.313, 2.313, 2.304	114.6, 114.7, 128.8	8.54	0.257
	AM05[b, c]	–	sp^2	2.309, 2.309, 2.303	114.1, 114.1, 129.3	9.58	0.282
	Exper.[d]			2.28, 2.32			

[a]calculated only at the PBE/PAW level; [b]spin excited state; [c]optimized using the trial wave functions obtained with PBEsol method; [d][42]

Table 11.2 Structural parameters of the Si(100) slab models optimized at different theory levels in comparison with experimental data, the distance notation being explained in Fig. 11.1e, f

Parameter	Experiment			Theory			
	[42]	[43]	[44]	[45][a]	PBE	PBEsol	PW91
A	2.28	2.26	2.67	2.29	2.304	2.304	2.302
U	2.34	–	–	2.35	2.380	2.369	2.378
L	2.32	–	–	2.31	2.324	2.311	2.322
y_u	3.63	3.74	3.84	3.60	4.172	4.172	4.172
x_s	3.55	3.64	3.70	3.60	3.728	3.728	3.727
y_l	4.02	3.94	3.84	4.08	4.172	4.172	4.172

[a]calculated for the (4 × 2) case

corresponds mainly to the measure of hybridization between the sp^2 and sp^3 types. The deviation from planarity can also be estimated using $\delta' = \sin(\delta)/\sin(\delta_T)$ [41], the angle $\delta_T = 35.26°$ corresponding to an ideal sp^3 hybridization ($\delta' = 1$). The δ and δ′ values calculated for both Si types are drastically different for the Si(sp^3) or Si(sp^2) atoms (Table 11.1). Finally, the differences between the Si(sp^2) atoms for the (110) and (100) surfaces are smaller regarding the Si-Si bond lengths and α angles (PBE level in Table 11.1) *versus* those between the Si(sp^3) ones.

The optimized Si(110) geometry with -(sp^2–sp^3)$_n$– chains (Fig. 11.1) in the top layer is pretty similar to the p(2 × 1)a structure with buckled dimers of asymmetric type for the (100) Si plane [43]. The relative formation energy for the Si(100) p (2 × 1)a SC type could compete with the one for complex p(2 × 2) and c(4 × 2) SCs, but their relative differences in total energies per dimer are small, around some meV, which prevents to make a proper selection [44, 45]. The accuracy of the PBE/PAW approach as well as of the other similar computational methods does not reach the meV scale; a chemical accuracy of 1 kcal/mol (~0.043 eV) seems to be the best estimate. The above mentioned latter two forms were observed with different experimental techniques, i.e., p(2 × 2) [46–48] and c(4 × 2) [49, 50]. According to [44], the flip between these forms and p(2 × 1) depends on the sample preparation conditions. That is why we admitted that a more detailed and precise geometry optimization for the Si(110) surface has no real influence to elucidate its stable final geometries with the most typical defects. Additionally, the energy variations between the packing models of the buckled dimers at Si(100) are also negligibly small *versus* the different energies of the products of the interactions with the O atoms along the deposition. The small deviations between the initial energies (and between the respective geometries) of the buckled dimers at the Si(100) plane or the -(sp^2-sp^3)$_n$- chains at the Si(110) surfaces are not representative due to the strong reconstructions of the successive systems along the reaction with the O atoms. As will be demonstrated below, a simultaneous reaction of Si(110) with 12 or 20 O atoms per SC destructs the regular upper Si(110) layer and strongly perturbs the second layer in the case with 20 deposited O atoms.

Relative to these deviations, we have also calculated other parameters regarding the Si atoms with the Bader partition scheme of the electron density for both (110) and (100) planes (Table 11.3). The obtained Bader charges for the Si(sp^3) or Si(sp^2) atoms possess opposite signs for both (110) and (100) planes. The different charges are coherent with the proposed origin of the band gap of 0.5 eV for a partially ionic Si-Si bond [45], which explains the non-metallic character of the Si (100) surface [51]. Small differences are observed between the Si(sp^2) Bader charges for the (110) and (100) surface but the Si(sp^3) charges coincide for both planes.

Following Fig. 11.1g, Si94(sp^3) contributes to the largest part of the PDOS at the "valence edge" in the system, while the PDOS decreases steeply *versus* its neighbor atoms. Edwards [41] pointed a partial shift of spin density from the DB Si atoms to their neighbor fully coordinated Si atoms. Taking into account such observation, we compared the α- and β-spin projected DOS of *s*- and *p*-orbitals of Si29(sp^2) and Si94(sp^3) for the optimized $H_{12}Si_{96}$(110) model (Fig. 11.3). The superposition of the α- and β-spin PDOS which shows a favored spin density for the Si94(sp^3) while both α- and β-spin projections are similar for Si29(sp^2) and for the deeper Si105 (fully coordinated). The obtained Si(110) model (Fig. 11.1a–c, g) will thus serve as starting point for the following oxidation schemes.

Table 11.3 Atomic Bader charges *q*(*e*) of Si(sp^n) atoms (n = 2, 3) and fully coordinated nearest Si atoms (noted * in the same columns below, i.e., Si79 is the closest neighbor of Si78, Si32 is the closest neighbor of Si29) in the slabs optimized at the PBE/PAW level. Atomic numbers are partly shown in Fig. 11.1b, c, e

Type	sp^3		sp^2	
	N	q	N	q
110	14	0.23	13	–0.19
	30	0.23	29	–0.19
	62	0.23	61	–0.19
	78	0.23	77	–0.19
	94	0.23	93	–0.19
	79*	0.08	32*	–0.10
100	17	0.23	9	–0.07
	33	0.23	25	–0.07
	49	0.23	41	–0.07
	65	0.23	57	–0.07
	50*	–0.07	58*	–0.06

Five and four atoms of the same type for (110) and (100), respectively, are given to show the minor computation errors for the charge. N corresponds to the atom label

11.3.2 Oxidation

11.3.2.1 Atomic Oxygen

A minimal number of 12 O atoms per SC is required to interact with the 6 Si(sp^3) and 6 Si(sp^2) types at the optimized Si(110) surface (Fig. 11.4). Two "+12 O" models were considered: with a rather "random" (v1-$H_{12}Si_{96}O_{12}$ in Fig. 11.4a) or more "ordered" (v2-$H_{12}Si_{96}O_{12}$ in Fig. 11.4b) location of O atoms over the surface at Si-O distances fixed as 1.8 Å (Table 11.4). The optimization of the v2-$H_{12}Si_{96}O_{12}$ model leads to an ordered structure with 8 three-coordinated Si defects (i.e., 3 Si atoms of Si_2O type or with two Si and one O neighbors, 3 Si atoms of O_3 type, and 2 Si atoms of SiO_2 type) and 3 mono-coordinated O defects with double bond *versus* the nearest Si atoms, i.e., >Si = O species with |Si = O| = 1.528 Å per SC. One should note that neither O atom, nor Si(sp^2) of the >Si = O group possesses a state in the mid band gap that was shown by the PDOS analysis for the $H_{12}Si_{96}O_{20}$ model constructed using an initial Si-O distance of 1.9 Å (see below). The first more stable v1-$H_{12}Si_{96}O_{12}$ model corresponds to only 2 Si three-coordinated defects with Si_3 and Si_2O neighbors. Both the v1- and v2-$H_{12}Si_{96}O_{12}$ models were then used for the consequent Al deposition resulting in the formation of the v1- and v2-Al_4O_{12} SCs of the VI class (Scheme 11.1).

In order to consider possible series of Si oxidation products with higher O concentration, we added more O atoms per SC; 18 O atoms per SC (Fig. 11.5) were deposited as "ordered" scheme and 20 O atoms per SC (Fig. 11.6) were deposited

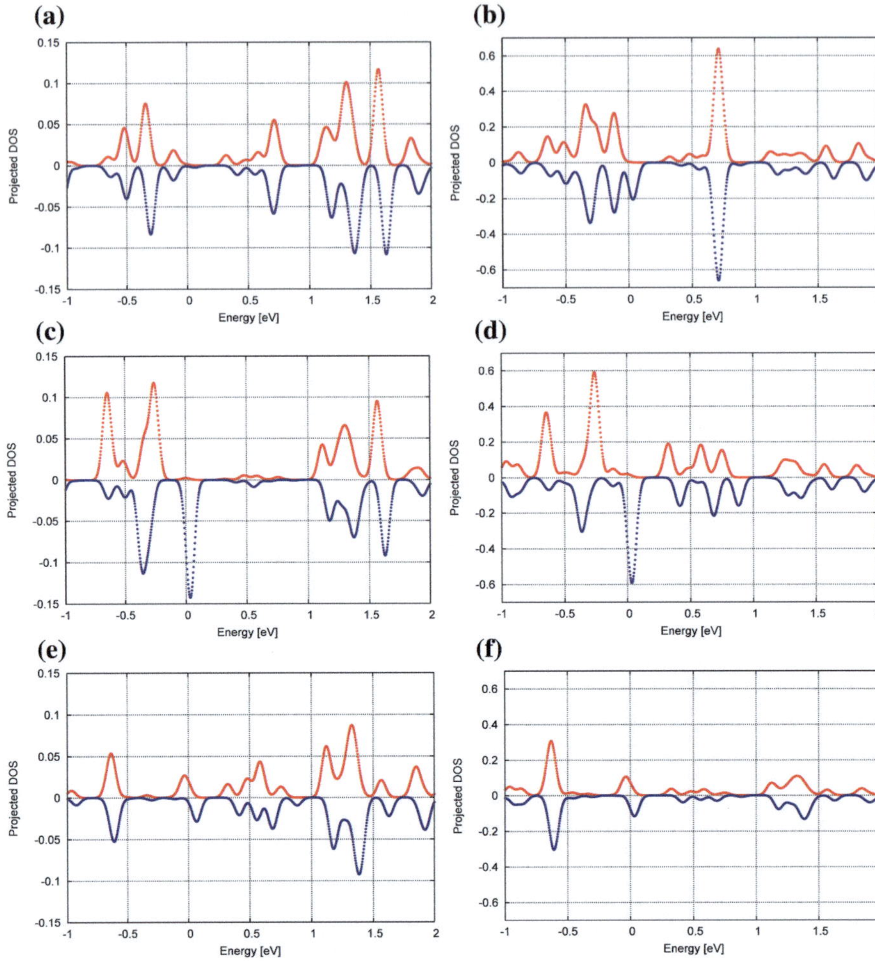

Fig. 11.3 Projected α- (*upper graph*) and β- (*lower graph*) spin DOS of *s*-type (**a**, **c**, **e**) and *p*-type (**b**, **d**, **f**) orbitals for Si29(sp^2) (**a**, **b**), Si94(sp^3) (**c**, **d**), and four-coordinated Si105 (**e**, **f**) atoms for the $H_{12}Si_{96}$ model. The atomic positions are shown in Fig. 11.1

as "random" scheme. In the "ordered" case, the 18 O atoms are arranged in one plane parallel above the (110) SC, the O…O distances ("x" and "y" in Fig. 11.5a) being selected according to the SC shape and dimensions. Despite the simplicity of this deposition scheme for 18 O atoms, the obtained band gap of 0.7 eV looks reasonable for a passivated Si slab (Fig. 11.5c). In the "random" case, the 20 O atoms were fixed at the same Si-O distance (1.7–1.9 Å) above the "defective" sp^2 or sp^3 Si atoms with random Si-O directions while avoiding strong O…O repulsion. The obtained final four $H_{12}Si_{96}O_{20}$ models considering the three different initial Si-O distances correspond to different stable products, the most stable being for the

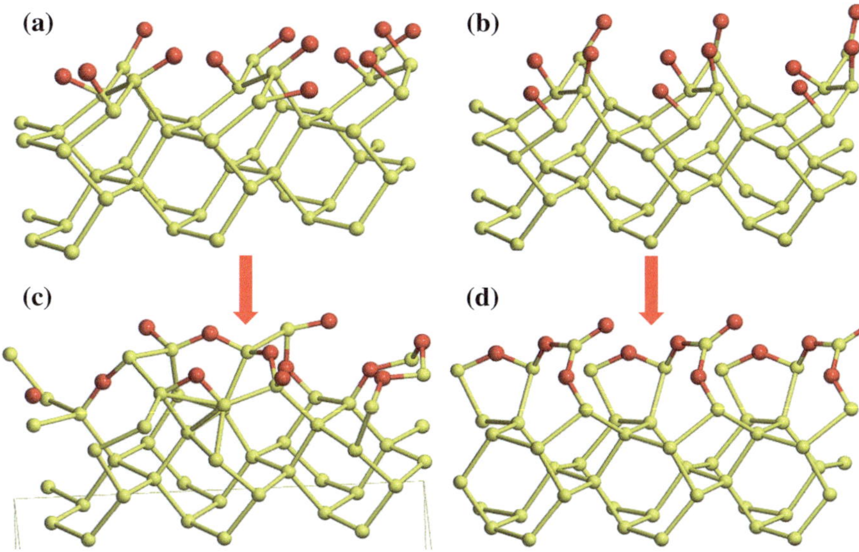

Fig. 11.4 Geometry configuration of the upper part of the initial (**a**, **b**) and final (**c**, **d**) structures for the v1-$H_{12}Si_{96}O_{12}$ (**a**, **c**) and v2-$H_{12}Si_{96}O_{12}$ (**b**, **d**) slabs obtained with the ordered and random O deposition schemes, respectively. Si, H, Al, and O atoms are in *yellow*, *grey*, *violet*, and *red*

longer distances, i.e., –3.950, –4.046/–4.197,[2] and –4.237 eV/Si for 1.7, 1.8, and 1.9 Å, respectively (Table 11.4). This trend is held irrespective of the different numbers of Si defects in the models obtained by O deposition at 1.7 (one sp^2), 1.8 (six sp^3)/1.8 (two sp^3), and 1.9 Å (one sp^3 and one sp^2). Additionally, the spatial hindrances caused by simultaneous deposition of many O atoms led to a smaller heat of oxidation. The destruction of the regular upper Si(110) layer along the addition of 12 (Fig. 11.6a) or 20 O atoms (Fig. 11.6b–d) led to energy losses. Nevertheless, the effect of oxidation can be calculated as stabilizing (exothermic) per O atom ($\Delta U/n_O$ in Table 11.4). The $\Delta U/n_O$ heat values herein are smaller than the maximal heat value estimated as –7.3 eV/O from the search of the optimal position for each additional O atom with conservation of the whole structure of the Si(100) surface [15]. However one should point out that such "atom-by-atom" addition with the following geometry optimization procedure for every atom can hardly be relevant for modeling general PA ALD deposition processes. The importance of the depth of the oxidized Si layer was emphasized by Dingemans et al. [53] for Al_2O_3 deposition as one of the principal parameters of the Si passivation. As well different thicknesses of SiO_2 layer were measured for the T ALD (0.5 ± 0.3 nm) and PA ALD (1.5 ± 0.5 nm) oxide slabs [54]. The thicker PA ALD oxide slabs can be assigned to the higher kinetic energy of the O atoms in O_2 PA ALD that leads to a deeper O penetration.

---

[2]For the second model see comment (c) in Table 11.4.

Table 11.4 Energy values of the reagent ($H_{12}Si_{96}$) and its oxidation products. Energy $\Delta U/n_O = U_C - U_T$ (eV), where $\Delta U = U_C - U_T$, n_O is the number of O atoms per SC, U_C is the value calculated at the PBE/PAW level, while U_T corresponds to the balances of Eq. (11.1). The energies of Si(bulk), $^3O_2$, Al_2O_3, and H_2 are –5.425, –9.858, –36.977, and –6.770 eV, respectively, at the PBE/PAW level

Slab	U_C	U_T	$\Delta U/n_O$[a]
$H_{12}Si_{96}$	–523.013	–561.420	–
v1-$H_{12}Si_{96}O_{12}$	–630.873	–582.165	–4.059
v2-$H_{12}Si_{96}O_{12}$	–627.455		–3.774
v1-$H_{12}Si_{96}Al_4O_{12}$	–645.980	–617.276	–2.392
v2-$H_{12}Si_{96}Al_4O_{12}$	–645.717		–2.370
v1-$H_{12}Si_{96}Al_{20}O_{42}$	–981.265	–933.385	–1.140
v2-$H_{12}Si_{96}Al_{20}O_{42}$	–974.210		–0.972
$H_{12}Si_{96}O_{18}$	–667.357	–611.576	–3.099
v1-$H_{12}Si_{96}Al_6O_{18}$	–707.267	–678.305	–1.609
v2-$H_{12}Si_{96}Al_6O_{18}$	–704.877		–1.476
$H_{12}Si_{96}Al_6O_{36}$	–823.739	–767.039	–1.575
$H_{12}Si_{96}Al_6O_{27}$	–780.935	–722.669	–2.158
$H_{12}Si_{96}Al_6O_{25}$	–770.701	–712.801	–2.316
$H_{12}Si_{96}Al_{22}O_{66}$	–1141.480	–1043.140	–1.490
v1-$H_{12}Si_{96}Al_{22}O_{48}$	–1051.490	–980.546	–1.478
v2-$H_{12}Si_{96}Al_{22}O_{48}$	–1053.183		–1.513
$H_{12}Si_{96}O_{20}$	–700.647[a]	–621.646	–3.950
	–702.550[b]; –704.163[b, c]; –705.588[b, c]		–4.045 –4.126 –4.197
	–706.341[d]		–4.238
v1-$H_{12}Si_{96}Al_7O_{20}$	–728.585	–682.905	–2.284
v2-$H_{12}Si_{96}Al_7O_{20}$	–724.328		–2.071
$H_{12}Si_{96}Al_{23}O_{44}$	–1015.978	–1004.934	–0.251
v1-$H_{12}Si_{96}Al_{23}O_{50}$	–1076.019	–1024.648	–1.027
v2-$H_{12}Si_{96}Al_{23}O_{50}$	–1069.266		–0.892
$H_{12}Si_{96}Al_{16}O_{30}$	–852.576	–839.136	–0.448
$H_{12}Si_{96}Al_{16}O_{38}$	–910.118	–866.290	–1.155

[a]model obtained starting the optimization with a Si…O distance of 1.7 Å for all three-coordinated Si atoms, resulting in one sp^2 defect; [b]model obtained starting the optimization with a Si…O distance of 1.8 Å for all three-coordinated Si atoms, resulting in six sp^3 defects; [c]optimization with NEB [52] reducing six initial Si defects to only two sp^3 defects; [d]model obtained starting the optimization with a Si…O distance of 1.9 Å for all three-coordinated Si atoms, resulting in one sp^3 and one sp^2 defects

One should note that our deposition scheme allows many ways to improve the quality of the passivation by using atomic transformations which require a TS search and analyses of the relative barriers. A complex structure of defects cannot however guarantee the obtaining of a passivated slab. A distorted rhombohedral Si66(O17O22)Si113 group (Fig. 11.6e, the view of total slab geometry is omitted)

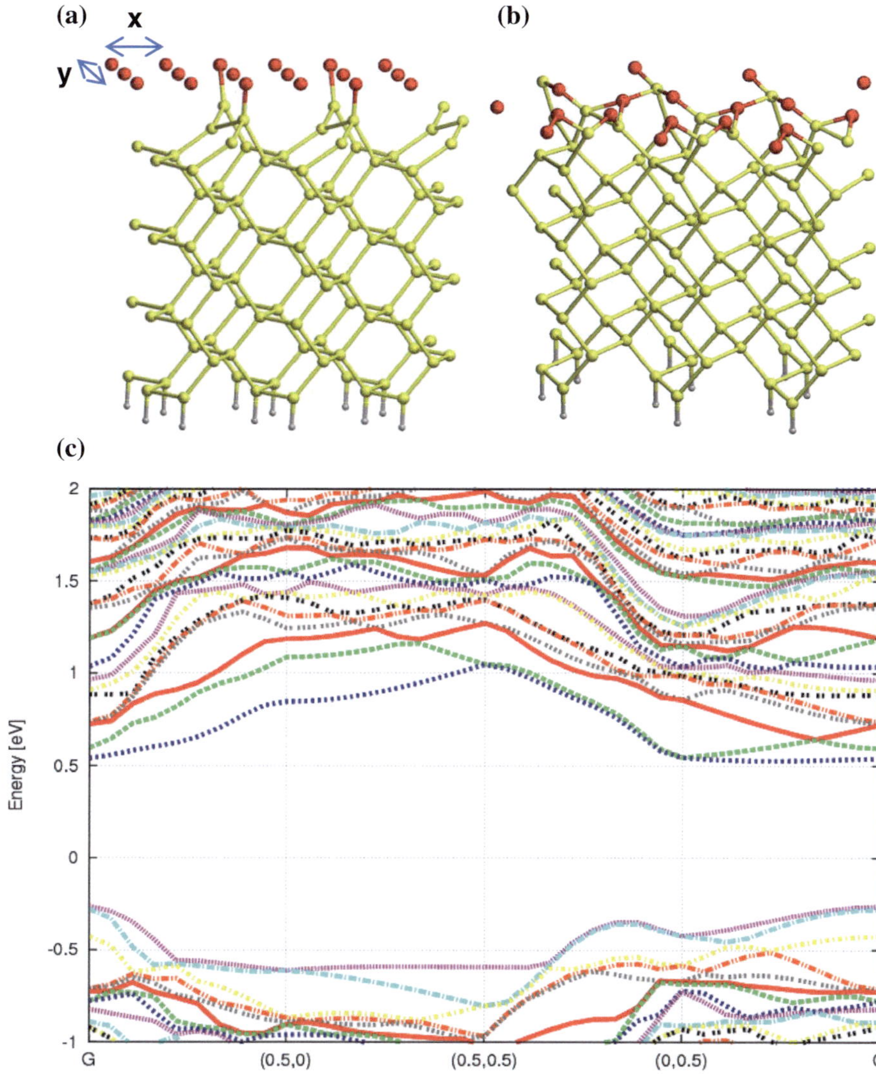

Fig. 11.5 Geometry (side view) of the initial (**a**), final (**b**), and band structure (**c**) for the $H_{12}Si_{96}O_{18}$ slab obtained with the ordered O deposition scheme. See Fig. 11.4 for *color* coding of atoms

was hence obtained as an intermediate stable structure with an energy of –704.163 eV (Table 11.4). This rhombohedral defect leads to a high *s*-type PDOS for the Si66 atom with an energy –2.5 eV (Fig. 11.6f). Despite of the close connections with the surface, this defect results in an unusual *k*-independent energy in the Brillouin zone (Fig. 11.6f) which is more typical for O_2 fragments weakly bonded to the framework. An analogous defect was obtained for the "+20 O" model, starting with an initial distance of 1.7 Å (the defect is not visible in

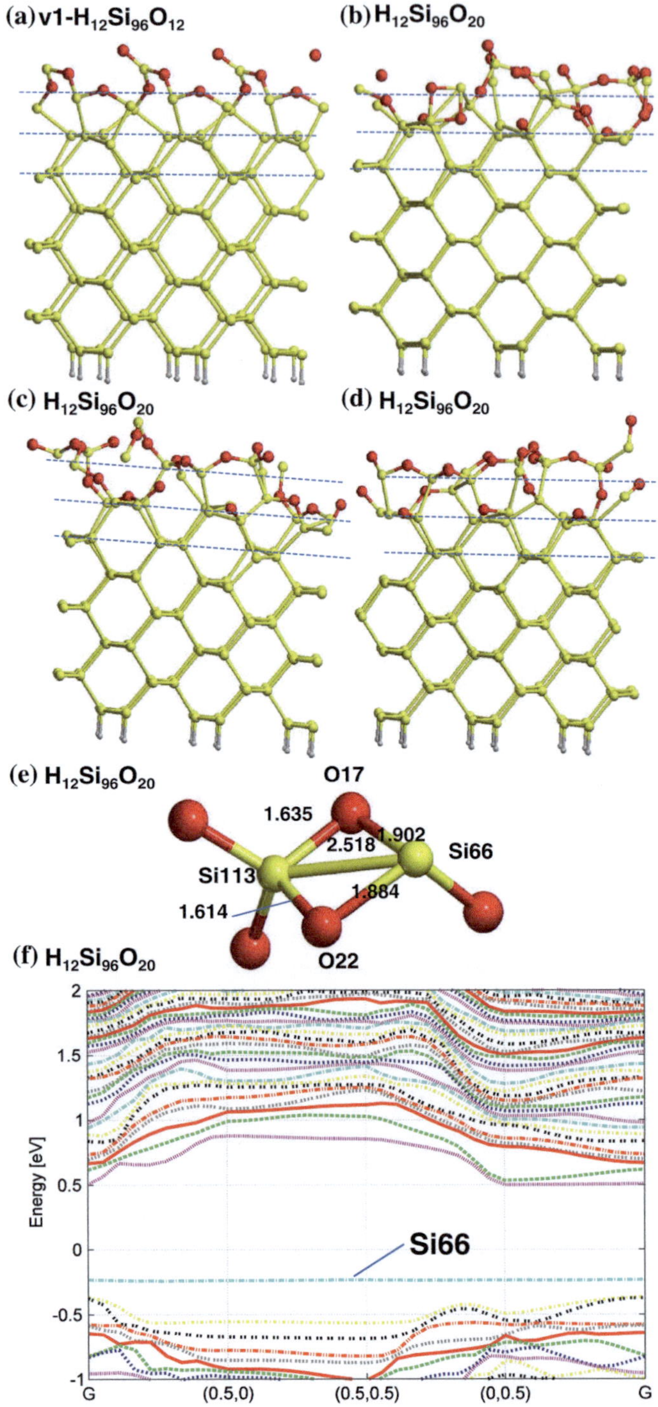
(a) v1-$H_{12}Si_{96}O_{12}$
(b) $H_{12}Si_{96}O_{20}$
(c) $H_{12}Si_{96}O_{20}$
(d) $H_{12}Si_{96}O_{20}$
(e) $H_{12}Si_{96}O_{20}$
O17
1.635
2.518
1.902
Si66
Si113
1.884
1.614
O22
(f) $H_{12}Si_{96}O_{20}$
Energy [eV]
2
1.5
1
0.5
0
-0.5
-1
Si66
G
(0.5,0)
(0.5,0.5)
(0,0.5)
G

Fig. 11.6 Geometry of the final structures for the v1-$H_{12}Si_{96}O_{12}$ (**a**) and $H_{12}Si_{96}O_{20}$ (**b-d**) slabs obtained with the O injection using different initial Si-O distances of 1.7 (**b**), 1.8 (**c**), and 1.9 Å (**d**). *Dashed lines* pass approximately through the three upper Si surface layers. Geometry of the defect (**e**) obtained for the $H_{12}Si_{96}O_{20}$ models (1.7 and 1.8 Å) with distances among atoms shown together with the band structure for the $H_{12}Si_{96}O_{20}$ (1.8 Å) case (**f**). See Fig. 11.4 for *color* coding of atoms

Fig. 11.6b) with a close Si…Si (2.520 Å) and shorter Si-O (1.675 – 1.790 Å) distances as compared to those in Si66(O17O22)Si113 defect (Fig. 11.6e).

11.3.2.2 Oxidation via Joining with $Al_{16}O_{30}$ with or Without Pre-oxidation Using Oxygen Atoms

Our first oxidation reaction of $H_{12}Si_{96}$ by the whole $Al_{16}O_{30}$ oxide fragment was modeled without pre-oxidation stage. The Si(110) – $Al_{16}O_{30}$ distance for the unique model of class I was chosen rather intuitively to allow the reaction between $H_{12}Si_{96}$ (Fig. 11.1b, c) and $Al_{16}O_{30}$ (Fig. 11.2). Formally, an insufficient oxygen concentration in the layer close to the Si atoms (6 O atoms at the lowest layer of $Al_{16}O_{30}$) does not allow a deep Si passivation. Nevertheless, this simple operation leads to a unique Si(sp^2) defect per SC which revives after the junction. The resulting $H_{12}Si_{96}Al_{16}O_{30}$ model (also noted $Al_{16}O_{30}$ for shortness) (Fig. 11.7a, b) contains one Si(sp^2) atom (Si75) of Si_3 neighbor type with a state in the mid band gap (Fig. 11.7e) which thus should be active in an electron spin resonance (ESR) spectra. We have tried to delete this Si75(sp^2) defect by addition of one O atom nearby and followed by an optimization. But together with curing of the Si (sp^2) defect and its transformation to a Si(sp^3) atom of Si_2O_2 neighbor type, a new Si(sp^2) atom ($\delta = 17.67°$) of the Si_3 neighbor type was obtained due to an angular switch of the nearest "O-Al" bond initially connected to the full coordinated Si. Despite the rather long initial distance of 4.772 Å between the old Si defect and new Si one (the final distance being 4.200 Å), the length of the "O-Al" bond is almost conserved (it varies from 1.720 to 1.727 Å only while switching) due to its angular motion towards the oxidized old Si(sp^2) defect including the Al atom. This example confirms the proposed revival way of Si(sp^2) defects beneath the oxide layer. The experimental ESR data relative to the (110) plane do not contradict that such Si(sp^2) defects can also be assigned to I_X type defects whose concentration are smaller (1/5) relative to the one of the main P_b ones [55]. The I_X type signal disappears when varying the thickness of the deleted SiO_2 layer (up to some μm) from the oxide surface and revives in a similar or a close geometry upon new oxidation [55]. The number of revival defects after oxidation can be evaluated from the ratio between the concentrations of the initial and final defects. The estimate leads to the conservation of one among 460 surface sp^2 defects as obtained by dividing the initial concentration per experimental occupancy (or final concentration) of I_X type (1×10^{12} cm^{-2} [55]). The initial concentration can be evaluated from the number of 6 total Si(sp^2) atoms per SC surface (Fig. 11.1b, c) as 6/130.8 Å^{-2} = 4.6×10^{14}

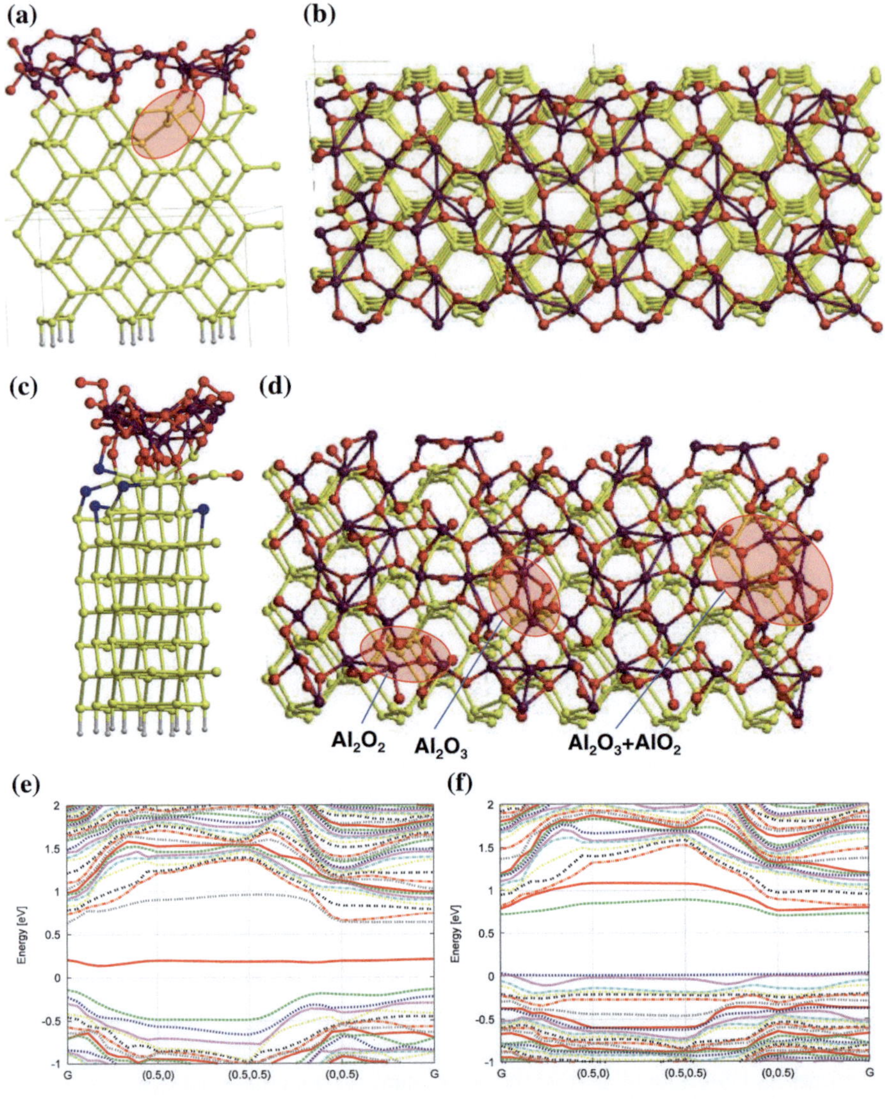

Fig. 11.7 Geometry (**a-d**) and band structure (**e**, **f**) of $H_{12}Si_{96}Al_{16}O_{30}$ (**a**, **b**, **e**) and $H_{12}Si_{96}Al_{16}O_{38}$ (**c**, **d**, **f**) models. Si(sp^3) and Si(sp^2) types of three-coordinated Si atoms are given *blue* (**c**), Si, H, Al, and O atoms are in *yellow*, *grey*, *violet*, and *red*. Al_2O_2 and Al_2O_3 defects are shown by ellipses (**d**) (see Fig. 11.8b, c, j for details)

cm^{-2}. The ESR signal of the sp^2 defect (Si_3 neighbor type) should not depend upon the temperature of oxidation in agreement with experiment as those of the I_X type [55] because the sp^2 defect is located immediately beneath the oxidized layer and does not interact with oxygen which oxidizes the Si atoms of the upper layer [55]. Formally, the initial concentration of Si(sp^3) atoms *per* SC surface is the same (6 Si

(sp^3) and 6 Si(sp^2) at the top layer, Fig. 11.1b, c) so that these assignments could be equally addressed to the sp^3 and sp^2 defects. But the main argument in favor of Si (sp^2) is conditioned by a similar ESR signal at different oxidation conditions. The Si (sp^3) atom can vary its angular geometry and bond lengths in a larger extent (Table 11.1) and hence modify the ESR signal while an Si(sp^2) can vary the bond lengths only. Regarding the small variation of bond lengths for one Si(sp^2) atom (Table 11.1), one could suspect less degrees of freedom with respect to the standard Si(sp^2) which would lead to an invariance of the ESR shape with a higher probability.

The conservation of the Si(sp^3) defects can be demonstrated for a similar merging of the $Al_{16}O_{30}$ fragment with a pre-oxidized Si(110) slab. In the first case above, we considered the junction with the lower plane of the $Al_{16}O_{30}$ slab which contains 6 O atoms. The opposite side of the slab contains 12 Al and 12 O atoms, this side being probably more rigid regarding the interaction with the Si plane. In order to induce a reaction between the Si(110) slab and the opposite side of the $Al_{16}O_{30}$ fragment, we rotated this last one by π radians (the rotation axis is shown in Fig. 11.2b). Usually, at the first step of PA ALD, oxygen is applied to form the initial oxidized layer over Si before the formation of an Al_XO_Y oxide layer [17]. While increasing the inter-surface distance by 0.86 Å between the inverted $Al_{16}O_{30}$ model and $H_{12}Si_{96}$ relative to the distance used in model I, we added 8 O atoms which simulates the initial stage of ALD deposition at the Si(110) surface (Fig. 11.7c, d, f). The content of the Al_XO_Y component was thus changed from X/Y = 16/30 = 0.53(3) to 16/38 = 0.421. Both ratios exceed the experimental value of 0.333 related to the large number of ALD cycles [37] but are smaller *versus* the ratio of 0.666 obtained in [17, 20, 38]. In such a way, we constructed the $H_{12}Si_{96}Al_{16}O_{38}$ slabs (model III, Fig. 11.7c, d, f). The reaction between the Si slab, 8 O atoms, and the inverted $Al_{16}O_{30}$ fragment leads to 5 three-coordinated Si(sp^2) and Si(sp^3) atoms of Si_3 (Si102, Si117), Si_2O (Si116), and SiO_2 (Si132, Si148, both with very short Si-Si distances lower than 2.3 Å) neighboring types (shown in blue in Fig. 11.7c). Only the *p*-PDOS of Si132 and Si148 results in a state near the conduction edge; the *s*- and *p*-type PDOS of Si102 and Si117 are below or near the Fermi energy (E_F) level (left columns in Table 11.5) while the contribution of the Si116 atom turns out to be very weak.

A very instructive aspect of this second variant of slab junction is the wide series of Al_XO_Y units which contain O_2 fragments with nearly *k*-independent *s*- and *p*-PDOS profiles. They were obtained essentially owing to the 6 O excessive atoms at the upper side of the $Al_{16}O_{30}$ fragment. Their *k*-independence points to a weak bonding with the slab and a possible easy detachment of O_2 or of other parts. The PDOS and energy values relative to the obtained Al_XO_Y species are given in the right columns of Table 11.5 while the geometry and atomic labels are presented in Fig. 11.8b, c, j. The most complex defect includes two Al_2O_3 moieties connected via one common Al atom (Fig. 11.8j). Only the O17 and O29 atoms (left part of Fig. 11.8j) reveal strong PDOS peaks in the mid band gap or at the edges as well as O14 and O26 in the isolated Al_2O_3 unit (Fig. 11.8c) and O13 and O22 in the isolated Al_2O_2 unit (Fig. 11.8b). Going forward, one should note that the "silent" O

Table 11.5 PDOS values (a.u.) and associated energy E (eV) of the *s*- and *p*-states of the Si(sp^2) (Si132, Si148) and Si(sp^3) (Si102, Si116, Si117) atoms with tetrahedral angle δ (°) (left columns) and of the Al_XO_Y type defects (right columns) in $Al_{16}O_{38}$ model

Type	N	δ	E	PDOS	Type	N	E	PDOS
sp^3	102*s*	47.4	–0.25	0.37	Al_2O_3[a]	14*p*	–0.2	8.5
	102*p*		–0.25	0.35			0.05	6.0
	117*p*	42.4	–0.4	0.24		26*p*	–0.2	1.0
			–0.2	0.31			0.05	6.2
			0.0	0.33	Al_2O_3[b]	17*p*	–0.2	8.6
	116*s*	36.2	–0.25	0.09			0.05	2.0
			0.0	0.15		29*p*	–0.2	3.5
	147*p*	–	–0.18	0.18			0.05	1.7
			–0.05	0.13		36*p*	0.05	0.1
sp^2	132*p*	5.0	–0.05	0.25	Al_2O_3[c]	13*p*	–0.4	1.75
			0.8	0.7			–0.2	0.45
	148*p*	13.9	–0.25	0.20		22*p*	–0.4	4.3
			0.9	0.85			–0.2	1.2

N corresponds to the atom label
[a]Al_2O_3 type shown in Fig. 11.8c; [b]Al_2O_3 type shown in Fig. 11.8j, with atomic labels; [c]Al_2O_2 in coherence with Fig. 11.8b

atoms (i.e., without PDOS peaks) as O15 and O27 (right part in Fig. 11.8j) can possess PDOS peaks in other structures (cf. the part about the Al_6O_{36} slab) and, on the opposite, the O13 and O22 type atoms in the isolated Al_2O_2 units can be silent (cf. the part about the $Al_{22}O_{55}$ and $Al_{22}O_{66}$ slabs). Let us also remind that Al_2O_2 moieties were usually considered as stable ones in zeolites ([56] and Refs. therein); where the stability of squared Al_2O_2 species was evaluated as favored as compared to the chain type fragment observed via our slab modeling.

11.3.3 Al Deposition

This part includes the various steps between the II and IV classes of Scheme I with different numbers of deposited Al atoms. Two Al_4O_{12} systems of type IV were constructed via successive deposition of O and Al atoms. The v1- and v2-Al_4O_{12} models were obtained from the v1- and v2-$H_{12}Si_{96}O_{12}$ slabs by injecting 4 Al atoms (Fig. 11.9a, b). Either two or one Al…Al bridge were obtained by optimization when injecting 4 Al atoms, respectively (Fig. 11.9c, d; bridges are shown by ellipses). The presence of states in the mid band gap in the systems depends on the Al…Al bond lengths, i.e., the formation of bridges, as well as on the Al coordination. For v2-Al_4O_{12} with one two-coordinated Al, one observes a state in the mid band gap (Fig. 11.9f), while such states are obtained only at the edges for v1-Al_4O_{12} (Fig. 11.9e), i.e., at the valence edge from the three-coordinated Si atoms and at the conduction edge from the Al atoms connected by a weak Al…Al

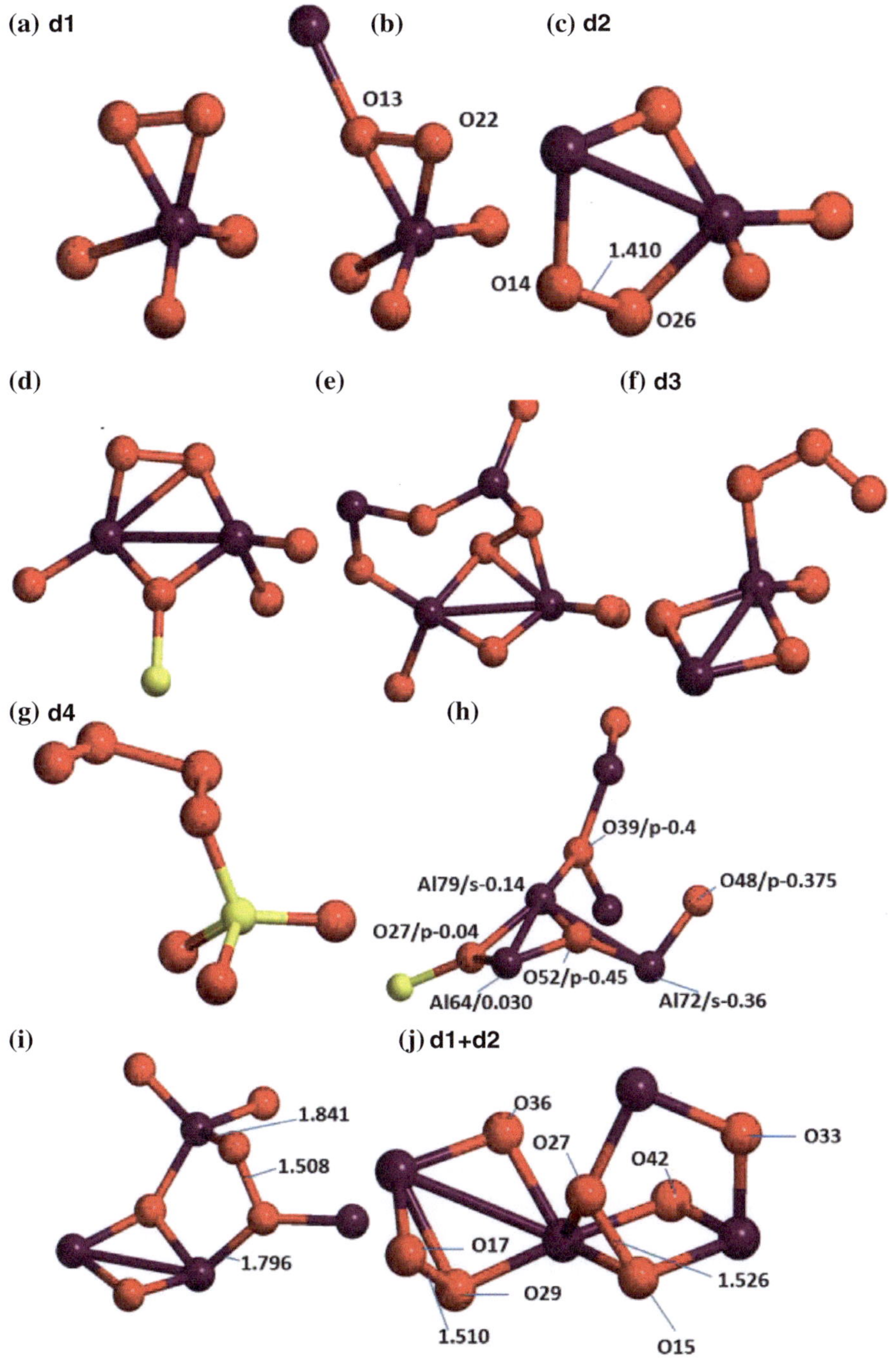
(a) d1
(b)
O13
O22
(c) d2
1.410
O14
O26
(d)
(e)
(f) d3
(g) d4
(h)
O39/p-0.4
Al79/s-0.14
O48/p-0.375
O27/p-0.04
O52/p-0.45
Al64/0.030
Al72/s-0.36
(i)
1.841
1.508
1.796
(j) d1+d2
O36
O27
O42
O33
O17
O29
1.526
1.510
O15

Fig. 11.8 Views of several defects AlO_2 (**a**), Al_2O_2 (**b**), Al_2O_3 (**c-e, i, j**), O_3 (**f**), O_4 (**g**), OAl_3 (**h**), and combination (**j**). The labels (d1–d4) are given for the band gap states in Fig. 11.11. More data about the geometries of d3 and d4 are given in Table 11.9, the band gap states are shown in Fig. 11.11e. The defects were obtained in $H_{12}Si_{96}Al_6O_{36}$ (**a, g**), $H_{12}Si_{96}Al_{22}O_{55}$ (**b**), $H_{12}Si_{96}Al_{16}O_{38}$ (**b, c, j**), $H_{12}Si_{96}Al_{22}O_{66}$ (**b, f, g**), $H_{12}Si_{96}Al_6O_{27}$ (**c, d**), v1-$H_{12}Si_{96}Al_{20}O_{42}$ (**e**), $H_{12}Si_{96}Al_{23}O_{44}$ (**h**), v2-$H_{12}Si_{96}Al_{20}O_{42}$ (**h**), and $H_{12}Si_{96}Al_{16}O_{30}$ (**i**) models. Maximum *s*- and *p*-PDOS values are given in (**h**)

bond (2.690 Å). Detailed PDOS analyses of the gap states (Fig. 11.9e, f) are given below (Fig. 11.9g, h).

v1-Al_4O_{12}. The slab contains 2 three-coordinated Si61 (of SiO_2 neighbor type, $\delta = 44.70°$) and Si77 (of Si_3 neighbor type, 46.70°) atoms, and 2 Al…Al bridges linked by rather long bonds of 2.521 and 2.670 Å. It is easy to verify that the three—coordinated Si atoms and the Al_2 groups interact. The Si77 is located close to the Al25…Al28 bond, the Al25 and Al28 atoms being quite similar (the Al25…Al28 bond equals 2.521 Å, the Al-O bonds equal 1.788 or 1.786 Å, and the Al–Si bonds equal 2.507 or 2.502 Å, respectively). The distances between Si77 and Al25 (2.990 Å) or Al28 (2.758 Å) are rather large but the PDOS profiles of these atoms are very similar at the valence edge and are nearly absent at the conduction edge. The PDOS is dominated by the profile of the Si77 atom (Fig. 11.9g) with a maximal value of 0.56 a.u. at $E = -0.38$ eV (to be compared with 0.24 and 0.28 a.u. for Al25 and Al28[3] in Table 11.6) and two small branches with peaks at –0.54 and –0.24 eV. The PDOS of Si77, with a similar shape to the PDOS of Al28 (Fig. 11.9g), is omitted in Fig. 11.9g to present all the PDOS values with a smaller scale. Hence, the Al25…Al28 group is involved in the state determined by the three-coordinated Si77. It is not the case for the second Al26…Al27 bridge of 2.670 Å with Al27, whose coordination is between two- and three-coordinated states, and leads to a state at the conduction edge.

The three-coordinated Si61 atom participates in a rhombohedral Si61(O14O15) Al26 group while the three-coordinated Al26 (O_3 neighbor type) is connected by a Al26…Al27 bond of 2.690 Å with Al27 (Si and O neighbors) whose coordination, between 2 and 3, is determined by the Al…Al bond. Only the influence of Al27 can explain the presence of the two PDOS maxima at the conduction edge for all three Al26, Al27, and Si61 (not shown) atoms at energies of 0.66 and 0.54 eV (Table 11.6). The relative PDOS maxima are larger for Al27 (Table 11.6, PDOS for Si61 is given in Fig. 11.9g). In parallel, Si61 manifests *s*- and *p*-DOS peaks of high intensity at the valence and both valence and conduction edges, respectively (Fig. 11.9g). Such behavior has been observed for Si atoms of the Si_4 and Si_3O neighbor types optimized using the Bethe models for different Si/SiO_2 interfaces (without *s/p* partition) [32]. (Several atoms which contribute to both sides of the gap with similar *s*- or *p*-type orbitals will be mentioned below). The participation of O14 and O15, which connect the Si61 and Al26 atoms in the rhombohedral Si61

[3]The puzzling aspect is the comparable PDOS values for the *s*- and *p*-PDOS of Al25 (Table 11.6) and a negligible *s*-PDOS of Al28 despite of the similarities between these Al atoms.

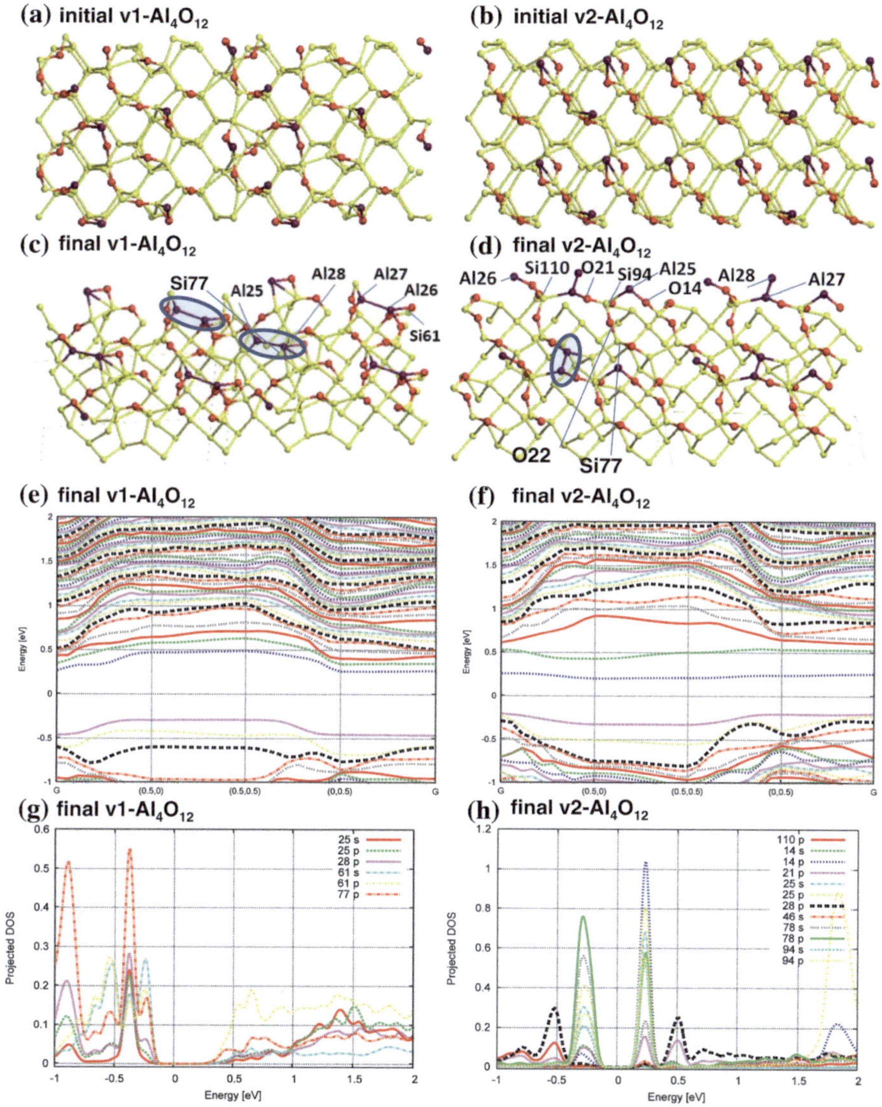

Fig. 11.9 Geometry (*top view*) of the (2 × 2) SC of the initial (**a**) and final (**c**) random v1-Al_4O_{12} and initial (**b**) and final (**d**) ordered v2-Al_4O_{12} models obtained with injection of 4 Al and corresponding band structures (**e**) and (**f**). The *s*- and *p*-PDOS values are given for v1-Al_4O_{12} (**g**) and v2-Al_4O_{12} (**h**). Al…Al bridges are shown by ellipses. See Fig. 11.4 for *color* coding of atoms

(O14O15)Al26 defect, in the state with maxima at 0.54 and 0.66 eV is less effective at the O15 atom (0.06 and 0.06 a.u.), which contributes to the state at the valence edge only similarly to Si61, and is negligible at O14 (0.02 and 0.04 a.u.). Nevertheless the non-zero contributions of O15 to the state at the conduction edge allow

Table 11.6 PDOS values (a.u.) and associated energy E (eV) of the *s*- and *p*-states of the Al(N) atoms in some models

N	v1-Al_4O_{12}		v2-Al_4O_{12}		N	v2-$Al_{20}O_{42}$	
	E	PDOS	E	PDOS		E	PDOS
25*s*	–0.38	0.24	–0.29 0.24	0.30 0.68	58*s*	0.5 –0.4	0.53 0.14
25*p*	–0.38	0.23	–0.29 0.24	0.40 **0.80**	58*p*	0.5	**0.68**
26*p*	0.54 0.66	0.16 0.11	–	–	55*p*	–0.4	0.18
27*p*	0.54 0.66	0.26 0.19	–0.56 0.53	0.22 0.30	56*p*	–0.6	0.11
28*p*	–0.38	0.27	–0.56 0.53	0.30 0.25	57*p*	0.5	0.10

N corresponds to the atom label

to confirm a delocalized state over the Al27...Al26-O15-Si61 chain. The weak Al26...Al27 bond of 2.670 Å clearly limits the position of the conduction edge (Fig. 11.9g) and thus determines the width of the band gap. All four Al atoms and Si77 and Si61 defects lead to a narrower gap in our model relative to the one of pure *c*-Si (1.15 eV). Its evolution after the junction with the Al_2O_3 fragment is presented with the v1-$Al_{20}O_{42}$ model in Sect 11.3.4.

v2-Al_4O_{12}. The band structure of this less stable system, by as much as 0.263 eV *versus* v1-Al_4O_{12} (Table 11.4), is more complex than the one of v1-Al_4O_{12}. The v2-Al_4O_{12} model contains a wide series of under-coordinated atoms, i.e., a mono-coordinated Al atom (as Al26 linked to an O at 1.732 Å), which does not show any state in the band gap, a two-coordinated Al atom (as Al25), three-coordinated Si atoms of Si_2O neighbor type (as Si77, $\delta = 45.98°$), Si78 of O_3 neighbor type (as Si78, $\delta = 31.73°$), Si93 of SiO_2 neighbor type ($\delta = 38.54°$), Si atoms of Si_3 neighbor type (as Si95, $\delta = 38.49°$), and Si atoms of Si_3 neighbor type, (as Si96, $\delta = 35.74°$). The DOS picture of v2-Al_4O_{12} is simplified by the two weak Si...Si bonds of 2.603 Å between Si77...Si95, and of 2.614 Å between Si93...Si96. As a result, the PDOS at these atoms has moderate maxima at the conduction edge only. An exception is Si77 (with a *p*-PDOS intensity of 0.08 a.u. in the mid band gap, i.e., at E = 0.24 eV) which is connected via O22 with the fully coordinated Si94 (Fig. 11.9d). The atomic geometry of Si94 is not strongly distorted, i.e., $\beta = 6.8°$, but its fourth bond of 2.554 Å is directed *versus* Al25, whose coordination varies between 2 and 1, if one counts this bond or not, respectively. The Al25 atom is probably the reason of an enhanced PDOS at both fully coordinated Si94 and O14 neighbors.

As a result of the delocalization of the state originated from the two-coordinated Al25, one observes a long (Al26-O17-Si110)-(Al27-O21-Si94)-(Al25-**O14**-Si78)-O15(O16) chain containing three repeating non-equivalent Al-O-Si fragments. The variation of the maximum DOS values, of the *s*- or *p*-type character, in the Al26-...-O15(O16) chain corresponds to the (0.0*s*-0.02*p*-0.12*p*)-(0.04*s*-0.14*p*-0.53*s*)-(0.68*s*-

1.03*p*-0.76*p*)-0.37*p*(0.20*p*) sequence. Both three- and four-coordinated Si78 and Si94, of SiO_3 neighbor type, possess states in the mid band gap and near the valence edge (Fig. 11.9f), like the one of Al25 (Table 11.6, Fig. 11.9h). The PDOS at O14 in the mid band gap (1.03*p* a.u. at 0.24 eV and 0.11*s* a.u. at –0.29 eV) is even larger than the ones at both its neighbors defect, Al25 and Si78 (Fig. 11.9h). The Al25 atom possesses both *s*- and *p*-type states in the mid band gap (0.8*p* a.u. at 0.24 eV and 0.4*p* a.u. at –0.29 eV; Fig. 11.9h and Table 11.6). Together with O14, other fully coordinated O neighbors of the three-coordinated Si78 also possesses valuable PDOS, i.e., O15 (0.37*p* a.u. at 0.24 eV and 0.54*p* a.u. at –0.29 eV) and O16 (0.20*p* a.u. at 0.24 eV and 0.29*p* a.u. at –0.29 eV). Owing to the weak Al27…Al28 bond of 2.612 Å, both Al atoms do not manifest *p*-DOS peaks (*s*-DOS peaks are minor ones) in the mid band gap but only near both edges (Al28 at –0.56 and 0.53 eV as shown in Fig. 11.9h, Al27 is not shown). Finally, the study of the slightly less stable v2-Al_4O_{12} system allows finding an interesting example of enhanced PDOS at the fully coordinated O14 atoms which reveals larger PDOS peaks *versus* the defective two-coordinated Al25 and three-coordinated Si78 neighbors. The v2-Al_4O_{12} model will serve for the following oxidation towards the v2-$Al_{20}O_{42}$ model using the $Al_{16}O_{30}$ fragment as described above.

Al_7O_{20}. Two v1- and v2-Al_7O_{20} models were constructed from the class II models (the $H_{12}Si_{96}O_{20}$ slab obtained with Si…O distances of 1.8 Å) by injecting 7 Al atoms (Fig. 11.10). The initial Al positions were targeted towards the O atoms so that the directions of Al-O vectors are nearly tilt for v1-Al_7O_{20} (Fig. 11.10a) and nearly perpendicular for v2-Al_7O_{20} (Fig. 11.10b) relative to the surface. Starting from different initial models, the two final optimized models led to an Al-clusterization. Two Al clusters were obtained for the two v1-Al_7O_{20} models (3 Al + 3 Al + 1 Al, shown by ellipses in Fig. 11.10c) and v2-Al_7O_{20} model (3 Al + 4 Al in Fig. 11.10d).

In the last v2-case, the largest PDOS (0.72 a.u., Figure 11.10f) belongs to Al34 (Fig. 11.10d), in the middle of the 4-atom chain (Al39-Al34-Al35-Al37) with decreasing values for its neighbors ranging from 0.42 a.u. for Si72 ($|$Si72-Al34$|$ = 2.645 Å), 0.34 a.u. for Al35 ($|$Al35-Al34$|$ = 2.478 Å) to 0.06 a.u. for Al37 ($|$Al37-Al34$|$ = 2.643 Å). By comparing the distances, one can outline that the bonds including Al37 and Si72 are extremely weak so that Al34 can be considered as a mono-coordinated defect. The other 3 Al atoms of the triangular group (Al33, Al36, Al38, not labelled in Fig. 11.10d) possess a state closer to the conduction edge at $E > 0.25$ eV.

A similar situation is observed for the obtained most stable v1-Al_7O_{20} model (Fig. 11.10c) with the largest PDOS belonging to Al33 (0.57 a.u., Figure 11.10e), which forms two typical bonds of 2.447 Å with Si41 and 2.367 Å with Si72 and one weak bond of 2.651 Å with Al34, thus leading to two-coordinated defects with a state in the mid band gap at nearly 0.1 eV (Fig. 11.10g).

Al_6O_{18}. Two v1- and v2-Al_6O_{18} models were constructed from the $H_{12}Si_{96}O_{18}$ slabs by injecting 6 Al atoms. Two grids with a parallelogram form in coherence with the SC were selected for the deposition of Al atoms with (Fig. 11.11a). The 6 Al atoms were positioned with different Al…Al distances in each of the two

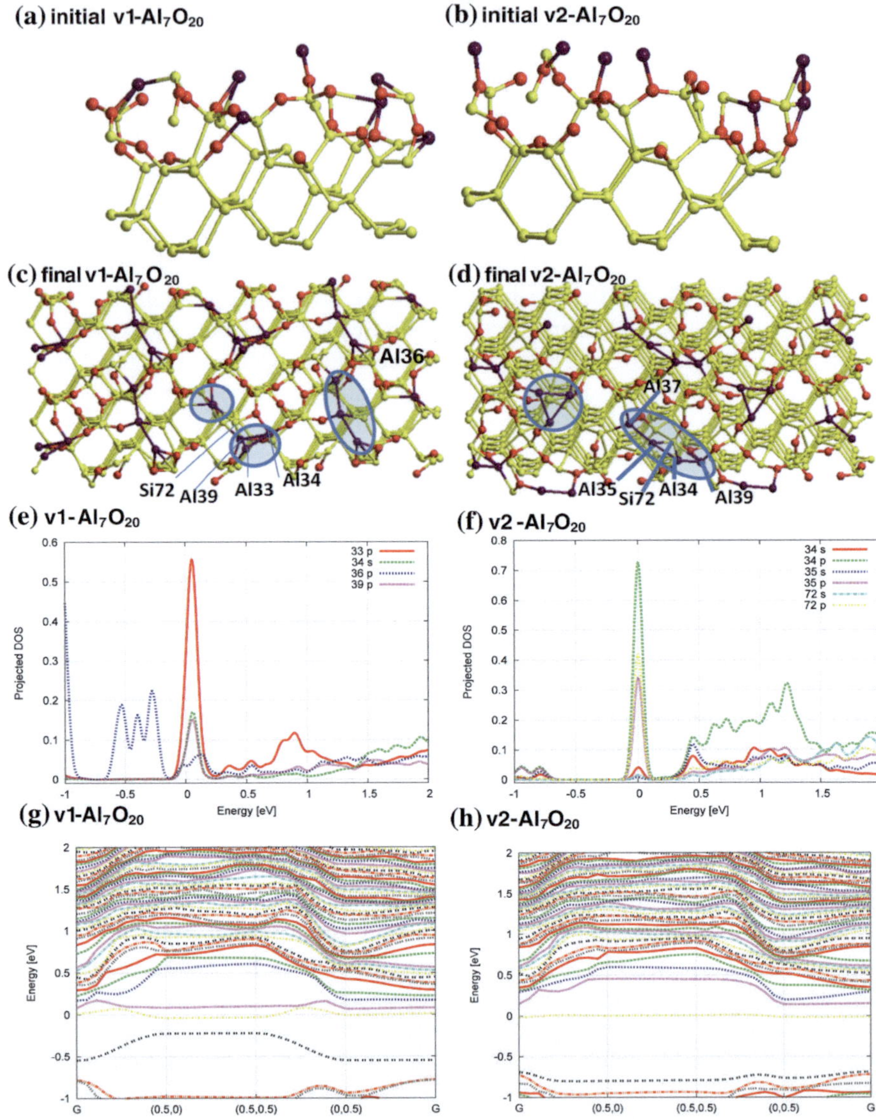

Fig. 11.10 Geometry of the initial (configuration of the upper part of one SC in (**a**), (**b**)) and of the final (2 × 2) SC (*top view*) v1-Al_7O_{20} (**c**), v2-Al_7O_{20} (**d**) models, obtained with the Al deposition schemes (**a**, **b**), corresponding *s*- and *p*-PDOS values (with atomic labels given in (**c**, **d**)) of selected atoms (**e**) and (**f**), and corresponding band structures (**g**) and (**h**). See Fig. 11.4 for *color* coding of atoms

directions corresponding to the sides of the parallelogram (Fig. 11.11a), i.e., three Al atoms located along the smaller and larger sides of the parallelogram in the v1 (Fig. 11.11a) and v2 variants (not shown here), respectively. For both cases, the larger side is oriented along the longer side of the SC (Fig. 11.11a). In the

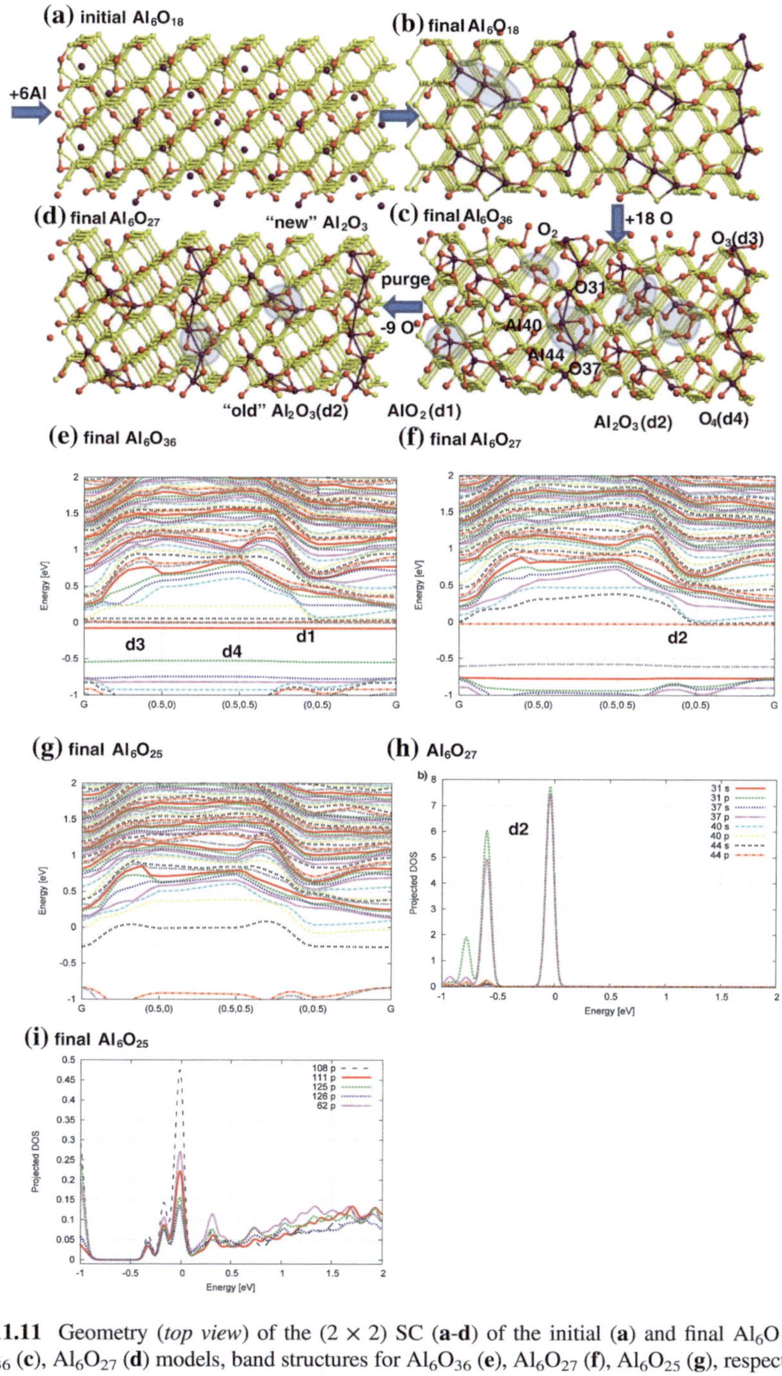

Fig. 11.11 Geometry (*top view*) of the (2 × 2) SC (**a-d**) of the initial (**a**) and final Al_6O_{18} (**b**), Al_6O_{36} (**c**), Al_6O_{27} (**d**) models, band structures for Al_6O_{36} (**e**), Al_6O_{27} (**f**), Al_6O_{25} (**g**), respectively, and *s*- and *p*-PDOS values for Al_6O_{27} (**h**) and Al_6O_{25} (**i**). The states of d1-d4 type defects are shown in (**e, f, h**). See Fig. 11.4 for *color* coding of atoms

Table 11.7 Al…X bond lengths (Å) (X = O, Al) in the Al_6O_{18} and Al_6O_{27} slabs as optimized at the PBE/PAW level

Atom	Al_6O_{18}	Al_6O_{27}
Al1	1.819, 2.640, 2.696	1.696, 1.722, 1.853, 1.875, 2.666, 2.831
Al2	1.817, 1.818, 2.640, 2.692	1.701, 1.702, 1.864, 1.937, 2.831, 2.843
Al3	1.780, 2.692, 2.696	1.707, 1.712, 1.835, 1.868, 2.666, 2.843

Table 11.8 Al…Al bond lengths (Å) in Al_2O_3 (bulk) and Al_6O_X slabs as optimized at the PBE/PAW level compared to the experimental values in boehmite [57]

Al_2O_3 (bulk)	Al_6O_{18}	Al_6O_{36}	Al_6O_{27}	Al_6O_{25}	Boehmite
2.793, 2.795, 2.801, 2.803, 2.809, 2.812, 2.835	2.640, 2.696, 2.692 (chain)	2.859, 2.717 (3-atom species)	2.831, 2.666, 2.843 (chain)	2.762, 2.809, 2.924, 2.852, 2.733 (two 3-atom species)	2.866, 2.868 (chain) [57]

v1-model, the grid with the initial (shorter) Al…Al distance of 3.752 Å leads to the formation of continuous chains containing three-, four-, and three-coordinated Al atoms (Al1, Al2, Al3), respectively (Table 11.7). The obtained Al…Al distances are shorter as compared to the experimental values of 2.868 and 2.866 Å observed in boehmite [57] or as compared to the Al…Al distances ranging from 2.793 to 2.812 Å in the SC of γ-Al_2O_3 with an $Al_{16}O_{24}$ content as optimized with VASP at the same computational level (Table 11.8). Three other Al atoms are linked together with also shorter Al…Al distances of 2.672 and 2.610 Å. One of the terminal Al atoms of this three-atom group forms two Al–Si bonds of 2.479 and 2.743 Å, but these bonds disappear at the next oxidation step. The short Al…Al bonds are the consequence of the low coordination numbers (3 or 4) for all deposited Al atoms.

Al_6O_{36}. The Al_6O_{36} model was constructed from the class IV model (the $H_{12}Si_{96}Al_6O_{18}$ slab) by injecting 18 O atoms (Fig. 11.11c) resulting in a product with the maximal O/Al ratio of 6. As a result, the Al chain is partly oxidized. The three-Al atom fragment decomposes to three separate Al atoms with Al-O-Al connections ($|Al\ldots Al| > 3$ Å) while the continuous Al-chain (Fig. 11.11b) is modified to a three-atom fragment (Table 11.8, Fig. 11.11c) with Al…Al distances closer to the ones in Al_2O_3 or in boehmite ($|Al\ldots Al| < 2.9$ Å). The Al-O-Al links were formed parallelly to each elongated Al…Al bond of the three-atom fragment. The difference with boehmite is the slightly lower coordination numbers for two of the three five-, six-, and five-coordinated Al atoms. The longer bonds are probably the consequence of the formation of Al-O-Al chains ($|Al\text{-}O| > 1.69$ Å) parallel to

Table 11.9 O…O bond lengths in the O_X ($X = 3$ as d3 in Fig. 11.8f and $X = 4$ as d4 in Fig. 11.8g) species stabilized as CaO_XCa, in the 8R or 6R + 4R clusters, CaMOR zeolite [58, 59], and coordinated as O_X to the Al or Si atoms of the Al_6O_{36} and $Al_{22}O_{66}$ slabs as optimized at different theory levels

System	Method	O…O distances	
		X = 3	X = 4
CaO_XCa(8R)	B3LYP/6-31G*	1.500, 1.503	1.340, 1.350, 1.998, 2.498
CaO_XCa(6R + 4R)	B3LYP/6-31G*	1.496, 1.498	1.358, 1.362, 1.928, 2.514
CaO_XCa(6R + 4R)	MP2/6-31G*	1.508, 1.509	1.439, 1.442, 2.276, 2.300
CaO_XCaMOR	PW91/PAW	1.500, 1.503	1.238, 1.358, 1.828, 2.414
$H_{12}Si_{96}Al_6O_{36}$	PBE/PAW	1.253, 1.560	1.239, 1.339, 1.831, 2.416
$H_{12}Si_{96}Al_{22}O_{66}$	PBE/PAW	1.283, 1.407	1.222, 1.263, 1.835, 3.260 1.232, 1.393, 1.782, 2.560

the Al…Al bond. The most important consequence of the oxidation with an O excess is the formation of the defect structures AlO_2, Al_2O_3, AlO_3, and AlO_4 (shown by ellipses in Fig. 11.11c, d) and O_2 molecules with gap states that are *k*-independent (Fig. 11.11e). Their domination in the defect states in the band gap is confirmed by the *s*- and *p*-PDOS analysis (Fig. 11.11h). The AlO_2 (Fig. 11.8a) and Al_2O_3 (Fig. 11.8c–e) are very similar to the ones obtained at the surface of the $Al_{16}O_{38}$ slab with essentially higher Al concentrations (Al/O = 0.421) than used herein (Al/O = 0.167). Let us stress that these AlO_2 and Al_2O_3 structures were obtained from deposited Al atoms and not from Al atoms of the oxide fragment as one could suspect after considering the part about the $Al_{16}O_{38}$ slab. It is easy to note that the AlO_2 and Al_2O_3 species form at the upper (non-reactive) side of the $Al_{16}O_{30}$ fragment while modeling the $Al_{16}O_{38}$ slab. It confirms the rather universal character of apparition of AlO_2 and Al_2O_3 fragments with an excess of oxygen, realized with O_2 PA ALD deposition.

It is also worth to mention that exothermic heats of formation for similar O_3 and O_4 type moieties were observed for the reactions of the MeO_XMe (X = 1–2) species with molecular oxygen (Table 11.9) [58, 59]. Close similarities are revealed between the O_4 geometries obtained in our slab and the ones noted in the cluster models and periodic models of zeolites despite of more variable geometries in the last case. The closeness between bond lengths (±0.02 Å) with the similar O_4 structures optimized at the PW91/PAW level for the Ca_2O_4 species in CaMOR zeolite is even surprising.

$Al_{22}O_{66}$. The attempt to finalize the formation of the $Si/SiO_X/Al_2O_3$ interface by merging the already deposited 18 O/6 Al/18 O slab (Al_6O_{36}) with the $Al_{16}O_{30}$ fragment (Fig. 11.2) without purging the Al_6O_{36} slab was not successful. The higher O concentration did not lead to the formation of new types of defects as compared to those obtained after injection of 18 O atoms per SC. The obtained O_3 or O_4 fragments did not interact with the atoms of the $Al_{16}O_{30}$ fragment even if the O_3 or O_4 species were located mostly at the upper side of the oxidized $Si/SiO_X/AlO_Y$ slab. Moreover, new O_4 species connected to the Si atom were obtained while the initial O_4 species switched their connection from the deposited Al atoms to Al ones of the $Al_{16}O_{30}$ oxide layer. An O1-O2 bond reduction from 1.339 to 1.263 Å in the O_4 species is noted for the part -Al-O1-O2-O3-O4 in the last O group. This shorter bond correlates with a weaker neighbor Al-O1 bond noted at the new position, i.e., 2.150 instead of 1.824 Å before. The other noted bond changes are minor but the longer distance between the terminal O1-O2-O3-O4 atoms increases up to 3.260 Å. It showed that the O_3 or O_4 moieties are relatively flexible and stable. The difference between the O_3 species in zeolites and in our slab, i.e., one shorter terminal O-O bond for the latter, allows to propose an easier O_2 deletion from the O_3 species. The question of which species could be desorbed or decomposed within the purging stage and how to desorb them is not trivial because O_2 or O_3 molecules can also be formed as soon as O_3 or O_4 fragments decompose. Their disappearance along the purging ALD stage could or not happen because the desirable state of passivation is achieved only after annealing. The possibility of the defects deletion should however be strictly confirmed by a precise TS evaluation. In this particular work, we assume the possibility of deletion of O_2 or O_3 fragments along the purging stage.

Al_6O_{27}. The Al_6O_{27} model was constructed from the class VI model (the $H_{12}Si_{96}Al_6O_{36}$ slab) by deleting 9 O atoms (Fig. 11.11d). The deletion of the defects from the Al_6O_{36} model was realized in two successive steps. First, we deleted the O_3 or O_4 fragments towards stable O_2 or O_3 molecules, respectively, and an O_2 molecule was cut from the AlO_2 moieties. The remaining Al-O bonds relax in the optimized Al_6O_{27} (or 18O-6Al-9O) model after desorption of O_2 or O_3. We thus conserved only one Al_2O_3 defect (shown by ellipses in Fig. 11.11c, d) which remains visible in the band gap. One supposes that the O atoms could be redistributed upon the loss of the O atoms by the neighboring Al atoms. However, the defect was kept in the chain connecting 3 six-coordinated Al atoms ("old" Al_2O_3 in Fig. 11.8c). Its deletion was realized in the Al_6O_{25} (18O-6Al-7O) model. Randomly, the new type of Al_2O_3 (or $Al_2O^*O_2$[4]) fragment appeared between two -(O_3) A1-O*-Al2(O_4)- atoms at the location of the deleted O_2 or O_3 fragments from Al1 and Al2 atoms, respectively ("new" Al_2O_3 in Fig. 11.8d). The 2 remaining O_1 and O_2 atoms of the O_3 or O_4 fragments form a new $Al_2O^*O_2$ fragment with another geometry with O_1 and O_2 atoms coordinated to one and two Al atoms, respectively, and with a longer $|O_1\text{-}O_2| = 1.522$ Å (Fig. 11.8d) than within the "old" Al_2O_3, $|O_1\text{-}O_2| = 1.403$ Å (Fig. 11.8c). One of two O atoms coordinates to one Al (1.746 Å) while the second possesses two non-equivalent Al-O bonds of 1.746 and 2.320 Å. The initial O* atom of the $Al_2O^*O_2$ moiety does not change of coordination within the transformation. The O atoms of the "new" Al_2O_3 group do not possess defect states in the mid band gap similarly to the defect in the v1-$H_{12}Si_{96}Al_{20}O_{42}$ model (Fig. 11.8e) discussed below.

[4]The * sign is used to depict the O* atom in the Al-O*-Al fragment which is conserved in the new Al_2O_3 defect.

Al_6O_{25}. The second stage of removing the defects from the Al_6O_{36} model consisted in preparing the Al_6O_{25} model from the Al_6O_{27} one. For this, the O_2 fragment of the "old" Al_2O_3 defect (Fig. 11.8c) was withdrawn in Al_6O_{27} and optimized, thus resulting in the complete deletion of three gas like states which are independent on the *k*-value (at the energies of –0.75, –0.6, and 0.0 eV in Fig. 11.11f). The remaining "new" Al_2O_3 defect (Fig. 11.11d, e) does not possess any state in the mid band gap. Despite the complete passivation of all added 6 Al atoms and O connected atoms, we obtained high density of states near the conduction edge and near the mid band gap. The last state near E_F (Fig. 11.11g) belongs mainly to the unique three-coordinated Si108 atom of the SiO_2 neighbor type ($\delta = 43.29°$). This atom is connected within the O28-Si108-Si111(Si62)-Si125-Si126 chain of Si_4 neighbor type (Si62, Si111,[5] Si126) which reveals higher *p*-PDOS values than the *s*-ones with the exception of Si125 of the Si_2O_2 neighbor type, i.e., 0.21*p*-0.43*p*-0.23*p*(0.23*p*)-0.18*s*-0.13*p*. The angular analysis does not confirm that the enhanced PDOS values are the result of structural Si distortions.

The Al…Al distances in the Al_6O_{25} model are presented in Table 11.8. Externally, Al_6O_{25} is pretty similar to the geometry of the Al_6O_{27} model (Fig. 11.11d) with a minor deviation. One of the Al…Al distances reaches 2.924 Å (Table 11.8), so that the chains which were continuous in Al_6O_{27} decompose into three-atom fragments in Al_6O_{25}. The Al_6O_{25} model includes 2 three-atom Al-fragments. This Al…Al elongation is not in contradiction with the stability of the Al-chains in the slabs observed at all preceding steps which have shown the possible boehmite formation at the Si(110) surface using O_2 PA ALD as discussed below (Sect. 11.4.2).

11.3.4 Junction with the $Al_{16}O_{30}$ Fragment After Al Deposition

v1-$Al_{20}O_{42}$. The v1-$Al_{20}O_{42}$ system of V type (Scheme 11.1) was obtained from the merging of the v1-Al_4O_{12} model (type IV) with the $Al_{16}O_{30}$ fragment. The lower part of $Al_{16}O_{30}$ reacts with the deposited Al atoms and three of the total of four Al atoms are trapped into the oxide layer initially located close to the upper Si plane. It could be the consequence either of the larger O/Al ratio in $Al_{16}O_{30}$, or of the necessary coverage of the deposited Al atoms by an O layer before such merging. Such bad example of junction accompanied by Al redistribution is nevertheless instructive leading to a wide series of Si defects (Table 11.10). If we accept as a conventional upper limit for a weak Si…Si bond 2.5 Å, then the final slab involves a two-coordinated Si91 of O_2 neighbor type (Si91 with two O20 and O23 neighbors), a Si of Si_2 neighbor type (Si94), a three-coordinated Si of SiO_2 neighbor type (Si107), a Si of Si_3 neighbor type (Si141), and a Si of O_3 neighbor

[5]The largest (fifth) Si111-Si125 distance of 2.507 Å allows to consider Si111 as a five-coordinated (Si_5) one.

Table 11.10 Si-X bond lengths (Å) (X = Si, Al, O), X-Si-X' bond angles (°), and *s*- or *p*-PDOS values (a.u.) of the fully and under-coordinated Si atoms, those being the neighbors of the dangling bond defects in the v1-$Al_{20}O_{42}$ (Si141) and v2-$Al_{20}O_{42}$ (Si92) models optimized at the PBE/PAW level

Model	Type	Si-X	X-Si-X'	PDOS
v1-$Al_{20}O_{42}$	Si91	1.643, 1.742, **2.604, 2.673**	89.7, 104.4, 105.0, 110.4, 116.4, 126.1	0.27*s*, 0.23*p*
	Si108	1.663, 1.668, 1.695, **2.673**	82.3, 103.4, 105.5, 107.4, 115.4, **132.5**	0.15 *s*, 0.13*p*
	Si141	2.340, 2.382, 2.435, **2.558**	91.5, 97.5, 98.2, 99.3, 124.1, **151.3**	**0.45*p***, 0.30*s*
	Si75	1.680, 2.267, 2.380, 2.383	95.8, 102.5, 106.0, 113.9, 115.1, 119.2	0.03*p*, 0.005*s*
	Si79	2.300, 2.399, 2.420, 2.435	**76.5**, 95.9, 111.1, 111.1, 118.6, **141.0**	0.18*p*, 0.07*s*
	Si94	2.342, 2.367, 2.558, **2.604**	**85.9**, 87.8, 104.8, 114.6, 120.9, **132.0**	**0.40*p***, 0.24*s*
	Si144	2.318, 2.340, 2.354, 2.355	104.9, 106.7, 107.2, 107.3, 111.9, 118.2	0.08*p*
v2-$Al_{20}O_{42}$	Si76[a]	1.724, 2.359, 2.400, **2.606**	92.5, 94.7, 99.7, 119.9, 122.3, 125.6	0.5*p*[b]
	Si92[a]	2.334, 2.414, **2.543*, 2.549**	82.8, 98.5, 101.0, 115.2, 125.3, 126.8	**1.14*p***, 0.09*s*
	Si107[a]	1.630, 1.703, 2.439, **2.606**	97.1, 101.8, 109.0, 109.8, 119.1, 121.3	0.8*p*[b] 0.27*s*[b]
	Si139[a]	1.664, 2.433, **2.601, 2.714**	102.8, 103.9, 104.8, 107.2, 112.4, 123.9	0.04*s,p*
	Si142[a]	2.388, 2.439, 2.465, **2.714**	96.2, 96.4, 102.8, 103.9, 120.4, 129.3	0.005*p*
	Si157[a]	2.400, 2.411, 2.473, **2.601**	97.7, 97.8, 101.5, 101.9, 108.7, **146.1**	0.03*p*

The * is for the X = Al neighbor atom. The larger (weak bonds) Si-X distances and X-Si-X' bond angles for each atom are given in bold. The largest PDOS is given in bold for each model. See also Figs. 11.12 and 11.16

[a]all PDOS profiles in the mid band gap are visible in Fig. 11.12f or g

[b]intensity of the PDOS values at the valence edge are visible in Fig. 11.12f or g

type (Si108) (Fig. 11.12a). Factually, all of them can be considered as fully coordinated ones with one (Si108, Si141) or two (Si91, Si94) weak bonds with the exception of Si107 which does not participate in any weak Si…Si bond (more precisely, |Si107…Si*| > 2.836 Å towards the nearest Si* atom). The angular deformation of each Si atom can be evaluated by the average absolute deviation of the α = X-Si-X angles (Table 11.10), X = Si or O, as:

$$\beta = \sum_{i=1,6} (|\alpha_i - 109.47°|)/6 \tag{11.2}$$

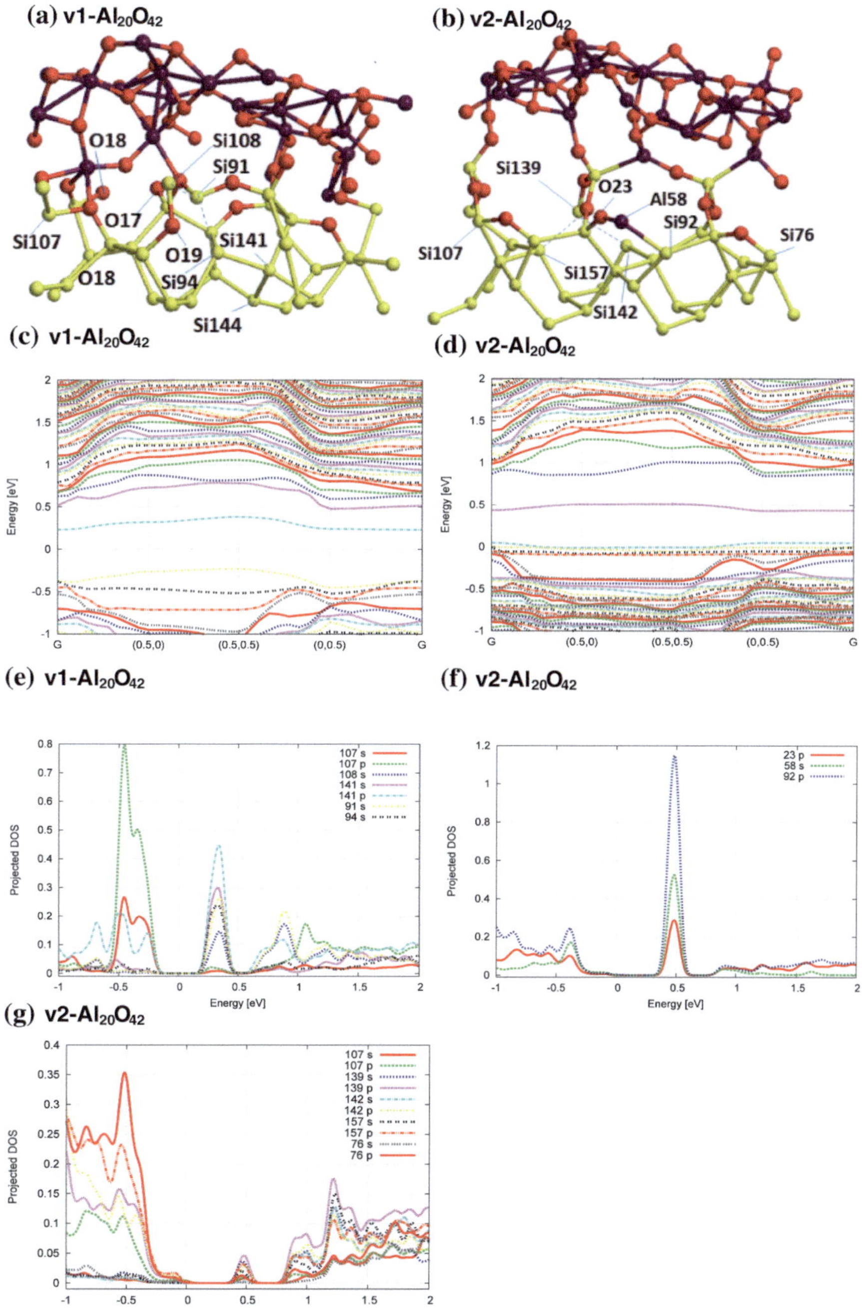
(a) v1-Al20O42
(b) v2-Al20O42
O18
Si108
Si91
O17
Si107
O19
Si141
O18
Si94
Si144
Si139
O23
Al58
Si92
Si76
Si107
Si157
Si142
(c) v1-Al20O42
(d) v2-Al20O42
Energy [eV]
G
(0.5,0)
(0.5,0.5)
(0,0.5)
G
(e) v1-Al20O42
(f) v2-Al20O42
107 s
107 p
108 s
141 s
141 p
91 s
94 s
23 p
58 s
92 p
Projected DOS
Energy [eV]
(g) v2-Al20O42
107 s
107 p
139 s
139 p
142 s
142 p
157 s
157 p
76 s
76 p
Projected DOS
Energy [eV]

Fig. 11.12 Geometry configuration of the upper part of one SC (**a**, **b**), band structure (**c**, **d**), and PDOS (**e-g**) of v1-$Al_{20}O_{42}$ (**a**, **c**, **e**) and v2-$Al_{20}O_{42}$ (**b**, **d**, **f**, **g**) models, the latter containing Si atoms linked by one (Si76, Si107, Si142, Si157) or two (Si139) weak Si...Si bonds (**b**), with the main Al58 and Si92 defects (**b**). See Fig. 11.4 for *color* coding of atoms

A maximum β parameter value of 17.8° is observed for Si141 among all Si atoms considered in this work. The difference between the twisting deformation per 60° required for a strong PDOS perturbation [32] and the deformation of β = 17.8° obtained herein could come from the different computational approaches (neither the tight binding approximation, nor the Bethe lattice for the Si part were considered herein) so that any further extent of Si distortion could be corrected by additional modeling. Regarding the PDOS profiles (Fig. 11.12c, e), we have added the fully Si_4 type coordinated Si79 and Si75 atoms (Table 11.10). The first Si79 atom is characterized by a strong angular distortion β = 15.1°. This specific geometry situation obtained in the system is related to the two weak Si94...Si91...Si108 bonds of 2.604 and 2.673 Å, respectively, with an angle of 126.1°. Usually the participation of under-coordinated atoms in a weak bond is sufficient to shift its state from the mid band gap to the edges [2]. However, the distribution of the PDOS peaks of the defects in the obtained v1-$Al_{20}O_{42}$ model seems to be more puzzling because the Si atoms (Si91, Si94) with two weak bonds are the intermediate ones in the order of atoms contributing to the mid band gap if one considers the roles at the extreme lower energies, i.e., Si141 > Si91 > Si94 > Si108, and at the conduction edge, i.e., then Si141 > Si91 > Si94, being absent at the extreme valence edge where the PDOS profiles of the Si141 and Si107 atoms are of the same order. In order to classify the influence of the bond weakening, we selected the largest Si...Si bond for each Si atom (upper part of Table 11.10) and calculated the upper Si...Si elongation (ΔR) relative to the equilibrium bulk Si-Si distance of 2.35 Å (Table 11.11 and Fig. 11.13b). The moderate correlation of the PDOS intensity *versus* ΔR (Fig. 11.13b) does not explain the unique dominance in the PDOS of the

Table 11.11 Maximum intensities of the *s*- and *p*-PDOS values (a.u.) of the Si atoms in the v1-$Al_{20}O_{42}$ model optimized at PBE/PAW theory level *versus* the Si...Si elongation of the longest bond relative to equilibrium bulk Si-Si distance of 2.35 Å (ΔR, Å) and average absolute deviation (β, °) of the α = X-Si-X angle (X = Si or O) expressed as $\beta = \Sigma_{i=1,6}\ (|\alpha_i\ -109.47°|)/6$, with number of weak bonds (N) and different neighbor types of each Si atom (see Fig. 11.13)

N	Atom	Type	*p*-PDOS	(*s* + *p*)-PDOS	ΔR	β
0	Si75	SiO_3	0.03	0.035	0.033	7.3
	Si79	Si_4	0.18	0.25	0.085	15.1
	Si144	Si_4	0.08	0.08	0.005	3.9
1	Si141	Si_4	0.45	0.75	0.208	17.8
	Si108	SiO_3	0.15	0.28	0.323	11.4
2	Si91	Si_2O_2	0.27	0.50	0.323	9.1
	Si94	Si_4	0.40	0.64	0.254	12.9

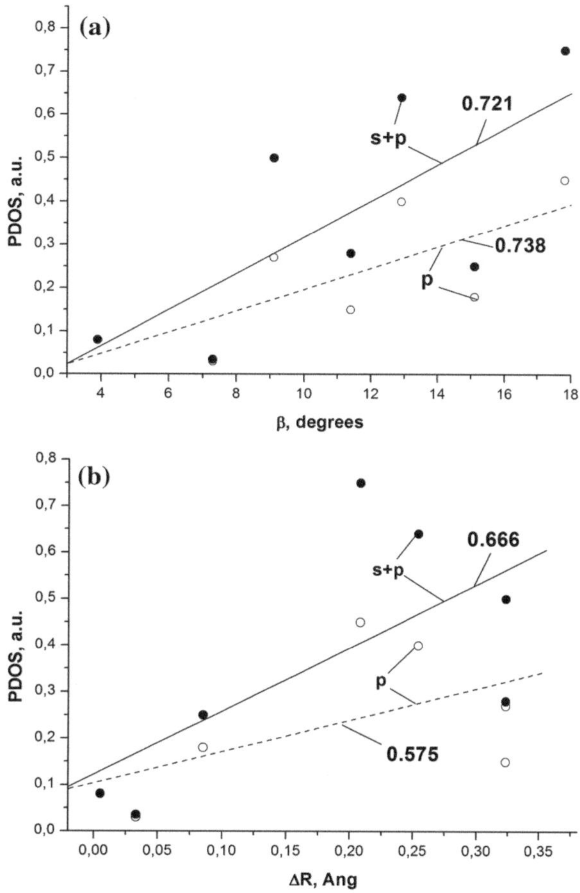

Fig. 11.13 PDOS maximum values (a.u.) for *p*-orbitals only (open circles) or (*s* + *p*)-orbitals together (closed circles) as a function (**a**) of the elongation of the longest Si…Si bond relative to 2.35 Å or *β* (°) bond angles (**b**) of the fully coordinated and under-coordinated Si atoms, those being the neighbors of the DB defects (Si141) in the v1-$Al_{20}O_{42}$ model optimized at the PBE/PAW level. The correlation *r* is given for the PDOS maximum values of *p*-orbitals only (*dashed lines*) or for the PDOS sum of (*s* + *p*)-orbitals together (*solid lines*). The type of neighbors for each Si atom and the PDOS values are given in Table 11.10

Si…Si elongation (ΔR in Table 11.11). A higher correlation was observed for the average absolute deviation β (Eq. 11.2) (Fig. 11.13a). Similar correlation coefficients (Fig. 11.13) were obtained for both the intensities of the *p*-PDOS (dashed lines in Fig. 11.13) or for the sum of *s*- and *p*-PDOS intensities (solid lines in Fig. 11.13). Regarding these results, we suppose that the PDOS could be accurately described by a more complex function depending on both ΔR and β.

v2-$Al_{20}O_{42}$. The second variant of V type (Scheme 11.1) was obtained from merging the v2-Al_4O_{12} model (type IV) with the $Al_{16}O_{30}$ fragment leading to a model that is less stable by 7.056 eV relative to the v1-variant (Table 11.4). Three of the total of four previously deposited Al atoms are trapped into the oxide layer but it is a consequence of their initial elevated positions above the surface in the starting v2-Al_4O_{12} model. Such location prevents a complete passivation and the isolation of the Al58 atom which was "conserved" at the Si slab surface after joining with $Al_{16}O_{30}$. The negligible *p*-PDOS of two of the Al atoms (Al55, Al56) in the mid band gap are similar to those of the Al atoms of the oxide slab while the one of Al57 is around 0.11, a consequence of the closeness to a chain of defects (see below). The Al57O_2 group (|O-O| = 1.620 Å) is similar to the defect model in Fig. 11.8b (with a second O atom also connected to the oxide layer) but it is inactive in the band gap.

The two-coordinated Al58 atom linked to Si92 (2.543 Å) and O23 (1.705 Å) (Fig. 11.12b) is an analogue to the Si94-Al25-O14 precursor in the initial v2-Al_4O_{12} model (Fig. 11.9c) but now it is not part of a long -(Al-O-Si)$_3$- chain. The Si92 participates in two weak bonds with Si93 and Si75 (2.414, 2.549 Å) in v2-$Al_{20}O_{42}$; as well as Si141 forms one weak bond (2.558 Å) in v1-$Al_{20}O_{42}$; and both Si92 and Si141atoms possess the most intensive PDOS profiles in the band gaps of both v2- and v1-$Al_{20}O_{42}$ models, respectively. For clarity, we partitioned the different groups of atoms in the v2-$Al_{20}O_{42}$ model into two graphs of different scales (the smaller PDOS peaks in the mid band gap in Fig. 11.12g and the largest ones in Fig. 11.12f). The lower extent of the angular distortion for Si92 as compared to the one for Si141 in the v1-$Al_{20}O_{42}$ model is expressed by $\beta = 14.8°$, but a cumulative influence of the angular Si92 distortion and, possibly, its two weak bonds result in higher PDOS intensities (1.14p a.u. in Fig. 11.12f) *versus* the one of Si141 (0.45p a.u. in Fig. 11.12e) in the mid band gap.[6] One also notes a state near the conduction edge (0.5 eV in Fig. 11.12d) which coincides with the DOS peaks of Si92 and Al58 (Fig. 11.12d) with a minor O23 participation. The two Al58 and Si92 atoms determine the mid band gap structure of the v2-$Al_{20}O_{42}$ variant. This state is more strongly delocalized along the Si75-(Si93)Si92-Al58-O23-Si91-O20-Si156-Al57-O21 chain (0.11p-(0.11p)1.14p-0.68p-0.29p-0.04p-0.13p-0.22p-0.11p-0.10p a.u.), the PDOS intensity being close to zero at the Si91 only, which can be assigned to a small β distortion of 3.8°. So, one observes a long chain for the state in the band gap similar to the one for the precursor v2-Al_4O_{12} model (Fig. 11.9d). The difference with the Si94-Al25-O14 group in v2-Al_4O_{12} (Fig. 11.9f) is related to the mid band gap states of the latter while the state of the Si92-Al58-O23 group is shifted to the conduction edge (Fig. 11.12f). The Si-Al-O angles in both groups differ rather slightly, i.e., 105.7° in v2-Al_4O_{12} and 116.3° in v2-$Al_{20}O_{42}$, which can hardly be the reason of a valuable deviation between these groups.

All other nearest neighbors (O or Si) of both Si92 and Al58 atoms such as the two-coordinated species (Si139 of SiO neighbor type), three-coordinated species, i.e., Si76 of Si_2O type, Si107 of SiO_2 type, Si142 of Si_3 type, and Si157 of Si_3 type, do not contribute significantly to the PDOS in the mid band gap (Fig. 11.12g). These "silent" atoms are of interest for the system due to numerous weak Si…Si bonds in the system (lower part of Table 11.10). According to the evaluations of [32], such Si…Si bond elongation has to be around 1.4 times the bulk $R_{Si\text{-}Si}$ value ($R_{Si\text{-}Si}$ = 2.35 Å) to lead to a substantial PDOS at the valence (and then at the conduction) edge. The results of the group of Drabold [12] showed the opposite trend for *a*-Si using a 512-atom model with much smaller bond changes (by 0.1 Å) using DFT and small atomic basis sets with the SIESTA code. More precisely, the

[6]We will see below that the state in the band gap can be obtained even at a smaller β value of 10°.

DOS increases nearly symmetrically at the valence and conduction edges upon the Si…Si shortening and elongation, respectively [12]. The first pair of the three-coordinated Si76 and Si107 atoms forms a weak Si…Si bond of 2.606 Å, which explains the absence of respective states in the band gap. Both other three-coordinated Si142 and Si157 atoms possess a sp^3 hybridization with δ angles around 40°[7] and elongated bond lengths compared to the one in the bulk, i.e., 2.388, 2.439, 2.465 Å and 2.400, 2.411, 2.473 Å for Si142 and Si157 atoms, respectively. Formally, these three-coordinated atoms have to reveal a state in the band gap.[8] However, the detailed analysis shows 2 weak Si…Si bonds of 2.601 Å (Si157…Si139) and 2.714 Å (Si142…Si139) centered around the two-coordinated Si139. Both bond lengths are also smaller than $1.4 \times R_{Si\text{-}Si}$ forming a Si142…Si139…Si157 angle of 102.8°. Their DOS intensities (Fig. 11.12g) are much smaller as compared to those of Al58 and Si92 (Fig. 11.12f). If "silent" cycled trimers with usual Si-Si distances (i.e., for bulk) were mentioned [15] in the third and fourth oxidized Si layers, one cannot speak about a cycled trimer of weak bonds because the third |Si142…Si157| distance in the trimer is rather long (4.154 Å).

v1-$Al_{23}O_{50}$. The v1-$Al_{23}O_{50}$ model (type V) was obtained via the successive deposition of 20 O and 7 Al atoms (from type IV) and junction with the $Al_{16}O_{30}$ fragment (Scheme 11.1). It results in very dense bond formations and conservation of 5 Al atoms of a total 7 at the oxidized surface (Fig. 11.14a). Its reduced heat of oxidation $\Delta U/n_O = -1.027$ eV per O is smaller (in absolute value) than the value $\Delta U/n_O = -2.284$ eV per O obtained at the previous stage of Al deposition according to respective balances:

$$H_{12}Si_{96} + 4.75 \times O_2 + 3.5 \times Al_2O_3 \rightarrow vl - H_{12}Si_{96}Al_7O_{20} \tag{11.3}$$

$$H_{12}Si_{96} + 7.75 \times O_2 + 11.5 \times Al_2O_3 \rightarrow vl - H_{12}Si_{96}Al_{23}O_{50} \tag{11.4}$$

Both values are smaller (in absolute value) than the maximal heat of Si(100) oxidation $\Delta U/n_O = -7.3$ eV/O [15]. Let us note that these partial (per O atom) values are more illustrative for the oxidation reactions without the Al_2O_3 fragment because only a part of its O atoms reacts. As a result, the $\Delta U/n_O$ value decreases in absolute value going from the reaction with O atoms only to the reaction with O atoms and the Al_2O_3 unit. The unique defect state belongs to the fully coordinated Si151 atom of Si_3O neighbor type (Fig. 11.14a) with two elongated Si-Si distances ("$Al_{23}O_{50}$" column 2 in Table 11.12) and moderate angular distortion β of 10°. According to [32] we should not obtain the DOS peaks for the Si atom of Si_3O neighbor type in the band gap (Fig. 2 of [32]). Regarding the (111) surface considered with the Bethe model, Sakurai and Sugano [2] showed with a TB approach that a weak Si-Si bond moves towards the valence band, while the weak Si-O bond

[7]If one calculates δ value as proposed in the preceding part (v1- $Al_{20}O_{42}$).

[8]While the absence of states in the mid band gap for Si_3O case has been noted (Fig. 2 in [31]), it is not the case for Si_2O.

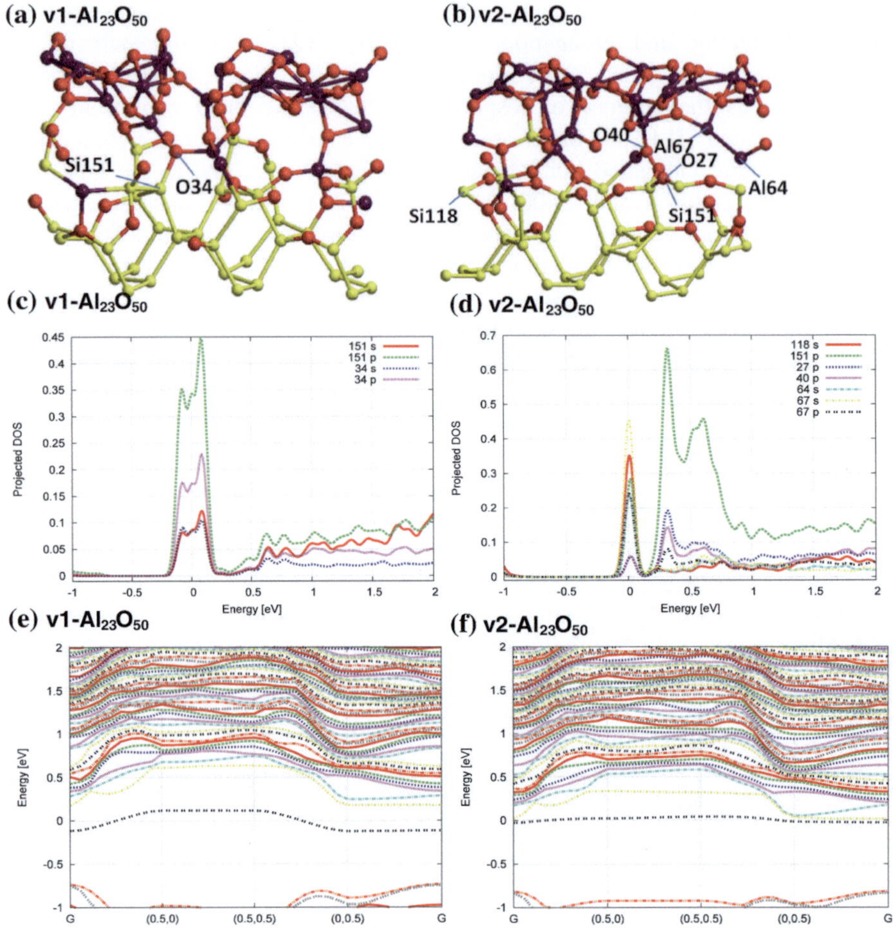

Fig. 11.14 Geometry configuration of the upper part of one SC—(**a**, **b**), band structure (**c**, **d**), and PDOS (**e**, **f**) of the v1-$Al_{23}O_{50}$ (**a**, **c**, **e**) and v2-$Al_{23}O_{50}$ (**b**, **d**, **f**) models. See Fig. 11.4 for *color* coding of atoms

produces a shift towards the conduction band. The Si151 is included in both weak Si-Si and weak Si-O bonds (Table 11.12) but probably the Si-O elongation by around 0.2 Å is more important than the Si-Si elongation by around 0.05–0.1 Å so that the state of Si151 shifts closer to the conduction edge in accordance with ref. [2]. The other 7 atoms of the same Si_3O neighbor type (but with Si-Si bond lengths shorter than 2.4 Å) are present in the model without any effective role near the E_F. Some of them are however useful to estimate the extent of the angular distortion which results in the band gap states. Neither Si121 ($\beta = 3.9°$), nor Si118 ($\beta = 7.3°$) atoms do not have states at the edges. Hence, we assigned the particular state of Si151 to the co-existence of both stretching (Si-O) and angular distortions β

Table 11.12 Type of neighbor atoms and Si-Y bond lengths (Å) for the atoms of the Si_3O (Si151, Si121) or Si_2O_2 (Si118) neighbor types in the initial models. See also Fig. 11.14

Type Y	Si151	Si121[a]	Si118[a]
	$Al_{23}O_{50}$	$Al_{23}O_{50}$	$Al_{23}O_{50}$
O35	1.880	1.753	1.628, 1.867
Si87	2.350	2.272	2.384
Si135	2.421	2.332	2.423
Si153	2.406	2.438	2.640

[a]The Si and O atomic numbers differ from the ones in the left column

(Eq. 11.2). Oxidizing the defect with one or two additional O atoms, we hope to cure the slab thus achieving a band gap without any defect states. The work is under progress.

v2-$Al_{23}O_{50}$. The second v2-$Al_{23}O_{50}$ variant deserves attention as the worse passivation case due to the Al64…Al67 bond length of 2.840 Å, meaning this connection cannot be counted as a coordination bond for both linked mono- (Al64) and two-coordinated (Al67) atoms (not including this Al…Al bond for both Al64 and Al67). This 2.840 Å bond length can be compared to 2.521 and 2.670 Å in the v1-Al_4O_{12} model and 2.612 Å in the v2-Al_4O_{12} model, where we evaluated 2.521 Å as weak Al…Al bond (Sect. 11.3.3). In the two last Al_4O_{12} models, Al atoms do not possess states in the band gap. Both Al64 and Al67 in the v2-$Al_{23}O_{50}$ possess *s*-(mainly) and *p*-type states with an energy exactly at the E_F level (Fig. 11.14d, f). Owing to the close states of the two-coordinated Si151 of the O_2 neighbor type, whose *p*-PDOS is the most important at the conduction edge, we have factually no band gap for this v2-variant. It is strictly localized so that the PDOS profiles at the 1st (O atoms) and 2nd (Al, Si) neighbors have no substantial values in the gap/edges.

11.4 Discussion

The Al_2O_3/Si boundary remains in the center of intensive theoretical [60–65] and experimental [66–73] studies partially owing to the rising interest to the Al_2O_3/III-V [60, 62, 63] and Al_2O_3/SiN_X [71–73] boundaries. While charged and neutral defects of α-Al_2O_3 [60, 61, 65], κ-Al_2O_3 [60], and θ-Al_2O_3 [60] forms in the bulk were properly considered, less attention was paid to the studies of the defects at the surface of the amorphous Al_2O_3 slab or at grain boundaries of Al_2O_3 [65]. Recent experimental results were presented in reviews [74, 75] with domination of ALD methods and few plasma enhanced chemical vapor deposition (PECVD) exceptions [72, 73]. That is why our Discussion parts are mainly oriented to the surface defects and structural aspects formed at the Al_2O_3/Si boundary in the course of PA ALD deposition together with the problems of accelerated formation of the interface layer, using an Al_2O_3 fragment.

11.4.1 Structural Distortion of the Defects

One of the interesting results is the alternation of Si(sp^2) (in blue in Fig. 11.1) with Si(sp^3) (in green in Fig. 11.1) neighbors in the upper (110) layers (Table 11.1). The Si(sp^2) and Si(sp^3) alternating atoms were also tentatively recorded using atomic resolved scanning tunneling microscopy in silicene nanosheets on Ag(111) [76]. However, these last authors evaluated very short Si-Si distances of 1.9 ± 0.1 Å, thus much shorter (about 17 %) than the one for Si bulk (2.35 Å), explaining them by the interaction with the Ag layer. The earlier works on the Si(sp^2) and Si(sp^3) hybridization did not note any drastic shortening of the Si-Si bonds for the sp^2 state [4, 5, 31, 40]. Such Si-Si bond decrease was indeed not observed with any of the computational approaches neither for the Si surface (Table 11.1), nor for silicene [77–79], thus possibly requiring further verification. We cannot guess any element which could demonstrate such wide single/double bond variation as 17 %. In the system with ordered O deposition of 12 O atoms, we obtained a Si = O double bound with |Si = O| = 1.528 Å (Sect. 11.3.2.1). Accepting a Si-O single bond length as 1.62 Å at the same computational level, we obtained a decrease of 5.1 %. It is much smaller than the suggested 17 %.

The simultaneous O deposition of many O atoms (as along a short ALD step) undertaken in our work deviates relatively to the consequent search of the most stable O geometries in [15]. We consider only surface trapping of oxygen and the following spontaneous reactions with Si atoms. As a result of these reactions, O atoms can penetrate to the deeper layers, so that we observe O diffusion into the deeper Si layers (second, third, …) with our approach. But a simultaneous attack of 12 (Fig. 11.6a) or 20 O atoms (Fig. 11.6b–d) leads to a serious destruction of the upper layer (1st of a total of 8) and penetration of the O atoms to the 2nd layer. It would therefore be useful to estimate the critical surface density of O atoms which corresponds to the favored penetration of O atoms into the deeper Si layers. According to the relative energies [15], the models with N oxygen atoms distributed between the two upper Si layers over the 2 × 2 SC of Si(100) (the SC surface of 59.6 Å^2) are preferred if $N > 8$ at nearly equilibrium reaction conditions (atom-by-atom). This leads to the critical surface O density of 8/59.6 = 0.13 Å^{-2} as compared to 0.153 Å^{-2} for the largest coverage of 20 O atoms over 2 × 3 SC of Si (110) (the SC surface of 130.8 Å^2) in our work. The coverages of 18 or 12 O atoms produce surface O densities of 0.138 or 0.092 Å^{-2}, respectively, which are slightly larger or smaller than the critical value of 0.13 Å^{-2} for the stabilization of O atoms in the second layer [15].

Some aspects which are important to develop future KMC approaches for the deposition of amorphous oxide were concerned above. The identification of fully coordinated defects in complex system, like we present above, should be done a priori on the basis of geometry and topology. But the quantitative parameters of fully coordinated Al and Si defects remain not clear. Some tips were however obtained in the presented results. The case of the "silent" Al25…Al28 (2.521 Å) bridge in the band gap and the limiting role of the Al26…Al27 (2.670 Å) bridge in the v1-Al_4O_{12} model is valid from both topological and geometrical points of view, providing an area for future estimations for weak bond lengths. Additionally, from the v1-$Al_{20}O_{53}$ model, one deduces a critical angular X-Si-X distortion β (Eq. 11.2) (X = Si, O) for the fully coordinated Si151 atom as $\beta > 10°$. Two other Si141 in v1-$Al_{20}O_{42}$ and Si92 in v-$Al_{20}O_{42}$ atoms (Table 11.10) with $\beta = 17.8$ and 14.8°, respectively, confirm this evaluation as illustrated in Fig. 11.13a.

11.4.2 Boehmite Formation

The tendency of forming Al chains has been reported for any number of Al atoms per SC surface, with an Al/O ratio of 3 [35]. Using different initial models with variable numbers of O (12, 18, 20) and Al (4, 6, 7) atoms over the SC of the Si(110) surface, we obtained a wide series of models each containing double, triple Al-atom groups or continuous chains. At the initial O/Al stage, the chain of low coordinated Al possesses too short bond lengths as compared to the ones in the experimental XRD data for boehmite (Table 11.8) [57]. These bonds elongate upon the following O/Al/O oxidation step due to the formation of an Al-O-Al linkage, parallel to Al-Al ones. Such Al…Al lengthening can vary in a wide range from 2.7 to 3.2 Å, which usually corresponds to strong Al-O bonds of 1.68–1.72 Å already at |Al…Al| = 2.9 Å. Hence, the chain decomposes to three-atom fragments (Fig. 11.11c). The deletion of the excess of oxygenated species as proposed at the next purging step of ALD, in the O_2 or O_3 forms, leads again to a decrease of the longest Al…Al bonds and the restoration of continuous chains. After this step, the similarity with a boehmite type becomes evident (Table 11.11). This purging involves the deletion of O_2 or O_3 molecules from AlO_2, Al_2O_3, O_3, or O_4 fragments. The problem is however to separate properly the steps for deletion of the defects upon purging and annealing.

We wish to emphasize that the stabilized boehmite-like chain is related to the Si/Al_2O_3 contact surface and not to a free one (with gas phase). At the higher O concentration provided by the usually longer ALD step of oxygen treatment and the necessary Si pre-oxidation ALD step, we believe that the Al coordination could be larger than at the free surface. The lower Al coordination at the Al_2O_3 surface is indeed well known. This result is in a partial contradiction with the transformation mechanism of T ALD deposited amorphous Al_2O_3 to crystalline Al_2O_3 along the annealing as suggested in [80]. The last work proposed a long-range ordering of the edge-sharing of tetrahedral and octahedral Al atoms without redistribution among their types as not visible from XPS. To our opinion, such conclusion [80] requires further verification because the concentration of octahedral Al atoms is very small, if any, in amorphous Al_2O_3 according to the experimental XRD or neutron data [81] as well as by modeling [82–84]. The deviation is less emphasized between the surfaces of crystalline γ-Al_2O_3 and amorphous Al_2O_3 [83–84] but such distinction of Al types at the surfaces exists. According to [85], "…the coordination number distribution and structural features, are different for amorphous and crystalline alumina surfaces, …". Hence, it is a problem to illustrate which kind of long-range transformation can lead to crystalline Al_2O_3 with dominating octahedral Al atoms from amorphous Al_2O_3 with dominating tetrahedral (and nearly absent octahedral) Al atoms [81] without local reordering.

The observed boehmite-like chain could agree with the mechanism of the Al_2O_3 transformation [80] provided that tetrahedral Al could be obtained in the same chain together with octahedral Al ones. It depends on the O/Al ratio which would require modeling with a smaller O/Al ratio than the one used herein (O/Al = 3). But, then the bond lengths should vary and probably the chain would not be stable and not boehmite-like. However, formally the mechanism proposed in [80] is related to T ALD deposition amorphous Al_2O_3 for which the mechanism of Al_2O_3 growth as proposed by us without products of TMA decomposition cannot be applied. These compounds containing methyl ligands obtained at the T ALD conditions are not oxidized by O_2 plasma and need to be considered. They could determine another way of oxide formation at every T ALD step.

The self-organization of boehmite from the extraframe work Al species was shown in HY zeolite in the form of $Al_{16}O_{13}(OH)_{22}(OH_2)_{11}$ clusters as observed via XRD already 30 years ago [80, 86].[9] The similarity between our observed chain and the octahedral linkage of boehmite is enforced by pre-oxidation of the Si(110) surface (O_2 plasma treatment) so that the Al–Si bonds are nearly absent after the first stage of Al deposition (Fig. 11.11b) with the exception of one bond of 2.479 Å which disappears at the next step. It is also possible that the formation of the Al-chains (and not of Al-clusters) is a more general topic which has also been studied recently at the Si(100) surface using a KMC approach [87].

11.4.3 *Joining Procedure:* Pros et Cons

The passivated oxide Al_2O_3/Si layer is relatively thin [17], but nevertheless exceeds 10 Å, so that any computational hints to accelerate the growing of the interface instead of the numerous steps of successive O/Al/O/Al/... depositions could be useful. It is why we have used the merging procedure with $Al_{16}O_{30}$ (joined via two opposite sides) or $Al_{16}O_{24}$ fragments which led to a passivation in most of the cases. The obtained defects are easily identified in terms of structural distortion or under-coordinated states. We distinguished two types of possible negative effects produced by joining the Si support and the oxide fragment. At small number of Al atoms (N_{Al} = 4) deposited on the oxidized Si slab, the oxide can attach the essential part of Al atoms thus redistributing them in favor of the oxide. It can be considered as a decomposition of the deposited interlayer. In such a case, the v1-$Al_{20}O_{42}$ model demonstrates the expansion of the band gap which is conditioned by the extraction of Al atoms from the oxidized Si surface. The wider band gap in the

[9]Randomly, the SC of crystalline γ-Al_2O_3 was selected by us with the same quantity of Al atoms ($Al_{16}O_{30}$).

latter is easier to obtain (Figs. 11.3 and 11.4) *versus* the similar width for the Al_xO_y models presented above. With the increase of the concentration of deposited Al atoms (as 7), the joining procedure for v1-$Al_{23}O_{50}$ does not result in the redistribution of Al atoms. In such a case, one observes a "true" passivation with the oxide layer.

The second type of negative effects of the proposed junction is related to the transformation of the OAl_2 and OAl_3 (Fig. 11.8h) groups of the Al_2O_3 oxide layer. The defective states of the OAl_2 groups could be previewed regarding three- or four-coordinated O atoms in the Al_2O_3 bulk, but the reasons of such transformation of OAl_3 groups are less understood and could be assigned tentatively to a geometry distortion. The two systems, among a total of ten, wherein we have observed this effect, unfortunately, do not allow to quantify the parameters which led to the O states located at the valence edge of the band gap or close to the E_F (Fig. 11.15). In the two cases, both v2-$Al_{20}O_{42}$ and $Al_{23}O_{44}$ models correspond to minimal oxidation effects obtained such as –0.972 eV per O and –0.251 eV per O, respectively. An example of such v2-$Al_{20}O_{42}$ system obtained from v2-Al_4O_{12} by junction with $Al_{16}O_{30}$ is presented in Fig. 11.15; the PDOS profiles for Si-O-O-Al, OAl_2, and OAl_3 type defects are shown in Fig. 11.15b–d; respective PDOS values are in Table 11.13.

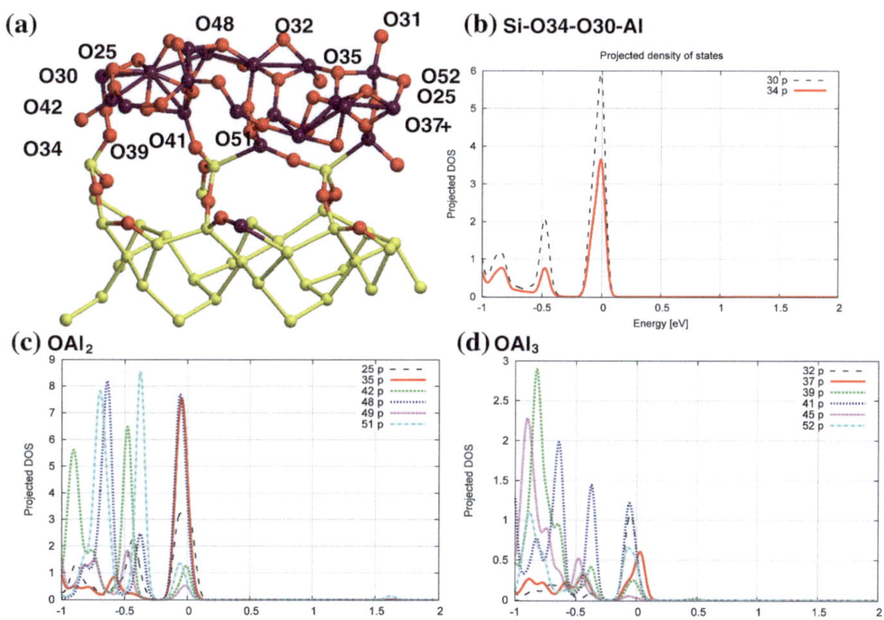

Fig. 11.15 Geometry configuration of the upper part of one SC (**a**) and PDOS (a.u.) (**b-d**) at the O atoms of the various defect groups in the v2-$Al_{20}O_{42}$ model. The band structure is given in Fig. 11.12d. See Fig. 11.4 for color coding of atoms

Table 11.13 *p*-PDOS values (a.u.) and associated energy E (eV) of several O(N) atoms of OT_X groups ($X = 2 - 3$, T = Al, Si), in the v2-$Al_{20}O_{42}$ and $Al_{23}O_{44}$ models optimized at the PBE/PAW level

Model	Type	N	O-T/O	T-O-T'/O	(E, PDOS)
v2-$Al_{20}O_{42}$	OAl_2	51	1.704, 1.693	130.08	–0.1, 1.4 –0.35, **8.6**
		48	1.702, 1.743	92.89	–0.1, 2.5 –0.35, **7.8**
		35	1.724, 1.754	126.4	–0.1, **7.7**
		25	1.776, 1.790	116.2	–0.1, 3.2 –0.4, 2.3
		49	1.654, 1.725	126.75	0.0, 0.5 –0.5, 1.8
	OAl_3	41	1.787, 1.864, 1.971	81.2, 112.0, 154.5	–0.1, 1.21 –0.35, 1.41
		37	1.787, 1.828, 1.878	94.4, 126.7, 127.8	0.0, 0.6 –0.5, 0.32
	O-O*-Al	30	1.476, 1.846	107.8	0.0, 6.0 –0.5, 2.0
	O-O*-Si	34	1.476, 1.675	110.3	0.0, 3.2
v1-$Al_{23}O_{44}$	OAl_2	48	1.697, 1.732	122.3	–0.3, 0.45
	OSi_2	24	1.662, 1.743	99.0	–0.1, 0.22
	OAl_3	52	1.792, 1.857, 1.927	94.2, 97.2, 144.3	–0.3, 0.37
		39	1.819, 1.848, 1.904	108.7, 121.9, 125.5	–0.3, 0.27
		33	1.839, 1.870, 1.893	93.1, 122.0, 138.7	–0.25, 0.15

N corresponds to the atom label

A similar effect was observed while replacing the $Al_{16}O_{30}$ fragment by the close $Al_{16}O_{24}$ one. The latter was applied because we tried to vary the O/Al ratio of the added fragment. One difference between $Al_{16}O_{24}$ and $Al_{16}O_{24}$ is the absence of the lowest layer containing 6 O atoms between the oxide and Si surfaces. A final v1-$Al_{23}O_{44}$ product was obtained from v1-Al_7O_{20} by junction with $Al_{16}O_{24}$. This v1-$Al_{23}O_{44}$ case corresponds to the much weaker peaks for the OAl_2 and OAl_3 groups in the PDOS relative to the ones for v2-$Al_{20}O_{42}$. The second difference between the v2-$Al_{20}O_{42}$ and v1-$Al_{23}O_{44}$ cases is the higher extent of delocalization in v1-$Al_{23}O_{44}$. From the relative maximum PDOS peaks for the v1-$Al_{23}O_{44}$ model (not shown), we observed that the O and Al atoms between the OAl_2 groups possess comparable PDOS values. For v2-$Al_{20}O_{42}$, we wish to point out …O48 (7.8)-Al60(0.04)-O51(8.6)… chain for two OAl_2 groups connected via the Al60 atom (with the PDOS maximum values in parentheses). This demonstrates more isolated states in the v2-$Al_{20}O_{42}$ slab. The third difference between v2-$Al_{20}O_{42}$ and v1-$Al_{23}O_{44}$ is related to similar orders of PDOS values for the OAl_2 and OAl_3

groups for v1-$Al_{23}O_{44}$ but not for v2-$Al_{20}O_{42}$ (Table 11.13, Fig. 11.15). But, as already mentioned above, at the moment, the sets of both defective OAl_3 and OAl_2 species are too limited to quantitatively conclude about the reasons of the presence of such defective groups in the deformed oxide layer.

The final set of defects including the OAl_X group, from the v1-$Al_{23}O_{44}$ model only, is also shown in Fig. 11.8. If some of them reveal usually several states in the gap (Fig. 11.8a, f, g), for others, the effect depends on the system and possibly on the exact geometry of the defect (Fig. 11.8b, c, d, g, h, i, j). The states corresponding to Al_2O_3 or AlO_2 are positioned near the valence edge and below the E_F similarly to those of OAl_2 and OAl_3 groups, while the states from the O_3 and O_4 groups can also be located closer to the conduction edge. All of them should be considered as the targets for studying the following decomposition reactions using NEB for example to evaluate all possible channels and to evaluate the relative stabilities.

11.5 Conclusions

Alternate O → Al → O surface reactions with the basic Si(110) surface were modeled theoretically at the Density Functional Theory PBE/PAW level in the course of O_2 plasma assisted atomic layer deposition (PA ALD) processes using periodic boundary conditions. After three first O/Al/O steps, numerous defects were obtained from atomic oxygen (O_2, O_3, O_4, AlO_2, and Al_2O_3) as determined by the *s*- and *p*-PDOS analyses and band structures. A limited series of defects which manifest states in the mid band gap was observed in the obtained models. Some of them, as Al_2O_2 and Al_2O_3, reveal states in the band gap depending of the obtained models and possibly of the exact defect geometry. The purging stage was then simulated by deleting excessive adsorbed O_2 and O_3 molecular species from the defects (O_4, AlO_2, Al_2O_3).

One observed the presence of states related to the weaken Si…Si bonds in the band gap if the respective Si atoms are connected with fully coordinated Si atoms possessing a distorted geometry. The evaluation of the extent of O-Si-O angular deformation was proposed as $\beta = \Sigma_{i=1,6}\ (|\alpha_i - 109.47°|)/6$, a $\beta > 10°$ revealing a defect state in the band gap. As a stronger angular distortion happens with a higher probability near the surface, where the defects are located, the obtained PDOS results from both a delocalization of the defect states and the deformation of atomic geometries.

The parallel addition of a series of O atoms was tested to simulate the ALD processes. It led to essentially lower but exothermic (negative) heats of oxidation *versus* the ones obtained via the successive steps of atom-by-atom optimization which can hardly be achieved during short ALD steps. Already after the injection of 12 O atoms, one could observe a strong reconstruction of the upper Si layer. It demonstrates that the search of the most stable geometry of trapped O atoms by the successive optimization of every O atom over a regular upper layer could not be relevant for ALD modeling.

An accelerated scheme for developing a γ-Al_2O_3 precursor has been proposed by joining the previously deposited O + Al layers with the above described γ-Al_2O_3 fragment (containing or not excessive oxygen with a total content of $Al_{16}O_{30}$ or $Al_{16}O_{24}$), the last one being strongly deformed in the course of the reaction. Some of the defects (as Al_2O_2 and Al_2O_3) were obtained via two different routes either at the O/Al deposited layer, or at the upper surface of the added $Al_{16}O_{30}$ fragment (the case of the $Al_{16}O_{38}$ slab). In two models out of a total of ten obtained by the merging, we noted defect states for the OAl_3 and OAl_2 types formed due to a deformation of the $Al_{16}O_{30}$ or $Al_{16}O_{24}$ oxide fragments added to the oxidized Si (110) surface. A sharp difference in the localization was observed in these two models. However, the reasons or other qualitative parameters which provoke such a distortion remain unclear. It shows that some care while using such a way to grow the oxide layer is needed. At small number of 4 or 6 Al atoms per SC surface (and 12 or 18 O atoms, respectively, deposited before Al), their largest part turns out to be trapped by Al_2O_3 forming more stable models. For the slab model with 7 deposited Al atoms, the larger part of Al atoms (5/7) remains at the surface in the most stable system. With the increase of the total number of deposited atoms (and nearly constant Al/O ratio 4/12, 6/18, and 7/20), a minor Al loss from the SiO_x surface was observed. Hence, this model with joined γ-Al_2O_3 fragment can serve as a precursor of an amorphous Al_2O_3 slab without the separation of the oxides at sufficiently high oxygen concentration at the oxidized Si surface.

It was also observed that long chains of 6–7 atoms possess defect states in the mid band gap with a steep drop of PDOS along them and DOS maxima at the fully coordinated Si atom (v2-$Al_{20}O_{42}$ models). The striking example of high PDOS value at fully coordinated atom relative to the ones at the nearest defects was observed. It corresponds to an O atom linking two under-coordinated neighbors, i.e., two-coordinated Al and three-coordinated Si atoms (as in the v2-$Al_{20}O_{42}$ model).

Boehmite-like chains with higher Al coordination ($N_{Al} = 6$) without Al–Si bonds were stabilized at the Si(110) surface. Traditionally, the lower Al coordination leads to a more stable free Al_2O_3 surface (with gas or air) compared to the coordination in the bulk of amorphous or crystalline oxides. These chains can be the precursor for the transition from amorphous to crystalline Al_2O_3 along the following annealing but such proposition requires further testing. These defects should be the topic of future search of transition states to evaluate their stabilities and possible transformation routes under the ALD conditions.

Acknowledgments The authors thank the financial support of Ministry of Education and Science of Russian Federation (Minobrnauka, GK No 07.514.11.4150) and Computer Complex SKIF of Moscow State University "Lomonosov" and "Chebyshev" for computational time [88]. This research used resources of the "Plateforme Technologique de Calcul Intensif (PTCI)" (http://www.ptci.unamur.be) located at the University of Namur, Belgium, which is supported by the F.R.S.-FNRS under the convention No. 2.4520.11. The PTCI is member of the "Consortium des Équipements de Calcul Intensif (CÉCI)" (http://www.ceci-hpc.be). The authors also thank the Interuniversity Attraction Poles Programmes 7/05: "Functional supramolecular systems'' initiated by the Belgian Science Policy Office for partial financial support.

References

1. Joannopoulos JD (1977) Phys Rev B 16:2764
2. Sakurai T, Sugano T (1981) J Appl Phys 52:2889
3. Brower KL (1983) Appl Phys Lett 43:1111
4. Carrico AS, Elliott RJ, Barrio RA (1986) Phys Rev B 34:8724
5. Carlos WE (1987) Appl Phys Lett 50:14505
6. Umeda T, Yamasaki S, Isoya J, Tanaka K (1999) Phys Rev B 59:4849
7. Schroeder T, Adelt M, Richter B, Naschitzki M, Bäumer M, Freund H-J (2000) Surf Rev Lett 07:7
8. Allan G, Delerue C, Lannoo M (1996) Phys Rev Lett 76: 2961
9. Dong J, Drabold DA (1998) Phys Rev Lett 80:1928
10. Neaton JB, Muller DA, Ashcroft NW (2000) Phys Rev Lett 85:1298
11. Ludlam JJ, Taraskin SN, Elliott SR, Drabold DA (2005) J Phys: Condens Matter 17:L321
12. Pan Y, Inam F, Zhang M, Drabold DA (2008) Phys Rev Lett 100:206403
13. Anderson PW (1958) Phys Rev 109:1492
14. Dahinden P, Broqvist P, Pasquarello A (2010) Phys Rev B 81:085331
15. Yamasaki T, Kato K, Uda T (2003) Phys Rev Lett 91:146102
16. Drabold DA (2009) Eur Phys J B 68:1
17. Hoex B, Schmidt J, Bock R, Altermatt PP, van de Sanden MCM, Kessels WMM (2007) Appl Phys Lett 91:112107
18. Mazaleyrat G, Estève A, Jeloaica L, Estève D, Djafari Rouhani M (2005) Rapport LAAS No 05332 HIKE Project EU-IST 29295, 13p
19. Mazaleyrat G, Estève A, Jeloaica L, Djafari-Rouhani M (2005) Comput Mater Sci 33:74
20. Tahir D, Kwon HL, Shin HC, Oh SK, Kang HJ, Heo S, Chung JG, Lee JC, Tougaard S (2010) J Phys D Appl Phys 43:255301
21. Hoex B, Heil SBS, Langereis E, van de Sanden MCM, Kessels WMM (2006) Appl Phys Lett 89:042112
22. Heil SBS, van Hemmen JL, van de Sanden MCM, Kessels WMM (2008) J Appl Phys 103:103302
23. Saint-Cast P, Kania D, Heller R, Kuehnhold S, Hofmann M, Rentsch J, Preu R (2012) Appl Surf Sci 258:8371
24. Miikkulainen V, Leskelä M, Ritala M, Puurunen RL (2013) J Appl Phys 113:021301
25. Elliott SD, Scarel G, Wiemer C, Fanciulli M, Pavia G (2006) Chem Mater 18:3764
26. Jeloaica L, Estève A, Djafari Rouhani M, Estève D (2003) Appl Phys Lett 83: 542
27. Halls MD, Raghavachari K, Frank MM, Chabal YJ (2003) Phys Rev B 68:161302
28. Heyman A, Musgrave CB (2004) J Phys Chem B 108:5718
29. Puurunen RL (2005) J Appl Phys 97:121301
30. Hu Z, Turner CH (2007) J Phys Chem C 111:5756
31. Longo RC, McDonnell S, Dick D, Wallace RM, Chabal YJ, Owen JHG, Ballard JB, Randall JN, Cho K (2014) J Vac Sci Technol B Microelectron Nanom Struct 32:03D112
32. Laughlin RB, Joannopoulos JD, Chadi DJ (1980) Phys Rev B 21:5733
33. Kresse G, Hafner J (1993) Phys Rev B 47:558
34. Kresse G, Furthmüller J (1996) Phys Rev B 54:11169
35. Paglia G, Buckley CE, Rohl AL, Hunter BA, Hart RD, Hanna JV, Byrne LT (2003) Phys Rev B 68:144110
36. Pinto HP, Nieminen RM, Elliott SD (2004) Phys Rev B 70:125402
37. Werner F, Veith B, Zielke D, Kühnemund L, Tegenkamp C, Seibt M, Brendel R, Schmidt J (2011) J Appl Phys 109:113701
38. Hoex B, Gielis JJH, van de Sanden MCM, Kessels WMM (2008) J Appl Phys 104:113703
39. Ugliengo P, Viterbo D, Chiari G (1993) Zeitschrift für Krist Cryst Mater 207:9
40. Larin AV, Milyaeva DV, Rybakov AA, Bezrukov DS, Trubnikov DN (2014) Mol Phys 112:956

41. Edwards AH (1987) Phys Rev B 36:9638
42. Abukawa T, Wei C, Yoshimura K, Kono S (2000) Phys Rev B 62:16069
43. Felici R, Robinson IK, Ottaviani C, Imperatori P, Eng P, Perfetti P (1997) Surf Sci 375:55
44. Gunnella R, Bullock EL, Patthey L, Natoli CR, Abukawa T, Kono S, Johansson LSO (1998) Phys Rev B 57:14739
45. Ramstad A, Brocks G, Kelly PJ (1995) Phys Rev B 51:14504
46. Tromp RM, Hamers RJ, Demuth JE (1985) Phys Rev Lett 55:1303
47. Hamers RJ, Tromp RM, Demuth JE (1986) Phys Rev B 34:5343
48. Wolkow RA (1992) Phys Rev Lett 68:2636
49. Tabata T, Aruga T, Murata Y (1987) Surf Sci 179:L63
50. Enta Y, Suzuki S, Kono S (1990) Phys Rev Lett 65:2704
51. Himpsel FJ, Eastman DE (1979) J Vac Sci Technol 16:1297
52. Henkelman G, Uberuaga BP, Jónsson H (2000) J Chem Phys 113:9901
53. Dingemans G, Terlinden NM, Verheijen MA, van de Sanden MCM, Kessels WMM (2011) J Appl Phys 110:093715
54. Hoex B, Bosman M, Nandakumar N, Kessels WMM (2013) Phys Status Solidi Rapid Res Lett 7:937
55. Keunen K, Stesmans A, Afanas'ev VV (2011) Phys Rev B 84: 085329
56. Larin AV, Zhidomirov GM (2014) J Struct Chem 55:583
57. Corbató CE, Tettenhorst RT, Christoph GG (1985) Clays Clay Miner 33:71
58. Zhidomirov GM, Larin AV, Trubnikov DN, Vercauteren DP (2009) J Phys Chem C 113:8258
59. Larin AV, Zhidomirov GM, Trubnikov DN, Vercauteren DP (2010) J Comp Chem 31:421
60. Weber JR, Janotti A, Van de Walle CG (2011) J Appl Phys 109:033715
61. Choi M, Janotti A, Van De Walle CG (2013) J Appl Phys 113:044501
62. Choi M, Lyons JL, Janotti A, Van De Walle CG (2013) Appl Phys Lett 102:142902
63. Kumagai Y, Oba F (2014) Phys Rev B 89:195205
64. Choi M, Janotti A, Van De Walle CG (2014) ACS Appl Mater Interfaces 6:4149
65. Ogawa T, Kuwabara A, Fisher CAJ, Moriwake H, Matsunaga K, Tsuruta K, Kitaoka S (2014) Acta Mater 69:365
66. Specht M, Städele M, Jakschik S, Schröder U (2004) Appl Phys Lett 84:3076
67. Yeh C-C, Ma TP, Ramaswamy N, Rocklein N, Gealy D, Graettinger T, Min K (2007) Appl Phys Lett 91:113521
68. Dingemans G, Beyer W, van de Sanden MCM, Kessels WMM (2010) Appl Phys Lett 97:152106
69. Dingemans G, Einsele F, Beyer W, van de Sanden MCM, Kessels WMM (2012) J Appl Phys 111:93713
70. Liao B, Stangl R, Mueller T, Lin F, Bhatia CS, Hoex B (2013) J Appl Phys 113:024509
71. Schuldis D, Richter A, Benick J, Saint-Cast P, Hermle M, Glunz SW (2014) Appl Phys Lett 105:231601
72. Töfflinger JA, Laades A, Korte L, Leendertz C, Montañez LM, Stürzebecher U, Sperlich H-P, Rech B (2014) Sol Energy Mater Sol Cells 135:49
73. Töfflinger JA, Laades A, Leendertz C, Montañez LM, Korte L, Stürzebecher U, Sperlich H-P, Rech B (2014) Energy Procedia 55:845
74. Dingemans G, Kessels WMM (2012) J Vac Sci Technol A Vac Surf. Film 30:040802
75. Robertson J, Wallace RM (2015) Mater Sci Eng R Rep 88:1
76. Lalmi B, Oughaddou H, Enriquez H, Kara A, Vizzini Ś, Ealet B, Aufray B (2010) Appl Phys Lett 97:223109
77. Cahangirov S, Topsakal M, Aktürk E, Şahin H, Ciraci S (2009) Phys Rev Lett 102:236804
78. Şahin H, Cahangirov S, Topsakal M, Bekaroglu E, Akturk E, Senger RT, Ciraci S (2009) Phys Rev B 80:155453
79. Lew Yan Voon LC, Sandberg E, Aga RS, Farajian AA (2010) Appl Phys Lett 97: 163114
80. Snijders PC, Jeurgens LPH, Sloof WG (2005) Surf Sci 589:98
81. Lamparter P, Kniep R (1997) Phys B 236:405
82. Chen FR, Davis JG, Fripiat JJ (1992) J Catal 133:263

83. Alvarez LJ, León LE, Sanz JF, Capitán MJ, Odriozola JA (1994) Phys Rev B 50:2561
84. Gutiérrez G, Johansson B (2002) Phys Rev B 65:104202
85. Adiga SP, Zapol P, Curtiss LA (2006) Phys Rev B 74:064204
86. Shannon RD, Gardner KH, Staley RH, Bergeret G, Gallezot P, Auroux A (1985) J Phys Chem 89:4778
87. Albia JR, Albao MA (2015) J Vac Sci Technol A Vac Surf. Film 33:021404
88. Voevodin VlV, Zhumatiy SA, Sobolev SI, Antonov AS, Bryzgalov PA, Nikitenko DA, Stefanov KS, Voevodin VadV (2012) Open Syst J Publ, Moscow, No.7. [http://www.osp.ru/os/2012/07/13017641/] (In Russian)

Chapter 12
Development of the Latest Tools for Building up "Nano-QSAR": Quantitative Features—Property/Activity Relationships (QFPRs/QFARs)

Andrey A. Toropov, Alla P. Toropova, Karel Nesmerak, Aleksandar M. Veselinović, Jovana B. Veselinović, Danuta Leszczynska and Jerzy Leszczynski

Abstract Computational studies of common compounds are already standard ways of their investigations. However, modeling properties of nanomaterials has been always a challenging task. This chapter reveals important differences between approaches applied to these two groups of species. The development of an optimal descriptor provides one of the efficient ways for the computational techniques to estimate endpoints related to nanospecies. Notably, the optimal descriptor can represent a translator of eclectic information into the endpoint prediction. Development of the optimal descriptor could start with consideration of a hybrid of topological indices calculated with the adjacency matrix of the molecular graph and application of additive scheme where a physicochemical parameter is modeled as the summation of contributions of molecular fragments. Further, the optimal descriptor might be advanced by taking into account contributions of various

A.A. Toropov (✉) · A.P. Toropova
IRCCS-Istituto di Ricerche Farmacologiche Mario Negri, Via La Masa 19, 20156 Milan, Italy
e-mail: andrey.toropov@marionegri.it

K. Nesmerak
Faculty of Science, Department of Analytical Chemistry, Charles University in Prague, Hlavova 8, 128 43 Prague 2, Czech Republic

A.M. Veselinović · J.B. Veselinović
Faculty of Medicine, Department of Chemistry, University of Niš, 18000 Niš, Serbia

D. Leszczynska
Interdisciplinary Nanotoxicity Center, Department of Civil and Environmental Engineering, Jackson State University, 1325 Lynch Street, Jackson, MS 39217-0510, USA

J. Leszczynski (✉)
Interdisciplinary Nanotoxicity Center, Department of Chemistry and Biochemistry, Jackson State University, 1400 J. R. Lynch Street, P.O. Box 17910, Jackson, MS 39217, USA
e-mail: jerzy@icnanotox.org

J. Leszczynski and M.K. Shukla (eds.), *Practical Aspects of Computational Chemistry IV*, DOI 10.1007/978-1-4899-7699-4_12

physicochemical conditions. Such contributions include presence/absence of defined chemical elements and/or defined kinds of covalent bonds, as well as different kinds of rings in the molecular system—factors which are able to modify the physicochemical (biochemical) behavior of a substance. Finally, the latest version of optimal descriptor involves the applications of eclectic data into building up model for endpoints related to nanomaterials. A recently acquired collection of models developed to predict various endpoints of nanomaterials is presented and discussed in this chapter.

12.1 Introduction

In the last few decades one notices strong influence and fast expansion of applications of various nanomaterials for industry, medicine, and many other areas that affect the everyday life. Unfortunately, current advantages of such materials can be accompanied by their unexpected and dangerous effects in the near future. These effects can include various types of toxicity together with hazardous ecological impacts. Obviously, possible hazardous aspects of nanomaterials require studies that would reveal their characteristics and provide guidelines to facilitate their safe applications. Predictive models for nanomaterials can be useful for both theoretical and practical reasons.

After introduction of the classical approaches to predict properties and activities (endpoints) of the typical chemical substances essential parts of this chapter are devoted to: (i) description of optimal descriptors which are translators of eclectic data into endpoint prediction; (ii) discussion of predictive potential of models based on the optimal descriptors; and (iii) discussion of possible ways to improve the optimal descriptors as a tool to build up QFPR/QFAR which are analogs of the traditional QSPR/QSAR for nanomaterials.

Quantitative Structure—Property/Activity Relationships (QSPR/QSAR) is one of the valuable tools of theoretical chemistry. To some extent, QSPR/QSAR analyses can be classified as investigations solely rely on and devoted to chemistry. However, in recent years more common situation emerges when QSPR/QSAR analysis accumulates and uses ideas and approaches adopted from two or more natural science areas. Unfortunately, at present these methods are only scantily involved in the nano-chemistry, nano-biology, and nano-ecology.

Milestones of QSPR have been established in works of Wiener [1–4]. Main idea of these studies is an application of a molecular graph, or more exactly a matrix of topological distances in calculation of special coefficients (descriptors) which can be correlated with thermodynamical properties of organic compounds.

The truth cannot be improved: mathematical predictions of any endpoints, hardly can be accomplished with high accuracy for large set of non-congeneric substances. As a rule, the numerical experimental data on an endpoint are sensitive to many factors besides the molecular structure. For instance, physicochemical properties can be modified by unknown impacts, such as electromagnetic field, ultrasound, etc. In the case of biochemical endpoints the variety of possible impacting effects considerably increases. For instance, the biochemical endpoints can be sensitive to atmospheric pressure, weather at the moment of the experimental definition of the endpoint, and many other factors. Though obviously, it is almost impossible to include all such factors in the predictive models. But a mathematical prediction of an endpoint can be useful at least as a preliminary estimation, especially, if the approach can be utilized for a group of compounds.

The typical approach involving a mathematical prediction of an endpoint can be expressed by the following scheme:

1. Using matrix of distances [1–4] for a set of congeneric compounds the sum of topological distances (elements of matrix of distances) or Wiener index are calculated. In fact the Wiener index is the first descriptor used for the QSPR analysis.
2. The numerical data on descriptors (e.g. Winer index) are compared with numerical data on the endpoint.

Structure	Matrix	Descriptor	Endpoint
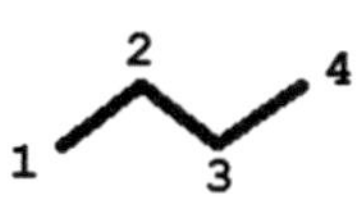		1 2 3 4 1	0 1 2 3 2	1 0 1 2 3	2 1 0 1 4	3 2 1 0	20	-0.5
		1 2 3 4 1	0 1 2 2 2	1 0 1 1 3	2 1 0 1 4	2 1 1 0	16	-11.7
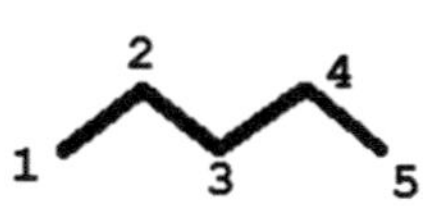		1 2 3 4 5 1	0 1 2 3 4 2	1 0 1 2 3 3	2 1 0 1 2 4	3 2 1 0 1 5	4 3 2 1 0	40	36.0
***	***	***	***

3. Using the least squares method one could develop the model: **Endpoint = F (descriptor)**. The calculation is based on the experimental data on endpoint and the numerical data on the descriptor.
4. One can use the equation: **Endpoint = F (descriptor)** for compounds which have not been examined in the experiment. However, the estimation of the endpoint is robust if these compounds are congeneric to compounds involved in building up model. In fact, the model as a rule is the following: **Endpoint = C_0 + C_1*Descriptor**.

It is to be noted that this scheme can be modified by means of using nonlinear model such as: **Endpoint = C_0 + C_1*Descriptor + C_2*Descriptor^2 + C_3*Descriptor^3 + …**

Also, one can utilize a group of descriptors. In the case of utilization of a group of descriptors the approach is well-known as multiple linear regression analysis (MLRA):

Endpoint = C_0 + C_1*Descriptor$_1$ + C_2*Descriptor$_2$ + C_3*Descriptor$_3$ + …

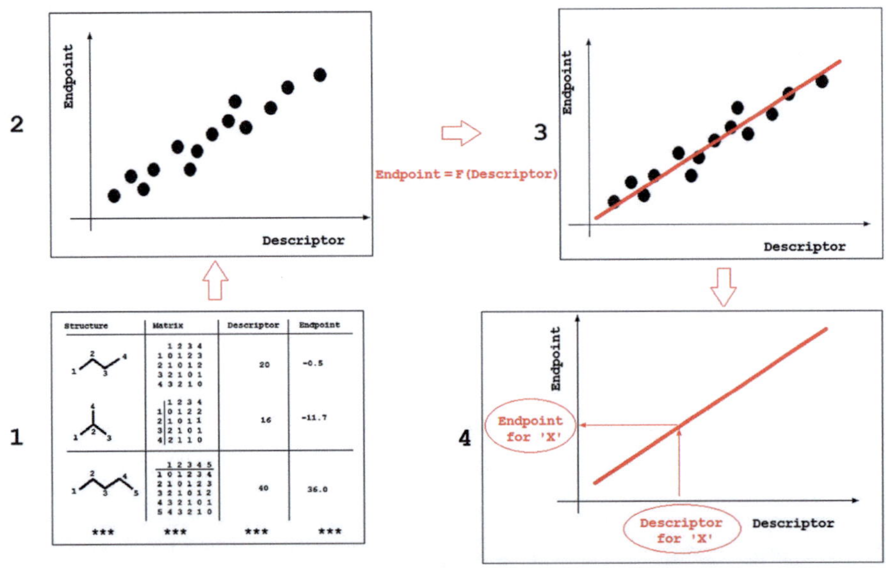

From the beginning of 1980s, a number of different descriptors conceptually analogical to Wiener number has significantly increased [5–13]. Most of these descriptors or indices were based on two special matrixes, named as the adjacency matrix and above mentioned the matrix of topological distances in a molecular graph.

On molecular graph of 2-methyl butane, with numbering of vertices of the adjacency matrix A(G) and matrix of topological distances D(G) are the following:

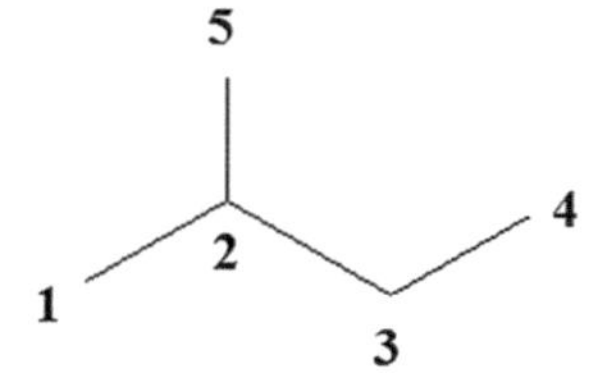

Numbering

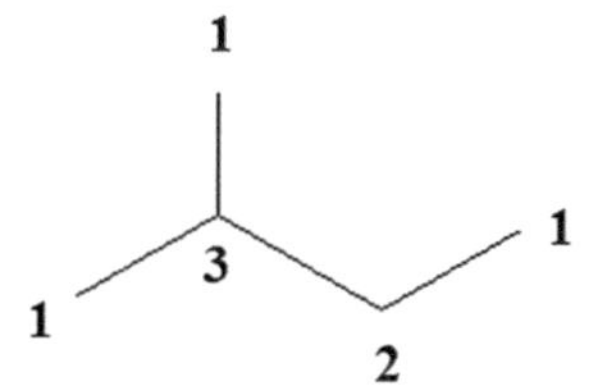

Vertex degree

A(G)

	1	2	3	4	5
1	0	1	0	0	0
2	1	0	1	0	1
3	0	1	0	1	0
4	0	0	1	0	0
5	0	1	0	0	0

D(G)

	1	2	3	4	5
1	0	1	2	3	2
2	1	0	1	2	1
3	2	1	0	1	2
4	3	2	1	0	3
5	2	1	2	3	0

In spite of existence of a large number of descriptors [9], the main idea of their calculation can be illustrated with Wiener number (***W***) and connectivity indices of zero-order ($^{0}\chi$) and first-order ($^{1}\chi$), the latter is also known as Randic index [8, 12–15].

General scheme of building up optimal descriptors with the hydrogen-suppressed graphs based approach can be demonstrated using example of ethyl isopropyl sulfide with the following numbering of atoms:

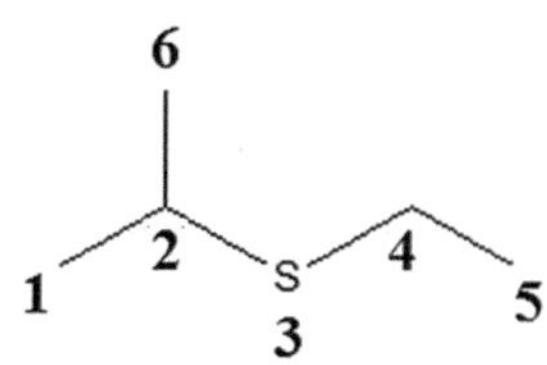

the Adjacency matrix and Modified Adjacency matrix for the compounds are the following:

Adjacency Matrix

	1	2	3	4	5	6	Row sum	
1	0	1	0	0	0	0	1	
2	1	0	1	0	0	1	3	
3	0	1	0	1	0	0	2	$^{1}\chi(0,0)$
4	0	0	1	0	1	0	2	
5	0	0	0	1	0	0	1	
6	0	1	0	0	0	0	1	

Modified Adjacency Matrix

	1	2	3	4	5	6	Row sum	
1	x	1	0	0	0	0	1+x	
2	1	x	1	0	0	1	3+x	
3	0	1	y	1	0	0	2+y	$^{1}\chi(x,y)$
4	0	0	1	x	1	0	2+x	
5	0	0	0	1	x	0	1+x	
6	0	1	0	0	0	x	1+x	

Optimal descriptor is a mathematical function of the molecular structure where instead of the rigid invariants (e.g. vertex degree, topological distances, etc.) specific coefficients are used. Thus, the traditional descriptor is in fact:

$$D = F(molecular\ structure) = F(I_1, I_2, \ldots I_m) \tag{12.1}$$

where $\boldsymbol{I_k}$ is the kth molecular invariant. It is a "rigid" version of descriptor. Formula (1) can be modified by replacing "rigid" components by flexible ones:

$$D = F(molecular\ structure) = F(CW(I_1),\ CW(I_2),\ \ldots CW(I_m)) \tag{12.2}$$

where $\boldsymbol{CW(I_k)}$ is correlation weight of the kth molecular invariant. It is flexible version of descriptor. These CWs, being some numerical coefficients, are used in calculation with Eq. (12.2).

Accordingly to a study published in one of the classical papers [10], application of x = +0.25 and y = −0.95 in calculation of the optimal connectivity index $^{1}\chi$(x,y) for correlation with normal boiling points of 21 sulfides gives considerable decrease of standard error, in comparison with the "inflexible" $^{1}\chi$ (0,0).

Correlation coefficient between descriptor calculated with Eq. (12.2) and property/activity (PA) of interest is also a mathematical function of the CWs,

$$R(PA, D) = R[PA, F(CW(I_1), CW(I_2), \ldots CW(I_m))] \tag{12.3}$$

where ***R(PA,D)*** is the correlation coefficient between the ***PA*** and ***D***, calculated with Eq. (12.2).

By means of an optimization procedure one can calculate the $\boldsymbol{CW^*(MI_1)}$, $\boldsymbol{CW^*(MI_2)}$, … $\boldsymbol{CW^*(MI_m)}$, which are being placed in Eq. (12.3) to provide maximum of the ***R(PA,D)*** for a training set of compounds under consideration. Rationality of such model can (and should) be tested with properly selected external set of species.

It is an important feature of the scheme based on the Eq. (12.3), that these correlation weights can be calculated not only for numerical invariant of molecular graph such as vertex degrees [16], extended connectivity of increasing orders [17], paths of length 2,3, or higher [9], valence shells of increasing orders [9], but also, for eclectic features of molecular structure, such as presence of different atoms, presence/absence of different rings, and so on. This option gives possibility to introduce and estimate measure of influence of given molecular attribute on property/activity of interest. This can be used as hint on the mechanism of phenomena under consideration—an important outcome of the QSAR approach.

It is to be noted that comparison of the hydrogen-suppressed graph based and the hydrogen-filled graph based optimal descriptors has been carried out in [18]. It has been shown that optimal descriptor based on hydrogen-filled graph improves accuracy prediction of normal boiling points of alkyl alcohols.

The graph of atomic orbitals (GAO) [16–22] also can be used as a basis for the optimal descriptors. The basic idea of the representation of the molecular structure by considering configuration of chemical elements is presented in the Table 12.1.

Figure 12.1 contains an example of translation of hydrogen suppressed molecular graph into the GAO. It is to be noted that there are two ways to translate the traditional molecular graph into the GAO: (i) translation of hydrogen suppressed graph into GAO; and (ii) translation of hydrogen filled graph into GAO [16–22]. For a training set of graphs of atomic orbitals, one can carry out the same optimization of correlation weights of the invariants using the same algorithms [23].

The scheme can be based not only on molecular graphs, but also on simplified Molecular Input Line Entry System (SMILES) [24–26]. SMILES contains an string of characters. These characters reflect molecular structure, or at least some part of attributes of the molecular structure. Based on such assumptions one can attempt to define a descriptor that represents a mathematical function of the SMILES characters. Figure 12.2 contains the scheme for building up optimal descriptors using SMILES.

Table 12.1 Groups of atomic orbitals, for some chemical elements, used in constructing the graph of atomic orbitals

Chemical element	Group of atomic orbitals
H	$1s^1$
C	$1s^2$, $2s^2$, $2p^2$
N	$1s^2$, $2s^2$, $2p^3$
O	$1s^2$, $2s^2$, $2p^4$
F	$1s^2$, $2s^2$, $2p^5$
P	$1s^2$, $2s^2$, $2p^6$, $3s^2$, $3p^3$
S	$1s^2$, $2s^2$, $2p^6$, $3s^2$, $3p^4$
Cl	$1s^2$, $2s^2$, $2p^6$, $3s^2$, $3p^5$
Br	$1s^2$, $2s^2$, $2p^6$, $3s^2$, $3p^6$, $3d^{10}$, $4\ s^2$, $4p^5$

Hydrogen suppressed graph:

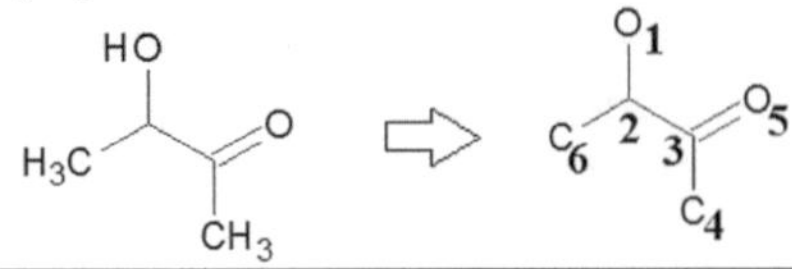

	O_1	C_2	C_3	C_4	O_5	C_6	$EC0_k$
O_1	0	1	0	0	0	0	1
C_2	1	0	1	1	0	0	3
C_3	0	1	0	0	0	0	1
C_4	0	1	0	0	1	1	3
O_5	0	0	0	1	0	0	1
C_6	0	0	0	1	0	0	1

Graph of atomic orbitals (GAO):

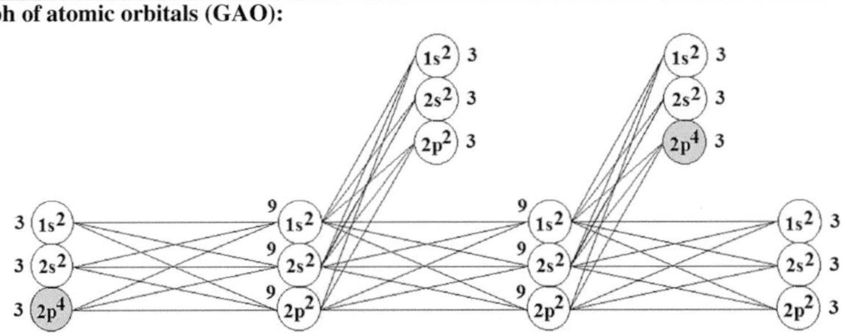

	O_1			C_2			C_3			C_4			O_5			C_6			
	$1s^2$	$2s^2$	$2p^4$	$1s^2$	$2s^2$	$2p^2$	$1s^2$	$2s^2$	$2p^2$	$1s^2$	$2s^2$	$2p^2$	$1s^2$	$2s^2$	$2p^4$	$1s^2$	$2s^2$	$2p^2$	$EC0_k$
$1s^2$	0	0	0	1	1	1	0	0	0	0	0	0	0	0	0	0	0	0	3
$2s^2$	0	0	0	1	1	1	0	0	0	0	0	0	0	0	0	0	0	0	3
$2p^4$	0	0	0	1	1	1	0	0	0	0	0	0	0	0	0	0	0	0	3
$1s^2$	1	1	1	0	0	0	1	1	1	1	1	1	0	0	0	0	0	0	9
$2s^2$	1	1	1	0	0	0	1	1	1	1	1	1	0	0	0	0	0	0	9
$2p^2$	1	1	1	0	0	0	1	1	1	1	1	1	0	0	0	0	0	0	9
$1s^2$	0	0	0	1	1	1	0	0	0	0	0	0	0	0	0	0	0	0	3
$2s^2$	0	0	0	1	1	1	0	0	0	0	0	0	0	0	0	0	0	0	3
$2p^2$	0	0	0	1	1	1	0	0	0	0	0	0	0	0	0	0	0	0	3
$1s^2$	0	0	0	1	1	1	0	0	0	0	0	0	1	1	1	1	1	1	9
$2s^2$	0	0	0	1	1	1	0	0	0	0	0	0	1	1	1	1	1	1	9
$2p^2$	0	0	0	1	1	1	0	0	0	0	0	0	1	1	1	1	1	1	9
$1s^2$	0	0	0	0	0	0	0	0	0	1	1	1	0	0	0	0	0	0	3
$2s^2$	0	0	0	0	0	0	0	0	0	1	1	1	0	0	0	0	0	0	3
$2p^4$	0	0	0	0	0	0	0	0	0	1	1	1	0	0	0	0	0	0	3
$1s^2$	0	0	0	0	0	0	0	0	0	1	1	1	0	0	0	0	0	0	3
$2s^2$	0	0	0	0	0	0	0	0	0	1	1	1	0	0	0	0	0	0	3
$2p^2$	0	0	0	0	0	0	0	0	0	1	1	1	0	0	0	0	0	0	3

Fig. 12.1 Example of the representation of Acetoin (CAS 513-86-0; and SMILES = "O = C(C)C (O)C") by means of (*i*) hydrogen suppressed graph; and (*ii*) Graph of atomic orbitals

Thus, there are four basic representation of the molecular structure which can be used as basis to build up the optimal descriptors (Fig. 12.3): (i) hydrogen suppressed graph; (ii) hydrogen filled graph; (iii) GAO; and (iv) SMILES. These representations also can be involved into hybrid version of the optimal descriptor where molecular features extracted from e.g. GAO and SMILES play the role of hybrid basis for a QSPR/QSAR predictions [27–32].

It is to be noted that structure of peptides represented by one-letter codes that reflect different amino acids is very similar to SMILES (Fig. 12.4). In fact, the representation of peptides can be used for their QSPR/QSAR analyses [33].

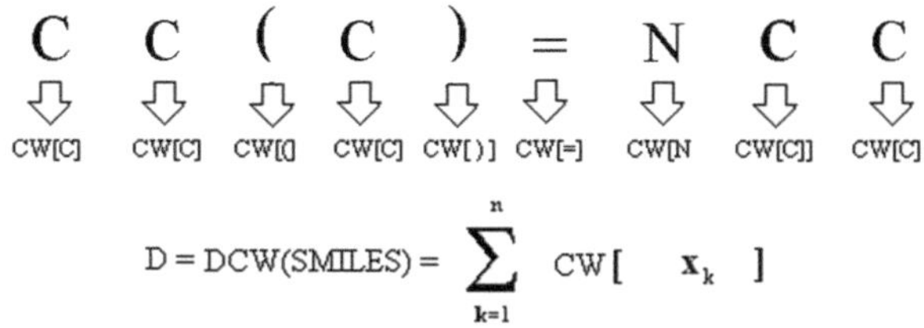

Fig. 12.2 The definition of optimal descriptors using representation of the molecular structure by SMILES

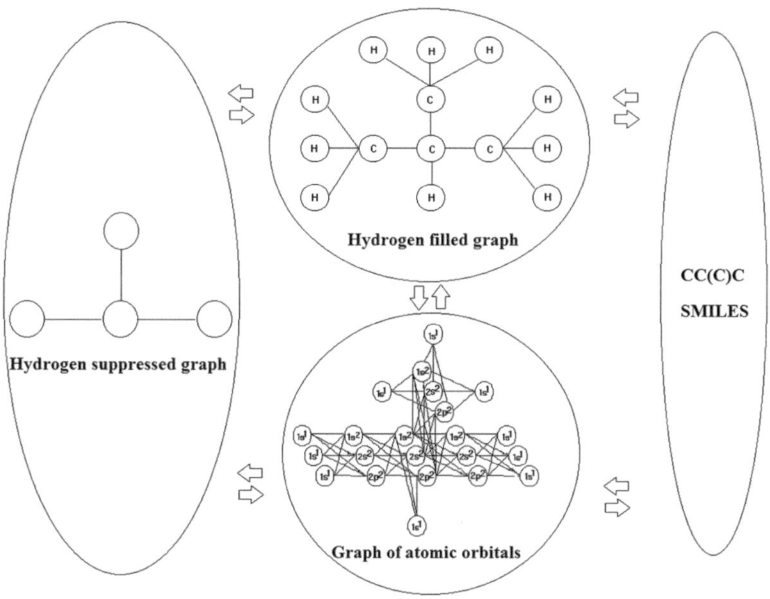

Fig. 12.3 The basic versions of the representations of the molecular structure which are used to build up optimal descriptors

Figure 12.5 contains the graphical representation of the Monte Carlo optimization. This approach is based on calculations of the correlation weights which give maximum correlation coefficient between experimental and predicted endpoint.

The above discussion provides summary of QSPR/QSAR approaches applied to classical, chemical compounds. However, an analysis devoted to nanomaterials having gigantic and complex molecular architecture lead to necessity of definition of new approaches for the predictive modelling, because the representation of their molecular structure by means of molecular graph and/or SMILES sometimes becomes very problematic (e.g. multi-walled carbon nanotubes [34], graphene [35]). In the first approximation, the optimal descriptors for such species should be a collector of all available data which are able to impact the physicochemical and/or biochemical behavior of nanomaterials. This concept is displayed in Fig. 12.6.

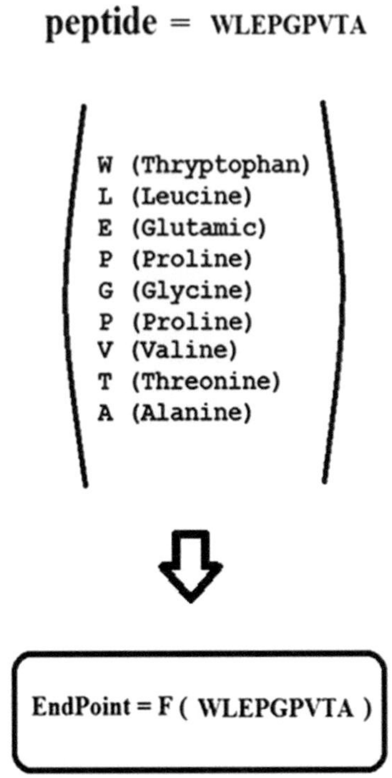

Fig. 12.4 The representation of peptides by means of symbols of amino acids. One notices its similarity with SMILES

It can be consider as a "quasi-SMILES" approach [36]. In contrast to traditional SMILES [24–26] one can use in such a case the quasi-SMILES [36] that reflect all available eclectic data on nanomaterials augmented by features (impacts, conditions), that have clear influence on the performance of nanomaterials.

Formally, the quasi-SMILES can replace the traditional SMILES in the above mentioned optimization procedure. However, in this case the new description should be used and the models should be named quantitative "features"—property/activity relationships (QFPRs/QFARs), since these models will be based on features which can be distant from the molecular structure (e.g. size, concentration, time exposure, etc.) that is used in classic QSPR/QSAR methods.

Figure 12.7 shows the evolution of concepts related to building up optimal descriptors. Interestingly, at the last stage one can see a new quality: the predicted details of molecular structure lost the advantage to be the only source of the information. The essence of difference between traditional descriptors and descriptors calculated with quasi-SMILES is depictured at the Fig. 12.8. Here eclectic features (impacts) partially replace data on the molecular structure.

In the cases of organic, inorganic, and organometallic compounds, as well as for various polymers the quantitative structure—property/activity relationships (QSPRs/QSARs) approaches represent efficient and available tools one can use in order to predict numerical data related to an endpoint of unknown substances. In order to accomplish it and develop QSPR/QSAR model such approaches require

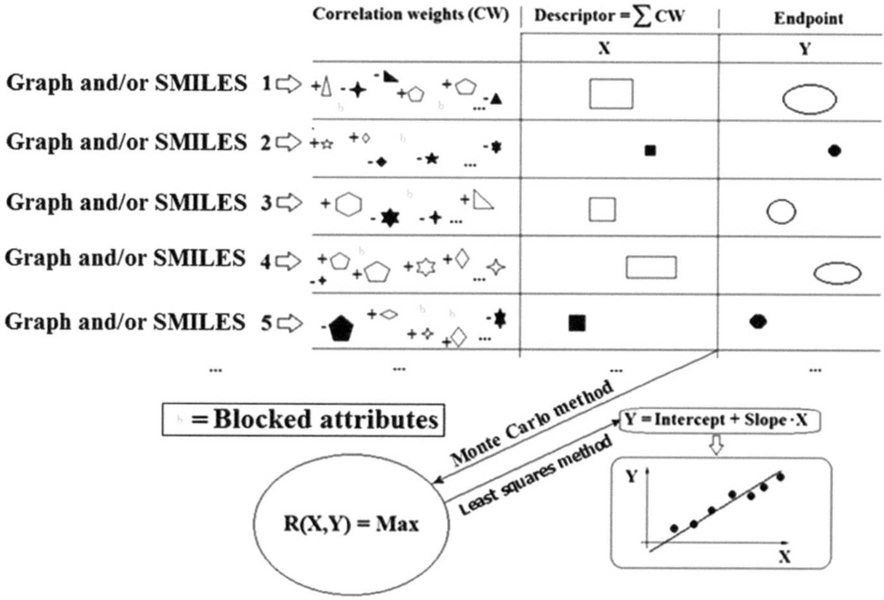

Fig. 12.5 The general scheme of the Monte Carlo optimization used as the basis of calculation of optimal descriptors. The row "Correlation weight" contains graphical images of various features (extracted from graph or SMILES) characterized by positive values of the correlation weights (they are indicated by *white color*) or by negative values of correlation weights (those are indicated by *black color*). Blocked (rare) features have correlation weights which are fixed to be equal to zero (indicated by *grey* "b"). The **R(X,Y)** is correlation coefficient between descriptor and endpoint

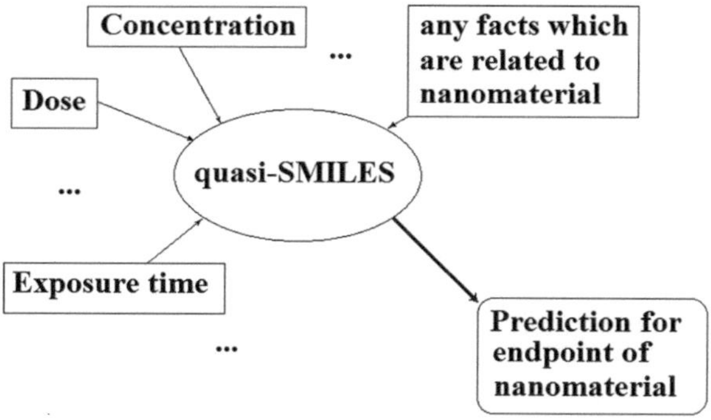

Fig. 12.6 The quasi-SMILES in fact is a collector of all available eclectic information related to some complex systems, such as nanomaterials, peptides, and micro electro mechanical systems

application of experimental data for this endpoint from analogical substances which have been investigated in the experimental studies.

There is considerable number of the literature sources dedicated to discussion traditional QSPR/QSAR analyses of nanomaterials based on selection of groups of

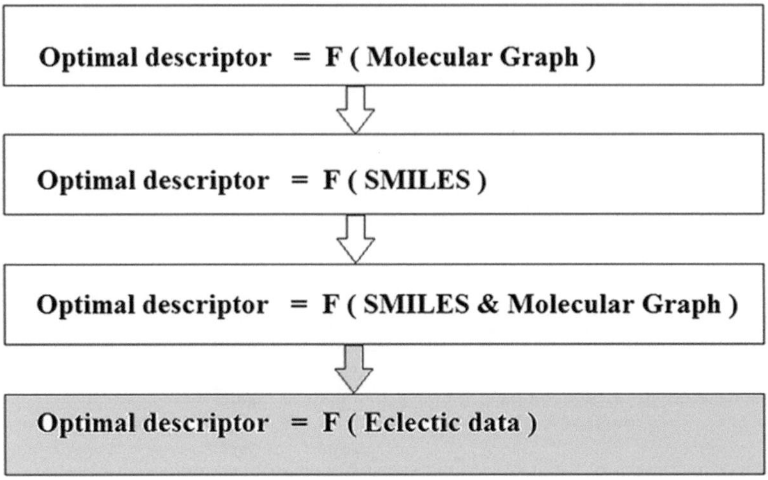

Fig. 12.7 The evolution of optimal descriptors: last stage represents descriptors of the new quality, since they become descriptors which are built up by means of analysis of features which at least partially are distinct from the molecular structure

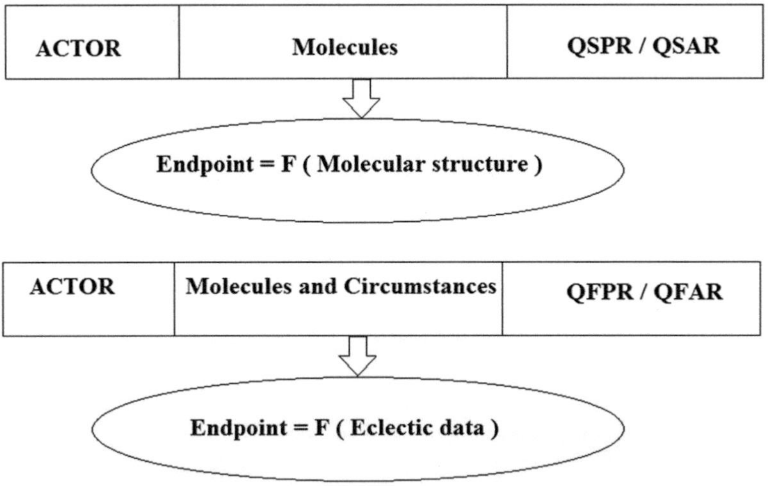

Fig. 12.8 The essence of difference between QSPR/QSAR and QFPR/QFAR

molecular descriptors [37–40]. However, unfortunately, the traditionally used QSPR/QSAR techniques, as a rule, do not provide convenient and efficient methodology to study nanomaterials, owing to a number of reasons. Firstly, molecular structure of majority of nanomaterials contains hundreds or even thousands of atoms. Of course, it generates a complex (but soluble technical problem) how exactly to save the data on the molecular structure in the form of the molecular graph and/or in the form of SMILES. Nevertheless, more important obstacle is due to the impossibility of comparison of such systems: they will be too similar (thousands of atoms are identical and only tenth of atoms are different) or too incomparable (thousands of atoms are different and only tenth are identical). In addition, modes of action of different nanomaterials in different biological systems (cell, organ, and organism) can be absolutely incomparable (different parts of gigantic molecule are acting in dissimilar environments). Also, in spite of manifold attempts to prepare and supply large databases on nanomaterials such collections of data still remain unsatisfactory and usually incomplete sources for the traditional QSPR/QSAR analyses.

Thus, in fact, endpoints related to traditional substances (i.e. non nanomaterials) represent phenomena basically defined by the molecular architecture which involves a "modest" number of atoms (e.g. less than 100). In contrast, endpoints related to nanomaterials are phenomena basically governed by many other factors besides the molecular characteristics of gigantic systems.

For the discussed above reasons, the application of the "classic" approaches to development of QSPR/QSAR models for nanomaterials becomes impossible, or at least very problematic. Consequently, for such species the paradigm "*Endpoint is a mathematical function of the molecular structure*" should be changed to the paradigm "*Endpoint is a mathematical function of all available eclectic data related to nanomaterials*" (Fig. 12.7).

Examples of the eclectic data include: (i) size of particles; (ii) porosity; (iii) condition of synthesis; (iv) irradiation (absence of the irradiation); (v) type of targets, e.g. cell, organ, organism, ecological system, etc.; (vi) electromagnetic field; (vii) presence of pollutants, and others. There is diversity and/or hierarchy of endpoints related to substances in general, and related to nanomaterials in particular. Different kinds of the acute toxicity are characterized by apparent and relatively fast effects. Skin sensitivity also represents a group of endpoints with apparent and relatively fast effects. Carcinogenicity and mutagenicity represent assemblies of endpoints characterized by effects which become visible only after considerable period of impact upon some targets (bacteria, cells, organisms). The experimental measurement of the above mentioned endpoints involve different costs, time, and equipment. Even for relatively simple endpoints such as density, normal boiling points, water solubility, etc., the computational prediction is an attractive alternative to a direct experiment. However, in the case of complex endpoints (e.g. carcinogenicity, mutagenicity, and therapeutic effects) applications of computational approaches become the necessity in both the case of traditional substances and in the case of nanomaterials.

The hierarchy of nanomaterials contains the considerable list of various groups, such as, (i) fullerenes and their derivatives, (ii) single wall carbon nanotubes (SWCN); (iii) multi walls carbon nanotubes (MWCNT); (iv) graphene; (v) micro-electro-mechanical systems (MEMS); (vi) metal nano oxides; (vii) quantum dots, and others. One can expect that in the near future this list will be extended owing to intensive work of experimentalists.

One of the valuable options for development and application of the computational approach that is devoted to estimate endpoints related to nanomaterials is the employment of the above mentioned optimal descriptor which can represent a translator of eclectic information into the endpoint prediction. As an initial form of the optimal descriptor one can apply a hybrid of topological indices calculated with the adjacency matrix of the molecular graph and additive scheme. In this form a physicochemical parameter is modeled as the summation of contributions of molecular fragments. Further, the optimal descriptor could be expanded by means of taking into account contributions of various physicochemical conditions (presence/absence of defined chemical elements and/or defined types of covalent bonds, as well as different types of rings in the molecular system) which are able to modify the physicochemical (biochemical) behavior of a substance. The next step of evolution of the optimal descriptor is the involvement of the above mentioned eclectic factors related to various modes of action of the nanomaterials (Fig. 12.8).

12.2 Methods

12.2.1 Data

12.2.1.1 Thermal Conductivity of Micro-Electro-Mechanical Systems

Experimental data on the thermal conductivity of Micro-Electro-Mechanical Systems (MEMS) have been posted recently at the Internet [41]. We used these data to building up a model of thermal conductivity for MEMS studied experimentally. The logarithm of the thermal conductivity expressed in [W/m/K] has been used as the endpoint. The collection of MEMS has been selected according to two principles (i) these MEMS have (partially) the same technological attributes (Table 12.2); and (ii) their number is as large as possible (Table 12.3). The development of suitable computational model involves rational application of the available experimental data. Here the experimental data for considered species were three times randomly split into the sub-training set (i.e. a group of MEMS which are "producer of model"), calibration set (i.e. a group of MEMS which are "critic of model"), test set (i.e. a group of MEMS which are "preliminary estimator of model"), and validation set (i.e. a group of MEMS which are "final estimator of model"). Table 12.4 contains the details of correlation weights for the model of MEMS.

12.2.1.2 Mutagenicity of Fullerene Under Various Conditions

Fullerene, that represents one of the first novel nanomaterials that was discover by interdisciplinary, multinational team 40 years ago, has been since meticulously investigated. Such investigations involve also interactions of fullerene with biological species. Recently bacterial reverse mutation caused by fullerene was tested

Table 12.2 Technological attributes and their codes, which are used for building up model of thermal conductivity for MEMS

Temperature, C	Code of the temperature
20	%1
25	%1
27	%1
80	%1
100	%1
127	%1
150	%1
200	%2
250	%2
273.1	%2
315	%3
350	%3
400	%4
425	%4
500	%5
540	%5
600	%6
650	%6
700	%7
800	%8
875	%9
1000	%10
1100	%11
1200	%12
1250	%12
1327	%13
1400	%14
1530	%15
1600	%16
2300	%17
Status of MEMS	*Code of status*
Ceramic	1
Single crystal	2
Cubic	3
CVD[a]	4
Glass	5

[a]CVD = Chemical Vapor Deposition

using *Salmonella typhimurium* strains TA100 in the presence and absence of metabolic activation under dark conditions and irradiation [42]. These data were used as experimental endpoints in our study. The data were split into the training, calibration, and validation sets according to the following principles: (i) the split is random; and (ii) the ranges of endpoint for the above-mentioned sets are similar. Table 12.5 displays list of features which are involved in building up the model for

Table 12.3 MEMS, their codes and data on the decimal logarithm of thermal conductivity, as well three splits of available data into the sub-training set (+), calibration set (–), test set (#), and validation set (*)

MEMS	Split			Codes for MEMS	lgTC
	1	2	3		
AlN-1	+	+	+	Al.N.%6	1.302
AlN-2	*	–	*	Al.N.%4	1.345
AlN-3	–	+	+	Al.N.%1	1.479
Al_2O_3-1	#	–	–	Al.Al.O.O.O.2	1.699
Al_2O_3-2	+	+	*	Al.Al.O.O.O.1.%14	0.735
Al_2O_3-3	–	–	+	Al.Al.O.O.O.1.%1	1.399
Al_2O_3-4	–	*	#	Al.Al.O.O.O.1.%3	1.189
Al_2O_3-5	*	*	+	Al.Al.O.O.O.1.%5	1.165
Al_2O_3-6	+	#	#	Al.Al.O.O.O.2.%1	1.634
Al_2O_3-7	*	#	+	Al.Al.O.O.O.2.%3	1.293
Al_2O_3-8	*	+	–	Al.Al.O.O.O.2.%8	1.084
BN-1	*	#	*	B.N.1.%3	1.458
BN-2	–	#	*	B.N.1.%7	1.431
BN-3	*	–	+	B.N.1.%10	1.425
Cd	–	+	–	Cd.%1	1.986
Cr	+	–	+	Cr.%1	1.956
CrB_2	–	#	–	Cr.B.B.%1	1.311
Cr_3C_3	–	+	–	Cr.Cr.Cr.C.C.C.1	2.278
GaAs	+	+	+	Ga.As.%1	1.663
Mo	–	+	*	Mo.%1	2.140
$MoSi_2$-1	#	*	–	Mo.Si.Si.%1.1	1.732
$MoSi_2$-2	#	+	#	Mo.Si.Si.%4.1	1.490
$MoSi_2$-3	#	–	#	Mo.Si.Si.%5.1	1.345
$MoSi_2$-4	–	#	+	Mo.Si.Si.%6.1	1.377
$MoSi_2$-5	#	–	–	Mo.Si.Si.%9.1	1.284
$MoSi_2$-6	+	*	#	Mo.Si.Si.%11.1	1.234
$(Al_2O_3)_3*(SiO_2)_2$-1	+	*	+	Al.Al.O.O.O.Al.Al.O.O.O.Al.Al.O.O.O.Si.O.O.Si.O.O.%1.1	0.782
$(Al_2O_3)_3*(SiO_2)_2$-2	#	#	#	Al.Al.O.O.O.Al.Al.O.O.O.Al.Al.O.O.O.Si.O.O.Si.O.O.%2.1	0.735
$(Al_2O_3)_3*(SiO_2)_2$-3	–	*	*	Al.Al.O.O.O.Al.Al.O.O.O.Al.Al.O.O.O.Si.O.O.Si.O.O.%4.1	0.663
$(Al_2O_3)_3*(SiO_2)_2$-4	+	#	–	Al.Al.O.O.O.Al.Al.O.O.O.Al.Al.O.O.O.Si.O.O.Si.O.O.%6.1	0.621
$(Al_2O_3)_3*(SiO_2)_2$-5	*	–	+	Al.Al.O.O.O.Al.Al.O.O.O.Al.Al.O.O.O.Si.O.O.Si.O.O.%8.1	0.599
$(Al_2O_3)_3*(SiO_2)_2$-6	#	*	*	Al.Al.O.O.O.Al.Al.O.O.O.Al.Al.O.O.O.Si.O.O.Si.O.O.%10.1	0.575
$(Al_2O_3)_3*(SiO_2)_2$-7	*	#	*	Al.Al.O.O.O.Al.Al.O.O.O.Al.Al.O.O.O.Si.O.O.Si.O.O.%12.1	0.575
$(Al_2O_3)_3*(SiO_2)_2$-8	–	+	*	Al.Al.O.O.O.Al.Al.O.O.O.Al.Al.O.O.O.Si.O.O.Si.O.O.%14.1	0.575
Ni	*	+	#	Ni.%1	1.957
Pt	#	+	–	Pt.%1	1.863
SiC-1	*	–	+	Si.C.3.4.%1	2.082
SiC-2	+	#	#	Si.C.3.4.%6	1.319
SiC-3	–	#	*	Si.C.3.4.%8	1.407
SiC-4	–	–	#	Si.C.3.4.%10	1.329
SiC-5	#	–	#	Si.C.3.4.%13	1.539
SiO_2-1	–	*	*	Si.O.O.1.%2	0.017
SiO_2-2	+	*	–	Si.O.O.1.%4	0.097

(continued)

Table 12.3 (continued)

MEMS	Split			Codes for MEMS	lgTC
	1	2	3		
SiO_2-3	#	–	#	Si.O.O.1.%8	0.223
SiO_2-4	#	#	–	Si.O.O.1.%12	0.320
SiO_2-5	#	+	*	Si.O.O.1.%16	0.400
SiO_2-6	*	–	+	Si.O.O.5.%1	0.140
SiO_2-7	*	*	–	Si.O.O.5.%2	0.107
SiO_2-8	+	#	–	Si.O.O.5.%3	0.134
SiO_2-9	#	*	#	Si.O.O.5.%4	0.176
SiO_2-10	*	#	+	Si.O.O.5.%7	0.255

Table 12.4 Lists of codes and their correlation weights calculated by the Monte Carlo method for three random splits of available data into training set (it includes sub-training, calibration, and test set) and external validation set

Split 1		Split 2			Split 3
Code	CW(Code)	Code	CW(Code)	Code	CW(Code)
%1	3.97200	%1	2.84600	%1	1.52100
%10	0.0	%10	1.29900	%10	0.0
%11	2.57800	%11	1.64050	%11	0.0
%12	0.0	%12	0.21350	%12	0.0
%13	0.0	%13	0.0	%13	0.0
%14	−0.21050	%14	0.0	%2	0.0
%16	0.0	%2	0.0	%3	0.0
%2	0.0	%3	0.0	%4	0.0
%3	1.35400	%4	0.0	%5	0.0
%4	0.22100	%5	2.32700	%6	0.0
%5	0.0	%6	0.77300	%7	0.0
%6	1.87800	%7	2.39900	%8	0.0
%7	0.0	%8	0.42400	%9	0.0
%8	0.0	%9	0.0	1	0.27900
%9	0.0	1	−1.80000	2	0.0
1	−0.38550	2	0.0	3	0.0
2	1.39700	3	−0.30400	4	0.0
3	2.32100	4	−0.33650	5	0.0
4	2.24600	5	−4.49700	B	0.0
5	−1.54250	B	1.46050	C	0.0
B	0.0	C	5.26050	Al	2.34800
C	1.81750	Al	1.20750	As	0.0
Al	1.34150	Cr	−0.80100	Cd	0.0
As	1.52500	N	1.88350	Cr	0.0
Cd	0.0	O	0.06450	Ga	0.0
Cr	2.72100	Mo	6.59900	N	−0.20200
Ga	1.23950	Ni	0.0	O	−2.07300
N	1.93350	Si	−1.40950	Mo	0.0
O	0.12700		−0.27900	Ni	0.0
Mo	7.07400			Pt	0.0
Pt	0.0			Si	−0.38350
Si	−2.26050				0.52700
.	−0.35950				

Table 12.5 List of attributes of fullerene C_{60} nanoparticles' exposure and their codes

Attribute	Codes of attributes (C_k) and their meaning
Dark or Irradiation	0 = Dark 1 = Irradiation
Mix S9	+ = with Mix S9 − = without Mix S9
Dose (g/plate)	A = 50 B = 100 C = 200 D = 400 E = 1000

Table 12.6 The list of fullerene C_{60} nanoparticles and data on the bacterial reverse mutation test TA100

No.	Set	Dark or irradiation	Mix s9	Dose	TA100
1	Validation	0	+	A	146
2	Training	0	+	B	141
3	Training	0	+	C	159
4	Validation	0	+	D	160
5	Training	0	+	E	177
6	Calibration	0	−	A	143
7	Training	0	−	B	139
8	Validation	0	−	C	169
9	Training	0	−	D	168
10	Training	0	−	E	152
11	Calibration	1	+	A	129
12	Training	1	+	B	131
13	Validation	1	+	C	138
14	Training	1	+	D	137
15	Calibration	1	+	E	160
16	Validation	1	−	A	136
17	Training	1	−	B	136
18	Training	1	−	C	138
19	Calibration	1	−	D	164
20	Calibration	1	−	E	172

fullerene mutagenicity. In addition, Table 12.6 contains the data set on the mutagenicity of fullerene under various conditions. The QFAR model is described in the recent work [43] (Table 12.7).

12.2.1.3 Prediction of Membrane Damage by Various TiO_2 Nanoparticles

Nano sized metal oxides are among the most important constituents of nanomaterials group. Titanium dioxide represents the largest tonnage metal oxide. It has various applications including cosmetics, paint and paper industry. Its characteristics make

Table 12.7 The QFAR for the mutagenicity of the fullerene C_{60} under various conditions

ID	Quasi-SMILES	DCW(1,3)	$TA100_{expr}$	$TA100_{calc}$	$\Delta TA100_{Expr}-TA100_{calc}$
Training set					
2	0 + B	2.93450	141.000	145.657	4.657
3	0 + C	3.37500	159.000	157.105	1.895
5	0 + E	3.68850	177.000	165.252	11.748
7	0 − B	2.87200	139.000	144.033	5.033
9	0 − D	3.49600	168.000	160.249	7.751
10	0 − E	3.62600	152.000	163.628	11.628
12	1 + B	2.30950	131.000	129.415	1.585
14	1 + D	2.93350	137.000	145.631	8.631
17	1 − B	2.24700	136.000	127.791	8.209
18	1 − C	2.68750	138.000	139.239	1.239
Calibration set					
6	0 − A	2.43550	143.000	132.690	10.310
11	1 + A	1.87300	129.000	118.072	10.928
15	1 + E	3.06350	160.000	149.010	10.990
19	1 − D	2.87100	164.000	144.007	19.993
20	1 − E	3.00100	172.000	147.386	24.614
Validation set					
1	0 + A	2.49800	146.000	134.314	11.686
4	0 + D	3.55850	160.000	161.873	−1.873
8	0 − C	3.31250	169.000	155.481	13.519
13	1 + C	2.75000	138.000	140.863	−2.863
16	1 − A	1.81050	136.000	116.448	19.552

this nanomaterial an important component of sunscreens protection against UV radiation. Due to large amount of TiO_2 that is manufactured an investigation of its interactions with biological species are important since the obtained data could shed a light on possible health and environmental effects of this very common nanomaterial. Recent experimental studies provide information on the physicochemical features of TiO_2 nanoparticles and their influence on the membrane damage [44]. These are (i) engineered size (nm); (ii) size in water suspension (nm); (iii) size in phosphate buffered saline (BPS, nm); (iv) concentration (mg/L); and (v) zeta potential (mV). The above-mentioned physicochemical features of TiO_2 nanoparticles [44] were used to build up computational model for membrane damage related to various TiO_2 nanoparticles (characterized by different physicochemical features). Three various splits of experimental data [44] into the training and test sets were examined in computational studies. These splits obey the following principles (i) they are random; and (ii) the ranges of the endpoint for the training and test sets are similar. Various features applied into building up the model for membrane damage by TiO_2 nanoparticles are displayed in the Table 12.8. Table 12.9 contains normalized representation of physicochemical features of TiO_2 nanoparticles. Experimental data

Table 12.8 Experimental data on physicochemical features of TiO_2 nanoparticles, and their denotations

ID	A Engineered size, nm	B Size in water, nm	C Size in PBS, nm	D Concentration, mg/L	E -Zeta potential, mV
1	30	125	1250	25	10
2	30	102	987	25	12
3	30	281	1543	50	15
4	30	101	1045	50	9
5	30	299	1754	100	11
6	30	134	961	100	11
7	30	600	1876	200	12
8	30	298	1165	200	12
9	45	129	2567	25	9
10	45	129	2309	25	10
11	45	201	2431	50	9
12	45	201	2987	50	11
13	45	451	2941	100	11
14	45	451	1934	100	9
15	45	876	1965	200	11
16	45	876	2109	200	10
17	125	136	3215	25	11
18	125	136	2667	25	10
19	125	149	3782	50	10
20	125	149	2144	50	15
21	125	343	3871	100	12
22	125	343	2890	100	9
23	125	967	3813	200	9
24	125	967	2671	200	8

allow developing a computational model that links characteristics of studied nanomaterials with membrane damage. Table 12.10 contains the Codes of Impacts and membrane damage values (MD, units/L) used to set-up definition of the quasi-SMILES. Based on this definition the quasi-SMILES parameters were selected. They are listed in the Table 12.11, together with the numerical data on three models for membrane damage obtained for three random splits of data into the visible training set and invisible validation set. The calculated values are in good agreement with experimental data. This provides a strong evidence of the applicability of the applied methodology to modeling environmental effects of nanomaterials.

12.2.1.4 Prediction of Membrane Damage by ZnO and TiO_2 Nanoparticles

In many cases rather than dealing with just a single nanomaterial one needs to consider a mixture of various species. Such a case was reported for investigations of

Table 12.9 Normalized representation of physicochemical features of TiO_2 nanoparticles

ID	A	B	C	D	E
	Engineered size, normalized	Size in water, normalized	Size in PBS, normalized	Concentration, normalized	-Zeta potential, normalized
1	0.39	0.21	0.46	0.22	0.78
2	0.39	0.19	0.40	0.22	0.87
3	0.39	0.36	0.52	0.33	1.00
4	0.39	0.19	0.42	0.33	0.74
5	0.39	0.37	0.56	0.56	0.83
6	0.39	0.22	0.40	0.56	0.83
7	0.39	0.66	0.59	1.00	0.87
8	0.39	0.37	0.44	1.00	0.87
9	0.48	0.22	0.73	0.22	0.74
10	0.48	0.22	0.68	0.22	0.78
11	0.48	0.28	0.70	0.33	0.74
12	0.48	0.28	0.82	0.33	0.83
13	0.48	0.52	0.81	0.56	0.83
14	0.48	0.52	0.60	0.56	0.74
15	0.48	0.91	0.61	1.00	0.83
16	0.48	0.91	0.64	1.00	0.78
17	1.00	0.22	0.86	0.22	0.83
18	1.00	0.22	0.75	0.22	0.78
19	1.00	0.23	0.98	0.33	0.78
20	1.00	0.23	0.64	0.33	1.00
21	1.00	0.42	1.00	0.56	0.87
22	1.00	0.42	0.80	0.56	0.74
23	1.00	1.00	0.99	1.00	0.74
24	1.00	1.00	0.75	1.00	0.70

the effect of metal oxides on membrane damage. Experimental data on the physicochemical features of TiO_2 and ZnO nanoparticles and their influence on the membrane damage (Table 12.11) were published recently [44]. The experimental study meticulously considered various features of nanoparticles that could affect final outcome. Such characteristics include: (i) engineered size (nm); (ii) size in water suspension (nm); (iii) size in phosphate buffered saline (PBS, nm); (iv) concentration (mg/L); and (v) zeta potential (mV). Such methodical study provides solid fundation for development of computational model. All measured characteristics of nanoparticles as well as presence of TiO_2 and/or ZnO in process that impacts on cells are the features which are applied in order to develop the quantitative features—activity relationships (QFAR). Membrane damage values related to TiO_2 and ZnO nanoparticles (characterized by different physicochemical features) were examined as the endpoint. Analysis of the results of three random splits of the experimental data into the sub-training, calibration, and test set and invisible

Table 12.10 Codes of impacts and membrane damage values (MD, units/L)

ID	Code for engineered size	Code for size in water	Code for size in PBS	Code for concentration	Code for zeta potential	MD, units/L
1	A3	B2	C4	D2	E7	0.90
2	A3	B1	C4	D2	E8	1.00
3	A3	B3	C5	D3	E9	0.75
4	A3	B1	C4	D3	E7	0.70
5	A3	B3	C5	D5	E8	1.04
6	A3	B2	C3	D5	E8	1.09
7	A3	B6	C5	D9	E8	1.15
8	A3	B3	C4	D9	E8	1.20
9	A4	B2	C7	D2	E7	0.90
10	A4	B2	C6	D2	E7	0.85
11	A4	B2	C7	D3	E7	0.75
12	A4	B2	C8	D3	E8	0.78
13	A4	B5	C8	D5	E8	1.40
14	A4	B5	C5	D5	E7	1.50
15	A4	B9	C6	D9	E8	1.35
16	A4	B9	C6	D9	E7	1.40
17	A9	B2	C8	D2	E8	1.25
18	A9	B2	C7	D2	E7	1.17
19	A9	B2	C9	D3	E7	1.00
20	A9	B2	C6	D3	E9	1.10
21	A9	B4	C9	D5	E8	1.50
22	A9	B4	C7	D5	E7	1.42
23	A9	B9	C9	D9	E7	1.60
24	A9	B9	C7	D9	E6	1.65

validation sets was performed in order to build up the QFAR approach. Table 12.12 contains the experimental data on features (impacts) of TiO_2 and ZnO nanoparticles, and their denotations. Table 12.13 contains the normalization of data on physicochemical properties and builds up codes of impacts for the quasi-SMILES. Table 12.14 presents the quasi-SMILES of representation of TiO_2 and ZnO nanoparticles and their experimental membrane damage. Table 12.15 contains the correlation weights of codes of impacts for calculation of the optimal descriptor for splits 1, 2, and 3 in the case of the TiO_2 and ZnO nanoparticles.

12.2.1.5 Prediction of Membrane Damage by Nano Metal Oxides

Due to a large pool of available nano metal oxides and their industrial applications numerous efforts have been carried out to link their characteristics to biological outcome. Such studies have been performed experimentally and augmented by details obtained using computational methods. Recently the experimental data on

Table 12.11 Numerical data on three models for membrane damage by TiO_2 nanoparticles under different conditions

Set	Quasi-SMILES	CW(2,20)	Expr	Calc	Expr-calc	ID
Split 1						
Training	A3B3C5D3E9	−0.52500	0.750	0.739	0.011	3
Training	A3B2C3D5E8	2.14600	1.090	1.079	0.011	6
Training	A3B3C4D9E8	3.33100	1.200	1.229	0.029	8
Training	A4B2C7D2E7	1.04700	0.900	0.939	0.039	9
Training	A4B2C7D3E7	−0.55100	0.750	0.735	0.015	11
Training	A4B9C6D9E7	4.46750	1.400	1.374	0.026	16
Training	A9B2C8D2E8	3.45650	1.250	1.245	0.005	17
Training	A9B2C7D2E7	2.58250	1.170	1.134	0.036	18
Training	A9B2C9D3E7	1.73050	1.000	1.026	0.026	19
Training	A9B4C7D5E7	4.91450	1.420	1.431	0.011	22
Test	A3B1C4D2E8	1.19900	1.000	0.958	0.042	2
Test	A4B2C8D3E8	0.32300	0.780	0.847	0.067	12
Test	A9B2C6D3E9	1.58250	1.100	1.007	0.093	20
Test	A9B4C9D5E8	5.78850	1.500	1.542	0.042	21
Test	A9B9C9D9E7	6.00300	1.600	1.570	0.030	23
Validation	A3B2C4D2E7	0.37850	0.900	0.854	0.046	1
Validation	A3B1C4D3E7	−0.52700	0.700	0.738	−0.038	4
Validation	A3B3C5D5E8	2.98850	1.040	1.186	−0.146	5
Validation	A3B6C5D9E8	3.18100	1.150	1.210	−0.060	7
Validation	A4B2C6D2E7	1.79300	0.850	1.034	−0.184	10
Validation	A4B5C8D5E8	4.25300	1.400	1.347	0.053	13
Validation	A4B5C5D5E7	4.12500	1.500	1.331	0.169	14
Validation	A4B9C6D9E8	4.59550	1.350	1.390	−0.040	15
Validation	A9B9C7D9E6	5.10900	1.650	1.456	0.194	24
Split 2						
Training	A3B1C4D2E8	2.04350	1.000	0.984	0.016	2
Training	A3B3C5D3E9	−0.00700	0.750	0.710	0.040	3
Training	A3B3C5D5E8	2.84400	1.040	1.091	0.051	5
Training	A3B2C3D5E8	2.49900	1.090	1.045	0.045	6
Training	A3B6C5D9E8	3.63250	1.150	1.196	0.046	7
Training	A4B2C7D2E7	1.94900	0.900	0.971	0.071	9
Training	A4B2C8D3E8	0.37800	0.780	0.761	0.019	12
Training	A4B5C8D5E8	5.67400	1.400	1.469	0.069	13
Training	A4B5C5D5E7	5.51500	1.500	1.448	0.052	14
Training	A4B9C6D9E7	4.87100	1.400	1.362	0.038	16
Training	A9B2C8D2E8	3.68100	1.250	1.203	0.047	17
Training	A9B2C9D3E7	2.54900	1.000	1.051	0.051	19
Training	A9B4C9D5E8	5.68850	1.500	1.471	0.029	21

(continued)

Table 12.11 (continued)

Set	Quasi-SMILES	CW(2,20)	Expr	Calc	Expr-calc	ID
Test	A3B1C4D3E7	1.06550	0.700	0.853	0.153	4
Test	A4B2C6D2E7	1.94900	0.850	0.971	0.121	10
Test	A9B2C6D3E9	2.18750	1.100	1.003	0.097	20
Test	A9B4C7D5E7	6.16350	1.420	1.534	0.114	22
Test	A9B9C9D9E7	6.72100	1.600	1.609	0.009	23
Validation	A3B2C4D2E7	1.46250	0.900	0.906	−0.006	1
Validation	A3B3C4D9E8	4.08450	1.200	1.256	−0.056	8
Validation	A4B2C7D3E7	0.69900	0.750	0.804	−0.054	11
Validation	A4B9C6D9E8	4.59900	1.350	1.325	0.025	15
Validation	A9B2C7D2E7	4.00200	1.170	1.245	−0.075	18
Validation	A9B9C7D9E6	6.35950	1.650	1.560	0.090	24
Split 3						
Training	A3B1C4D2E8	2.70350	1.000	0.957	0.043	2
Training	A3B3C5D3E9	0.72325	0.750	0.705	0.045	3
Training	A3B1C4D3E7	1.00225	0.700	0.741	0.041	4
Training	A3B3C5D5E8	4.08150	1.040	1.133	0.093	5
Training	A3B2C3D5E8	3.44475	1.090	1.052	0.038	6
Training	A4B2C7D2E7	2.54800	0.900	0.938	0.038	9
Training	A4B2C8D3E8	1.38075	0.780	0.789	0.009	12
Training	A4B5C8D5E8	6.52100	1.400	1.444	0.044	13
Training	A4B5C5D5E7	6.62900	1.500	1.458	0.042	14
Training	A4B9C6D9E7	5.82700	1.400	1.356	0.044	16
Training	A9B4C9D5E8	6.52100	1.500	1.444	0.056	21
Training	A9B9C9D9E7	8.19575	1.600	1.658	0.058	23
Training	A9B9C7D9E6	8.04050	1.650	1.638	0.012	24
Test	A3B2C4D2E7	1.66875	0.900	0.826	0.074	1
Test	A3B3C4D9E8	3.65450	1.200	1.079	0.121	8
Test	A4B9C6D9E8	6.41375	1.350	1.431	0.081	15
Test	A9B2C8D2E8	4.22525	1.250	1.152	0.098	17
Test	A9B2C6D3E9	2.47600	1.100	0.928	0.172	20
Test	A9B4C7D5E7	5.88100	1.420	1.363	0.057	22
Validation	A3B6C5D9E8	3.98150	1.150	1.120	0.030	7
Validation	A4B2C6D2E7	1.96250	0.850	0.863	−0.013	10
Validation	A4B2C7D3E7	1.43350	0.750	0.796	−0.046	11
Validation	A9B2C7D2E7	4.27800	1.170	1.158	0.012	18
Validation	A9B2C9D3E7	3.21675	1.000	1.023	−0.023	19

cell membrane damage measured by propidium iodide (PI) uptake were reported for 24 nano metal-oxides (ZrO_2, ZnO, Yb_2O_3, Y_2O_3, WO_3, TiO_2, SnO_2, SiO_2, Sb_2O_3, NiO, Ni_2O_3, MnO_3, La_2O_3, In_2O_3, HfO_2, Gd_2O_3, Fe_3O_4, Fe_2O_3, CuO, Cr_2O_3, CoO,

Table 12.12 Experimental data on features (impacts) of TiO_2 and ZnO nanoparticles, and their denotations

ID	NPs	Engineered size, nm	Size in water, nm	Size in PBS, nm	Concentration, mg/L	-Zeta potential, mV
		A	B	C	D	E
1.	TiO_2	30	125	1250	25	10
2.	–	30	102	987	25	12
3.	–	30	281	1543	50	15
4.	–	30	101	1045	50	9
5.	–	30	299	1754	100	11
6.	–	30	134	961	100	11
7.	–	30	600	1876	200	12
8.	–	30	298	1165	200	12
9.	–	45	129	2567	25	9
10.	–	45	129	2309	25	10
11.	–	45	201	2431	50	9
12.	–	45	201	2987	50	11
13.	–	45	451	2941	100	11
14.	–	45	451	1934	100	9
15.	–	45	876	1965	200	11
16.	–	45	876	2109	200	10
17.	–	125	136	3215	25	11
18.	–	125	136	2667	25	10
19.	–	125	149	3782	50	10
20.	–	125	149	2144	50	15
21.	–	125	343	3871	100	12
22.	–	125	343	2890	100	9
23.	–	125	967	3813	200	9
24.	–	125	967	2671	200	8
25.	ZnO	50	55	158	25	55
26.	–	60	68	208	25	45
27.	–	70	71	198	25	50
28.	–	50	56	258	50	50
29.	–	60	78	386	50	50
30.	–	70	95	279	50	50
31.	–	50	168	314	100	25
32.	–	60	151	385	100	30
33.	–	70	172	354	100	29
34.	–	1000	1245	1319	25	44
35.	–	1200	1268	1325	25	33
36.	–	1500	1198	1381	25	25
37.	–	1000	1268	1459	50	30
38.	–	1200	1301	1587	50	32

(continued)

Table 12.12 (continued)

ID	NPs	Engineered size, nm	Size in water, nm	Size in PBS, nm	Concentration, mg/L	-Zeta potential, mV
		A	B	C	D	E
39.	–	1500	1283	1523	50	29
40.	–	1000	1243	1925	100	20
41.	–	1200	1124	1805	100	21
42.	–	1500	1269	2109	100	21

Table 12.13 The normalization of data on physicochemical properties and build up codes of impacts for the quasi-SMILES

ID	NPs	A	B	C	D	E
1	T	A0	B1	C3	D2	E2
2	T	A0	B1	C2	D2	E3
3	T	A0	B2	C4	D3	E3
4	T	A0	B1	C2	D3	E2
5	T	A0	B2	C4	D5	E3
6	T	A0	B1	C2	D5	E3
7	T	A0	B4	C5	D9	E3
8	T	A0	B2	C3	D9	E3
9	T	A0	B1	C6	D2	E2
10	T	A0	B1	C6	D2	E2
11	T	A0	B1	C6	D3	E2
12	T	A0	B1	C7	D3	E3
13	T	A0	B3	C7	D5	E3
14	T	A0	B3	C5	D5	E2
15	T	A0	B6	C5	D9	E3
16	T	A0	B6	C5	D9	E2
17	T	A1	B1	C8	D2	E3
18	T	A1	B1	C7	D2	E2
19	T	A1	B1	C9	D3	E2
20	T	A1	B1	C5	D3	E3
21	T	A1	B2	C9	D5	E3
22	T	A1	B2	C7	D5	E2
23	T	A1	B7	C9	D9	E2
24	T	A1	B7	C7	D9	E2
25	Z	A0	B0	C0	D2	E9
26	Z	A0	B0	C0	D2	E8
27	Z	A0	B0	C0	D2	E9
28	Z	A0	B0	C1	D3	E9
29	Z	A0	B0	C1	D3	E9
30	Z	A0	B1	C1	D3	E9
31	Z	A0	B1	C1	D5	E5

(continued)

Table 12.13 (continued)

ID	NPs	A	B	C	D	E
32	Z	A0	B1	C1	D5	E6
33	Z	A0	B1	C1	D5	E5
34	Z	A6	B9	C3	D2	E8
35	Z	A8	B9	C3	D2	E6
36	Z	A9	B9	C3	D2	E5
37	Z	A6	B9	C4	D3	E6
38	Z	A8	B9	C4	D3	E6
39	Z	A9	B9	C4	D3	E5
40	Z	A6	B9	C5	D5	E4
41	Z	A8	B8	C4	D5	E4
42	Z	A9	B9	C5	D5	E4
	Minimum	30	55	158	25	8
	Maximum	1500	1301	3871	200	55

Co_3O_4, CeO_2, Al_2O_3) [45]. These data were used to develop computational models. The experimental results on this endpoint related to four doses (50, 100, 150, and 200 μg/mL) and seven exposure time (from 1 to 7 h) for all 24 nano metal-oxides were examined. Figure 12.9 contains an example of the representation for (i) nano metal-oxide; and (ii) conditions of its impact upon cell membrane. The data set contains the numerical values of cell membrane damage which are characterized by dispersion $<10\ \%$ [46].

12.2.2 Principles of Building up Optimal Descriptors

An experience acquired during developments of QSAR/QSPR methods applied to various nanomaterials allows establishing useful principles applied to obtaining optimal descriptors. Few key rules are given below:

(i) One should collect information on all features (impacts) which are able to influence biochemical behavior of a complex system;
(ii) One should define preliminary hierarchy of the above features;
(iii) The traditional paradigm of the QSPR/QSAR analysis "Endpoint = Mathematical function of the molecular structure" must be replaced by paradigm "Endpoint = Mathematical function of all available eclectic data". It is to be noted, that the "molecular structure" can represent a "specific eclectic data";
(iv) Rational discrimination of features of the complex system in accordance with their frequencies (in the visible training set and invisible validation set) and their information contributions (correlation weights) should be carried out by means of comparison of series of Monte Carlo optimizations

Table 12.14 The quasi-SMILES of representation of TiO_2 and ZnO nanoparticles and their experimental membrane damage

ID	quasi-SMILES	Membrane damage (units/L)
1	TA0B1C3D2E2	0.90
2	TA0B1C2D2E3	1.00
3	TA0B2C4D3E3	0.75
4	TA0B1C2D3E2	0.70
5	TA0B2C4D5E3	1.04
6	TA0B1C2D5E3	1.09
7	TA0B4C5D9E3	1.15
8	TA0B2C3D9E3	1.20
9	TA0B1C6D2E2	0.90
10	TA0B1C6D2E2	0.85
11	TA0B1C6D3E2	0.75
12	TA0B1C7D3E3	0.78
13	TA0B3C7D5E3	1.40
14	TA0B3C5D5E2	1.50
15	TA0B6C5D9E3	1.35
16	TA0B6C5D9E2	1.40
17	TA1B1C8D2E3	1.25
18	TA1B1C7D2E2	1.17
19	TA1B1C9D3E2	1.00
20	TA1B1C5D3E3	1.10
21	TA1B2C9D5E3	1.50
22	TA1B2C7D5E2	1.42
23	TA1B7C9D9E2	1.60
24	TA1B7C7D9E2	1.65
25	ZA0B0C0D2E9	1.10
26	ZA0B0C0D2E8	1.03
27	ZA0B0C0D2E9	1.08
28	ZA0B0C1D3E9	1.00
29	ZA0B0C1D3E9	0.92
30	ZA0B1C1D3E9	0.99
31	ZA0B1C1D5E5	1.12
32	ZA0B1C1D5E6	1.25
33	ZA0B1C1D5E5	1.19
34	ZA6B9C3D2E8	1.58
35	ZA8B9C3D2E6	1.69
36	ZA9B9C3D2E5	1.59
37	ZA6B9C4D3E6	0.92
38	ZA8B9C4D3E6	0.95
39	ZA9B9C4D3E5	0.84
40	ZA6B9C5D5E4	1.25
41	ZA8B8C4D5E4	1.39
42	ZA9B9C5D5E4	1.45

Table 12.15 Correlation weights of codes of impacts for calculation of the optimal descriptor for splits 1, 2, and 3 in the case of the TiO_2 and ZnO nanoparticles

C_k	$CW(C_k)$	Frequency of C_k in training set	Frequency of C_k in calibration set	Frequency of C_k in test set
Split 1				
A0	0.49800	6	11	3
A1	1.36250	5	1	1
A6	1.18950	3	0	0
A8	1.70300	1	1	1
A9	0.92600	2	1	0
B0	0.68250	1	2	0
B1	0.50500	6	4	3
B2	0.54600	2	1	1
B3	0.0	0	2	0
B4	0.0	0	1	0
B6	1.50400	1	1	0
B7	2.00500	1	1	0
B8	0.0	0	1	0
B9	1.07100	6	1	1
C0	0.90400	1	1	0
C1	0.54900	1	3	0
C2	0.0	0	2	0
C3	2.00200	4	0	1
C4	0.55200	3	2	0
C5	1.45300	3	4	0
C6	0.0	0	0	3
C7	1.55500	3	2	0
C9	1.27900	2	0	1
D2	1.18450	5	2	3
D3	0.50300	6	2	1
D5	2.00300	3	7	1
D9	1.52100	3	3	0
E2	1.00100	5	3	3
E3	0.93000	4	5	1
E4	1.09500	1	2	0
E5	1.45400	3	1	0
E6	0.87400	2	0	1
E8	1.05100	1	1	0
E9	1.27800	1	2	0
T	1.10950	9	8	4
Z	0.98450	8	6	1
Split 2				
A0	−0.13550	11	8	3
A1	2.05500	4	2	2

(continued)

Table 12.15 (continued)

C_k	$CW(C_k)$	Frequency of C_k in training set	Frequency of C_k in calibration set	Frequency of C_k in test set
A6	0.0	2	0	0
A8	0.0	0	2	0
A9	0.0	1	2	0
B0	0.0	2	1	1
B1	−0.90000	7	5	3
B2	0.0	2	2	1
B3	0.0	1	1	0
B6	0.0	2	0	0
B7	0.0	1	1	0
B8	0.0	0	1	0
B9	0.40400	3	3	0
C0	0.0	1	0	1
C1	0.63650	3	2	1
C2	0.0	1	1	1
C3	0.99150	3	1	0
C4	0.0	1	4	0
C5	0.74500	4	2	0
C6	0.0	1	1	0
C7	0.0	2	3	0
C8	0.0	0	0	1
C9	0.0	2	0	1
D2	1.79700	7	1	2
D3	0.05950	4	5	1
D5	2.53950	4	6	2
D9	2.84700	3	2	0
E2	0.90100	8	2	1
E3	0.98750	3	6	2
E4	0.0	1	2	0
E5	0.0	2	1	1
E6	0.0	0	2	0
E8	0.0	2	0	0
E9	0.0	2	1	1
T	0.40100	11	8	3
Z	1.91150	7	6	2
Split 3				
A0	−0.20300	13	4	6
A1	0.0	0	6	0
A6	1.49900	3	0	0
A8	2.57600	1	0	1
A9	1.75200	2	1	0

(continued)

Table 12.15 (continued)

C_k	$CW(C_k)$	Frequency of C_k in training set	Frequency of C_k in calibration set	Frequency of C_k in test set
B0	0.10400	4	0	1
B1	−0.30300	5	6	3
B2	0.0	0	3	1
B3	3.04900	2	0	0
B4	1.10300	1	0	0
B6	2.38450	1	0	1
B7	0.0	0	1	0
B8	2.33350	1	0	0
B9	1.21550	5	1	1
C0	0.34900	3	0	0
C1	2.05400	4	0	2
C2	−0.29700	1	1	0
C3	0.72400	1	3	0
C4	−0.29600	3	0	2
C5	1.80100	5	1	1
C6	0.64250	1	0	2
C7	1.40300	1	3	0
C8	0.0	0	1	0
C9	0.0	0	2	0
D2	1.72500	5	3	1
D3	−0.21250	5	3	4
D5	3.50200	7	3	1
D9	3.20300	2	2	1
E2	1.52500	4	4	2
E3	1.14900	2	6	2
E4	−0.22100	3	0	0
E5	0.68650	2	1	1
E6	1.14900	2	0	1
E8	3.02900	2	0	0
E9	2.52900	4	0	1
T	0.85400	6	10	4
Z	1.69700	13	1	3

with various distributions of the data into the visible training set and invisible validation set;

(v) The validation of the predictive potential of the preferable rational model should be carried out with quasi-SMILES of the validation set. This quasi-SMILES are invisible during building up the model.

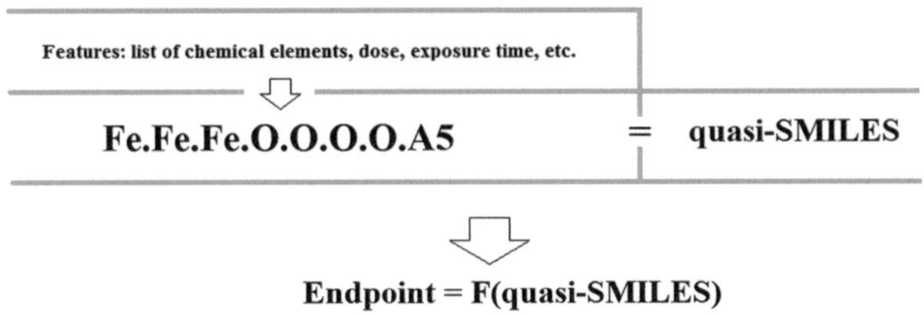

Fig. 12.9 Building up the representation of impact of Fe_3O_4 upon cell membrane according to codes of Table 12.16; dose is 200 μg/mL ("A") and exposure time is five hours ("5")

12.2.3 Calculation of the Optimal Descriptors

The general scheme of calculation of optimal descriptors is the following:

$$\mathbf{DCW}(\mathbf{quasi-SMILES, Threshold, N_{epoch}}) = \sum \mathbf{CW}(\mathbf{C_k}) \tag{12.4}$$

where **Threshold** is coefficient for classification of various features extracted from quasi-SMILES into two classes: (i) active (in this case correlation weight is involved in the modeling process); and (ii) rare (in this case correlation weight is not involved in the modeling process); the $\mathbf{N_{epoch}}$ is the number of epochs of the Monte Carlo optimization which gives the best statistical quality for the calibration set; $\mathbf{C_k}$ is a symbol of quasi-SMILES which is representation of a feature taken into account for this model. Having numerical data on correlation weights of features which give the preferable statistics for the calibration set, one can calculate (using the training set) the model

$$\mathbf{Endpoint = C_0 + C_1{*}DCW}(\mathbf{quasi-SMILES, Threshold, N_{epoch}}) \tag{12.5}$$

The predictive potential of the model should be checked up with external validation set (quasi-SMILES which are invisible during build up the model).

12.3 Results and Discussion

12.3.1 QFPR for Thermal Conductivity of MEMS

Thermal conductivity [W/m/K] of nanomaterials was examined as endpoint in recent works [47, 48] where approach similar to described here has been used. Technological attributes (Table 12.2) are the basis of the models. Table 12.3 contains the codes for the technological attributes, in fact these are basis for the quasi-SMILES. Table 12.4 contains the correlation weights for technological attributes which were obtained for three different splits into the sub-training, calibration, test, and validation sets. The models of decimal logarithm of thermal conductivity (lgTC) are the following:

$$LgTC = 0.7914(\pm 0.009) + 0.1556(\pm 0.0030)*DCW(1, 44) \quad (12.6)$$

$$n = 11,\ r^2 = 0.9752,\ q^2 = 0.9628,\ s = 0.097,\ F = 355 \text{(sub-training set)}$$
$$n = 14,\ r^2 = 0.7977,\ s = 0.373 \text{ (calibration set)}$$
$$n = 13,\ r^2 = 0.9601,\ s = 0.419 \text{ (test set)}$$
$$n = 13,\ r^2 = 0.8229,\ s = 0.351 \text{ (validation set)}$$

$$LgTC = 0.8914(\pm 0.010) + 0.1380(\pm 0.0017)*DCW(1, 55) \quad (12.7)$$

$$n = 12,\ r^2 = 0.9598,\ q^2 = 0.9490,\ s = 0.120,\ F = 239 \text{ (sub-training set)}$$
$$n = 15,\ r^2 = 0.8505,\ s = 0.239 \text{ (calibration set)}$$
$$n = 12,\ r^2 = 0.8811,\ s = 0.317 \text{ (test set)}$$
$$n = 12,\ r^2 = 0.9295,\ s = 0.456 \text{ (validation set)}$$

$$LgTC = 0.8700(\pm 0.0284) + 0.2283(\pm 0.0145)*DCW(3, 57) \quad (12.8)$$

$$n = 14,\ r^2 = 0.6742,\ q^2 = 0.5570,\ s = 0.327,\ F = 49 \text{ (sub-training set)}$$
$$n = 13,\ r^2 = 0.8463,\ s = 0.379 \text{ (calibration set)}$$
$$n = 12,\ r^2 = 0.8508,\ s = 0.247 \text{ (test set)}$$
$$n = 12,\ r^2 = 0.6751,\ s = 0.321 \text{ (validation set)}$$

The calculations were carried out for three various splits into the sub-training, calibration, test, and external (invisible) validation sets. The statistical quality of these models is different, but can be estimated as satisfactory for all three splits.

12.3.2 QFAR for Mutagenicity (TA100) of Fullerene

Table 12.5 contains the eclectic data related to mutagenicity of fullerene under different conditions. Table 12.6 contains quasi-SMILES which were used as the basis of the QFAR for the mutagenicity of fullerene. The model calculated with these quasi-SMILES are the following:

$$TA100 = 69.3980(\pm 7.2079) + 25.9872(\pm 2.5129)*DCW(1, 3) \quad (12.9)$$

$$n=10,\ r^2=0.7549,\ q^2=0.5709,\ s=7.67,\ F=25\ \text{(Training set)}$$
$$n=5,\ r^2=0.8987,\ s=18.4\ \text{(Calibration set)}$$
$$n=5,\ r^2=0.6968,\ s=10.9\ \text{(Validation set)}$$

Table 12.7 contains the details related to the model calculated with Eq. 12.9.

Unfortunately the estimation of the approach with different splits into the training set and calibration set together with external validation set is impossible since available data are limited. However, the approach can be extended to other similar data when those will become available.

12.3.3 QFAR for Membrane Damage by Various TiO_2 Nanoparticles

Table 12.8 contains experimental data related to membrane damage by TiO_2 nanoparticles. In order to build up quasi-SMILES for this eclectic system these data were normalized and discriminated into nine categories. The formula for the normalization is the following:

$$Norm\,(X_k)=\frac{\min X_k+X_k}{\min X_k+\max X_k} \tag{12.10}$$

where X_k is a physicochemical parameter (Table 12.8).

Table 12.9 contains the normalized data on the physicochemical parameters. Table 12.10 contains basis to build up quasi-SMILES.

Predictive models for the membrane damage by TiO_2 nanoparticles (MD, units/L) calculated for three random splits are the following:

$$MD=0.8054(\pm 0.0044)+0.1273(\pm 0.0014)*DCW(2,20) \tag{12.11}$$

$$n=10,\ r^2=0.9893,\ q^2=0.9845,\ RMSE=0.025,\ F=741\ \text{(training set)}$$
$$n=5,\ r^2=0.9647,\ RMSE=0.066\ \text{(test set)}$$
$$n=9,\ r^2=0.8679,\ RMSE=0.115\ \text{(validation set)}$$

$$MD=0.7108(\pm 0.0084)+0.1336(\pm 0.0025)*DCW(2,19) \tag{12.12}$$

$$n=13,\ r^2=0.9639,\ q^2=0.9495,\ RMSE=0.049,\ F=294\ \text{(training set)}$$
$$n=5,\ r^2=0.9263,\ RMSE=0.123\ \text{(test set)}$$
$$n=6,\ r^2=0.9748,\ RMSE=0.054\ \text{(validation set)}$$

$$MD=0.6128(\pm 0.0080)+0.1275(\pm 0.0016)*DCW(2,21) \tag{12.13}$$

$n = 13$, $r^2 = 0.9792$, $q^2 = 0.9718$, RMSE = 0.049, F = 518 (training set)

$n = 6$, $r^2 = 0.8959$, RMSE = 0.118 (test set)

$n = 5$, $r^2 = 0.9925$, RMSE = 0.025 (validation set)

Table 12.11 contains the numerical data related to models calculated with Eqs. 12.11–12.13 together with the distribution into the training set, test set, and validation set.

12.3.4 QFAR for Membrane Damage by Means of Various TiO_2 and ZnO Nanoparticles

The scheme of building up quasi-SMILES for the case TiO_2 together with ZnO is in fact the same scheme which is used for solely TiO_2 nanoparticles (Tables 12.12, 12.13 and 12.14), but with special symbols ('T' and 'Z' in Table 12.14) in order to define two kinds of the nanoparticles (i.e. TiO_2 and ZnO). The translation of the scheme of building up the model for TiO_2 (solely) into the scheme of building up the model for TiO_2 and ZnO is demonstration of the possibility of connecting of two eclectic systems into one generalized model. The QFAR models for the membrane damage (MD) for the case TiO_2 together with ZnO, calculated with three random splits are the following:

Split 1

$$MD = -0.3708(\pm 0.0499) + 0.2378(\pm 0.0081)*DCW(1,5) \qquad (12.14)$$

$n = 17$, $r^2 = 0.8131$, $q^2 = 0.7582$, $s = 0.114$, F = 65 (sub-training set)

$n = 14$, $r^2 = 0.8062$, $s = 0.256$ (calibration set)

$n = 5$, $r^2 = 0.9952$, $s = 0.241$, (test set)

$n = 6$, $r^2 = 0.8362$, $s = 0.244$, (validation set)

Split 2

$$MD = 0.5784(\pm 0.0187) + 0.1596(\pm 0.0051)*DCW(3,19) \qquad (12.15)$$

$n = 18$, $r^2 = 0.7845$, $q^2 = 0.7308$, $s = 0.129$, F = 58 (sub-training set)

$n = 14$, $r^2 = 0.8197$, $s = 0.110$ (calibration set)

$n = 5$, $r^2 = 0.9873$, $s = 0.056$, (test set)

$n = 5$, $r^2 = 0.7159$, $s = 0.157$, (validation set)

Split 3

$$MD = 0.4544(\pm 0.0209) + 0.0974(\pm 0.0027) * DCW(1, 13) \tag{12.16}$$

$n = 19$, $r^2 = 0.9255$, $q^2 = 0.8932$, $s = 0.068$, $F = 211$ (sub-training set)
$n = 11$, $r^2 = 0.6693$, $s = 0.354$ (calibration set)
$n = 7$, $r^2 = 0.8866$, $s = 0.102$, (test set)
$n = 5$, $r^2 = 0.6763$, $s = 0.318$, (validation set)

Table 12.15 contains the correlation weights for the models calculated with Eqs. 12.14–12.16.

12.3.5 QFAR for Cellular Membrane Damage (CMD) by Metal Oxide Nanoparticles

Table 12.16 contains the eclectic data used as basis for the QFAR model. Table 12.17 contains the correlation weights for development of the model. Table 12.18 contains available data represented by quasi-SMILES and distribution

Table 12.16 The list of C_k which are used for representation of nano metal-oxides and conditions of their action. For example, the "Zn.O.A2" means (*i*) nano metal-oxide ZnO; (*ii*) dose is 200 μg/mL; and (*iii*) exposure time is 2 h

C_k	Comment
Chemical elements	
Al, Ce, Co, Cr, Cu, Fe, Gd, Hf, O, In, La, Mn, Ni, W, Sb, Si, Y, Sn, Ti, Yb, Zn, and Zr	
Separator for chemical elements (dot)	
"."	
Doses	
"A"	200 μg/mL
"B"	150 μg/mL
"C"	100 μg/mL
"D"	50 μg/mL
Exposure time	
"1"	1 h
"2"	2 h
"3"	3 h
"4"	4 h
"5"	5 h
"6"	6 h
"7"	7 h

Table 12.17 Correlation weights, $\boldsymbol{CW(C_k)}$, for calculation of $DCW(T,N_{epoch})$ with Eq. 12.17

C_k	$CW(C_k)$				
	Split 1	Split 2	Split 3	Split 4	Split 5
Al	1.37800	1.25100	1.31450	1.19250	1.12900
Ce	3.12300	2.80850	2.49500	3.00200	2.56550
Co	0.18850	−0.00300	0.18350	0.00200	0.00200
Cr	0.80950	0.81750	0.74700	0.37800	0.68650
Cu	4.62100	4.81050	3.99600	5.18250	4.06650
Fe	0.43950	0.37200	0.37200	0.12700	0.25100
Gd	1.06650	0.87300	0.93250	0.80850	0.81050
Hf	3.81650	3.50400	3.12300	3.50300	3.12700
O	0.49600	0.81450	0.24700	1.00200	0.56650
In	1.37300	1.12900	1.31450	1.12500	1.06550
La	0.87100	0.56350	0.68850	0.55950	0.68750
Mn	2.56450	1.93650	2.37600	2.00400	1.93950
Ni	0.99700	1.99500	0.99500	2.62100	1.87500
W	1.49800	1.12600	1.87500	0.87300	1.49700
Sb	0.56750	0.44250	0.62500	0.18950	0.56750
Si	3.05950	2.81450	2.62500	2.93550	2.43550
Y	1.37900	1.18250	1.25400	1.00100	0.99800
Sn	3.31750	3.06350	2.68550	3.18750	2.68850
Ti	3.99600	3.69050	3.50200	3.62800	3.00500
Yb	1.25200	1.06250	1.19150	0.94250	1.00100
Zn	5.94150	6.12300	4.87200	6.37400	4.99600
Zr	3.25000	3.18850	2.63000	3.31150	2.69050
	0.12700	0.05950	0.12500	0.12600	0.12300
A	1.55950	1.75300	1.55850	0.94150	1.74700
B	1.31650	1.43450	0.94150	0.62700	1.49600
C	1.30950	1.49900	1.06250	0.44150	1.56750
D	1.55850	1.81050	1.37200	0.74900	1.62500
1	1.93750	0.99600	0.75400	1.31250	1.24600
2	1.87100	1.00100	0.87600	1.43850	1.44150
3	2.00500	1.37300	1.12100	1.74900	1.69050
4	1.56250	0.87600	0.62600	1.06650	1.25200
5	1.93650	1.31150	0.93550	1.62300	1.56050
6	1.50100	0.75000	0.56350	1.19150	1.12200
7	2.25400	1.68850	1.37300	1.87700	1.68250

into the sub-training set, calibration set, and validation set (Split 1). In order to validate the approach five random splits were examined. The models are the following:

Table 12.18 The model for split 1: (i) representation of the nano metal-oxides; (ii) distribution into the sub-training, calibration and validation sets; (iii) the numerical data on the DCW(1,10); and (iv) numerical data for cell membrane damage (CMD) [1]

ID	Set	Quasi-SMILES	DCW(1,10)	CMD_{Expr}	CMD_{Calc}
3	Sub-training	Zn.O.A2	10.12200	−2.125	−2.101
4	Sub-training	Yb.Yb.O.O.O.A1	8.12400	−3.500	−2.978
5	Sub-training	Y.Y.O.O.O.A1	8.37800	−2.750	−2.867
6	Sub-training	Y.Y.O.O.O.A3	8.44550	−3.000	−2.837
7	Sub-training	Y.Y.O.O.O.A5	8.37700	−3.000	−2.867
9	Sub-training	Y.Y.O.O.O.A7	8.69450	−3.000	−2.728
11	Sub-training	Sn.O.O.A4	7.81250	−3.000	−3.115
12	Sub-training	Sn.O.O.A5	8.18650	−3.000	−2.951
14	Sub-training	Si.O.O.A2	7.86300	−3.250	−3.093
15	Sub-training	Si.O.O.A3	7.99700	−3.125	−3.034
17	Sub-training	Si.O.O.A5	7.92850	−3.125	−3.064
19	Sub-training	Sb.Sb.O.O.O.A2	6.68850	−3.125	−3.609
20	Sub-training	Sb.Sb.O.O.O.A4	6.38000	−2.875	−3.744
21	Sub-training	Sb.Sb.O.O.O.A7	7.07150	−3.125	−3.440
22	Sub-training	La.La.O.O.O.A7	7.67850	−2.625	−3.174
23	Sub-training	Hf.O.O.A6	8.25000	−2.875	−2.923
24	Sub-training	Gd.Gd.O.O.O.A2	7.68650	−3.000	−3.170
26	Sub-training	Gd.Gd.O.O.O.A4	7.37800	−3.125	−3.306
27	Sub-training	Gd.Gd.O.O.O.A5	7.75200	−3.125	−3.142
29	Sub-training	Gd.Gd.O.O.O.A7	8.06950	−2.750	−3.002
31	Sub-training	Fe.Fe.Fe.O.O.O.O.A5	7.68750	−3.500	−3.170
32	Sub-training	Cu.O.A1	8.86800	−3.125	−2.652
33	Sub-training	Cr.Cr.O.O.O.A1	7.23900	−2.875	−3.367
34	Sub-training	Cr.Cr.O.O.O.A2	7.17250	−3.000	−3.396
35	Sub-training	Cr.Cr.O.O.O.A6	6.80250	−4.000	−3.559
37	Sub-training	Co.O.A7	4.75200	−5.000	−4.459
38	Sub-training	Co.Co.Co.O.O.O.O.A6	6.49900	−3.000	−3.692
39	Sub-training	Ce.O.O.A4	7.61800	−4.500	−3.200
40	Sub-training	Yb.Yb.O.O.O.B6	7.44450	−2.875	−3.277
42	Sub-training	Y.Y.O.O.O.B2	8.06850	−3.000	−3.003
43	Sub-training	Y.Y.O.O.O.B5	8.13400	−2.750	−2.974
44	Sub-training	Y.Y.O.O.O.B7	8.45150	−3.125	−2.835
45	Sub-training	W.O.O.O.B1	6.74800	−3.125	−3.582
46	Sub-training	W.O.O.O.B2	6.68150	−3.500	−3.612
47	Sub-training	W.O.O.O.B6	6.31150	−4.575	−3.774
48	Sub-training	W.O.O.O.B7	7.06450	−3.000	−3.443
51	Sub-training	Sn.O.O.B4	7.56950	−3.125	−3.222
53	Sub-training	Si.O.O.B6	7.25000	−3.000	−3.362

(continued)

Table 12.18 (continued)

ID	Set	Quasi-SMILES	DCW(1,10)	CMD_{Expr}	CMD_{Calc}
55	Sub-training	Sb.Sb.O.O.O.B1	6.51200	−4.625	−3.686
56	Sub-training	Sb.Sb.O.O.O.B2	6.44550	−4.375	−3.715
58	Sub-training	Gd.Gd.O.O.O.B7	7.82650	−3.550	−3.109
59	Sub-training	Fe.Fe.Fe.O.O.O.O.B7	7.76200	−3.125	−3.137
61	Sub-training	Cu.O.B1	8.62500	−2.875	−2.758
62	Sub-training	Cu.O.B2	8.55850	−3.000	−2.788
63	Sub-training	Cu.O.B5	8.62400	−1.500	−2.759
65	Sub-training	Zr.O.O.C2	7.80350	−3.125	−3.119
66	Sub-training	Zr.O.O.C3	7.93750	−3.125	−3.060
67	Sub-training	Zr.O.O.C6	7.43350	−3.000	−3.281
68	Sub-training	Zr.O.O.C7	8.18650	−3.000	−2.951
70	Sub-training	Y.Y.O.O.O.C7	8.44450	−2.875	−2.838
71	Sub-training	W.O.O.O.C6	6.30450	−4.500	−3.777
74	Sub-training	Ti.O.O.C3	8.68350	−2.875	−2.733
75	Sub-training	Ti.O.O.C4	8.24100	−2.875	−2.927
76	Sub-training	Ti.O.O.C6	8.17950	−2.875	−2.954
78	Sub-training	Sn.O.O.C7	8.25400	−3.000	−2.921
79	Sub-training	Si.O.O.C7	7.99600	−3.125	−3.035
80	Sub-training	Ni.O.C1	4.99400	−3.550	−4.352
83	Sub-training	Ni.O.C5	4.99300	−4.750	−4.353
84	Sub-training	Ni.O.C6	4.55750	−4.875	−4.544
85	Sub-training	In.In.O.O.O.C2	8.04950	−3.000	−3.011
89	Sub-training	In.In.O.O.O.C6	7.67950	−3.000	−3.173
92	Sub-training	Gd.Gd.O.O.O.C1	7.50300	−3.000	−3.251
93	Sub-training	Fe.Fe.O.O.O.C2	6.18250	−3.625	−3.831
94	Sub-training	Fe.Fe.O.O.O.C3	6.31650	−3.375	−3.772
96	Sub-training	Cu.O.C1	8.61800	−3.000	−2.761
97	Sub-training	Cu.O.C6	8.18150	−3.250	−2.953
98	Sub-training	Cr.Cr.O.O.O.C6	6.55250	−4.000	−3.668
100	Sub-training	Co.Co.Co.O.O.O.O.C4	6.31050	−4.000	−3.774
101	Sub-training	Ce.O.O.C3	7.81050	−2.750	−3.116
102	Sub-training	Ce.O.O.C6	7.30650	−2.875	−3.337
104	Sub-training	Zr.O.O.D7	8.43550	−3.000	−2.842
105	Sub-training	Yb.Yb.O.O.O.D1	8.12300	−2.875	−2.979
106	Sub-training	Yb.Yb.O.O.O.D2	8.05650	−3.000	−3.008
107	Sub-training	Yb.Yb.O.O.O.D7	8.43950	−2.875	−2.840
109	Sub-training	Y.Y.O.O.O.D4	8.00200	−2.750	−3.032
110	Sub-training	Y.Y.O.O.O.D5	8.37600	−3.000	−2.868
111	Sub-training	Y.Y.O.O.O.D6	7.94050	−2.875	−3.059
112	Sub-training	W.O.O.O.D7	7.30650	−2.750	−3.337

(continued)

Table 12.18 (continued)

ID	Set	Quasi-SMILES	DCW(1,10)	CMD_{Expr}	CMD_{Calc}
113	Sub-training	Sn.O.O.D4	7.81150	−3.125	−3.116
114	Sub-training	Sn.O.O.D5	8.18550	−3.000	−2.951
119	Sub-training	Sb.Sb.O.O.O.D2	6.68750	−3.500	−3.609
120	Sub-training	Mn.O.O.O.D5	8.05550	−3.000	−3.008
121	Sub-training	La.La.O.O.O.D2	7.29450	−4.150	−3.343
122	Sub-training	La.La.O.O.O.D3	7.42850	−3.675	−3.284
123	Sub-training	La.La.O.O.O.D6	6.92450	−3.125	−3.505
124	Sub-training	La.La.O.O.O.D7	7.67750	−2.875	−3.174
125	Sub-training	In.In.O.O.O.D2	8.29850	−2.875	−2.902
126	Sub-training	In.In.O.O.O.D3	8.43250	−3.000	−2.843
127	Sub-training	In.In.O.O.O.D4	7.99000	−2.875	−3.037
128	Sub-training	In.In.O.O.O.D5	8.36400	−3.125	−2.873
130	Sub-training	Hf.O.O.D1	8.68550	−2.750	−2.732
131	Sub-training	Gd.Gd.O.O.O.D7	8.06850	−3.525	−3.003
133	Sub-training	Co.Co.Co.O.O.O.O.D4	6.55950	−3.875	−3.665
134	Sub-training	Ce.O.O.D2	7.92550	−2.875	−3.066
135	Sub-training	Ce.O.O.D3	8.05950	−2.750	−3.007
137	Sub-training	Al.Al.O.O.O.D2	8.30850	−2.875	−2.897
16	Calibration	Si.O.O.A4	7.55450	−3.125	−3.228
25	Calibration	Gd.Gd.O.O.O.A3	7.82050	−3.050	−3.112
28	Calibration	Gd.Gd.O.O.O.A6	7.31650	−3.000	−3.333
30	Calibration	Fe.Fe.Fe.O.O.O.O.A2	7.62200	−2.875	−3.199
36	Calibration	Co.O.A5	4.43450	−4.375	−4.598
41	Calibration	Yb.Yb.O.O.O.B7	8.19750	−2.875	−2.946
50	Calibration	Sn.O.O.B3	8.01200	−3.125	−3.028
57	Calibration	Ni.O.B3	5.06850	−3.575	−4.320
69	Calibration	Yb.Yb.O.O.O.C7	8.19050	−2.875	−2.949
72	Calibration	W.O.O.O.C7	7.05750	−3.000	−3.447
77	Calibration	Ti.O.O.C7	8.93250	−2.500	−2.623
81	Calibration	Ni.O.C3	5.06150	−3.800	−4.323
82	Calibration	Ni.O.C4	4.61900	−4.500	−4.517
86	Calibration	In.In.O.O.O.C3	8.18350	−3.125	−2.952
87	Calibration	In.In.O.O.O.C4	7.74100	−2.875	−3.147
88	Calibration	In.In.O.O.O.C5	8.11500	−3.000	−2.982
90	Calibration	Hf.O.O.C3	8.50400	−2.800	−2.812
91	Calibration	Hf.O.O.C7	8.75300	−2.900	−2.702
115	Calibration	Sn.O.O.D6	7.75000	−3.000	−3.143
132	Calibration	Fe.Fe.Fe.O.O.O.O.D7	8.00400	−2.625	−3.031
136	Calibration	Ce.O.O.D7	8.30850	−3.000	−2.897
1	Validation	Zr.O.O.A7	8.43650	−3.000	−2.841

(continued)

Table 12.18 (continued)

ID	Set	Quasi-SMILES	DCW(1,10)	CMD_{Expr}	CMD_{Calc}
2	Validation	Zn.O.A1	10.18850	−2.125	−2.072
8	Validation	Y.Y.O.O.O.A6	7.94150	−3.125	−3.058
10	Validation	Ti.O.O.A7	9.18250	−2.875	−2.514
13	Validation	Sn.O.O.A6	7.75100	−3.125	−3.142
18	Validation	Si.O.O.A7	8.24600	−2.875	−2.925
49	Validation	Sn.O.O.B2	7.87800	−3.250	−3.086
52	Validation	Sn.O.O.B7	8.26100	−3.125	−2.918
54	Validation	Si.O.O.B7	8.00300	−3.125	−3.031
60	Validation	Fe.Fe.O.O.O.B6	5.81950	−4.750	−3.990
64	Validation	Cr.Cr.O.O.O.B2	6.92950	−3.875	−3.503
73	Validation	Ti.O.O.C2	8.54950	−3.000	−2.792
95	Validation	Fe.Fe.O.O.O.C5	6.24800	−4.125	−3.802
99	Validation	Co.Co.Co.O.O.O.O.C3	6.75300	−4.000	−3.580
103	Validation	Ce.O.O.C7	8.05950	−3.000	−3.007
108	Validation	Y.Y.O.O.O.D3	8.44450	−3.000	−2.838
116	Validation	Sn.O.O.D7	8.50300	−3.125	−2.812
117	Validation	Si.O.O.D1	7.92850	−3.500	−3.064
118	Validation	Si.O.O.D5	7.92750	−3.250	−3.065
129	Validation	In.In.O.O.O.D6	7.92850	−3.175	−3.064

Split 1

$$CMD = -6.5448(\pm 0.0455) + 0.4390(\pm 0.0057) * \boldsymbol{DCW(1, 10)} \qquad (12.17)$$

$$n = 96,\ r^2 = 0.5213,\ q^2 = 0.4947,\ s = 0.392,\ F = 102\ (\text{sub-training set})$$
$$n = 21,\ r^2 = 0.8424,\ s = 0.290\ (\text{calibration set})$$
$$n = 20,\ r^2 = 0.9174,\ s = 0.280\ (\text{validation set})$$

Split 2

$$CMD = -6.3034(\pm 0.0411) + 0.4261(\pm 0.0053) * \boldsymbol{DCW(1, 8)} \qquad (12.18)$$

$$n = 96,\ r^2 = 0.5093,\ q^2 = 0.4837,\ s = 0.393,\ F = 98\ (\text{sub-training set})$$
$$n = 21,\ r^2 = 0.8628,\ s = 0.296\ (\text{calibration set})$$
$$n = 20,\ r^2 = 0.8110,\ s = 0.315\ (\text{validation set})$$

Split 3

$$\mathrm{CMD} = -5.6821(\pm 0.0336) + 0.4539(\pm 0.0058)^{*}\boldsymbol{DCW}(\boldsymbol{1},\boldsymbol{9}) \tag{12.19}$$

$n = 96,\ r^2 = 0.5014,\ q^2 = 0.4754,\ s = 0.393,\ F = 95$ (sub-training set)
$n = 21,\ r^2 = 0.8235,\ s = 0.352$ (calibration set)
$n = 20,\ r^2 = 0.6979,\ s = 0.255$ (validation set)

Split 4

$$\mathrm{CMD} = -6.3770(\pm 0.0422) + 0.4297(\pm 0.0055)^{*}\boldsymbol{DCW}(\boldsymbol{1},\boldsymbol{9}) \tag{12.20}$$

$n = 91,\ r^2 = 0.5026,\ q^2 = 0.4767,\ s = 0.389,\ F = 90$ (sub-training set)
$n = 26,\ r^2 = 0.7010,\ s = 0.250$ (calibration set)
$n = 20,\ r^2 = 0.9268,\ s = 0.402$ (validation set)

Split 5

$$\mathrm{CMD} = -7.5325(\pm 0.0495) + 0.6161(\pm 0.0068)^{*}\boldsymbol{DCW}(\boldsymbol{1},\boldsymbol{8}) \tag{12.21}$$

$n = 91,\ r^2 = 0.5437,\ q^2 = 0.5215,\ s = 0.394,\ F = 106$ (sub-training set)
$n = 24,\ r^2 = 0.8824,\ s = 0.214$ (calibration set)
$n = 22,\ r^2 = 0.7809,\ s = 0.348$ (validation set)

The technical details for these models are available in the literature [46]. The comparison of statistical quality for model calculated with Eqs. 12.17–12.21 shows that the distribution of available data into visible training set (i.e. sub-training set and calibration set) and invisible validation set influences the predictive potential of these models.

12.4 Conclusions

Though details of computational investigation of properties and activities of nanomaterials are still being developed the quantitative "features"—property/activity relationships (QFPRs/QFARs) methods discussed in this chapter provide efficient tool to predict various characteristics of nanospecies. They are developed following a set of rules: (i) All examined eclectic systems of data are translated into quasi-SMILES with further prediction of the above mentioned endpoints by the same approach; (ii) building up these models using the quasi-SMILES can be carried out by the same algorithm (software) which is used in the case of the traditional SMILES [49–56]; (iii) eclectic systems can be similar in order for their implementation into common generalized model; (iv) the suggested model has mechanistic interpretation in terms of revealing promoters of increase and decrease for an endpoint. Thus, one concludes that the proposed approach provides models for nanomaterials, in accordance with the OECD principles [57].

Acknowledgments The authors are grateful to the EU project PROSIL funded under the LIFE program (Project LIFE12 ENV/IT/000154), the EC FP7 project NanoPUZZLES (Project Reference: 309837) and EU FP7 project PreNanoTox (contract 309666). D.L. and J.L. acknowledge support from the National Science Foundation (NSF/CREST HRD-0833178), and EPSCoR (Award #: 362492-190200-01/NSFEPS-090378).

References

1. Wiener H (1947) J Am Chem Soc 69(1):17
2. Wiener H (1947) J Am Chem Soc 69(11):2636
3. Wiener H (1948) J Phys Chem Soc 52(6):1082
4. Wiener H (1948) J Phys Chem Soc 52(2):425
5. Hosoya H (1972) J Chem Doc 12:181
6. Amidon GL, Anik ST (1976) J Pharm Sci 65:801
7. Bonchev D, Balaban AT, Mekenyan O (1980) J Chem Inf Comput Sci 20:106
8. Chemical application of topology and graph theory: a collection of papers from a symposium held at the University of Georgia, Athens, Georgia, U.S.A, 18–22 April 1983
9. Randic M (2001) J Chem Inf Comput Sci 41(3):627
10. Randic M, Basak SC (2001) J Chem Inf Comput Sci 41:614
11. Randic M, Plavsic D, Lers N (2001) J Chem Inf Comput Sci 41(3):657
12. Roy K, Leonard TJ (2004) Bioorg Med Chem 12:745
13. Jalbout AF, Li X (2003) J Mol Struct (Theochem) 663:19
14. Visco DP Jr, Poppale RS, Rintoul MD, Faulon J-L (2002) J Mol Graph Mod 20:429
15. http://www.esi.umontreal-a/accelrys/life/cerius46/qsar/theory_descriptors.html. Accessed 22 Jan 2015
16. Toropov AA, Toropova AP (1988) Russ J Coord Chem 24:81
17. Toropov AA, Toropova AP (2003) J Mol Struct (Theochem) 637:1
18. Krenkel G, Castro EA, Toropov AA (2001) J Mol Struct (Theochem) 542:107
19. Toropov AA, Toropova AP (2004) J Mol Struct (Theochem) 711:173
20. Toropov AA, Gutman I, Furtula B (2005) J Serb Chem Soc 70:669
21. Gutman I, Toropov AA, Toropova AP (2005) MATCH Commun Math Comput Chem 53:215
22. Gutman I, Furtula B, Toropov AA, Toropova AP (2005) MATCH Commun Math Comput Chem 53:225
23. Toropov AA, Schultz TW (2003) J Chem Inf Comput Sci 43:560
24. Weininger D (1988) J Chem Inf Comput Sci 28:31
25. Weininger D, Weininger A, Weininger JL (1989) J Chem Inf Comput Sci 29:97
26. Weininger D (1990) J Chem Inf Comput Sci 30:237
27. Achary PGR (2014) SAR QSAR Environ Res 25(6):507
28. Toropov AA, Toropova AP, Martyanov SE, Benfenati E, Gini G, Leszczynska D, Leszczynski J (2011) Chemometr Intell Lab Syst 109(1):94
29. Toropova AP, Toropov AA, Martyanov SE, Benfenati E, Gini G, Leszczynska D, Leszczynski J (2013) Mol Inf 32(2):145

30. Toropova AP, Toropov AA, Rasulev BF, Benfenati E, Gini G, Leszczynska D, Leszczynski J (2012) Struct Chem 23(6):1873
31. Toropov AA, Toropova AP, Martyanov SE, Benfenati E, Gini G, Leszczynska D, Leszczynski J (2012) Chemom Intell Lab Syst 112:65
32. Toropova AP, Toropov AA, Martyanov SE, Benfenati E, Gini G, Leszczynska D, Leszczynski J (2012) Chemom Intell Lab Syst 110(1):177
33. Toropov AA, Toropova AP, Raska I Jr, Benfenati E, Gini G (2012) Struct Chem 23:1891
34. Apul OG, Wang Q, Shao T, Rieck JR, Karanfil T (2013) Environ Sci Technol 47(5):2295
35. Yousefinejad S, Hemmateenejad B (2014) Colloids Surf A 441:766
36. Toropov AA, Toropova AP (2015) Chemosphere 124:40
37. Melagraki G, Afantitis A (2014) RSC Adv 4(92):50713
38. Petrova T, Rasulev BF, Toropov AA, Leszczynska D, Leszczynski J (2011) J Nanopart Res 13 (8):3235
39. Toropov AA, Rasulev BF, Leszczynska D, Leszczynski J (2008) Chem Phys Lett 457 (4–6):332
40. Toropov AA, Rasulev BF, Leszczynska D, Leszczynski J (2007) Chem Phys Lett 444 (1–3):209
41. MEMSnet: https://www.memsnet.org/material. Accessed 19 Feb 2013
42. Shinohara N, Matsumoto K, Endoh S, Maru J, Nakanishi J (2009) Toxicol Lett 191:289
43. Toropov AA, Toropova AP (2014) Chemosphere 104:262
44. Sayes C, Ivanov I (2010) Risk Anal 30:1723
45. Patel T, Low-Kam C, Ji ZH, Zhang H, Xia T, Nel AE, Zinc JI, Telesca D (2012) COBRA preprint series 2012, Working Paper 101. http://biostats.bepress.com/cobra/art101
46. Toropova AP, Toropov AA, Benfenati E, Korenstein R, Leszczynska D, Leszczynski J (2015) Environ Sci Pollut Res 22:745
47. Toropov AA, Leszczynska D, Leszczynski J (2007) Mater Lett 61(26):4777
48. Toropova AP, Toropov AA, Puzyn T, Benfenati E, Leszczynska D, Leszczynski J (2013) J Math Chem 51(8):2230
49. Veselinović AM, Milosavljević JB, Toropov AA, Nikolić GM (2013) Eur J Pharm Sci 48:532
50. Veselinović AM, Milosavljević JB, Toropov AA, Nikolić GM (2013) Arch Pharm 346:134
51. Nesměrák K, Toropov AA, Toropova AP (2014) Struct Chem 25(1):311
52. Nesměrák K, Toropov AA, Toropova AP, Kohoutova P, Waisser K (2013) Eur J Med Chem 67:111
53. Worachartcheewan A, Nantasenamat C, Isarankura-Na-Ayudhya C, Prachayasittikul V (2014) Lett Drug Des Discovery 11(4):420
54. Masand VH, Toropov AA, Toropova AP, Mahajan DT (2014) Curr Comput Aided Drug Des 10(1):75
55. Achary PGR (2014) SAR QSAR Environ Res 25(1):73
56. Mullen LMA, Duchowicz PR, Castro EA (2011) Chemom Intell Lab Syst 107(2):269
57. OECD: Organisation for Economic Co-operation and Development (2007) Guidance document on the validation of (Quantitative) Structure-Activity Relationships ((Q)SAR) models, OECD, Paris. http://www.oecd.org/dataoecd/55/35/38130292.pdf

Index

J. Leszczynski and M.K. Shukla (eds.), *Practical Aspects of Computational Chemistry IV*, DOI 10.1007/978-1-4899-7699-4

MIX
Papier aus verantwortungsvollen Quellen
Paper from responsible sources
FSC® C105338

If you have any concerns about our products,
you can contact us on
ProductSafety@springernature.com

In case Publisher is established outside the EU,
the EU authorized representative is:
Springer Nature Customer Service Center GmbH
Europaplatz 3, 69115 Heidelberg, Germany

Printed by Libri Plureos GmbH
in Hamburg, Germany